W9-AEY-731

Theory of Point Estimation

The Wadsworth & Brooks/Cole Statistics/Probability Series

Series Editors

O. E. Barndorff-Nielsen, Aarhus University
Peter J. Bickel, University of California, Berkeley
William S. Cleveland, AT&T Bell Laboratories
Richard M. Dudley, Massachusetts Institute of Technology

R. Becker, J. Chambers, A. Wilks, *The New S Language: A Programming Environment for Data Analysis and Graphics*
P. Bickel, K. Doksum, J. Hodges, Jr., *A Festschrift for Erich L. Lehmann*
G. Box, *The Collected Works of George E. P. Box, Volumes I and II*, G. Tiao, editor-in-chief
L. Breiman, J. Friedman, R. Olshen, C. Stone, *Classification and Regression Trees*
G. Casella, R. Berger, *Statistical Inference*
J. Chambers, W. S. Cleveland, B. Kleiner, P. Tukey, *Graphical Methods for Data Analysis*
W. S. Cleveland, M. McGill, *Dynamic Graphics for Statistics*
K. Dehnad, *Quality Control, Robust Design, and the Taguchi Method*
R. Durrett, *Lecture Notes on Particle Systems and Percolation*
R. Durrett, *Probability: Theory and Examples*
F. Graybill, *Matrices with Applications in Statistics, Second Edition*
L. Le Cam, R. Olshen, *Proceedings of the Berkeley Conference in Honor of Jerzy Neyman and Jack Kiefer, Volumes I and II*
E. Lehmann, *Testing Statistical Hypotheses, Second Edition*
E. Lehmann, *Theory of Point Estimation*
P. Lewis, E. Orav, *Simulation Methodology for Statisticians, Operations Analysts, and Engineers*
H. J. Newton, *TIMESLAB*
J. Rawlings, *Applied Regression Analysis*
J. Rice, *Mathematical Statistics and Data Analysis*
J. Romano, A. Siegel, *Counterexamples in Probability and Statistics*
J. Tanur, F. Mosteller, W. Kruskal, E. Lehmann, R. Link, R. Pieters, G. Rising, *Statistics: A Guide to the Unknown, Third Edition*
J. Tukey, *The Collected Works of J. W. Tukey*, W. S. Cleveland, editor-in-chief
Volume I: *Time Series: 1949–1964*, edited by D. Brillinger
Volume II: *Time Series: 1965–1984*, edited by D. Brillinger
Volume III: *Philosophy and Principles of Data Analysis: 1949–1964*, edited by L. Jones
Volume IV: *Philosophy and Principles of Data Analysis: 1965–1986*, edited by L. Jones
Volume V: *Graphics: 1965–1985*, edited by W. S. Cleveland
Volume VI: *More Mathematical: 1938–1984*, edited by C. Mallows

Theory of Point Estimation

E. L. LEHMANN
Professor of Statistics
University of California, Berkeley

W Wadsworth & Brooks/Cole Advanced Books & Software
Pacific Grove, California

Wadsworth & Brooks/Cole Advanced Books & Software
A Division of Wadsworth, Inc.

Printed in the United States of America
10 9 8 7 6 5 4 3 2

Library of Congress Cataloging-in-Publication Data
Lehmann, E. L. (Erich Leo), [date]–
Theory of point estimation / E.L. Lehmann.
p. cm. — (The Wadsworth & Brooks/Cole statistics/probability series)
Reprint. Originally published: New York : Wiley, c1983.
Includes bibliographical references and indexes.
ISBN 0-534-15978-8
1. Fix-point estimation. I. Title. II. Series.
QA276.8.L43 1991 90-20739
519.5'44—dc20 CIP

Sponsoring Editor: *John Kimmel*
Editorial Assistant: *Jennifer Kehr*
Production Coordinator: *Dorothy Bell*
Cover Design: *Lisa Thompson*
Printing and Binding: *Arcata Graphics/Fairfield*

To Stephen, Barbara, Fia

Preface

This book is concerned with point estimation in Euclidean sample spaces. The first four chapters deal with exact (small-sample) theory, and their approach and organization parallel those of the companion volume, *Testing Statistical Hypotheses* (TSH). Optimal estimators are derived according to criteria such as unbiasedness, equivariance, and minimaxity, and the material is organized around these criteria. The principal applications are to exponential and group families, and the systematic discussion of the rich body of (relatively simple) statistical problems that fall under these headings constitutes a second major theme of the book.

A theory of much wider applicability is obtained by adopting a large-sample approach. The last two chapters are therefore devoted to large-sample theory, with Chapter 5 providing a fairly elementary introduction to asymptotic concepts and tools. Chapter 6 establishes the asymptotic efficiency, in sufficiently regular cases, of maximum likelihood and related estimators, and of Bayes estimators, and presents a brief introduction to the local asymptotic optimality theory of Hajek and LeCam. Even in these two chapters, however, attention is restricted to Euclidean sample spaces, so that estimation in sequential analysis, stochastic processes, and function spaces, in particular, is not covered.

The text is supplemented by numerous problems. These and references to the literature are collected at the end of each chapter. The literature, particularly when applications are included, is so enormous and spread over the journals of so many countries and so many specialties that complete coverage did not seem feasible. The result is a somewhat inconsistent coverage which, in part, reflects my personal interests and experience.

It is assumed throughout that the reader has a good knowledge of calculus and linear algebra. Most of the book can be read without more advanced mathematics (including the sketch of measure theory which is presented in Section 1.2 for the sake of completeness) if the following conventions are accepted.

1. A central concept is that of an integral such as $\int f\,dP$ or $\int f\,d\mu$. This covers both the discrete and continuous case. In the discrete case $\int f\,dP$ becomes $\Sigma f(x_i)P(x_i)$ where $P(x_i) = P(X = x_i)$ and $\int f\,d\mu$ becomes $\Sigma f(x_i)$. In the continuous case, $\int f\,dP$ and $\int f\,d\mu$ become, respectively, $\int f(x)p(x)\,dx$ and $\int f(x)\,dx$. Little is lost (except a unified notation and some generality) by always making these substitutions.

2. When specifying a probability distribution P, it is necessary to specify not only the sample space $\mathscr{X}$, but also the class $\mathscr{A}$ of sets over which P is to be defined. In nearly all examples $\mathscr{X}$ will be a Euclidean space and $\mathscr{A}$ a large class of sets, the so-called Borel sets, which in particular includes all open and closed sets. The references to $\mathscr{A}$ can be ignored with practically no loss in the understanding of the statistical aspects.

A forerunner of this book appeared in 1950 in the form of mimeographed lecture notes taken by Colin Blyth during a course I taught at Berkeley; they subsequently provided a text for the course until the stencils gave out. Some sections were later updated by Michael Stuart and Fritz Scholz. Throughout the process of converting this material into a book, I greatly benefited from the support and advice of my wife, Juliet Shaffer. Parts of the manuscript were read by Rudy Beran, Peter Bickel, Colin Blyth, Larry Brown, Fritz Scholz, and Geoff Watson, all of whom suggested many improvements. Sections 6.7 and 6.8 are based on material provided by Peter Bickel and Chuck Stone, respectively. Very special thanks are due to Wei-Yin Loh, who carefully read the complete manuscript at its various stages and checked all the problems. His work led to the correction of innumerable errors and to many other improvements. Finally, I should like to thank Ruth Suzuki for her typing, which by now is legendary, and Sheila Gerber for her expert typing of many last-minute additions and corrections.

E. L. LEHMANN

Berkeley, California,
March 1983

Comments for Instructors

The two companion volumes, *Testing Statistical Hypotheses* (TSH) and *Theory of Point Estimation* (TPE), between them provide an introduction to classical statistics from a unified point of view. Different optimality criteria are considered, and methods for determining optimum procedures according to these criteria are developed. The application of the resulting theory to a variety of specific problems as an introduction to statistical methodology constitutes a second major theme.

On the other hand, the two books are essentially independent of each other. (As a result, there is some overlap in the preparatory chapters; also, each volume contains cross-references to related topics in the other.) They can therefore be taught in either order. However, TPE is somewhat more discursive and written at a slightly lower mathematical level and, for this reason, may offer the better starting point.

The material of the two volumes combined somewhat exceeds what can be comfortably covered in a year's course meeting 3 hours a week, thus providing the instructor with some choice of topics to be emphasized. A one-semester course covering both estimation and testing, can be obtained, for example, by deleting all large-sample considerations, all nonparametric material, the sections concerned with simultaneous estimation and testing, the minimax chapter of TSH, and some of the applications. Such a course might consist of the following sections: TPE: Chap. 2, § 1 and a few examples from §§ 2, 3; Chap. 3, §§1–3; Chap. 4, §§ 1–4. TSH: Chap. 3, §§ 1–3, 5, 7 (without proof of Theorem 6); Chap. 4, §§1–6; Chap. 5, §§ 1–7; Chap. 6, §§1–6; Chap. 7, §§1–3, 5–9, together with material from the preparatory chapters (TSH Chap. 1, 2; TPE Chap. 1) as it is needed.

Contents

Theory of Point Estimation

CHAPTER 1

Preparations

1. THE PROBLEM

Statistics is concerned with the collection of data and with their analysis and interpretation. We shall not consider the problem of data collection in this book but shall take the data as given and ask what they have to tell us. The answer depends not only on the data, on what is being observed, but also on background knowledge of the situation; the latter is formalized in the assumptions with which the analysis is entered. We shall distinguish between three principal lines of approach.

Data analysis. Here the data are analyzed on their own terms, essentially without extraneous assumptions. The principal aim is the organization and summarization of the data in ways that bring out their main features and clarify their underlying structure.

Classical inference and decision theory. The observations are now postulated to be the values taken on by random variables which are assumed to follow a joint probability distribution, P, belonging to some known class $\mathcal{P}$. Frequently, the distributions are indexed by a parameter, say θ (not necessarily real-valued), taking values in a set, Ω, so that

$$\mathcal{P} = \{P_\theta, \theta \in \Omega\}. \tag{1}$$

The aim of the analysis is then to specify a plausible value for θ (this is the problem of point estimation), or at least to determine a subset of Ω of which we can plausibly assert that it does, or does not, contain θ (estimation by confidence sets or hypothesis testing). Such a statement about θ can be viewed as a summary of the information provided by the data and may be used as a guide to action.

Bayesian analysis. In this approach, it is assumed in addition that θ is itself a random variable (though unobservable) with a *known* distribution. This prior distribution (specified prior to the availability of the data) is modified in light of the data to determine a posterior distribution (the conditional distribution of θ given the data), which summarizes what can be said about θ on the basis of the assumptions made *and the data*.

These three methods of approach permit increasingly strong conclusions, but they do so at the price of assumptions which are correspondingly more detailed and hence less reliable. It is often desirable to use different formulations in conjunction, for example, by planning a study (e.g., determining sample size) under rather detailed assumptions but performing the analysis under a weaker set which appears more trustworthy. In practice, it is often useful to model a problem in a number of different ways. One may then be satisfied if there is reasonable agreement among the conclusions; in the contrary case, a closer examination of the different sets of assumptions will be indicated.

In this book we shall be concerned principally with the second formulation. (A book-length treatment of the first is Tukey's *Exploratory Data Analysis* and of the third, Lindley's *Introduction to Probability and Statistics from a Bayesian Viewpoint*.) In this approach, one tries to specify what is meant by a "best" statistical procedure for a given problem and to develop methods for determining such best procedures.

This program encounters two difficulties. First, there is typically no unique, convincing definition of optimality. Various optimality criteria with which we shall be concerned in Chapters 2 through 4 will be discussed later in this section. Perhaps even more serious is the fact that the optimal procedure and its properties may depend very heavily on the precise nature of the assumed probability model (1), which often rests on rather flimsy foundations. It therefore becomes important to consider the *robustness* of the proposed solution under deviations from the model. Some aspects of robustness will be taken up in Chapter 5.

The discussion so far has been quite general; let us now specialize to point estimation. In terms of the model (1), suppose that g is a real-valued function defined over Ω and that we would like to know the value of $g(\theta)$ (which may, of course, be θ itself). Unfortunately, θ, and hence $g(\theta)$, is unknown. However, the data can be used to obtain an estimate of $g(\theta)$, a value that one hopes will be close to $g(\theta)$.

Point estimation is one of the most common forms of statistical inference. One measures a physical quantity in order to estimate its value; surveys are conducted to estimate the proportion of voters favoring a candidate or watching a television program; agricultural experiments are carried out to estimate the effect of a new fertilizer, and clinical experiments to estimate the improved life expectancy or cure rate resulting from a medical treatment. As a prototype of such an estimation problem, consider the determination of an unknown quantity by measuring it.

Example 1.1. The measurement problem. A number of measurements are taken of some quantity, for example, a distance (or temperature), in order to obtain an estimate of the quantity θ being measured. If the n measured values are $x_1, \ldots, x_n$, a common recommendation is to estimate θ by their mean

$$\bar{x} = \frac{(x_1 + \cdots + x_n)}{n}.$$

The idea of averaging a number of observations to obtain a more precise value seems so commonplace today that it is difficult to realize it has not always been in use. It appears to have been introduced only toward the end of the seventeenth century (see Plackett, 1958). But why should the observations be combined in just this way? The following are two properties of the mean, which were used in early attempts to justify this procedure.

(i) An appealing approximation to the true value being measured is the value a, for which the sum of squared differences $\Sigma(x_i - a)^2$ is a minimum. That this *least squares* estimate of θ is $\bar{x}$ is seen from the identity

$$\Sigma(x_i - a)^2 = \Sigma(x_i - \bar{x})^2 + n(\bar{x} - a)^2 \tag{2}$$

since the first term on the right side does not involve a and the second term is minimized by $a = \bar{x}$. (For the history of least squares, see Eisenhart, 1964, Plackett, 1972, Harter 1974–1976: and Stigler 1981. Least squares estimation will be discussed in a more general setting in Section 3.4.)

(ii) The least squares estimate defined in (i) is the value minimizing the sum of the squared residuals, the residuals being the differences between the observations x_i and the estimated value. Another approach is to ask for the value a for which the sum of the residuals is zero, so that the positive and negative residuals are in balance. The condition on a is

$$\Sigma(x_i - a) = 0, \tag{3}$$

and this again immediately leads to $a = \bar{x}$. (That the two conditions lead to the same answer is, of course, obvious since (3) expresses that the derivative of (2) with respect to a is zero.)

These two principles clearly belong to the first (data analytic) level mentioned at the beginning of the section. They derive the mean as a reasonable descriptive measure of the center of the observations, but they cannot justify $\bar{x}$ as an estimate of the true value θ since no explicit assumption has been made connecting the observations x_i with θ. To establish such a connection, let us now assume that the x_i are the observed values of n independent random variables which have a common distribution depending on θ. Eisenhart (1964) attributes the crucial step of introducing such probability models for this purpose to Simpson (1755).

More specifically, we shall assume that $X_i = \theta + U_i$, where the measurement error U_i is distributed according to a distribution F symmetric about 0 so that the X_i are symmetrically distributed about θ with distribution

$$P(X_i \leqslant x) = F(x - \theta). \tag{4}$$

In terms of this model, can we now justify the idea that the mean provides a more precise value than a single observation? If the X's have a finite variance σ^2, the variance of the mean $\bar{X}$ is σ^2/n; the expected squared difference between $\bar{X}$ and θ is therefore only $1/n$ of what it is for a single observation. However, if the X's have a Cauchy distribution, the distribution of $\bar{X}$ is the same as that of a single X_i (Problem 1.8), so that nothing is gained by taking several measurements and then averaging them. Whether $\bar{X}$ is a reasonable estimator of θ thus depends on the nature of the X_i.

As illustrated in this example, formalization of an estimation problem involves two basic ingredients:

(i) A real-valued function g defined over a parameter space Ω, whose value at θ is to be estimated; we shall call $g(\theta)$ the *estimand*. [In Example 1.1, $g(\theta) = \theta$].

(ii) A *random observable* X (typically vector-valued) taking on values in a sample space $\mathcal{X}$ according to a distribution P_θ, which is known to belong to a family $\mathcal{P}$ as stated in (1). [In Example 1.1, $X = (X_1, \ldots, X_n)$, where the X_i are independently, identically distributed (iid) and their distribution is given by (4)]. The observed value x of X constitutes the *data*.

The problem is the determination of a suitable *estimator*, that is, a real-valued function δ defined over the sample space, of which it is hoped that $\delta(X)$ will tend to be close to the unknown $g(\theta)$. The value $\delta(x)$ taken on by $\delta(X)$ for the observed value x of X is then the *estimate* of $g(\theta)$, which will be our "educated guess" for the unknown value. One could adopt a slightly more restrictive definition. In applications it is often desirable to restrict δ to possible values of $g(\theta)$, for example, to be positive when g takes on only positive values, to be integer-valued when g is, and so on. For the moment, however, it is more convenient not to impose this additional restriction.

The estimator δ is to be close to $g(\theta)$ and, since $\delta(X)$ is a random variable, we shall interpret this to mean that it will be close on the average. To make this requirement precise, it is necessary to specify a measure of the average closeness of (or distance from) an estimator to $g(\theta)$. Examples of such measures are

$$P(|\delta(X) - g(\theta)| < c) \quad \text{for some} \quad c > 0, \tag{5}$$

and

$$E|\delta(X) - g(\theta)|^p \quad \text{for some} \quad p > 0. \tag{6}$$

(Of these, we want the first to be large and the second to be small.) If g and

δ take on only positive values, one may be interested in

$$E\left|\frac{\delta(X)}{g(\theta)} - 1\right|^p,$$

which suggests generalizing (6) to

$$\kappa(\theta) E|\delta(X) - g(\theta)|^p. \tag{7}$$

Quite generally, suppose that the consequences of estimating $g(\theta)$ by a value d are measured by $L(\theta, d)$. Of the *loss function* L, we shall assume that

$$L(\theta, d) \geqslant 0 \qquad \text{for all} \quad \theta, d \tag{8}$$

and

$$L[\theta, g(\theta)] = 0 \qquad \text{for all} \quad \theta, \tag{9}$$

so that the loss is zero when the correct value is estimated. The accuracy, or rather inaccuracy, of an estimator δ is then measured by the *risk function*

$$R(\theta, \delta) = E_\theta\{L[\theta, \delta(X)]\}, \tag{10}$$

the long-term average loss resulting from the use of δ. One would like to find a δ which minimizes the risk for all values of θ.

As stated, this problem has no solution. For by (9), it is possible to reduce the risk at any given point θ_0 to zero by making $\delta(x)$ equal to $g(\theta_0)$ for all x. There thus exists no *uniformly best* estimator, that is, no estimator which simultaneously minimizes the risk for all values of θ, except in the trivial case that $g(\theta)$ is constant.

One way of avoiding this difficulty is to restrict the class of estimators by ruling out estimators that too strongly favor one or more values of θ at the cost of neglecting other possible values. This can be achieved by requiring the estimator to satisfy some condition which enforces a certain degree of impartiality. One such condition requires that the *bias* $E_\theta[\delta(X)] - g(\theta)$, sometimes called the systematic error, of the estimator δ, be zero, that is, that

$$E_\theta[\delta(X)] = g(\theta) \qquad \text{for all} \quad \theta \in \Omega. \tag{11}$$

This condition of *unbiasedness* ensures that, in the long run, the amounts by which δ over- and underestimates $g(\theta)$ will balance, so that the estimated value will be correct "on the average." A somewhat similar condition is obtained by considering not the amount but only the frequency of over- and underestimation. This leads to the condition

$$P_\theta[\delta(X) < g(\theta)] = P_\theta[\delta(X) > g(\theta)] \tag{12}$$

or slightly more generally to the requirement that $g(\theta)$ be a median of $\delta(X)$ for all values of θ. To distinguish it from this condition of *median-unbiasedness*, (11) is called *mean-unbiasedness* if there is a possibility of confusion. A more general unbiasedness concept, of which the above two are special cases, will be discussed in Section 3.1. The theory of mean-unbiased estimation is the topic of Chapter 2.

A different impartiality condition can be formulated when symmetries are present in a problem. It is then natural to require a corresponding symmetry to hold for the estimator. The resulting condition of *equivariance* will be explored in Chapter 3.

In the next two chapters it will be seen that in many important problems the conditions of unbiasedness and equivariance lead to estimators that are uniformly best among the estimators satisfying these restrictions. Nevertheless, the applicability of both conditions is limited. There is an alternative approach which is more generally applicable. Instead of asking for an estimator which minimizes the risk uniformly in θ, one can more modestly ask that the risk function be low only in some overall sense. Two natural global measures of the size of the risk are the average

$$\int R(\theta, \delta) w(\theta)\, d\theta \tag{13}$$

for some weight function w and the maximum of the risk function

$$\sup_{\Omega} R(\theta, \delta). \tag{14}$$

The estimator minimizing (13) (discussed in Section 4.1) formally coincides with that obtained from the Bayes approach, mentioned at the beginning of the section when θ is assumed to be a random variable with probability density w. Minimizing (14) leads to the *minimax* estimator, which will be considered in Section 4.2.

The formulation of an estimation problem in a concrete situation along the lines described in this section requires specification of the probability model (1) and of a measure of inaccuracy $L(\theta, d)$. In the measurement problem of Example 1.1 and its generalizations to linear models, it is frequently reasonable to assume that the measurement errors are approximately normally distributed (but see Section 5.3). In other situations, the assumptions underlying a binomial or Poisson distribution may be appropriate. Thus, knowledge of the circumstances and previous experience with similar situations will often suggest a particular parametric family $\mathcal{P}$ of distributions. If such information is not available, one may instead adopt a nonparametric model, which requires only very general assumptions such as independence or symmetry but does not lead to a particular parametric family of distributions. As a compromise between these two approaches, one may be willing to assume that the true distribution, though not exactly following a particular parametric form, lies within a stated distance of some parametric family. Such a neighborhood model will be considered in Section 5.3.

The choice of an appropriate model requires judgment and utilizes experience; it is also affected by considerations of convenience. Analogous considerations for choice of the loss function L appear to be much more difficult. The most common fate of a point estimate (for example, of the distance of a star or the success probability of an operation) is to wind up in a research report or paper. It is likely to be used on different occasions and in various settings for a variety of purposes which cannot be foreseen at the time the estimate is made. Under these circumstances one wants the estimator to be accurate, but just what measure of accuracy should be used is fairly arbitrary.

This was recognized very clearly by Laplace (1820) and Gauss (1821), who compared the estimation of an unknown quantity, on the basis of observations with random errors, with a game of chance and the error in the estimated value with the loss resulting from such a game. Gauss proposed the square of the error as measure of loss or inaccuracy. Should someone object to this specification as arbitrary, he writes, he is in complete agreement. He defends his choice by an appeal to mathematical simplicity and convenience. Among the infinite variety of possible functions for the purpose, the square is the simplest and is therefore preferable.

When estimates are used to make definite decisions (for example, to determine the amount of medication to be given a patient or the size of an order that a store should place for some goods), it is sometimes possible to specify the loss function by the consequences of various errors in the estimate. A general discussion of the distinction between inference and decision problems is given by Blyth (1970).

Actually, it turns out that much of the general theory does not require a detailed specification of the loss function but applies to large classes of such functions, in particular to loss functions $L(\theta, d)$, which are convex in d. [For example, this includes (7) with $p \geqslant 1$ but not with $p < 1$. It does not include (5).] We shall here develop the theory for suitably general classes of loss functions whenever the cost in complexity is not too high. However, in applications to specific examples—and these form a large part of the subject—the choice of squared error as loss has the twofold advantage of ease of computation and of leading to estimators that can be obtained explicitly. For these reasons, in the examples we shall typically take the loss to be squared error.

2. MEASURE THEORY AND INTEGRATION

A convenient framework for theoretical statistics is measure theory in abstract spaces. The present section will sketch (without proofs) some of the principal concepts, results, and notational conventions of this theory. Such a sketch should provide sufficient background for a comfortable understanding of the ideas and results and the essentials of most of the proofs, in this book. A fuller account of measure theory can be found in many standard books, for example, Billingsley (1979), Halmos (1950), and Rudin (1966).

The most natural example of a "measure" is that of the length, area, or volume of sets in one, two, or three-dimensional Euclidean space. As in these special cases, a measure assigns non-negative (not necessarily finite) values to sets in some space $\mathcal{X}$. A measure μ is thus a set function; the value it assigns to a set A will be denoted by $\mu(A)$.

In generalization of the properties of length, area, and volume, a measure will be required to be *additive*, that is, to satisfy

$$\mu(A \cup B) = \mu(A) + \mu(B) \qquad \text{when } A, B \text{ are disjoint,} \tag{1}$$

where $A \cup B$ denotes the union of A and B. From (1), it follows immediately by induction that additivity extends to any finite union of disjoint sets. The measures with which we shall be concerned will be required to satisfy the stronger condition of *sigma-additivity*, namely that

$$\mu\left(\bigcup_{i=1}^{\infty} A_i\right) = \sum_{i=1}^{\infty} \mu(A_i) \tag{2}$$

for any countable collection of disjoint sets.

The domain over which a measure μ is defined is a class of subsets of $\mathcal{X}$. It would seem easiest to assume that this is the class of all subsets of $\mathcal{X}$. Unfortunately, it turns out that typically it is not possible to give a satisfactory definition of the measures of interest for all subsets of $\mathcal{X}$ in such a way that (2) holds. (Such a negative statement holds in particular for length, area, and volume (see, for example, Halmos, p. 70) but not for the measure μ of Example 2.1 below). It is therefore necessary to restrict the definition of μ to a suitable class of subsets of $\mathcal{X}$. This class should contain the whole space $\mathcal{X}$ as a member, and for any set A also its *complement* $\mathcal{X} - A$. In view of (2), it should also contain the union of any countable collection of disjoint sets of the class. A class of sets satisfying these conditions is called a *σ-field* or *σ-algebra*. It is easy to see that if $A_1, A_2, \ldots$ are members of a σ-field $\mathcal{A}$, then so are their union and intersection (Problem 2.1).

If $\mathcal{A}$ is a σ-field of subsets of a space $\mathcal{X}$, then $(\mathcal{X}, \mathcal{A})$ is said to be a *measurable space* and the sets A of $\mathcal{A}$ to be *measurable*. A *measure* μ is a non-negative set function defined over a σ-field $\mathcal{A}$ and satisfying (2). If μ is a measure defined over a measurable space $(\mathcal{X}, \mathcal{A})$, the triple $(\mathcal{X}, \mathcal{A}, \mu)$ is called a *measure space*.

A measure is *σ-finite* if there exist sets A_i in $\mathcal{A}$ whose union is $\mathcal{X}$ and such that $\mu(A_i) < \infty$. All measures with which we shall be concerned in this book are σ-finite, and we shall therefore use the term *measure* to mean a σ-finite measure.

The following are two important examples of measure spaces.

Example 2.1. Counting measure. Let $\mathcal{X}$ be countable and $\mathcal{A}$ the class of all subsets of $\mathcal{X}$. For any A in $\mathcal{A}$, let $\mu(A)$ be the number of points of A if A is finite, and $\mu(A) = \infty$ otherwise. This measure μ is called *counting measure*. That μ is σ-finite is obvious.

Example 2.2. Lebesgue measure. Let $\mathcal{X}$ be n-dimensional Euclidean space E_n, and let $\mathcal{A}$ be the smallest σ-field containing all open rectangles

$$A = \{(x_1, \ldots, x_n) : a_i < x_i < b_i\}, \qquad -\infty < a_i < b_i < \infty. \tag{3}$$

We shall then say that $(\mathcal{X}, \mathcal{A})$ is *Euclidean*. The members of $\mathcal{A}$ are called *Borel sets*. This is a very large class which contains, among others, all open and all closed subsets of $\mathcal{X}$. There exists a (unique) measure μ, defined over $\mathcal{A}$, which assigns to (3) the measure

$$\mu(A) = (b_1 - a_1) \cdots (b_n - a_n), \tag{4}$$

that is, its volume; μ is called *Lebesgue measure*.

The intuitive meaning of measure suggests that any subset of a set of measure zero should again have measure zero. If $(\mathcal{X}, \mathcal{A}, \mu)$ is a measure space, it may, however, happen that a subset of a set in $\mathcal{A}$ which has measure zero is not in $\mathcal{A}$ and hence not measurable. This difficultly can be remedied by the process of completion. Consider the class $\mathcal{B}$ of all sets $B = A \cup C$ where A is in $\mathcal{A}$ and C is a subset of a set in $\mathcal{A}$ having measure zero. Then $\mathcal{B}$ is a σ-field (Problem 2.7). If μ' is defined over $\mathcal{B}$ by $\mu'(B) = \mu(A)$, μ' agrees with μ over $\mathcal{A}$, and $(\mathcal{X}, \mathcal{B}, \mu')$ is called the *completion* of the measure space $(\mathcal{X}, \mathcal{A}, \mu)$.

When the process of completion is applied to Example 2.2 so that $\mathcal{X}$ is Euclidean and $\mathcal{A}$ is the class of Borel sets, the resulting larger class $\mathcal{B}$ is the class of Lebesgue measurable sets. The measure μ' defined over $\mathcal{B}$, which agrees with Lebesgue measure over the Borel sets, is also called Lebesgue measure.

A third principal concept needed in addition to σ-field and measure is that of the integral of a real-valued function f with respect to a measure μ. However, before defining this integral, it is necessary to specify a suitable class of functions f. This will be done in three steps.

First, consider the class of real-valued functions s called *simple*, which take on only a finite number of values, say $a_1, \dots, a_m$ and for which the sets

$$A_i = \{x: s(x) = a_i\} \tag{5}$$

belong to $\mathcal{A}$. An important special case of a simple function is the *indicator* I_A of a set A in $\mathcal{A}$, defined by

$$I_A(x) = \begin{cases} 1 & \text{if} \quad x \in A \\ 0 & \text{if} \quad x \notin A. \end{cases} \tag{6}$$

Second, let $s_1, s_2, \dots$ be a nondecreasing sequence of non-negative simple functions and let

$$f(x) = \lim_{n \to \infty} s_n(x). \tag{7}$$

Note that this limit exists since for every x the sequence $s_1(x), s_2(x), \dots$ is nondecreasing but that $f(x)$ may be infinite. A function with domain $\mathcal{X}$ and range $[0, \infty)$, that is, non-negative and finite valued, will be called $\mathcal{A}$-measurable or, for short, *measurable* if there exists a nondecreasing sequence of non-negative simple functions such that (7) holds for all $x \in \mathcal{X}$.

Third, for an arbitrary function f, define its *positive and negative part* by

$$f^+(x) = \max(f(x), 0), \qquad f^-(x) = -\min(f(x), 0),$$

so that f^+ and f^- are both non-negative and

$$f = f^+ - f^-.$$

Then a function with domain $\mathcal{X}$ and range $(-\infty, \infty)$ will be called *measurable* if both its positive and its negative parts are measurable. The measurable functions constitute a very large class which has a simple alternative characterization.

It can be shown that a real-valued function f is $\mathcal{A}$-measurable if and only if, for every Borel set B on the real line, the set

$$\{x: f(x) \in B\}$$

is in $\mathcal{A}$. It follows from the definition of Borel sets that it is enough to check that $\{x: f(x) < b\}$ is in $\mathcal{A}$ for every b. This shows in particular that if $(\mathcal{X}, \mathcal{A})$ is Euclidean and f continuous, then f is measurable. As another important class, consider functions taking on a countable number of values. If f takes on distinct values $a_1, a_2, \ldots$ on sets $A_1, A_2, \ldots$ it is measurable if and only if $A_i \in \mathcal{A}$ for all i.

The integral can now be defined in three corresponding steps.

(i) For a non-negative simple function s taking on values a_i on the sets A_i, define

$$\int s\, d\mu = \Sigma a_i \mu(A_i) \tag{8}$$

where $a\mu(A)$ is to be taken as zero when $a = 0$ and $\mu(A) = \infty$.

(ii) For a non-negative measurable function f given by (7), define

$$\int f\, d\mu = \lim_{n \to \infty} \int s_n\, d\mu. \tag{9}$$

Here the limit on the right side exists since the fact that the functions s_n are nondecreasing implies the same for the numbers $\int s_n\, d\mu$. The definition (9) is meaningful because it can be shown that if $\{s_n\}$ and $\{s'_n\}$ are two nondecreasing sequences with the same limit function, their integrals also will have the same limit. Thus, the value of $\int f\, d\mu$ is independent of the particular sequence used in (7).

The definitions (8) and (9) do not preclude the possibility that $\int s\, d\mu$ or $\int f\, d\mu$ is infinite. A non-negative measurable function is *integrable* (with respect to μ) if $\int f\, d\mu < \infty$.

(iii) An arbitrary measurable function f is said to be integrable if its positive and negative parts are integrable, and its integral is then defined by

$$\int f\, d\mu = \int f^+\, d\mu - \int f^-\, d\mu. \tag{10}$$

Important special cases of this definition are obtained by taking for μ the measures defined in Examples 2.1 and 2.2.

Example 2.1. (*Continued*). If $\mathcal{X} = \{x_1, x_2, \ldots\}$ and μ is counting measure, it is easily seen from (8) through (10) that

$$\int f\, d\mu = \Sigma f(x_i).$$

Example 2.2. (*Continued*). If μ is Lebesgue measure, then $\int f\, d\mu$ exists whenever the Riemann integral (the integral taught in calculus courses) exists and the two agree. However, the integral defined in (8) through (10) exists for many functions for which the Riemann integral is not defined. A simple example is the function f for which $f(x) = 1$ or 0, as x is rational or irrational. It follows from (22) below that the integral of f with respect to Lebesgue measure is zero; on the other hand, f is not Riemann integrable (Problem 2.12).

In analogy with the customary notation for the Riemann integral, it will frequently be convenient to write the integral (10) as $\int f(x)\, d\mu(x)$. This is especially true when f is given by an explicit formula.

The integral defined above has the properties one would expect of it. In particular, for any real numbers $c_1, \ldots, c_m$ and any integrable functions $f_1, \ldots, f_m$, $\Sigma c_i f_i$ is also integrable and

$$\int (\Sigma c_i f_i)\, d\mu = \Sigma c_i \int f_i\, d\mu. \tag{11}$$

Also, if f is measurable and g integrable and if $0 \leqslant f \leqslant g$, then f is also integrable, and

$$\int f\, d\mu \leqslant \int g\, d\mu. \tag{12}$$

We shall often be dealing with statements that hold except on a set of measure zero. If a statement holds for all x in $\mathcal{X} - N$ where $\mu(N) = 0$, the statement is said to hold a.e. (almost everywhere) μ (or a.e. if the measure μ is clear from the context).

It is sometimes required to know when $f(x) = \lim f_n(x)$ or more generally when

$$f(x) = \lim f_n(x) \qquad (\text{a.e. } \mu) \tag{13}$$

implies that

$$\int f\, d\mu = \lim \int f_n\, d\mu. \tag{14}$$

The following is a simple sufficient condition known as the *dominated convergence theorem*.

If the f_n are measurable and satisfy (13), and if there exists an integrable function g such that

$$|f_n(x)| \leqslant g(x) \qquad \text{for all} \quad x, \tag{15}$$

then the f_n and f are integrable and (14) holds.

The following is another useful result concerning integrals of sequences of functions.

Fatou's Lemma. *If* $\{f_n\}$ *is a sequence of non-negative measurable functions, then*

$$\int \left(\liminf_{n\to\infty} f_n \right) d\mu \leqslant \liminf_{n\to\infty} \int f_n \, d\mu \tag{16}$$

with the reverse inequality holding for lim sup. (*Recall that the lim inf and lim sup of a sequence of numbers are, respectively, the smallest and largest limit points that can be obtained through subsequences. See Problems* 2.5 *and* 2.6)

As a last extension of the concept of integral, define

$$\int_A f \, d\mu = \int I_A f \, d\mu \tag{17}$$

when the integral on the right exists. It follows in particular from (8) and (17) that

$$\int_A d\mu = \mu(A). \tag{18}$$

Obviously such properties as (11) and (12) continue to hold when $\int$ is replaced by $\int_A$.

It is often useful to know under what conditions an integrable function f satisfies

$$\int_A f \, d\mu = 0. \tag{19}$$

This will clearly be the case when either

$$f = 0 \quad \text{on } A \tag{20}$$

or

$$\mu(A) = 0. \tag{21}$$

More generally, it will be the case whenever

$$f = 0 \quad \text{a.e. on } A, \tag{22}$$

that is, f is zero except on a subset of A having measure zero.

Conversely, if f is a.e. non-negative on A,

$$\int_A f\,d\mu = 0 \Rightarrow f = 0 \qquad \text{a.e. on } A, \tag{23}$$

and if f is a.e. positive on A, then

$$\int_A f\,d\mu = 0 \Rightarrow \mu(A) = 0. \tag{24}$$

Note that, as a special case of (22), if f and g are integrable functions differing only on a set of measure zero, that is, if $f = g$ (a.e. μ), then

$$\int f\,d\mu = \int g\,d\mu.$$

It is a consequence that functions can never be determined by their integrals uniquely but at most up to sets of measure zero.

For a non-negative integrable function f, let us now consider

$$\nu(A) = \int_A f\,d\mu \tag{25}$$

as a set function defined over $\mathcal{A}$. Then ν is non-negative, σ-finite, and σ-additive and hence a measure over $(\mathcal{X}, \mathcal{A})$.

If μ and ν are two measures defined over the same measurable space $(\mathcal{X}, \mathcal{A})$, it is a question of central importance whether there exists a function f such that (25) holds for all $A \in \mathcal{A}$. By (21), a necessary condition for such a representation is clearly that

$$\mu(A) = 0 \Rightarrow \nu(A) = 0. \tag{26}$$

When (26) holds, ν is said to be *absolutely continuous* with respect to μ. It is a surprising and basic fact known as the *Radon-Nikodym* theorem that (26) is not only necessary but also sufficient for the existence of a function f satisfying (25) for all $A \in \mathcal{A}$. The resulting function f is called the *Radon-Nikodym derivative* of ν with respect to μ. This f is not unique because it can be changed on a set of μ-measure zero without affecting the integrals (25). However, it is *unique a.e.* μ in the sense that if g is any other integrable function satisfying (25), then $f = g$ (a.e. μ). It is a useful consequence of this result that

$$\int_A f\, d\mu = 0 \qquad \text{for all} \quad A \in \mathcal{A}$$

implies that $f = 0$ (a.e. μ).

The last theorem on integration we require is a form of Fubini's theorem which essentially states that in a repeated integral of a non-negative function, the order of integration is immaterial. To make this statement precise, define the Cartesian product of any two sets A, B as the set of all ordered pairs (a, b) with $a \in A$, $b \in B$. Let $(\mathcal{X}, \mathcal{A}, \mu)$ and $(\mathcal{Y}, \mathcal{B}, \nu)$ be two measure spaces, and define $\mathcal{A} \times \mathcal{B}$ to be the smallest σ-field containing all sets $A \times B$ with $A \in \mathcal{A}$ and $B \in \mathcal{B}$. Then there exists a unique measure λ over $\mathcal{A} \times \mathcal{B}$ which to any product set $A \times B$ assigns the measure $\mu(A) \cdot \nu(B)$. The measure λ is called the *product measure* of μ and ν and is denoted by $\mu \times \nu$.

Example 2.3. Borel sets. If $\mathcal{X}$ and $\mathcal{Y}$ are Euclidean spaces E_m and E_n and $\mathcal{A}$ and $\mathcal{B}$ the σ-fields of Borel sets of $\mathcal{X}$ and $\mathcal{Y}$, respectively, then $\mathcal{X} \times \mathcal{Y}$ is Euclidean space E_{m+n}, and $\mathcal{A} \times \mathcal{B}$ is the class of Borel sets of $\mathcal{X} \times \mathcal{Y}$. If, in addition, μ and ν are Lebesgue measure on $(\mathcal{X}, \mathcal{A})$ and $(\mathcal{Y}, \mathcal{B})$, then $\mu \times \nu$ is Lebesgue measure on $(\mathcal{X} \times \mathcal{Y}, \mathcal{A} \times \mathcal{B})$.

An integral with respect to a product measure generalizes the concept of a double integral. The following theorem, which is one version of Fubini's theorem, states conditions under which a double integral is equal to a repeated integral and under which it is permitted to change the order of integration in a repeated integral.

Theorem (Fubini). *Let $(\mathcal{X}, \mathcal{A}, \mu)$ and $(\mathcal{Y}, \mathcal{B}, \nu)$ be measure spaces and let* f *be a non-negative $\mathcal{A} \times \mathcal{B}$-measurable function defined on $\mathcal{X} \times \mathcal{Y}$. Then*

$$\text{(27)} \qquad \int_{\mathcal{X}}\left[\int_{\mathcal{Y}} f(x, y)\, d\nu(y)\right] d\mu(x) = \int_{\mathcal{Y}}\left[\int_{\mathcal{X}} f(x, y)\, d\mu(x)\right] d\nu(y)$$

$$= \int_{\mathcal{X}\times\mathcal{Y}} f\, d(\mu \times \nu).$$

Here the first term is the repeated integral in which f is first integrated for fixed x with respect to ν, and then the result with respect to μ. The inner integrals of the first two terms in (27) are, of course, not defined unless $f(x, y)$, for fixed values of either variable is a measurable function of the other. Fortunately, under the assumptions of the theorem, this is always the case. Similarly, existence of the outer integrals requires the inner integrals to be measurable functions of the variable that has not been integrated. This condition is also satisfied.

The remainder of this section is concerned with the specialization of these concepts to probability theory. A measure P defined over a measure space $(\mathcal{X}, \mathcal{A})$ satisfying

$$P(\mathcal{X}) = 1 \tag{28}$$

is a *probability measure* (or *probability distribution*), and the value $P(A)$ it assigns to A is the *probability* of A. If P is absolutely continuous with respect to a measure μ with Radon–Nikodym derivative p, so that

$$P(A) = \int_A p\, d\mu, \tag{29}$$

p is called the *probability density* of P with respect to μ. Such densities are, of course, determined only up to sets of μ-measure zero.

We shall here be concerned only with situations in which $\mathcal{X}$ is Euclidean, and typically the distributions will either be discrete (in which case μ can be taken to be counting measure) or absolutely continuous with respect to Lebesgue measure.

Statistical problems are concerned not with single probability distributions but with families of such distributions

$$\mathcal{P} = \{P_\theta, \theta \in \Omega\} \tag{30}$$

defined over a common measurable space $(\mathcal{X}, \mathcal{A})$. When all the distributions of $\mathcal{P}$ are absolutely continuous with respect to a common measure μ, as will usually be the case, the family $\mathcal{P}$ is said to be *dominated* (by μ).

Most of the examples with which we shall deal belong to one or the other of the following two cases.

(i) *The discrete case.* Here $\mathcal{X}$ is a countable set, $\mathcal{A}$ the class of subsets of $\mathcal{X}$, and the distributions of $\mathcal{P}$ are dominated by counting measure.

(ii) *The absolutely continuous case.* Here $\mathcal{X}$ is a Borel subset of a Euclidean space, $\mathcal{A}$ the class of Borel subsets of $\mathcal{X}$, and the distributions of $\mathcal{P}$ are dominated by Lebesgue measure over $(\mathcal{X}, \mathcal{A})$.

It is one of the advantages of the general approach of this section that it includes both these cases, as well as mixed situations such as those arising with censored data.

When dealing with a family $\mathcal{P}$ of distributions, the most relevant null-set concept is that of a *$\mathcal{P}$-null set*, that is, of a set N satisfying

$$P(N) = 0 \quad \text{for all} \quad P \in \mathcal{P}. \tag{31}$$

If a statement holds except on a set N satisfying (31), we shall say that the statements holds (a.e. $\mathscr{P}$). If $\mathscr{P}$ is dominated by μ, then

$$\mu(N) = 0 \tag{32}$$

implies (31). When the converse also holds, μ and $\mathscr{P}$ are said to be *equivalent*.

To bring the customary probabilistic framework and terminology into consonance with that of measure theory, it is necessary to define the concepts of random variable and random vector. A random variable is the mathematical representation of some real-valued aspect of an experiment with uncertain outcome. The experiment may be represented by a space $\mathscr{E}$, the full details of its possible outcomes by the points e of $\mathscr{E}$. The frequencies with which outcomes can be expected to fall into different subsets E of $\mathscr{E}$ (assumed to form a σ-field $\mathscr{B}$) are given by a probability distribution over $(\mathscr{E}, \mathscr{B})$. A *random variable* is then a real-valued function X defined over $\mathscr{E}$. Since we wish the probabilities of the events $X \leqslant a$ to be defined, the function X must be measurable and the probability

$$F_X(a) = P(X \leqslant a) \tag{33}$$

is simply the probability of the set $\{e\colon X(e) \leqslant a\}$. The function F_X defined through (33) is the *cumulative distribution function* (*cdf*) of X.

It is convenient to digress here briefly in order to define another concept of absolute continuity. A real-valued function f on $(-\infty, \infty)$ is said to be *absolutely continuous* if given any $\varepsilon > 0$ there exists $\delta > 0$ such that for each finite collection of disjoint bounded open intervals (a_i, b_i),

$$\Sigma(b_i - a_i) < \delta \qquad \text{implies} \quad \Sigma|f(b_i) - f(a_i)| < \varepsilon. \tag{34}$$

A connection with the earlier concept of absolute continuity of one measure with respect to another is established by the fact that a cdf F on the real line is absolutely continuous if and only if the probability measure it generates is absolutely continuous with respect to Lebesgue measure. Any absolutely continuous function is continuous (Problem 2.13), but the converse does not hold. In particular, there exist continuous cumulative distribution functions, which are not absolutely continuous and therefore do not have a probability density with respect to Lebesgue measure. Such distributions are rather pathological and play little role in statistics.

If not just one but n real-valued aspects of an experiment are of interest, these are represented by a measurable vector-valued function $(X_1, \ldots, X_n)$

defined over $\mathcal{E}$, with the joint cdf

$$(35) \qquad F_X(a_1,\ldots,a_n) = P[X_1 \leqslant a_1,\ldots,X_n \leqslant a_n]$$

being the probability of the event

$$(36) \qquad \{e\colon X_1(e) \leqslant a_1,\ldots,X_n(e) \leqslant a_n\}.$$

The cdf (35) determines the probabilities of $(X_1,\ldots,X_n)$ falling into any Borel set A, and these agree with the probabilities of the events

$$\{e\colon [X_1(e),\ldots,X_n(e)] \in A\}.$$

From this description of the mathematical model, one might expect the starting point for modeling a specific situation to be the measurable space $(\mathcal{E}, \mathcal{B})$ and a family $\mathcal{P}$ of probability distributions defined over it. However, the statistical analysis of an experiment is typically not based on a full description of the experimental outcome (which would, for example, include the minutest details concerning all experimental subjects) represented by the points e of $\mathcal{E}$. More often, the starting point is a set of observations, represented by a random vector $X = (X_1,\ldots,X_n)$, with all other aspects of the experiment being ignored. The specification of the model will therefore begin with X, the *data*; the measurable space $(\mathcal{X}, \mathcal{A})$ in which X takes on its values, the *sample space*; and a family $\mathcal{P}$ of probability distributions to which the distribution of X is known to belong. Real-valued or vector-valued measurable functions $T = (T_1,\ldots,T_k)$ of X are called *statistics*; in particular, estimators are statistics.

The change of starting point from $(\mathcal{E}, \mathcal{B})$ to $(\mathcal{X}, \mathcal{A})$ requires clarification of two definitions: (1) In order to avoid reference to $(\mathcal{E}, \mathcal{B})$ it is convenient to require T to be a measurable function over $(\mathcal{X}, \mathcal{A})$ rather than over $(\mathcal{E}, \mathcal{B})$. Measurability with respect to the original $(\mathcal{E}, \mathcal{B})$ is then an automatic consequence (Problem 2.14). (2) Analogously, the expectation of a real-valued integrable T is originally defined as

$$\int T[X(e)]\, dP(e).$$

However, it is legitimate to calculate it instead from the formula

$$E(T) = \int T(x)\, dP_X(x)$$

where P_X denotes the probability distribution of X.

As a last concept, we mention the *support* of a distribution P on $(\mathcal{X}, \mathcal{A})$. It is the set of all points x for which $P(A) > 0$ for all open rectangles A [defined by (3)] which contain x.

Example 2.4. Let X be a random variable with distribution P and cdf F, and suppose the support of P is a finite interval I with end points a and b. Then I must be the closed interval $[a, b]$ and F is strictly increasing on $[a, b]$ (Problem 2.15).

If P and Q are two probability measures on $(\mathcal{X}, \mathcal{A})$ and are equivalent (i.e., each is absolutely continuous with respect to the other), then they have the same support; however, the converse need not be true (Problems 2.17, 2.18).

Having outlined the mathematical foundation on which the statistical developments of the later chapters are based, we shall from now on ignore it as far as possible and instead concentrate on the statistical issues. In particular, we shall pay little or no attention to two technical difficulties that occur throughout.

(i) The estimators that will be derived are statistics and hence need to be measurable. However, we shall not check that this requirement is satisfied. In specific examples, it is usually obvious. In more general constructions, it will be tacitly understood that the conclusion holds only if the estimator in question is measurable. In practice, the sets and functions in these constructions usually turn out to be measurable although verification of their measurability can be quite difficult.

(ii) Typically, the estimators are also required to be integrable. This condition will not be as universally satisfied in our examples as measurability and will therefore be checked when it seems important to do so. In other cases it will again be tacitly assumed.

3. GROUP FAMILIES

The two principal families of models with which we shall be concerned in this book are *exponential families* and *group families*. Between them, these families cover many of the more common statistical models. In this and the next section, we shall discuss these families and some of their properties together with some of the more important special cases. More details about these and other special distributions can be found in the four-volume reference work on statistical distributions by Johnson and Kotz (1969–1972).

One of the main reasons for the central role played by these two families in statistics is that in each of them it is possible to effect a great simplification of the data. In an exponential family there exists a fixed (usually rather small) number of statistics to which the data can be reduced without loss of information, regardless of the sample size. In a group family, the simplification stems from the fact that the different distributions of the family play a highly symmetric role. This symmetry in the basic structure again leads essentially to a reduction of the dimensionality of the data since it is then natural to impose a corresponding symmetry requirement on the estimator.

A *group family* of distributions is obtained by subjecting a random variable with a fixed distribution to a family of transformations.

Example 3.1. Location-scale families. Let U be a random variable with a fixed distribution F. If a constant a is added to U, the resulting variable

$$X = U + a \tag{1}$$

has distribution

$$P(X \leqslant x) = F(x - a). \tag{2}$$

The totality of distributions (2), for fixed F and as a varies from $-\infty$ to ∞, is said to constitute a *location family*.

Analogously a *scale family* is generated by the transformations

$$X = bU, \qquad b > 0 \tag{3}$$

and has the form

$$P(X \leqslant x) = F(x/b). \tag{4}$$

Combining these two types of transformations into

$$X = a + bU, \qquad b > 0, \tag{5}$$

one obtains the *location-scale* family

$$P(X \leqslant x) = F\left(\frac{x-a}{b}\right). \tag{6}$$

In applications of these families, F usually has a density f with respect to Lebesgue measure. The density of (5) is then given by

$$\frac{1}{b} f\left(\frac{x-a}{b}\right). \tag{7}$$

Table 3.1 exhibits several such densities, which will be used in the sequel.

In each of (1), (3), and (5), the class of transformations has the following two properties.

(i) *Closure under composition.* Application of a 1 : 1 transformation g_1 from $\mathscr{X}$ to $\mathscr{X}$ followed by another, g_2, results in a new such transformation

Table 3.1. Location-Scale Families[a]

Density	Name	Notation
$\frac{1}{\sqrt{2\pi}\,b} e^{-(x-a)^2/2b^2}$	Normal	$N(a, b^2)$
$\frac{1}{2b} e^{-\lvert x-a\rvert/b}$	Double exponential	$DE(a, b)$
$\frac{b}{\pi} \frac{1}{b^2 + (x-a)^2}$	Cauchy	$C(a, b)$
$\frac{1}{b} \frac{e^{-(x-a)/b}}{[1 + e^{-(x-a)/b}]^2}$	Logistic	$L(a, b)$
$\frac{1}{b} e^{-(x-a)/b} u(a, x)$	Exponential	$E(a, b)$
$\frac{1}{b} u\left(a - \frac{b}{2}, x\right) u\left(x, a + \frac{b}{2}\right)$	Uniform	$U\left(a - \frac{b}{2}, a + \frac{b}{2}\right)$

[a]Here $u(x, y) = \begin{cases} 1 & \text{if } x < y \\ 0 & \text{otherwise.} \end{cases}$

called the composition of g_1 with g_2 and denoted by $g_2 \cdot g_1$. For the transformation (1), addition first of a_1 and then of a_2 results in the addition of $a_1 + a_2$. For (3), multiplication by b_1 and then by b_2 is equivalent to multiplication by $b_2 \cdot b_1$. The composition rule (5) is slightly more complicated. First transforming u to $x = a_1 + b_1 u$ and then the result to $y = a_2 + b_2 x$ results in the transformation

$$(8) \qquad y = a_2 + b_2(a_1 + b_1 u) = (a_2 + b_2 a_1) + b_2 b_1 u.$$

A class $\mathfrak{T}$ of transformations is said to be *closed* under composition if $g_1 \in \mathfrak{T}$, $g_2 \in \mathfrak{T}$ implies that $g_2 \cdot g_1 \in \mathfrak{T}$. We have just shown that the three classes of transformations,

$$(9) \qquad \begin{array}{ll} (1) & \text{with } -\infty < a < \infty, \\ (3) & \text{with } 0 < b, \\ (5) & \text{with } -\infty < a < \infty, 0 < b, \end{array}$$

are all closed with respect to composition. On the other hand, the class (1) with $|a| < 1$ is not, since $U + \frac{1}{2}$ and $U + \frac{2}{3}$ are both members of the class but their composition is not.

(ii) *Closure under inversion.* Given any $1:1$ transformation $x' = gx$, let g^{-1}, the *inverse* of g, denote the transformation which undoes what g

did, that is, takes x' back to x so that $x = g^{-1}x'$. For the transformation which adds a, the inverse subtracts a; the inverse in (3) of multiplication by b is division by b; and the inverse of $a + bu$ is $(x - a)/b$. A class $\mathfrak{T}$ is said to be closed under inversion if $g \in \mathfrak{T}$ implies $g^{-1} \in \mathfrak{T}$. The three classes listed in (9) are all closed under inversion. On the other hand, the class (1) with $0 \leqslant a$ is not.

A class $\mathfrak{T}$ of transformations is called a *transformation group* if it is closed under both composition and inversion. The *identity transformation* $x \equiv x$ is a member of any transformation group G since $g \in G$ implies $g^{-1} \in G$ and hence $g^{-1}g \in G$, and by definition $g^{-1}g$ is the identity. Note also that the inverse $(g^{-1})^{-1}$ of g^{-1} is g, so that gg^{-1} is also the identity.

A transformation group G which satisfies

$$g_2 \cdot g_1 = g_1 \cdot g_2$$

for all $g_1, g_2 \in G$ is called *commutative*. The first two groups of (9) are commutative, but the third is not.

Example 3.1. (*Continued*). The families (2), (4), and (6) generalize easily to the case that $\underline{U}$ is a vector $\underline{U} = (U_1, \ldots, U_n)$, if one defines

$$\underline{U} + a = (U_1 + a, \ldots, U_n + a) \text{ and } b\underline{U} = (bU_1, \ldots, bU_n). \tag{10}$$

This covers in particular the case that $X_1, \ldots, X_n$ are iid according to one of the previous families, for example one of the densities of Table 3.1. Larger families are obtained in the same way by letting

$$\underline{U} + \underline{a} = (U_1 + a_1, \ldots, U_n + a_n) \quad \text{and} \quad \underline{b}\underline{U} = (b_1U_1, \ldots, b_nU_n). \tag{11}$$

Example 3.2. Multivariate normal distribution. As a more special but very important example, suppose next that $\underline{U} = (U_1, \ldots, U_p)$ where the U_i are independently distributed as $N(0, 1)$ and let

$$\begin{pmatrix} X_1 \\ \vdots \\ X_p \end{pmatrix} = \begin{pmatrix} a_1 \\ \vdots \\ a_p \end{pmatrix} + B \begin{pmatrix} U_1 \\ \vdots \\ U_p \end{pmatrix} \tag{12}$$

where B is any nonsingular $p \times p$ matrix. The resulting family of distributions in p-space is the family of nonsingular *p-variate normal distributions*. If the three columns of (12) are denoted by $\underline{X}$, $\underline{a}$, and $\underline{U}$, respectively,* (12) can be written as

$$\underline{X} = \underline{a} + B\underline{U} \tag{13}$$

From this equation, it is seen that the expectation and covariance matrix $\underline{\Sigma}$ of $\underline{X}$ are

*When it is not likely to cause confusion, we shall use $\underline{U}$, and so on, to denote both the vector and the column with elements U_i.

given by

$$(14) \qquad E(\underset{\sim}{X}) = \underset{\sim}{a} \quad \text{and} \quad \underset{\sim}{\Sigma} = E[(\underset{\sim}{X} - \underset{\sim}{a})(\underset{\sim}{X} - \underset{\sim}{a})'] = BB'.$$

To obtain the density of $\underset{\sim}{X}$, write the density of $\underset{\sim}{U}$ as

$$\frac{1}{(\sqrt{2\pi})^p} e^{-(1/2)\underset{\sim}{u}'\underset{\sim}{u}}.$$

Now $\underset{\sim}{U} = B^{-1}(\underset{\sim}{X} - \underset{\sim}{a})$ and the Jacobian of the linear transformation (13) is just the determinant $|B|$ of B. Thus, by the usual formula for transforming densities, the density of $\underset{\sim}{X}$ is seen to be

$$(15) \qquad \frac{|B|^{-1}}{(\sqrt{2\pi})^p} e^{-(\underset{\sim}{x} - \underset{\sim}{a})'\underset{\sim}{\Sigma}(\underset{\sim}{x} - \underset{\sim}{a})/2}.$$

For the case $p = 2$, this reduces to (Problem 3.5)

$$(16) \qquad \frac{1}{2\pi\sigma\tau\sqrt{1 - \rho^2}} e^{-[(x-\xi)^2/\sigma^2 - 2\rho(x-\xi)(y-\eta)/\sigma\tau + (y-\eta)^2/\tau^2]/2(1-\rho^2)}$$

where we write (x, y) for (x_1, x_2) and (ξ, η) for (a_1, a_2), and where $\sigma^2 = \text{var}(X)$, $\tau^2 = \text{var}(Y)$, $\rho\sigma\tau = \text{cov}(X, Y)$.

There is a difference between the transformation groups (1), (3), and (5), on the one hand, and (13), on the other. In the first three cases, different transformations of the group lead to different distributions. This is not true of (13) since the distributions of

$$\underset{\sim}{a}_1 + B_1\underset{\sim}{U} \quad \text{and} \quad \underset{\sim}{a}_2 + B_2\underset{\sim}{U}$$

coincide provided $\underset{\sim}{a}_1 = \underset{\sim}{a}_2$ and $B_1B_1' = B_2B_2'$. This occurs when $\underset{\sim}{a}_1 = \underset{\sim}{a}_2$ and $(B_2^{-1}B_1)(B_2^{-1}B_1)'$ is the identity matrix, that is, when $B_2^{-1}B_1$ is orthogonal. The same family of distributions can therefore be generated by restricting the matrices B in (13) to belong to a smaller group. In particular, it is enough to let G be the group of lower triangular matrices, in which all elements above the main diagonal are zero (Problems 3.6–3.8).

Example 3.3. Linear model. Let us next consider a different generalization of a location-scale family. As before, let $\underset{\sim}{U} = (U_1, \ldots, U_n)$ have a fixed joint distribution and consider the transformations

$$(17) \qquad X_i = a_i + bU_i, \qquad i = 1, \ldots, n$$

where the translation vector $\underset{\sim}{a} = (a_1, \ldots, a_n)$ is restricted to lie in some s-dimensional linear subspace Ω of n-space, that is, to satisfy a set of linear equations

$$(18) \qquad a_i = \sum_{j=1}^{s} d_{ij}\beta_j \qquad (i = 1, \dots, n).$$

Here the d_{ij} are fixed (without loss of generality the matrix $D = (d_{ij})$ is assumed to be of rank s) and the β_j are arbitrary.

The most important case of this model is that in which the U's are iid as $N(0, 1)$. The joint distribution of the X's is then given by

$$(19) \qquad \frac{1}{(\sqrt{2\pi}\, b)^n} \exp\left[-\frac{1}{2b^2}\Sigma(x_i - a_i)^2\right]$$

with $\underline{a}$ ranging over Ω.

We shall next consider a number of models in which the groups (and hence the resulting families of distributions) are much larger than in the situations discussed so far.

Example 3.4. A nonparametric iid family. Let $U_1, \dots, U_n$ be n independent random variables with a fixed continuous common distribution, say $N(0, 1)$, whose support is the whole real line, and let G be the class of all transformations

$$(20) \qquad X_i = g(U_i)$$

where g is any continuous, strictly increasing function satisfying

$$(21) \qquad \lim_{u \to -\infty} g(u) = -\infty, \qquad \lim_{u \to \infty} g(u) = \infty.$$

This class constitutes a group. The X_i are again iid with common distribution, say F_g. The class $\{F_g: g \in G\}$ is the class of all continuous distributions whose support is $(-\infty, \infty)$, that is, the class of all distributions whose cdf is continuous and strictly increasing on $(-\infty, \infty)$.

In this example one may wish to impose on g the additional restriction of differentiability for all u. The resulting family of distributions will be as before but restricted to have probability density with respect to Lebesgue measure.

Many variations of this basic example are of interest, we shall mention only a few.

Example 3.5. Symmetric distributions. Consider the situation of Example 3.4 but with g restricted to be odd, that is, to satisfy $g(-u) = -g(u)$ for all u. This leads to the class of all distributions whose support is the whole real line and which are symmetric with respect to the origin. If instead we let $X_i = g(u_i) + a$, $-\infty < a < \infty$, the resulting class is that of all distributions whose support is the real line and which are symmetric without the point a of symmetry being specified.

Example 3.4. (*Continued*). In Example 3.4, replace $N(0, 1)$ as initial distribution of the U_i with the uniform distribution on $(0, 1)$, and let G be the class of all strictly increasing continuous functions g on (0, 1) satisfying $g(0) = 0$, $g(1) = 1$. If, then, $X_i = a + bg(U_i)$ with $-\infty < a < \infty$, $0 < b$, the resulting group family is that of all continuous distributions whose support is an interval.

The examples of group families considered so far are of two types. In Examples 3.1 through 3.3, the distributions within a family were naturally indexed by a relatively small number of parameters (a and b in Example 3.1; the elements of the matrix B and the vector $\underline{a}$ in Example 3.2; the quantities b and $\beta_1, \ldots, \beta_s$ in Example 3.3). On the other hand, in Examples 3.4 and 3.5 the distribution of the X_i was fairly unrestricted, subject only to conditions such as independence, identity of distribution, nature of support, continuity, and symmetry. The next example is the prototype of a third kind of model arising in survey sampling.

Example 3.6. Sampling from a finite population. To motivate this model, consider a finite population of N elements (or subjects) to each of which is attached a real number (for example, the age or income of the subject) and an identifying label. A random sample of n elements drawn from this population constitutes the observations. Let the observed values and labels be $(X_1, J_1), \ldots, (X_n, J_n)$. The following group family provides a possible model for this situation.

Let $v_1, \ldots, v_N$ be any fixed N real numbers, and let $(U_1, J_1), \ldots, (U_n, J_n)$ be n of the pairs $(v_1, 1), \ldots, (v_N, N)$ selected at random, that is, in such a way that all $\binom{N}{n}$ possible choices of n pairs are equally likely. Finally, let G be the group of transformations

$$X_1 = U_1 + a_{J_1}, \ldots, X_n = U_n + a_{J_n} \tag{22}$$

where $(a_1, \ldots, a_N)$ ranges over all possible N-tuples $-\infty < a_1, a_2, \ldots, a_N < \infty$. If we put $y_i = v_i + a_i$, then $(X_1, J_1), \ldots, (X_n, J_n)$ is a random sample from the population $(y_1, 1), \ldots, (y_N, N)$, the y values being arbitrary.

This example can be extended in a number of ways. In particular, the sampling method, reflecting some knowledge concerning the population of y values may be more complex. In *stratified sampling*, for instance, the population of N is divided into, say, s subpopulations of $N_1, \ldots, N_s$ members ($\Sigma N_i = N$) and a sample of n_i is drawn at random from the ith subpopulation (Problem 3.11). This and some other sampling schemes will be considered in Section 3.6. A different modification places some restrictions on the y's such as $0 < y_i < \infty$, or $0 < y_i < 1$ (Problem 3.10).

It was stated at the beginning of the section that in a group family, the different members of the family play a highly symmetric role. However, the general construction of such a family $\mathcal{P}$ as the distributions of gU, where U has a fixed distribution P_0 and g ranges over a group G of transformations, appears to single out the distribution P_0 of U (which is a member of $\mathcal{P}$ since the identity transformation is a member of G) as the starting point of the construction. This asymmetry is only apparent. For let P_1 be any distribution of $\mathcal{P}$ other than P_0 and consider the family $\mathcal{P}'$ of distributions of gV as g ranges over G where V has distribution P_1. Since P_1 is an element of $\mathcal{P}$, there exists an element g_0 of G for which g_0U is distributed according to P_1. Thus g_0U can play the role of V and $\mathcal{P}'$ is the family of distributions of gg_0U as g ranges over G. However, as g ranges over G, so does gg_0 (Problem 3.4), so that the family of distributions of gg_0U, $g \in G$ is the same as the family $\mathcal{P}$ of gU, $g \in G$. A group family is thus independent of which of its members is taken as starting distribution.

If one cannot find a group generating a given family $\mathcal{P}$ of distributions, the question arises whether such a group exists, that is, whether $\mathcal{P}$ is a group family. In principle, the answer is easy. For the sake of simplicity, suppose that $\mathcal{P}$ is a family of univariate distributions with continuous and strictly increasing cumulative distribution functions. Let F_0 and F be two such c.d.f.s and suppose that U is distributed according to F_0. Then if g is strictly increasing, $g(U)$ is distributed according to F if and only if $g = F^{-1}(F_0)$ (Problem 3.13). Thus, the transformations generating the family must be the transformations

$$\{F^{-1}(F_0),\ F \in \mathcal{P}\}. \tag{23}$$

The family $\mathcal{P}$ will be a group family if and only if the transformations (23) form a group. That is, are closed under composition and inversion. In specific situations, the calculations needed to check this requirement may not be easy. For an important class of problems, the question has been settled by Borges and Pfanzagl (1965).

4. EXPONENTIAL FAMILIES

A family $\{P_\theta\}$ of distributions is said to form an s-parameter exponential family if the distributions P_θ have densities of the form

$$p_\theta(x) = \exp\left[\sum_{i=1}^{s} \eta_i(\theta)T_i(x) - B(\theta)\right]h(x) \tag{1}$$

with respect to some common measure μ. Here the η_i and B are real-valued functions of the parameters and the T_i are real-valued statistics. Table 4.1 exhibits several such densities with $s = 1$ or 2, which will be used in the sequel. Frequently, it is more convenient to use the η_i as the parameters and write the density in the *canonical form*

$$p(x, \eta) = \exp\left[\sum_{i=1}^{s} \eta_i T_i(x) - A(\eta)\right]h(x). \tag{2}$$

Table 4.1. Some One- and Two-Parameter Exponential Families

Density[a]	Name	Notation	Support
$\frac{1}{\Gamma(g)b^g}x^{g-1}e^{-x/b}$	Gamma	$\Gamma(g, b)$	$0 < x < \infty$
$\frac{1}{\Gamma\left(\frac{f}{2}\right)2^{f/2}}x^{f/2-1}e^{-x/2}$	Chi-square	χ_f^2	$0 < x < \infty$
$\frac{\Gamma(a+b)}{\Gamma(a)\Gamma(b)}x^{a-1}(1-x)^{b-1}$	Beta	$B(a, b)$	$0 < x < 1$
$\binom{n}{x}p^x(1-p)^{n-x}$	Binomial	$b(p, n)$	$x = 0, 1, \ldots, n$
$\frac{1}{x!}\lambda^x e^{-\lambda}$	Poisson	$P(\lambda)$	$x = 0, 1, \ldots$
$\binom{m+x-1}{m-1}p^m q^x$	Negative binomial	$Nb(p, m)$	$x = 0, 1, \ldots$

[a] The density of the first three distributions is with respect to Lebesgue measure, and that of the last three with respect to counting measure.

It should be noted that this form is not unique. We can, for example, multiply η_i by a constant c if at the same time T_i is replaced by T_i/c. More generally, we can make linear transformations of the η's and T's.

Both (1) and (2) are redundant in that the factor $h(x)$ could be absorbed into μ. The reason for not doing so is that it is then usually possible to take μ to be either Lebesgue measure or counting measure rather than having to define a more elaborate measure.

The function p given by (2) is non-negative and is therefore a probability density with respect to the given μ, provided its integral with respect to μ equals 1. A constant $A(\eta)$ for which this is the case exists if and only if

$$\int e^{\Sigma \eta_i T_i(x)} h(x)\, d\mu(x) < \infty. \tag{3}$$

The set Ξ of points $\eta = (\eta_1, \ldots, \eta_s)$ for which this is the case is called the *natural parameter space* of the family (2). It is not difficult to see that Ξ is convex (TSH, p. 51). In most applications, it turns out to be open, but this need not be the case (Problem 4.1). In the parameterization (1), the natural parameter space is the set of θ values for which $[\eta_1(\theta), \ldots, \eta_s(\theta)]$ is in Ξ.

If the statistics $T_1, \ldots, T_s$ satisfy a linear constraint, the number s of terms in the exponent of (1) can be reduced. Unless this is done, the parameters η_i are statistically meaningless; they are "unidentifiable"* (Problem 4.2). A reduction is also possible when the η's satisfy a linear constraint. In the latter case, the natural parameter space will be a convex set which lies in a linear subspace of dimension less than s. We shall assume without loss of generality that the representation (2) is minimal in the sense that neither the T's nor the η's satisfy a linear constraint. The natural parameter space will then be a convex set in E_s containing an open s-dimensional rectangle.† If (2) is minimal and the parameter space contains an s-dimensional rectangle, the family (2) is said to be of *full rank*.

Example 4.1. Multinomial. In n independent trials with $s + 1$ possible outcomes, let the probability of the ith outcome be p_i in each trial. If X_i denotes the number of trials resulting in outcome i ($i = 0, 1, \ldots, s$), then the joint distribution of the X's is the *multinomial distribution* $M(p_0, \ldots, p_s; n)$

$$P(X_0 = x_0, \ldots, X_s = x_s) = \frac{n!}{x_0! \cdots x_s!} p_0^{x_0} \cdots p_s^{x_s}, \tag{4}$$

which can be rewritten as

$$\exp(x_0 \log p_0 + \cdots + x_s \log p_s) h(x).$$

Since the x_i add up to n, this can be reduced to

$$\exp[n \log p_0 + x_1 \log(p_1/p_0) + \cdots + x_s \log(p_s/p_0)] h(x) \tag{5}$$

This is an s-parameter exponential family with

$$\eta_i = \log(p_i/p_0), \qquad A(\eta) = -n \log p_0 = n \log\left[1 + \sum_{i=1}^{s} e^{\eta_i}\right]. \tag{6}$$

The natural parameter space is the set of all $(\eta_1, \ldots, \eta_s)$ with $-\infty < \eta_i < \infty$.

Let X and Y be independently distributed according to s-parameter exponential families with densities

$$\exp[\Sigma \eta_i T_i(x) - A(\eta)] h(x) \quad \text{and} \quad \exp[\Sigma \eta_i U_i(y) - C(\eta)] k(y) \tag{7}$$

with respect to measures μ and ν over $(\mathcal{X}, \mathcal{A})$ and $(\mathcal{Y}, \mathcal{B})$, respectively. Then the joint distribution of X, Y is again an exponential family, and by

*If X is distributed according to p_θ, then θ is said to be unidentifiable on the basis of X if there exist $\theta \neq \theta'$ for which $P_\theta = P_{\theta'}$.

†A more detailed account of such minimal representations is given in Barndorff–Nielsen (1978), Section 8.1.

induction the result extends to the joint distribution of more than two factors. The most important special case is that of iid random variables X_i, each distributed according to (1): The exponential structure is preserved under random sampling. The joint density of $X = (X_1, \dots, X_n)$ is

$$\exp[\Sigma\eta_i(\theta)T_i'(x) - nB(\theta)]h(x_1)\cdots h(x_n) \tag{8}$$

with $T_i'(x) = \Sigma_{j=1}^n T_i(x_j)$.

Example 4.2. ***Normal sample.*** Let X_i $(i = 1, \dots, n)$ be iid according to $N(\xi, \sigma^2)$. Then the joint density of the sample $X_1, \dots, X_n$ with respect to Lebesgue measure in E_n is

$$\exp\left(\frac{\xi}{\sigma^2}\Sigma x_i - \frac{1}{2\sigma^2}\Sigma x_i^2 - \frac{n}{2\sigma^2}\xi^2\right)\cdot\frac{1}{(\sqrt{2\pi}\,\sigma)^n}. \tag{9}$$

This constitutes a two-parameter exponential family with natural parameters $(\xi/\sigma^2, -1/2\sigma^2)$.

Example 4.3. ***Bivariate normal.*** Suppose that (X_i, Y_i), $i = 1, \dots, n$ is a sample from the bivariate normal density (3.16). Then it is seen that the joint density of n pairs is a five-parameter exponential density with statistics

$$T_1 = \Sigma X_i, \quad T_2 = \Sigma X_i^2, \quad T_3 = \Sigma X_i Y_i, \quad T_4 = \Sigma Y_i, \quad T_5 = \Sigma Y_i^2$$

This example easily generalizes to the p-variate case (Problem 4.3).

A useful property of exponential families is given by the following theorem, which is proved, for example, in TSH (Chap. 2, Theorem 9) and in Barndorff-Nielsen (1978 Section 7.1).

Theorem 4.1. *For any integrable function* f *and any* η *in the interior of* Ξ, *the integral*

$$\int f(x)\,exp[\Sigma\eta_i T_i(x)]h(x)\,d\mu(x) \tag{10}$$

is continuous and has derivatives of all orders with respect to the η*'s, and these can be obtained by differentiating under the integral sign.*

As an application, differentiate the identity

$$\int \exp[\Sigma\eta_i T_i(x) - A(\eta)]h(x)\,d\mu(x) = 1$$

with respect to η_j to find

$$(11) \qquad E_\eta(T_j) = \frac{\partial}{\partial \eta_j} A(\eta).$$

Differentiating (11), in turn, with respect to η_k leads to

$$(12) \qquad \operatorname{cov}(T_j, T_k) = \frac{\partial^2}{\partial \eta_j \partial \eta_k} A(\eta).$$

(For the corresponding formulas in terms of (1), see Johnson, Ladalla, and Liu, 1979).

Example 4.1. (*Continued*). From (6), (11), and (12), one easily finds for the multinomial variables of Example 4.1 that (Problem 4.13)

$$(13) \qquad E(X_i) = np_i, \qquad \operatorname{cov}(X_j, X_k) = \begin{cases} np_j(1 - p_j) & \text{if } k = j \\ -np_j p_k & \text{if } k \neq j. \end{cases}$$

As will be discussed in the next section, in an exponential family the statistics $T = (T_1, \ldots, T_s)$ carry all the information about η or θ contained in the data, so that all statistical inferences concerning these parameters will be based on the T's. For this reason, we shall frequently be interested in calculating not only the first two moments of the T's given by (11) and (12) but also some of the higher moments

$$(14) \qquad \alpha_{r_1, \ldots, r_s} = E(T_1^{r_1} \cdots T_s^{r_s})$$

and central moments

$$(15) \qquad \mu_{r_1, \ldots, r_s} = E\{[T_1 - E(T_1)]^{r_1} \cdots [T_s - E(T_s)]^{r_s}\}.$$

A tool that often facilitates such calculations is the *moment generating function*

$$(16) \qquad M_T(u_1, \ldots, u_s) = E(e^{u_1 T_1 + \cdots + u_s T_s}).$$

If M_T exists in some neighborhood $\Sigma u_i^2 < \delta$ of the origin, then all moments $\alpha_{r_1, \ldots, r_s}$ exist and are the coefficients in the expansion of M_T as a power series

$$M_T(u_1, \ldots, u_s) = \sum_{(r_1, \ldots, r_s)} \alpha_{r_1, \ldots, r_s} u_1^{r_1} \cdots u_s^{r_s} / r_1! \cdots r_s! \tag{17}$$

As an alternative, it is sometimes more convenient to calculate instead the *cumulants* $\kappa_{r_1, \ldots, r_s}$, defined as the coefficients in the expansion of the *cumulant generating function*

$$\begin{aligned} K_T(u_1, \ldots, u_s) &= \log M_T(u_1, \ldots, u_s) \\ &= \sum_{(r_1, \ldots, r_s)} \kappa_{r_1, \ldots, r_s} u_1^{r_1} \cdots u_s^{r_s} / r_1! \cdots r_s! \end{aligned} \tag{18}$$

From the cumulants, the moments can be determined by formal comparison of the two power series (see, for example, Cramér, 1946, p. 186, or Kendall and Stuart, Vol. 1, 1977, p. 70 and 86). For $s = 1$, one finds, for example (Problem 4.5),

$$\begin{aligned} \alpha_1 = \kappa_1, \qquad \alpha_2 = \kappa_2 + \kappa_1^2, \qquad & \alpha_3 = \kappa_3 + 3\kappa_1\kappa_2 + \kappa_1^3, \\ \alpha_4 = \kappa_4 + 3\kappa_2^2 + 4\kappa_1\kappa_3 & + 6\kappa_1^2\kappa_2 + \kappa_1^4. \end{aligned} \tag{19}$$

For exponential families, the moment and cumulant generating functions can be expressed rather simply as follows.

Theorem 4.2. *If* X *is distributed with density* (2), *then for any* η *in the interior of* Ξ, *the moment and cumulant generating functions* $\mathrm{M_T(u)}$ *and* $\mathrm{K_T(u)}$ *of the* T*'s exist in some neighborhood of the origin and are given by*

$$\mathrm{K_T(u) = A(\eta + u) - A(\eta)} \tag{20}$$

and

$$\mathrm{M_T(u) = e^{A(\eta+u)}/e^{A(\eta)}} \tag{21}$$

respectively.

Frequently, the calculation of moments becomes particularly easy when they can be represented as the sum of independent terms. We shall illustrate two approaches for the case $s = 1$.

Suppose $X = X_1 + \cdots + X_n$ where the X_i are independent with moment and cumulant generating functions $M_{X_i}(u)$ and $K_{X_i}(u)$, respectively. Then

$$M_X(u) = E[e^{(X_1 + \cdots + X_n)u}] = M_{X_1}(u) \cdots M_{X_n}(u)$$

and therefore

$$K_X(u) = \sum_{i=1}^{n} K_{X_i}(u).$$

From the definition of cumulants, it then follows that

$$\kappa_r = \sum_{i=1}^{n} \kappa_{ir} \tag{22}$$

where κ_{ir} is the rth cumulant of X_i.

The situation is also very simple for low central moments. If $\xi_i = E(X_i)$, $\sigma_i^2 = \text{var}(X_i)$ and the X_i are independent, one easily finds (Problem 4.5)

$$\text{var}(\Sigma X_i) = \Sigma\sigma_i^2, \qquad E[\Sigma(X_i - \xi_i)]^3 = \Sigma E(X_i - \xi_i)^3 \tag{23}$$

$$E[\Sigma(X_i - \xi_i)]^4 = \Sigma E(X_i - \xi_i)^4 + 6\sum_{i<j}\sigma_i^2\sigma_j^2.$$

For the case of identical components with $\xi_i = \xi$, $\sigma_i^2 = \sigma^2$ this reduces to

$$\text{var}(\Sigma X_i) = n\sigma^2, \qquad E(\Sigma(X_i - \xi))^3 = nE(X_1 - \xi)^3, \tag{24}$$

$$E(\Sigma(X_i - \xi))^4 = nE(X_1 - \xi)^4 + 3n(n-1)\sigma^4.$$

The following are a few of the many important special cases of exponential families and some of their moments. Additional examples are given in the problems. For applications of exponential families to directional data see Beran (1979).

Example 4.4. Binomial. Let X have the binomial distribution $b(p, n)$ so that

$$P(X = x) = \binom{n}{x} p^x q^{n-x} \qquad (0 < p < 1;\ x = 0, 1, \ldots, n;\ q = 1 - p) \tag{25}$$

This is the special case of the multinomial distribution (4) with $s = 1$. The probability (25) can be rewritten as

$$\binom{n}{x} e^{x \log(p/q) + n \log q},$$

which defines an exponential family, with μ being counting measure over the points

$x = 0, 1, \ldots, n$ and with

$$\eta = \log(p/q), \qquad A(\eta) = n\log(1 + e^{\eta}). \tag{26}$$

From (21) and (26), one easily finds that (Problem 4.6)

$$M_X(u) = (q + pe^u)^n. \tag{27}$$

An easy way to obtain the expectation and the first three central moments of X is to use the fact that X arises as the number of successes in n binomial trials with success probability p, and hence that $X = \Sigma X_i$ where X_i is 1 or 0, as the ith trial is or is not a success. From (24) and the moments of X_i, one then finds (Problem 4.6)

$$\begin{aligned} E(X) &= np \qquad E(X - np)^3 = npq(q - p) \\ \text{var}(X) &= npq \qquad E(X - np)^4 = 3(npq)^2 + npq(1 - 6pq). \end{aligned} \tag{28}$$

Example 4.5. ***Poisson.*** A random variable X has the Poisson distribution $P(\lambda)$ if

$$P(X = x) = \frac{\lambda^x}{x!}e^{-\lambda}, \qquad x = 0, 1, \ldots; \lambda > 0. \tag{29}$$

Writing this as an exponential family in canonical form, we find

$$\eta = \log\lambda, \qquad A(\eta) = \lambda = e^{\eta} \tag{30}$$

and hence

$$K_X(u) = \lambda(e^u - 1), \qquad M_X(u) = e^{\lambda(e^u - 1)}, \tag{31}$$

so that, in particular, $\kappa_r = \lambda$ for all r. The expectation and first three central moments are given by (Problem 4.7)

$$\begin{aligned} E(X) &= \lambda \qquad E(X - \lambda)^3 = \lambda \\ \text{var}(X) &= \lambda \qquad E(X - \lambda)^4 = \lambda + 3\lambda^2. \end{aligned} \tag{32}$$

Example 4.6. ***Normal.*** Let X have the normal distribution $N(\xi, \sigma^2)$ with density

$$\frac{1}{\sqrt{2\pi}\,\sigma}e^{-(x-\xi)^2/2\sigma^2} \tag{33}$$

with respect to Lebesgue measure. For fixed σ, this is a one-parameter exponential family with

$$(34) \qquad \eta = \xi/\sigma^2 \quad \text{and} \quad A(\eta) = \eta^2\sigma^2/2 + \text{constant}.$$

It is thus seen that

$$(35) \qquad M_X(u) = e^{\xi u + (1/2)\sigma^2 u^2},$$

and hence in particular that

$$(36) \qquad E(X) = \xi.$$

Since the distribution of $X - \xi$ is $N(0, \sigma^2)$, the central moments μ_r of X are simply the moments α_r of $N(0, \sigma^2)$, which are obtained from the moment-generating function

$$M_0(u) = e^{\sigma^2 u^2/2}$$

to be

$$(37) \qquad \mu_{2r+1} = 0, \qquad \mu_{2r} = 1 \cdot 3 \cdot \cdots \cdot (2r-1)\sigma^{2r}, \qquad r = 1, 2, \ldots$$

Example 4.7. ***Gamma.*** A random variable X has the *gamma distribution* $\Gamma(\alpha, b)$ if its density is

$$(38) \qquad \frac{1}{\Gamma(\alpha) b^\alpha} x^{\alpha - 1} e^{-x/b}, \qquad x > 0, \alpha > 0, b > 0,$$

with respect to Lebesgue measure on $(0, \infty)$. Here b is a scale parameter, whereas α is called the *shape parameter* of the distribution. For $\alpha = f/2$ (f an integer), $b = 2$, this is the χ^2-distribution χ_f^2 with f degrees of freedom. For fixed shape parameter α, (38) is a one-parameter exponential family with $\eta = -1/b$ and

$$A(\eta) = \alpha \log b = -\alpha \log(-\eta).$$

Thus, the moment- and cumulant-generating functions are seen to be

$$(39) \qquad M_X(u) = (1 - bu)^{-\alpha} \quad \text{and} \quad K_X(u) = -\alpha \log(1 - bu), u < 1/b.$$

From the first of these formulas one finds

$$(40) \qquad E(X^r) = \alpha(\alpha + 1) \cdots (\alpha + r - 1) b^r = \frac{\Gamma(\alpha + r)}{\Gamma(\alpha)} b^r$$

and hence (Problem 4.15)

$$E(X) = \alpha b \qquad E(X-\alpha b)^3 = 2\alpha b^3 \tag{41}$$

$$\operatorname{var}(X) = \alpha b^2 \qquad E(X-\alpha b)^4 = (3\alpha^2 + 6\alpha)b^4.$$

Not only the moments of the statistics T_i appearing in (1) and (2) are of interest but also the family of distributions of the T's. This turns out again to be an exponential family.

Theorem 4.3. *If* X *is distributed according to an exponential family with density* (1) *with respect to a measure* μ *over* $(\mathcal{X}, \mathcal{A})$, *then* $\mathrm{T} = (\mathrm{T}_1, \ldots, \mathrm{T}_s)$ *is distributed according to an exponential family with density*

$$exp[\Sigma\eta_i t_i - A(\eta)]k(t) \tag{42}$$

with respect to a measure ν *over* E_s.

For a proof, see, for example, TSH, p. 52.

Let us now apply this theorem to the case of two independent exponential families with densities (7). Then it follows from Theorem 4.3 that $(T_1 + U_1, \ldots, T_s + U_s)$ is also distributed according to an s-parameter exponential family, and by induction this result extends to the sum of more than two independent terms. In particular, let $X_1, \ldots, X_n$ be independently distributed, each according to a one-parameter exponential family with density

$$\exp[\eta T_i(x_i) - A_i(\eta)]h_i(x_i). \tag{43}$$

Then the sum $\Sigma_{i=1}^n T_i(X_i)$ is again distributed according to a one-parameter exponential family. In fact, the sum of independent Poisson or normal variables again has a distribution of the same type, and the same is true for a sum of independent binomial variables with common p, or a sum of independent gamma variables $\Gamma(\alpha_i, b)$ with common b.

The normal distributions $N(\xi, \sigma^2)$ for fixed σ constitute both a one-parameter exponential family (Example 4.6) and a location family (Table 3.1). It is natural to ask whether there are any other families that enjoy this double advantage. Another example is obtained by putting $X = \log Y$ where Y has the gamma distribution $\Gamma(\alpha, b)$ given by (38), and where the location parameter θ is $\theta = \log b$. Since multiplication of a random variable by a constant $c \neq 0$ preserves both the exponential and location structure, a more general example is provided by the random variable $c \log Y$ for any $c \neq 0$. It was shown by Dynkin (1951) and Ferguson (1962, 1963) that the cases in which X is normal or is equal to $c \log Y$ with Y being gamma, provide the only examples of exponential location families. The extension of this result to more general group families is discussed, for example, by Barndorff-Nielsen, Blaesild, Jensen, and Jørgensen (1982).

5. SUFFICIENT STATISTICS

The starting point of a statistical analysis, as formulated in the preceding sections, is a random observable X taking on values in a sample space $\mathcal{X}$, and a family of possible distributions of X. It often turns out that some part of the data carries no information about the unknown distribution and that X can therefore be replaced by some statistic $T = T(X)$ (not necessarily real-valued) without loss of information. A statistic T is said to be *sufficient* for X, or for the family $\mathcal{P} = \{P_\theta, \theta \in \Omega\}$ of possible distributions of X, or for θ, if the conditional distribution of X given $T = t$ is independent of θ for all t.

This definition is not quite precise and we shall return to it later in this section. However, consider first in what sense a sufficient statistic T contains all the information about θ contained in X. For that purpose suppose that an investigator reports the value of T, but on being asked for the full data admits that they have been discarded. In an effort at reconstruction, one can use a random mechanism (such as a table of random numbers) to obtain a random quantity X' according to the conditional distribution of X given t. (This would not be possible, of course, if the conditional distribution depended on the unknown θ.) Then the unconditional distribution of X' is the same as that of X, that is,

$$P_\theta(X' \in A) = P_\theta(X \in A) \qquad \text{for all} \quad A,$$

regardless of the value of θ. Hence from a knowledge of T alone, it is possible to construct a quantity X' which is completely equivalent to the original X. Since X and X' have the same distribution for all θ, they provide exactly the same information about θ (for example, the estimators $\delta(X)$ and $\delta(X')$ have identical distributions for any θ).

In this sense, a sufficient statistic provides a reduction of the data without loss of information. This property holds, of course, only as long as attention is restricted to the model $\mathcal{P}$ and no distributions outside $\mathcal{P}$ are admitted as possibilities. Thus, in particular, restriction to T is not appropriate when testing the validity of $\mathcal{P}$.

The construction of X' is in general effected with the help of an independent random mechanism. An estimator $\delta(X')$ depends, therefore, not only on T but also on this mechanism. It is thus not an estimator as defined in Section 1, but a randomized estimator. Quite generally, if X is the basic random observable, a *randomized estimator* of $g(\theta)$ is a rule which assigns to each possible outcome x of X a random variable $Y(x)$ with a known distribution. When $X = x$, an observation of $Y(x)$ will be taken and will constitute the estimate of $g(\theta)$. The risk of the resulting estimator is then $E_\theta[M(\theta, X)]$ where

$$M(\theta, x) = EL[\theta, Y(x)].$$

The operational significance of sufficiency can now be stated formally as follows.

Theorem 5.1. *Let* X *be distributed according to* $P_\theta \in \mathcal{P}$ *and let* T *be sufficient for* $\mathcal{P}$. *Then for any estimator* $\delta(X)$ *of* $g(\theta)$ *there exists a (possibly randomized) estimator based on* T *which has the same risk function as* $\delta(X)$.

Example 5.1. Let X_1, X_2 be independent Poisson variables with common expectation λ, so that their joint distribution is

$$P(X_1 = x_1, X_2 = x_2) = \frac{\lambda^{x_1+x_2}}{x_1!x_2!}e^{-2\lambda}.$$

Then the conditional distribution of X_1 given $X_1 + X_2 = t$ is given by

$$P(X_1 = x_1 | X_1 + X_2 = t) = \frac{\lambda^t e^{-2\lambda}/x_1!(t-x_1)!}{\sum_{y=0}^{t} \lambda^t e^{-2\lambda}/y!(t-y)!}$$

$$= \frac{1}{x_1!(t-x_1)!} \Big/ \frac{1}{\sum_{y=0}^{t} 1/y!(t-y)!}$$

Since this is independent of λ, so is the conditional distribution given t of $(X_1, X_2 = t - X_1)$, and hence $T = X_1 + X_2$ is a sufficient statistic for λ. To see how to reconstruct (X_1, X_2) from T, note that

$$\Sigma \frac{1}{y!(t-y)!} = \frac{1}{t!}2^t$$

so that

$$P(X_1 = x_1 | X_1 + X_2 = t) = \binom{t}{x_1}\left(\frac{1}{2}\right)^{x_1}\left(\frac{1}{2}\right)^{t-x_1},$$

that is, the conditional distribution of X_1 given t is the binomial distribution $b(\frac{1}{2}, t)$ corresponding to t trials with success probability 1/2. Let X_1' and $X_2' = t - X_1'$ be, respectively, the number of heads and the number of tails in t tosses with a fair coin. Then the joint conditional distribution of (X_1', X_2') given t is the same as that of (X_1, X_2) given t.

Example 5.2. ***Uniform.*** Let $X_1, \ldots, X_n$ be independently distributed according to the uniform distribution $U(0, \theta)$. Let T be the largest of the n X's, and consider the conditional distribution of the remaining $n - 1$ X's given t. Thinking of the n variables as n points on the real line, it is intuitively obvious and not difficult to see formally (Problem 5.2) that the remaining $n - 1$ points (after the largest is fixed at t) behave like $n - 1$ points selected at random from the interval $(0, t)$. Since this conditional distribution is independent of θ, T is sufficient. Given only $T = t$, it is obvious how to reconstruct the original sample: Select $n - 1$ points at random on $(0, t)$.

Example 5.3. Suppose that X is normally distributed with mean zero and unknown variance σ^2. Then the distribution of X is symmetric about the origin. Given that $|X| = t$, the only two possible values of X are $\pm t$, and by symmetry the conditional probability of each is 1/2. The conditional distribution of X given t is thus independent of σ and $T = |X|$ is sufficient. In fact, a random variable X' with the same distribution as X can be obtained from T by tossing a fair coin and letting $X' = T$ or $-T$ as the coin falls heads or tails.

The definition of sufficiency given at the beginning of the section depends on the concept of conditional probability, and this unfortunately is not capable of a treatment which is both general and elementary. Difficulties arise when $P_\theta(T = t) = 0$ so that the conditioning event has probability zero. The definition of conditional probability can then be changed at one or more values of t (in fact, at any set of t-values which has probability zero) without affecting the distribution of X, the result of combining the distribution of T with the conditional distribution of X given T.

In elementary treatments of probability theory, the conditional probability $P(X \in A|t)$ is considered for fixed t as defining the conditional distribution of X given $T = t$. A more general approach can be obtained by a change of viewpoint, namely, by considering $P(X \in A|t)$ for fixed A as a function of t, defined in such a way that in combination with the distribution of T it leads back to the distribution of X. (See TSH, Chapter 2, Section 4 for details.) This provides a justification, for instance, of the assignment of conditional probabilities in Example 5.3 and Example 5.6.

In the same way, the conditional expectation $\eta(t) = E[\delta(X)|t]$ can be defined in such a way that

$$E\eta(T) = E\delta(X) \tag{1}$$

that is, so that the expected value of the conditional expectation is equal to the unconditional expectation.

Conditional expectation essentially satisfies the usual laws of expectation. However, since it is only determined up to sets of probability zero, these laws can only hold a.e. More specifically, we have with probability 1

$$E[af(X) + bg(X)|t] = aE[f(X)|t] + bE[g(X)|t]$$

(2) and

$$E[b(T)f(X)|t] = b(t)E[f(X)|t].$$

As just discussed, the functions $P(A|t)$ are not uniquely defined, and the question arises whether determinations exist which for each fixed t define a conditional probability distribution. It turns out that this is not always possible (see Blackwell and Ryll–Nardzewsky, 1963) but that it is possible when the sample space is Euclidean (see TSH, Chapter 2, Section 5), as will be the case throughout most of this book. When this is the case, a statistic T can be defined to be sufficient if there exists a determination of the conditional distribution functions of X given t which is independent of θ.

The determination of sufficient statistics by means of the definition is inconvenient since it requires, first, guessing a statistic T that might be sufficient and, then, checking whether the conditional distribution of X given t is independent of θ. However, for dominated families, that is, when the distributions have densities with respect to a common measure, there is a simple criterion for sufficiency.

Theorem 5.2. (Factorization criterion). *A necessary and sufficient condition for a statistic* T *to be sufficient for a family* $\mathcal{P} = \{P_\theta, \ \theta \in \Omega\}$ *of distributions of* X *dominated by a σ-finite measure* μ, *is that there exist non-negative functions* g_θ *and* h *such that the densities* p_θ *of* P_θ *satisfy*

$$p_\theta(x) = g_\theta[T(x)]h(x) \qquad (\text{a.e. } \mu). \tag{3}$$

For a proof, see TSH, Theorem 2.8 and Corollary 2.1.

Example 5.4. Poisson. In generalization of Example 5.1 suppose that $X_1, \ldots, X_n$ are iid according to a Poisson distribution with expectation λ. Then

$$P_\lambda(X_1 = x_1, \ldots, X_n = x_n) = \lambda^{\Sigma x_i} e^{-n\lambda} / \Pi(x_i!).$$

This satisfies (3) with $T = \Sigma X_i$, which is therefore sufficient.

Example 5.5. Normal. Let $X_1, \ldots, X_n$ be iid as $N(\xi, \sigma^2)$ so that their joint density is

$$p_{\xi,\sigma}(x) = \frac{1}{(\sqrt{2\pi}\,\sigma)^n} \exp\left[-\frac{1}{2\sigma^2}\Sigma x_i^2 + \frac{\xi}{\sigma^2}\Sigma x_i - \frac{n}{2\sigma^2}\xi^2\right]. \tag{4}$$

Then it follows from the factorization criterion that $T = (\Sigma X_i^2, \Sigma X_i)$ is sufficient for $\theta = (\xi, \sigma^2)$. Sometimes it is more convenient to replace T by the equivalent statistic $T' = (\bar{X}, S^2)$ where $\bar{X} = \Sigma X_i/n$ and $S^2 = \Sigma(X_i - \bar{X})^2 = \Sigma X_i^2 - n\bar{X}^2$. The two representations are equivalent in that they identify the same points of the sample space, that is, $T(x) = T(y)$ if and only if $T'(x) = T'(y)$.

Example 5.2. (*Continued*). The joint density of a sample $X_1, \ldots, X_n$ from $U(0, \theta)$ is

$$p_\theta(x) = \frac{1}{\theta^n} \prod_{i=1}^{n} u(0, x_i) u(x_i, \theta)$$

where

$$u(a, b) = \begin{cases} 1 & \text{if } a < b \\ 0 & \text{otherwise.} \end{cases} \tag{5}$$

Now

$$\Pi u(0, x_i) u(x_i, \theta) = u(x_{(n)}, \theta) \Pi u(0, x_i)$$

where $x_{(n)}$ is the largest of the x values. It follows from Theorem 5.2 that $X_{(n)}$ is sufficient, as had been shown directly in Example 5.2.

As a final illustration, consider Example 5.3 from the present point of view.

Example 5.3. (*Continued*). If X is distributed as $N(0, \sigma^2)$, the density of X is

$$\frac{1}{\sqrt{2\pi}\,\sigma} e^{-x^2/2\sigma^2}$$

which depends on x only through x^2, so that (3) holds with $T(x) = x^2$. As always, of course, there are many equivalent statistics such as $|X|$, X^4 or e^{X^2}.

Quite generally, two statistics, $T = T(X)$ and $T' = T'(X)$, will be said to be equivalent (with respect to a family $\mathcal{P}$ of distributions of X) if each is a function of the other a.e. $\mathcal{P}$, that is, if there exists a $\mathcal{P}$-null set N and functions f and g such that $T(x) = f[T'(x)]$ and $T'(x) = g[T(x)]$ for all $x \notin N$. Two such statistics carry the same amount of information.

Example 5.6. Order statistics. Let $X = (X_1, \ldots, X_n)$ be iid according to an unknown continuous distribution F and let $T = (X_{(1)}, \ldots, X_{(n)})$ where $X_{(1)} < \cdots < X_{(n)}$ denote the ordered observations, the so-called *order statistics*. By the continuity assumptions, the X's are distinct with probability 1. Given T, the only possible values for X are the $n!$ vectors $(X_{(i_1)}, \ldots, X_{(i_n)})$ and by symmetry each of these has conditional probability $1/n!$ The conditional distribution is thus independent of F, and T is sufficient. In fact, a random vector X' with the same distribution as X can be obtained from T by labeling the n coordinates of T at random. Equivalent to T is the statistic $U = (U_1, \ldots, U_n)$ where $U_1 = \Sigma X_i$, $U_2 = \Sigma X_i X_j$ $(i \neq j), \ldots, U_n = X_1 \cdots X_n$, and also the statistic $V = (V_1, \ldots, V_n)$ where $V_k = X_1^k + \cdots + X_n^k$ (Problem 5.9).

Equivalent forms of a sufficient statistic reduce the data to the same extent. There may, however, also exist sufficient statistics which provide different degrees of reduction.

Example 5.7. Different sufficient statistics. Let $X_1, \ldots, X_n$ be iid as $N(0, \sigma^2)$ and consider the statistics

$$T_1(X) = (X_1, \ldots, X_n)$$

$$T_2(X) = (X_1^2, \ldots, X_n^2)$$

$$T_3(X) = (X_1^2 + \cdots + X_m^2, X_{m+1}^2 + \cdots + X_n^2)$$

$$T_4(X) = X_1^2 + \cdots + X_n^2.$$

These are all sufficient (Problem 5.5), with T_i providing increasing reduction of the data as i increases.

It follows from the interpretation of sufficiency given at the beginning of this section that if T is sufficient and $T = H(U)$, then U is also sufficient. Knowledge of U implies knowledge of T and hence permits reconstruction of the original data. Furthermore, T provides a greater reduction of the data than U unless H is 1 : 1, in which case T and U are equivalent. A sufficient statistic T is said to be *minimal* if of all sufficient statistics it provides the greatest possible reduction of the data, that is, if for any sufficient statistic U there exists a function H such that $T = H(U)$ (a.e. $\mathcal{P}$). Minimal sufficient statistics can be shown to exist under weak assumptions (see, for example, Bahadur, 1954), but exceptions are possible (Landers and Rogge, 1972, Pitcher, 1957). Minimal sufficient statistics exist in particular if the basic measurable space is Euclidean in the sense of Example 2.2 and the family $\mathcal{P}$ of distributions is dominated (Bahadur, 1957).

It is typically fairly easy to construct a minimal sufficient statistic. For the sake of simplicity, we shall restrict attention to the case that the distributions of $\mathcal{P}$ all have the same support (but see Problems 5.11–5.17).

Theorem 5.3. *Let $\mathcal{P}$ be a finite family with densities p_i, $i = 0, 1, \ldots, k$ all having the same support. Then the statistic*

$$T(X) = \left(\frac{p_1(X)}{p_0(X)}, \frac{p_2(X)}{p_0(X)}, \ldots, \frac{p_k(X)}{p_0(X)} \right) \tag{6}$$

is minimal sufficient.

The proof is an easy consequence of the following corollary of Theorem 5.2 (Problem 5.6).

Corollary 5.1. *Under the assumptions of Theorem 5.2, a necessary and sufficient condition for a statistic* U *to be sufficient is that for any fixed θ and θ_0 the ratio* $p_\theta(x)/p_{\theta_0}(x)$ *is a function of* U(x).

Proof of Theorem 5.3. The corollary states that U is a sufficient statistic for $\mathcal{P}$ if and only if T is a function of U, and this proves T to be minimal.

Theorem 5.3 immediately extends to the case that $\mathcal{P}$ is countable. Generalizations to uncountable families are also possible (see Lehmann and Scheffé, 1950, Dynkin, 1951 and Barndorff–Nielsen, Hoffmann–Jørgensen, and Pedersen, 1976), but must contend with measure theoretic difficulties. In most applications, minimal sufficient statistics can be obtained for uncountable families by combining Theorem 5.3 with the following lemma.

Lemma 5.1. *If $\mathcal{P}$ is a family of distributions with common support and $\mathcal{P}_0 \subset \mathcal{P}$, and if* T *is minimal sufficient for $\mathcal{P}_0$ and sufficient for $\mathcal{P}$, it is minimal sufficient for $\mathcal{P}$.*

Proof. If U is sufficient for $\mathcal{P}$, it is also sufficient for $\mathcal{P}_0$, and hence T is a function of U.

Example 5.8. Location families. As an application, let us now determine minimal sufficient statistics for a sample $X_1, \ldots, X_n$ from a location family $\mathcal{P}$, that is, when

$$p_\theta(x) = f(x_1 - \theta) \cdots f(x_n - \theta), \tag{7}$$

where f is assumed to be known. By Example 5.6, sufficiency permits the rather trivial reduction to the order statistics for all f. However, this reduction uses only the iid assumption and neither the special structure (7) nor the knowledge of f. To illustrate the different possibilities that arise when this knowledge is utilized, we shall take for f the six densities of Table 3.1, each with $b = 1$.

(i) *Normal.* If $\mathcal{P}_0$ consists of the two distributions $N(\theta_0, 1)$ and $N(\theta_1, 1)$ it follows from Theorem 5.3 that the minimal sufficient statistic for $\mathcal{P}_0$ is $T(x) = p_{\theta_1}(X)/p_{\theta_0}(X)$, which is equivalent to $\bar{X}$. Since $\bar{X}$ is sufficient for $\mathcal{P} = \{N(\theta, 1), -\infty < \theta < \infty\}$ by the factorization criterion, it is minimal sufficient.

(ii) *Exponential.* If the X's are distributed as $E(\theta, 1)$, it is easily seen that $X_{(1)}$ is minimal sufficient (Problem 5.17).

(iii) *Uniform.* For a sample from $U(\theta - \frac{1}{2}, \theta + \frac{1}{2})$, the minimal sufficient statistic is $(X_{(1)}, X_{(n)})$ (Problem 5.16).

In these three instances, sufficiency was able to reduce the original n-dimensional data to one or two dimensions. Such extensive reductions are not possible for the remaining three distributions of Table 3.1.

(iv) *Logistic*. The joint density of a sample from $L(\theta, 1)$ is

$$p_\theta(x) = \exp[-\Sigma(x_i - \theta)]/\Pi\{1 + \exp[-(x_i - \theta)]\}^2. \tag{8}$$

Consider a subfamily $\mathcal{P}_0$ consisting of the distributions (8) with $\theta_0 = 0$ and $\theta_1, \dots, \theta_k$. Then by Theorem 5.3, the minimal sufficient statistic for $\mathcal{P}_0$ is $T(X) = [T_1(X), \dots, T_k(X)]$, where

$$T_j(x) = e^{n\theta_j} \prod_{i=1}^{n} \left(\frac{1 + e^{-x_i}}{1 + e^{-x_i + \theta_j}} \right)^2. \tag{9}$$

We shall now show that for $k = n + 1$, $T(X)$ is equivalent to the order statistics, that is, that $T(x) = T(y)$ if and only if $x = (x_1, \dots, x_n)$ and $y = (y_1, \dots, y_n)$ have the same order statistics, which means that one is a permutation of the other. The equation $T_j(x) = T_j(y)$ is equivalent to

$$\Pi \left(\frac{1 + \exp(-x_i)}{1 + \exp(-x_i + \theta_j)} \right)^2 = \Pi \left(\frac{1 + \exp(-y_i)}{1 + \exp(-y_i + \theta_j)} \right)^2$$

and hence $T(x) = T(y)$ to

$$\prod_{i=1}^{n} \frac{1 + \xi u_i}{1 + u_i} = \prod_{i=1}^{n} \frac{1 + \xi v_i}{1 + v_i} \qquad \text{for} \quad \xi = \xi_1, \dots, \xi_{n+1}, \tag{10}$$

where $\xi_j = e^{\theta_j}$, $u_i = e^{-x_i}$ and $v_i = e^{-y_i}$. Now the left and right-hand sides of (10) are polynomials in ξ of degree n which agree for $n + 1$ values of ξ if and only if the coefficients of ξ^r agree for all $r = 0, 1, \dots, n$. For $r = 0$, this implies $\Pi(1 + u_i) = \Pi(1 + v_i)$, so that (10) reduces to $\Pi(1 + \xi u_i) = \Pi(1 + \xi v_i)$ for $\xi = \xi_1, \dots, \xi_{n+1}$, and hence for all ξ. It follows that $\Pi(\eta + u_i) = \Pi(\eta + v_i)$ for all η, so that these two polynomials in η have the same roots. Since this is equivalent to the x's and y's having the same order statistics, the proof is complete.

Similar arguments show that in the Cauchy and double exponential cases, too, the order statistics are minimal sufficient (Problem 5.10). This is, in fact, the typical situation for location families, examples (i) through (iii) being happy exceptions.

As a second application of Theorem 5.3 and Lemma 5.1, let us determine minimal sufficient statistics for exponential families.

Example 5.9. Exponential families. Let X be distributed with density (4.2). We will then show that $T = (T_1, \dots, T_s)$ is minimal sufficient provided the family (4.2) is of full rank.

That T is sufficient follows immediately from Theorem 5.2. To prove minimality, let $\mathcal{P}_0$ be a subfamily consisting of $s + 1$ distributions $\eta^{(j)} = (\eta_1^{(j)}, \dots, \eta_s^{(j)})$, $j = 0, 1, \dots, s$. Then the minimal sufficient statistic for $\mathcal{P}_0$ is equivalent to

$$\Sigma(\eta_i^{(1)} - \eta_i^{(0)})T_i(X), \dots, \Sigma(\eta_i^{(s)} - \eta_i^{(0)})T_i(X),$$

which is equivalent to $T = [T_1(X), \ldots, T_s(X)]$ provided the $s \times s$ matrix $\|\eta_i^{(j)} - \eta_i^{(0)}\|$ is nonsingular. A subfamily $\mathcal{P}_0$ for which this condition is satisfied exists under the assumption of full rank. (It exists, in fact, under the much weaker assumption that the parameter space contains $s + 1$ points $\eta^{(j)}$ $(j = 0, \ldots, s)$, which span E_s, in the sense that they do not belong to a proper affine subspace of E_s). This completes the proof.

It is seen from this result that the sufficient statistics T of Examples 5.4 and 5.5 are minimal.

Let $X_1, \ldots, X_n$ be iid each with density (4.2), assumed to be of full rank. Then the joint distribution of the X's is again full-rank exponential, with $T = (T_1^*, \ldots, T_s^*)$ where $T_i^* = \sum_{j=1}^n T_i(X_j)$. This shows that in a sample from an exponential family, the data can be reduced to an s-dimensional sufficient statistic, regardless of the sample size.

The reduction of a sample to a smaller number of sufficient statistics greatly simplifies the statistical analysis, and it is therefore interesting to ask what other families permit such a reduction. The dimensionality of a sufficient statistic is a property which differs from those considered so far in that it depends not only on the sets of points of the sample space for which the statistic takes on the same value but also on these values; that is, the dimensionality may not be the same for different representations of a sufficient statistic (see, for example, Denny, 1964). To make the concept of dimensionality meaningful, let us call T a *continuous s-dimensional sufficient statistic* over a Euclidean sample space $\mathcal{X}$ if the assumptions of Theorem 5.2 hold, if $T(x) = [T_1(x), \ldots, T_s(x)]$ where T is continuous, and if the factorization (3) holds not only a.e. but for all $x \in \mathcal{X}$.

Theorem 5.4. *Suppose* $X_1, \ldots, X_n$ *are real-valued iid according to a distribution with density* $f_\theta(x_i)$ *with respect to Lebesgue measure, which is continuous in* x_i *and whose support for all* θ *is an interval* I. *Suppose that for the joint density of* $X = (X_1, \ldots, X_n)$

$$p_\theta(x) = f_\theta(x_1) \cdots f_\theta(x_n)$$

there exists a continuous k*-dimensional sufficient statistic. Then*

(i) *if* $k = 1$, *there exist functions* η_1, B *and* h *such that* (4.1) *holds*;

(ii) *if* $k > 1$, *and if the densities* $f_\theta(x_i)$ *have continuous partial derivatives with respect to* x_i, *then there exist functions* η_i, B *and* h *such that* (4.1) *holds with* $s \leq k$.

For a proof of this result, see Barndorff–Nielsen and Pedersen (1968). A corresponding problem for the discrete case is considered by Andersen (1970).

This theorem states essentially that among "smooth" absolutely continuous families of distributions with fixed support, exponential families are the only ones that permit dimensional reduction of the sample through sufficiency. It is crucial for this result that the support of the distributions P_θ is independent of θ. In the contrary case, a simple example of a family possessing a one-dimensional sufficient statistic for any sample size is provided by the uniform distribution (Example 5.2).

The Dynkin–Ferguson theorem mentioned at the end of the last section and Theorem 5.4 state roughly that (a) the only location families which are one-parameter exponential families are the normal and log of gamma distributions, and (b) only exponential families permit reduction of the data through sufficiency. Together, these results appear to say that the only location families with fixed support in which a dimensional reduction of the data is possible are the normal and log of gamma families. This is not quite correct, however, because a location family—though it is a one-parameter family—may also be what in definition (4.1) has been called an s-parameter exponential family, with $s > 1$.

Example 5.10. Let $X_1, \ldots, X_n$ be iid with joint density (with respect to Lebesgue measure)

$$(11)\quad C\exp\left[-\sum_{i=1}^{n}(x_i - \theta)^4\right]$$

$$= C\exp(-n\theta^4)\exp\left(4\theta^3\Sigma x_i - 6\theta^2\Sigma x_i^2 + 4\theta\Sigma x_i^3 - \Sigma x_i^4\right).$$

According to (4.1) this is a three-parameter exponential family, and it provides an example of a location family with a three-dimensional sufficient statistic satisfying all the assumptions of Theorem 5.4.

Exponential families such as (11), in which the parameter space forms a curved subset of s-space, are called *curved exponential families* by Efron (1975, 1978).

The tentative conclusion, which had been reached just before Example 5.10 and which was contradicted by this example, is nevertheless basically correct. Typically, a location family with fixed support $(-\infty, \infty)$ will not constitute even a curved exponential family and will, therefore, not permit a dimensional reduction of the data without loss of information.

Example 5.8 shows that the degree of reduction that can be achieved through sufficiency is extremely variable, and an interesting question is what characterizes the situations in which sufficiency leads to a substantial reduction of the data. The ability of a sufficient statistic to achieve such a reduction appears to be related to the amount of ancillary information it contains. A statistic $V(X)$ is said to be *ancillary* if its distribution does not depend on θ and *first-order ancillary* if its expectation $E_\theta[V(X)]$ is constant, independent of θ. An ancillary statistic by itself contains no information about θ, but minimal sufficient statistics may still contain much ancillary material. In Example 5.8(iv), for instance, the differences $X_{(n)} - X_{(i)}$ ($i = 1, \ldots, n-1$) are ancillary despite the fact that they are functions of the minimal sufficient statistics ($X_{(1)}, \ldots, X_{(n)}$). A sufficient statistic T appears to be most successful in reducing the data if no nonconstant function of T is ancillary or even first-order ancillary, that is, if $E_\theta[f(T)] = c$ for all $\theta \in \Omega$ implies $f(t) = c$ (a.e. $\mathcal{P}$). By subtracting c, this condition is seen to be equivalent to

$$(12) \qquad E_\theta[f(T)] = 0 \qquad \text{for all} \quad \theta \in \Omega \text{ implies } f(t) = 0 \text{ (a.e. } \mathcal{P})$$

where $\mathcal{P} = \{P_\theta, \theta \in \Omega\}$. A statistic T satisfying (12) is said to be *complete*. As will be seen later, completeness brings with it substantial simplifications of the statistical situation.

Since complete sufficient statistics are particularly effective in reducing the data, it is not surprising that a complete sufficient statistic is always minimal. Proofs are given in Lehmann and Scheffé (1950) and Bahadur (1957); see also Problem 5.24.

What happens to the ancillary statistics when the minimal sufficient statistics is complete is shown by the following result.

Theorem 5.5. **(Basu's theorem).** *If* T *is a complete sufficient statistic for the family* $\mathcal{P} = \{P_\theta, \theta \in \Omega\}$, *then any ancillary statistic* V *is independent of* T.

Proof. If V is ancillary, the probability $p_A = P(V \in A)$ is independent of θ for all A. Let $\eta_A(t) = P(V \in A | T = t)$. Then $E_\theta[\eta_A(T)] = p_A$ and hence, by completeness,

$$\eta_A(t) = p_A \text{(a.e. } \mathcal{P})$$

This establishes the independence of V and T.

We conclude this section by giving some examples of completeness.

Theorem 5.6. *If* X *is distributed according to the exponential family* (4.2) *and is of full rank, then* $T = [T_1(X), \ldots, T_s(X)]$ *is complete.*

For a proof, see TSH Section 4.3, Theorem 1 or Barndorff–Nielsen (1978), Lemma 8.2.

Example 5.11. Theorem 5.6 proves completeness of X for the binomial family $\{b(p, n), 0 < p < 1\}$ and the Poisson family $\{P(\lambda), 0 < \lambda\}$; also for $(\bar{X}, S^2)$ of Example 5.5 in the normal family $\{N(\xi, \sigma^2), -\infty < \xi < \infty, 0 < \sigma\}$.

Example 5.12. (i) Let $X_1, \ldots, X_n$ be iid according to the uniform distribution $U(0, \theta)$, $0 < \theta$. It was seen in Example 5.2 that $T = X_{(n)}$ is sufficient for θ. To see that T is complete, note that

$$P(T \leqslant t) = t^n/\theta^n, \qquad 0 < t < \theta$$

so that T has probability density

$$(13) \qquad p_\theta(t) = nt^{n-1}/\theta^n, \qquad 0 < t < \theta.$$

Suppose $E_\theta f(T) = 0$ for all θ, and let f^+ and f^- be its positive and negative parts. Then

$$\int_0^\theta t^{n-1} f^+(t)\, dt = \int_0^\theta t^{n-1} f^-(t)\, dt$$

for all θ. It follows that

$$\int_A t^{n-1} f^+(t)\, dt = \int_A t^{n-1} f^-(t)\, dt$$

for all Borel sets A, and this implies $f = 0$ a.e.

(ii) Let $Y_1, \ldots, Y_n$ be iid according to the exponential distribution $E(\eta, 1)$. If $X_i = e^{-Y_i}$, $\theta = e^{-\eta}$, then $X_1, \ldots, X_n$ are iid as $U(0, \theta)$ (Problem 5.23), and it follows from (i) that $X_{(n)}$ or, equivalently, $Y_{(1)}$ is sufficient and complete.

Example 5.13. Let $X_1, \ldots, X_n$ be iid according to $U(\theta - \frac{1}{2}, \theta + \frac{1}{2})$, $-\infty < \theta < \infty$. Here $T = (X_{(1)}, X_{(n)})$ is minimal sufficient (Problem 5.16). On the other hand, T is not complete since $X_{(n)} - X_{(1)}$ is ancillary. For example, $E_\theta[X_{(n)} - X_{(1)} - (n-1)/(n+1)] = 0$ for all θ.

Example 5.14. Let $X_1, \ldots, X_n$ be iid according to the exponential distribution $E(a, b)$, $-\infty < a < \infty$, $0 < b$, and let $T_1 = X_{(1)}$, $T_2 = \Sigma[X_i - X_{(1)}]$. Then (T_1, T_2) are independently distributed as $E(a, b/n)$ and $\frac{1}{2}b\chi^2_{2n-2}$, respectively (Problem 5.18), and they are jointly sufficient and complete. Sufficiency follows from the factorization criterion. To prove completeness suppose that

$$E_{a,b}[f(T_1, T_2)] = 0 \qquad \text{for all} \quad a, b.$$

Then if

$$(14) \qquad g(t_1, b) = E_b f(t_1, T_2)$$

we have that for any fixed b

$$\int_a^\infty g(t_1, b) e^{-nt_1/b} dt_1 = 0 \qquad \text{for all} \quad a.$$

It follows from Example 5.12(ii) that

$$g(t_1, b) = 0$$

except on a set N_b of t_1 values which has Lebesgue measure zero and which may depend on b. Then by Fubini's theorem, for almost all t_1 we have

$$g(t_1, b) = 0 \text{ a.e. in } b.$$

Since the densities of T_2 constitute an exponential family, $g(t_1, b)$ by (14) is a continuous function of b for any fixed t_1. It follows that for almost all t_1, $g(t_1, b) = 0$, not only a.e. but for all b. Applying completeness of T_2 to (14) we see that for almost all t_1, $f(t_1, t_2) = 0$ a.e. in t_2. Thus, finally, $f(t_1, t_2) = 0$ a.e. with respect to Lebesgue measure in the (t_1, t_2) plane. [For measurability aspects which have been ignored in this proof, see Lehmann and Scheffé (1955, Theorem 7.1).]

6. CONVEX LOSS FUNCTIONS

The estimation problem outlined in Section 1 simplifies in a number of ways when the loss function $L(\theta, d)$ is a convex function of d.

Definition. *A real-valued function* ϕ *defined over an open interval* $I = (a, b)$ *with* $-\infty \leqslant a < b \leqslant \infty$ *is convex if for any* $a < x < y < b$ *and any* $0 < \gamma < 1$

$$\phi[\gamma x + (1 - \gamma)y] \leqslant \gamma\phi(x) + (1 - \gamma)\phi(y). \tag{1}$$

The function is said to be strictly convex if strict inequality holds in (1) *for all indicated values of* x, y, *and* γ. *A function* ϕ *is concave on* (a, b) *if* $-\phi$ *is convex.*

Convexity is a very strong condition which implies, for example, that ϕ is continuous in (a, b) and has a left and right derivative at every point of (a, b). Proofs of these properties and of the other properties of convex functions stated in the following without proof can be found, for example, in Hardy, Littlewood, and Polya (1934), in Roberts and Varberg (1973), or in Rudin (1966).

Determination of whether or not a function is convex is often easy with the help of the following two criteria.

Theorem 6.1. (i) *If* ϕ *is defined and differentiable on* (a, b), *then a necessary and sufficient condition for* ϕ *to be convex is that*

$$\phi'(x) \leqslant \phi'(y) \qquad \textit{for all} \quad a < x < y < b. \tag{2}$$

The function is strictly convex if and only if the inequality (2) *is strict for all* $x < y$.

(ii) *If, in addition,* ϕ *is twice differentiable, then the necessary and sufficient condition* (2) *is equivalent to*

$$\phi''(x) \geqslant 0 \qquad \textit{for all} \quad a < x < b \tag{3}$$

with strict inequality sufficient (but not necessary) for strict convexity.

Example 6.1. From these criteria it is easy to see that the following functions are convex over the indicated intervals:

Function ϕ	Interval (a, b)
(i) $\lvert x \rvert$	$-\infty < x < \infty$
(ii) x^2	$-\infty < x < \infty$
(iii) $x^p, p \geq 1$	$0 < x$
(iv) $1/x^p, p > 0$	$0 < x$
(v) e^x	$-\infty < x < \infty$
(vi) $-\log x$	$0 < x < \infty$

In all these cases, ϕ is strictly convex, except in (i) and (iii) with $p = 1$.

In general, a convex function is not strictly convex if and only if it is linear over some subinterval of (a, b) (Problems 6.1 and 6.6).

A basic property of convex functions is contained in the following theorem.

Theorem 6.2. *Let ϕ be a convex function defined on* I = (a, b) *and let* t *be any fixed point in* I. *Then there exists a straight line*

$$y = L(x) = c(x - t) + \phi(t) \tag{4}$$

through the point $[t, \phi(t)]$ *such that*

$$L(x) \leq \phi(x) \quad \textit{for all } x \textit{ in } I. \tag{5}$$

By definition, a function ϕ is convex if the value of the function at the weighted average of two points does not exceed the weighted average of its values at these two points. By induction, this is easily generalized to the average of any finite number of points (Problem 6.7). In fact, the inequality also holds for the weighted average of any infinite set of points and in this general form is known as Jensen's inequality.

The weighted average of ϕ with respect to the weight function Λ is represented by

$$\int_I \phi \, d\Lambda \tag{6}$$

where Λ is a measure with $\Lambda(I) = 1$. In the particular case that Λ assigns measure γ and $1 - \gamma$ to the points x and y, respectively, this reduces to the right side of (1). It is convenient to interpret (6) as the expected value of $\phi(X)$, where X is a random variable taking on values in I according to the probability distribution Λ.

Theorem 6.3. (Jensen's inequality). *If ϕ is a convex function defined over an open interval* I, *and* X *is a random variable with* $P(X \in I) = 1$ *and finite expectation, then*

$$\phi[E(X)] \leqslant E[\phi(X)]. \tag{7}$$

If ϕ is strictly convex, the inequality is strict unless X *is a constant with probability* 1.

Proof. Let $y = L(x)$ be the equation of the line which satisfies (5) and for which $L(t) = \phi(t)$ when $t = E(X)$. Then

$$E[\phi(X)] \geqslant E[L(X)] = L[E(X)] = \phi[E(X)], \tag{8}$$

which proves (7). If ϕ is strictly convex, the inequality in (5) is strict for all $x \neq t$, and hence the inequality in (8) is strict unless $\phi(X) = E[\phi(X)]$ with probability 1.

Note that the theorem does not exclude the possibility that $E[\phi(X)] = \infty$.

Corollary 6.1. *If* X *is a nonconstant positive random variable with finite expectation, then*

$$\frac{1}{E(X)} < E\left(\frac{1}{X}\right) \tag{9}$$

and

$$E(\log X) < \log[E(X)]. \tag{10}$$

In Theorem 5.1, it was seen that if T is a sufficient statistic, then for any statistical procedure there exists an equivalent procedure (i.e., having the same risk function) based only on T. We shall now show that in estimation with a strictly convex loss function, a much stronger statement is possible: Given any estimator $\delta(X)$ which is not a function of T, there exists a *better* estimator depending only on T.

Theorem 6.4. (Rao–Blackwell theorem). *Let* X *be a random observable with distribution* $P_\theta \in \mathcal{P} = \{P_{\theta'}, \theta' \in \Omega\}$, *and let* T *be sufficient for* $\mathcal{P}$. *Let* δ *be an estimator of an estimand* $g(\theta)$, *and let the loss function* $L(\theta, d)$ *be a strictly convex function of* d. *Then, if* δ *has finite expectation and risk,*

$$R(\theta, \delta) = EL[\theta, \delta(X)] < \infty$$

and if

$$\eta(t) = E[\delta(X)|t], \tag{11}$$

the risk of the estimator $\eta(T)$ *satisfies*

$$R(\theta, \eta) < R(\theta, \delta) \tag{12}$$

unless $\delta(X) = \eta(T)$ *with probability* 1.

Proof. In Theorem 6.3, let $\phi(d) = L(\theta, d)$, let $\delta = \delta(X)$, and let X have the conditional distribution $P^{X|t}$ of X given $T = t$. Then

$$L[\theta, \eta(t)] < E\{L[\theta, \delta(X)]|t\}$$

unless $\delta(X) = \eta(T)$ with probability 1. Taking expectation on both sides of this inequality yields (12), unless $\delta(X) = \eta(T)$ with probability 1.

Some points concerning this result are worth noting.

1. Sufficiency of T is used in the proof only to ensure that $\eta(T)$ does not depend on θ and hence is an estimator.
2. If the loss function is convex but not strictly convex, the theorem remains true provided the inequality sign in (12) is replaced by $\leqslant$. The theorem still provides information even in that case beyond the results of Section 5 because it shows that the particular estimator $\eta(T)$ is at least as good as $\delta(X)$.
3. The theorem is not true if the convexity assumption is dropped. Examples illustrating this fact will be given in Chapters 2 and 4.

In Section 5, randomized estimators were introduced, and such estimators may be useful, for example, in reducing the maximum risk (see Ch. 4, Example 2.2), but this can never be the case when the loss function is convex.

Corollary 6.2. *Given any randomized estimator of* $g(\theta)$, *there exists a nonrandomized estimator which is uniformly better if the loss function is strictly convex and at least as good when it is convex.*

Proof. Note first that a randomized estimator can be obtained as a nonrandomized estimator $\delta^*(X, U)$, where X and U are independent and U is uniformly distributed on $(0, 1)$. This is achieved by observing $X = x$ and then using U to construct the distribution of Y given $X = x$, where $Y = Y(x)$ is the random variable employed in the definition of a randomized estimator (Problem 6.8). To prove the theorem, we therefore need to show that given any estimator $\delta^*(X, U)$ of $g(\theta)$ there exists an estimator $\delta(X)$, depending on X only, which has uniformly smaller risk. However, this is an immediate consequence of the Rao–Blackwell theorem since for the observations (X, U), the statistic X is sufficient. For $\delta(X)$, one can therefore take the conditional expectation of $\delta^*(X, U)$ given X.

An estimator δ is said to be *inadmissible* if there exists another estimator δ' which *dominates* it (that is, such that $R(\theta, \delta') \leqslant R(\theta, \delta)$ for all θ, with strict inequality for some θ) and *admissible* if no such estimator δ' exists. If the loss function L is strictly convex, it follows from Corollary 6.2 that every admissible estimator must be nonrandomized. Another property of admissible estimators in the strictly convex case is provided by the following uniqueness result.

Theorem 6.5. *If* L *is strictly convex and* δ *is an admissible estimator of* $g(\theta)$, *and if* δ' *is another estimator with the same risk function that is, satisfying* $R(\theta, \delta) = R(\theta, \delta')$ *for all* θ, *then* $\delta' = \delta$ *with probability* 1.

Proof. If $\delta^* = \frac{1}{2}(\delta + \delta')$, then

$$R(\theta, \delta^*) < \tfrac{1}{2}[R(\theta, \delta) + R(\theta, \delta')] = R(\theta, \delta) \tag{13}$$

unless $\delta = \delta'$ with probability 1, and (13) contradicts the admissibility of δ.

The preceding considerations can be extended to the situation in which the estimand $g(\theta) = [g_1(\theta), \dots, g_k(\theta)]$ and the estimator $\delta(X) = [\delta_1(X), \dots, \delta_k(X)]$ are vector-valued.

For any two points $\underline{x} = (x_1, \dots, x_k)$ and $\underline{y} = (y_1, \dots, y_k)$ in E_k define $\gamma\underline{x} + (1 - \gamma)\underline{y}$ to be the point with coordinates $\gamma x_i + (1 - \gamma)y_i$, $i = 1, \dots, k$. Then a set S in E_k is *convex* if for any $\underline{x}$, $\underline{y} \in S$ the points

$$\gamma\underline{x} + (1 - \gamma)\underline{y}, \qquad 0 < \gamma < 1$$

are also in S. Geometrically, this means that the line segment connecting any two points in S lies in S. Furthermore, a real-valued function ϕ defined over an open convex set S in E_k is *convex* if (1) holds with x and y replaced by $\underline{x}$ and $\underline{y}$; it is strictly convex if the inequality is strict for all $\underline{x}$ and $\underline{y}$.

***Example** 6.2.* If ϕ_j is a convex function of a real variable defined over an interval I_j for each $j = 1, \dots, k$, then for any positive constants $a_1, \dots, a_k$

$$\phi(\underline{x}) = \Sigma a_j \phi_j(x_j) \tag{14}$$

is a convex function defined over the k-dimensional rectangle with sides $I_1, \dots, I_k$; it is strictly convex provided $\phi_1, \dots, \phi_k$ are all strictly convex. This example implies, in particular, that the loss function

$$L(\theta, \underline{d}) = \Sigma a_i [d_i - g_i(\theta)]^2 \tag{15}$$

is strictly convex.

A useful criterion to determine whether a given function ϕ is convex is the following generalization of (3).

Theorem 6.6. *Let ϕ be defined over an open convex set* S *in* E_k *and twice differentiable in* S. *Then a necessary and sufficient condition for ϕ to be convex is that the* $k \times k$ *matrix with* ij*th element* $\partial^2\phi(x_1, \ldots, x_k)/\partial x_i \partial x_j$, *the so-called Hessian matrix, is positive semidefinite; if the matrix is positive definite, then ϕ is strictly convex.*

Example 6.3. Consider the loss function

$$L(\theta, \underline{d}) = \Sigma\Sigma a_{ij}[d_i - g_i(\theta)][d_j - g_j(\theta)]. \tag{16}$$

Since $\partial^2 L/\partial d_i \partial d_j = a_{ij}$, L is strictly convex provided the matrix $\|a_{ij}\|$ is positive definite.

Theorem 6.2 generalizes to the following supporting hyperplane theorem for convex functions.

Theorem 6.7. *Let ϕ be a convex function defined over an open convex set* S *in* E_k *and let* $\underline{t}$ *be any point in* S. *Then there exists a hyperplane*

$$y = L(\underline{x}) = \Sigma c_i(x_i - t_i) + \phi(\underline{t}) \tag{17}$$

through the point $[\underline{t}, \phi(\underline{t})]$ *such that*

$$L(\underline{x}) \leqslant \phi(\underline{x}) \qquad \textit{for all} \quad \underline{x} \in S. \tag{18}$$

Jensen's inequality (Theorem 6.3) generalizes in the obvious way. The only changes that are needed are replacement of the interval I by an open convex set S, of the random variable X by a random vector $\underline{X}$ satisfying $P(\underline{X} \in S) = 1$, and of the expectation $E(X)$ by the expectation vector $E(\underline{X}) = [E(X_1), \ldots, E(X_k)]$. For the resulting modification of the inequality (7) to be meaningful, it is necessary to know that $E(\underline{X})$ is in S so that $\phi[E(\underline{X})]$ is defined.

Lemma 6.1. *If* $\underline{X}$ *is a random vector with* $P(\underline{X} \in S) = 1$ *where* S *is an open convex set in* E_k, *and if* $E(\underline{X})$ *exists, then* $E(\underline{X}) \in S$.

A formal proof is given by Ferguson (1967, p. 74). Here we shall give only a sketch. Suppose that $k = 2$, and suppose that $\xi = E(\underline{X})$ is not in S. Then a theorem similar to Theorem 6.2 guarantees the existence of a line $a_1x_1 + a_2x_2 = b$ through the point (ξ_1, ξ_2) such that S lies entirely on one side of the line. By a rotation of the plane, it can be assumed without loss of generality that the equation of the line is $x_2 = \xi_2$ and that S lies above this line so that $P(X_2 > \xi_2) = 1$. It follows that $E(X_2) > \xi_2$, which is a contradiction.

As a last consequence of the adoption of a convex loss function, we shall consider a somewhat more special problem that arises in location models. In

Section 1, it was pointed out that there exists a unique number a minimizing $\Sigma(x_i - a)^2$, namely $\bar{x}$, and that the minimizing value of $\Sigma_{i=1}^n |x_i - a|$ is either unique (when n is odd) or the minimizing values constitute an interval. This interval structure of the minimizing values does not hold, for example, when minimizing $\Sigma\sqrt{|x_i - a|}$. In the case $n = 2$, for instance, there exist two minimizing values, $a = x_1$ and $a = x_2$ (Problem 6.10). This raises the general question of the set of values a minimizing $\Sigma\rho(x_i - a)$, which in turn is a special case of the following problem. Let X be a random variable and $L(\theta, d) = \rho(d - \theta)$ a loss function, with ρ even. Then what can be said about the set of values a minimizing $E[\rho(X - a)]$? This specializes to the earlier case if X takes on the values $x_1, \ldots, x_n$ with probabilities $1/n$ each.

Theorem 6.8. *Let ρ be a convex function defined on $(-\infty, \infty)$ and X a random variable such that $\phi(a) = E[\rho(X - a)]$ is finite for some a. If ρ is not monotone, $\phi(a)$ takes on its minimum value and the set on which this value is taken is a closed interval. If ρ is strictly convex, the minimizing value is unique.*

The proof is based on the following lemma.

Lemma 6.2. *Let ϕ be a convex function on $(-\infty, \infty)$ which is bounded below and suppose that ϕ is not monotone. Then ϕ takes on its minimum value; the set S on which this value is taken on is a closed interval and is a single point when ϕ is strictly convex.*

Proof. Since ϕ is convex and not monotone, it tends to ∞ as $x \to \pm\infty$. Since ϕ is also continuous, it takes on its minimizing value. That S is an interval follows from convexity and that it is closed from continuity.

Proof of Theorem 6.8. By the lemma, it is enough to prove that $\phi(a)$ is (strictly) convex and not monotone. That ϕ is not monotone follows from the fact that $\phi(a) \to \infty$ as $a \to \pm\infty$. This latter property of ϕ is a consequence of the facts that $X - a$ tends in probability to $\mp\infty$ as $a \to \pm\infty$, and that $\rho(t) \to \infty$ as $t \to \pm\infty$. (Strict) convexity of ϕ follows from the corresponding property of ρ.

Example 6.4. Squared error. Let $\rho(t) = t^2$ and suppose that $E(X^2) < \infty$. Since ρ is strictly convex, it follows that $\phi(a)$ has a unique minimizing value. If $E(X) = \mu$, which by assumption is finite, we have in fact

$$\phi(a) = E(X - a)^2 = E(X - \mu)^2 + (\mu - a)^2, \tag{19}$$

which shows that $\phi(a)$ is a minimum if and only if $a = \mu$.

Example 6.5. Absolute error. Let $\rho(t) = |t|$ and suppose that $E|X| < \infty$. Since ρ is convex but not strictly convex, it follows from Theorem 6.8 that $\phi(a)$ takes on its minimum value and that the set S is a closed interval. The set S is, in fact, the set of *medians* of X (Problems 1.9, 1.10).

The following is a useful consequence of Theorem 6.8.

Corollary 6.3. *Under the assumptions of Theorem* 6.8, *suppose that* ρ *is even and* X *is symmetric about* μ. *Then* ϕ(a) *attains its minimum at* a = μ.

Proof. By Theorem 6.8 the minimum is taken on. If $\mu + c$ is a minimizing value, so is $\mu - c$ and so, therefore, are all values a between $\mu - c$ and $\mu + c$, which includes $a = \mu$.

Now consider an example in which ρ is not convex.

Example 6.6. Let $\rho(t) = 1$ if $|t| \geqslant k$ and $\rho(t) = 0$ otherwise. Minimizing $\phi(a)$ is then equivalent to maximizing $\psi(a) = P(|X - a| < k)$. Consider the following two special cases (Problem 6.17).

(i) The distribution of X has a probability density (with respect to Lebesgue measure) which is continuous, unimodal, and such that $f(x)$ decreases strictly as x moves away from the mode in either direction. Then there exists a unique value a for which $f(a - k) = f(a + k)$, and this is the unique maximizing value of $\psi(a)$.

(ii) Suppose that f is even and U-shaped with $f(x)$ attaining its maximum at $x = \pm A$ and $f(x) = 0$ for $|x| > A$. Then $\psi(a)$ attains its maximum at the two points $a = -A + k$ and $a = A - k$.

Convex loss functions have been seen to lead to a number of simplifications of estimation problems. One may wonder, however, whether such loss functions are likely to be realistic. If $L(\theta, d)$ represents not just a measure of inaccuracy but a real (for example, financial) loss, one may argue that all such losses are bounded: once you have lost all, you cannot lose any more. On the other hand, if d can take on all values in $(-\infty, \infty)$ or $(0, \infty)$ no nonconstant bounded function can be convex (Problem 6.13). Unfortunately, bounded loss functions with unbounded d, lead to completely unreasonable estimators (see, for example, Theorem 2.1.3). The reason is roughly that arbitrarily large errors can then be committed with essentially no additional penalty and their leverage used to unfair advantage. Perhaps convex loss functions result in more reasonable estimators because the large penalties they exact for large errors compensate for the unrealistic assumption of unbounded d: they make such values so expensive that the estimator will try hard to avoid them.

The most widely used loss function is squared error

$$L(\theta, d) = [d - g(\theta)]^2 \tag{20}$$

or slightly more generally weighted squared error

$$L(\theta, d) = w(\theta)[d - g(\theta)]^2. \tag{21}$$

Since these are strictly convex in d, the simplifications represented by Theorem 6.4, Corollary 6.2, and Theorem 6.5 are valid in these cases. The most slowly growing even convex loss function is absolute error

$$L(\theta, d) = |d - g(\theta)|. \tag{22}$$

The faster the loss function increases, the more attention it pays to extreme values of the estimators and hence to outlying observations, so that the performance of the resulting estimators is strongly influenced by the tail behavior of the assumed distribution of the observable random variables. As a consequence, fast-growing loss functions lead to estimators that tend to be sensitive to the assumptions made about this tail behavior, and these assumptions typically are based on little information and thus are not very reliable.

It turns out that the estimators produced by squared error loss often are uncomfortably sensitive in this respect. On the other hand, absolute error appears to go too far in leading to estimators which discard all but the central observations. For many important problems, the most appealing results are obtained from the use of loss functions which lie between (20) and (22). One interesting class of such loss functions, due to Huber (1964), puts

$$L(\theta, d) = \begin{cases} [d - g(\theta)]^2 & \text{if } |d - g(\theta)| \leqslant k \\ 2k|d - g(\theta)| - k^2 & \text{if } |d - g(\theta)| \geqslant k. \end{cases} \tag{23}$$

This agrees with (20) for $|d - g(\theta)| \leqslant k$, but above k and below $-k$ it replaces the parabola with straight lines joined to the parabola so as to make the function continuous and continuously differentiable (Problem 6.16).

The Huber loss functions are convex but not strictly convex. An alternative family, which also interpolates between (20) and (22) and which is strictly convex, is

$$L(\theta, d) = |d - g(\theta)|^p, \qquad 1 < p < 2. \tag{24}$$

It is a disadvantage of both (23) and (24) that the resulting estimators, even in fairly simple problems, cannot be obtained in closed form and hence are more difficult to grasp intuitively and to interpret. This may account at least in part for the fact that squared error is by far the most commonly used loss function or measure of accuracy and that the classic estimators in most situations are the ones derived on this basis. As indicated at the end of

Section 1, we shall here develop the theory under the more general assumption of convex loss functions (which, in practice does not appear to be a serious limitation) but work most examples for the conventional squared error loss. The issue of the robustness of the resulting estimators, which requires going outside the assumed model, will be taken up in Chapter 5.

7. PROBLEMS

Section 1

1.1 If $(x_1, y_1), \ldots, (x_n, y_n)$ are n points in the plane, determine the best fitting line $y = \alpha + \beta x$ in the least squares sense, that is, determine the values α, β that minimize $\Sigma[y_i - (\alpha + \beta x_i)]^2$.

1.2 Let $X_1, \ldots, X_n$ be uncorrelated random variables with common expectation θ and variance σ^2. Then among all linear estimators $\Sigma\alpha_i X_i$ of θ satisfying $\Sigma\alpha_i = 1$, the mean $\bar{X}$ has the smallest variance.

1.3 In the preceding problem, minimize the variance of $\Sigma\alpha_i X_i$ $(\Sigma\alpha_i = 1)$

(i) When the variance of X_i is σ^2/a_i (a_i known).

(ii) When the X_i have common variance σ^2 but are correlated with common correlation coefficient ρ.

(For generalizations of these results, see, for example, Watson, 1967, and Kruskal, 1968.)

1.4 Let X, Y have common expectation θ, variances σ^2 and τ^2, and correlation coefficient ρ. Determine the conditions on σ, τ, and ρ under which

(i) $\text{var}(X) < \text{var}[(X + Y)/2]$.

(ii) The value of α that minimizes $\text{var}[\alpha X + (1 - \alpha)Y]$ is negative.

Give an intuitive explanation of your results.

1.5 (i) If two estimators δ_1, δ_2 have continuous symmetric densities $f_i(x - \theta)$, $i = 1, 2$, and $f_1(0) > f_2(0)$, then

$$P[|\delta_1 - \theta| < c] > P[|\delta_2 - \theta| < c] \quad \text{for some} \quad c > 0 \tag{1}$$

and hence δ_1 will be closer to θ than δ_2 with respect to the measure (1.5).

(ii) Let X, Y be independently distributed with common continuous symmetric density f, and let $\delta_1 = X$, $\delta_2 = (X + Y)/2$. Then (1) will hold provided

$$2\int f^2(x)\,dx < f(0) \tag{2}$$

(Edgeworth, 1883, Stigler, 1980).

1.6 (i) Let $f(x) = \frac{1}{2}(k - 1)/(1 + |x|)^k$, $k \geqslant 2$. Show that f is a probability density and that all its moments of order $< k - 1$ are finite.

(ii) The density of (i) satisfies (2).

1.7 Let X_i $(i = 1, 2)$ be independently distributed according to the Cauchy densities $C(a_i, b_i)$. Then $X_1 + X_2$ is distributed as $C(a_1 + a_2, b_1 + b_2)$. [*Hint:* Transform to new variables $Y_1 = X_1 + X_2$, $Y_2 = X_2$.]

1.8 If $X_1, \dots, X_n$ are iid as $C(a, b)$, the distribution of $\bar{X}$ is again $C(a, b)$. [*Hint:* Prove by induction, using Problem 1.7.]

1.9 A *median* of X is any value m such that

$$P(X \leqslant m) \geqslant \tfrac{1}{2} \quad \text{and} \quad P(X \geqslant m) \geqslant \tfrac{1}{2}. \tag{3}$$

(i) Show that (3) is equivalent to

$$P(X < m) \leqslant \tfrac{1}{2} \quad \text{and} \quad P(X > m) \leqslant \tfrac{1}{2}. \tag{4}$$

(ii) Show that the set of medians is always a closed interval

$$m_0 \leqslant m \leqslant m_1.$$

1.10 If $\phi(a) = E|X - a| < \infty$ for some a, show that $\phi(a)$ is minimized by any median of X. [*Hint:* If $m_0 \leqslant m \leqslant m_1$ (in the notation of Problem 1.9) and $m_1 < c$, then

$$E|X - c| - E|X - m| = (c - m)[P(X \leqslant m) - P(X > m)]$$

$$+ 2\int_{m<x<c} (c - x)\, dP(x)].$$

1.11 (i) The median of any set of distinct real numbers $x_1, \dots, x_n$ is defined to be the middle one of the ordered x's when n is odd, and any value between the two middle ordered x's when n is even. Show that this is also the median of the random variable X which takes on each of the values $x_1, \dots, x_n$ with probability $1/n$.

(ii) For any set of distinct real numbers $x_1, \dots, x_n$, the sum of absolute deviations $\Sigma|x_i - a|$ is minimized by any median of the x's.

(iii) For n given points (x_i, y_i), $i = 1, \dots, n$, determine the value b that minimizes $\Sigma|y_i - bx_i|$. [*Hint:* Reduce the problem to a special case of Problem 1.10.]

1.12 (i) If X is binomial $b(p, n)$, then

$$E\left|\frac{X}{n} - p\right| = 2\binom{n-1}{k-1} p^k (1-p)^{n-k+1} \quad \text{for} \quad \frac{k-1}{n} \leqslant p \leqslant \frac{k}{n}.$$

(ii) Graph the risk function of part (i) for $n = 4$ and $n = 5$.

[(i) Use the identity

$$\binom{n}{x}(x-np) = n\left[\binom{n-1}{x-1}(1-p) - \binom{n-1}{x}p\right], \quad 1 \leq x \leq n.$$

(Johnson, 1957–1958; and Blyth, 1980)].

1.13* Let f be a unimodal density symmetric about 0, and let $L(\theta, d) = \rho(d-\theta)$ be a loss function with ρ nondecreasing on $(0, \infty)$ and symmetric about 0.

(i) The function φ defined in Theorem 6.8 takes on its minimum at 0.
(ii) If

$$S_a = \{x: [\rho(x+a) - \rho(x-a)][f(x+a) - f(x-a)] \neq 0\}$$

then $\varphi(a)$ takes on its unique minimum value at $a = 0$ if and only if there exists a_0 such that $\varphi(a_0) < \infty$, and $\mu(S_a) > 0$ for all a. [*Hint:* Note that $\varphi(0) \leq \frac{1}{2}[\varphi(2a) + \varphi(-2a)]$, with strict inequality holding if and only if $\mu(S_a) > 0$ for all a.]

1.14* (i) Suppose that f and ρ satisfy the assumptions of Problem 1.13 and that f is strictly decreasing on $[0, \infty)$. Then if $\varphi(a_0) < \infty$ for some a_0, $\varphi(a)$ has a unique minimum at zero unless there exists $c \leq d$ such that

$$\rho(0) = c \quad \text{and} \quad \rho(x) = d \qquad \text{for all} \quad x \neq 0.$$

(ii) If ρ is symmetric about 0, strictly increasing on $[0, \infty)$, and $\varphi(a_0) < \infty$ for some a_0, then $\varphi(a)$ has a unique minimum at 0 for all symmetric unimodal f.

Section 2

2.1 If $A_1, A_2, \ldots$ are members of a σ-field $\mathcal{A}$ (the A's need not be disjoint), so are their union and intersection.

2.2 For any $a < b$, the following sets are Borel sets: (i) $\{x: a < x\}$: (ii) $\{x: a \leq x \leq b\}$.

2.3 Under the assumptions of Problem 2.1, let

$$\underline{A} = \liminf A_n = \{x: x \in A_n \text{ for all except a finite number of } n\text{'s}\}$$

$$\bar{A} = \limsup A_n = \{x: x \in A_n \text{ for infinitely many } n\}.$$

Then $\underline{A}$ and $\bar{A}$ are in $\mathcal{A}$.

2.4 (i) If $A_1 \subset A_2 \subset \cdots$, then $\underline{A} = \bar{A} = \cup A_n$.
(ii) If $A_1 \supset A_2 \supset \cdots$, then $\underline{A} = \bar{A} = \cap A_n$.

*Problems 1.13 and 1.14 were communicated to me by Dr. W. Y. Loh.

2.5 For any sequence of real numbers $a_1, a_2, \ldots$, show that the set of all limit points of subsequences is closed. The smallest and largest such limit point (which may be infinite) are denoted by $\liminf a_k$ and $\limsup a_k$, respectively.

2.6 Under the assumptions of Problems 2.1 and 2.3,

$$I_{\underline{A}}(x) = \liminf I_{A_k}(x) \quad \text{and} \quad I_{\bar{A}}(x) = \limsup I_{A_k}(x)$$

where $I_A(x)$ denotes the indicator of the set A.

2.7 Let $(\mathcal{X}, \mathcal{A}, \mu)$ be a measure space and let $\mathcal{B}$ be the class of all sets $A \cup C$ with $A \in \mathcal{A}$ and C a subset of a set $A' \in \mathcal{A}$ with $\mu(A') = 0$. Show that $\mathcal{B}$ is a σ-field.

2.8 If f and g are measurable functions, so are (i) $f + g$; and (ii) $\max(f, g)$.

2.9 If f is integrable with respect to μ, so is $|f|$, and

$$|\int f \, d\mu| \leqslant \int |f| \, d\mu.$$

[*Hint:* Express $|f|$ in terms of f^+ and f^-.]

2.10 Let $\mathcal{X} = \{x_1, x_2, \ldots\}$, μ = counting measure on $\mathcal{X}$, and f integrable. Then

$$\int f \, d\mu = \Sigma f(x_i).$$

[*Hint:* Suppose, first, that $f \geqslant 0$ and let $s_n(x)$ be the simple function, which is $f(x)$ for $x = x_1, \ldots, x_n$, and 0 otherwise.]

2.11 Let X have a standard normal distribution and $Y = 2X$. Determine whether

(i) the cdf $F(x, y)$ of (X, Y) is continuous,

(ii) the distribution of (X, Y) is absolutely continuous with respect to Lebesgue measure in the (x, y) plane.

2.12 Let $f(x) = 1$ or 0 as x is rational or irrational. Show that the Riemann integral of f does not exist.

2.13 Show that any function f which satisfies (2.34) is continuous.

2.14 Let X be a *measurable transformation* from $(\mathcal{E}, \mathcal{B})$ to $(\mathcal{X}, \mathcal{A})$ (i.e., such that for any $A \in \mathcal{A}$, the set $\{e\colon X(e) \in A\}$ is in $\mathcal{B}$), and let Y be a measurable transformation from $(\mathcal{X}, \mathcal{A})$ to $(\mathcal{Y}, \mathcal{C})$. Then $Y[X(e)]$ is a measurable transformation from $(\mathcal{E}, \mathcal{B})$ to $(\mathcal{Y}, \mathcal{C})$.

2.15 In Example 2.4, show that the support of P is $[a, b]$ if and only if F is strictly increasing on $[a, b]$.

2.16 Let S be the support of a distribution on a Euclidean space $(\mathcal{X}, \mathcal{A})$. Then (1) S is closed; (2) $P(S) = 1$; (3) S is the intersection of all closed sets C with $P(C) = 1$.

2.17 If P, Q are two probability measures over the same Euclidean space which are equivalent, then they have the same support.

2.18 Let P and Q assign probabilities

$$P: P\left(X = \frac{1}{n}\right) = p_n > 0, \qquad n = 1, 2, \ldots \qquad (\Sigma p_n = 1)$$

$$Q: P(X = 0) = \tfrac{1}{2}; \quad P\left(X = \frac{1}{n}\right) = q_n > 0; \quad n = 1, 2, \ldots \quad (\Sigma q_n = \tfrac{1}{2}).$$

Then P and Q have the same support but are not equivalent.

Section 3

3.1 If the distributions of a positive random variable X form a scale family, show that the distributions of $\log X$ form a location family.

3.2 If X is distributed according to the uniform distribution $U(0, \theta)$, determine the distribution of $-\log X$.

3.3 Let U be uniformly distributed on $(0, 1)$ and consider the variables $X = U^\alpha$, $0 < \alpha$. Show that this defines a group family, and determine the density of X.

3.4 If g_0 is any element of a group G, show that as g ranges over G so does gg_0.

3.5 Show that for $p = 2$ the density (3.15) specializes to (3.16).

3.6 Show that the family of transformations (3.12) with B nonsingular and lower triangular form a group G.

3.7 Show that the totality of nonsingular multivariate normal distributions can be obtained by the subgroup G of (3.12) described in Problem 3.6.

3.8 In the preceding problem, show that G can be replaced by the subgroup G_0 of lower triangular matrices $B = (b_{ij})$, in which the diagonal elements $b_{11}, \ldots, b_{pp}$ are all positive, but that no proper subgroup of G_0 will suffice.

3.9 Show that the family of all continuous distributions whose support is an interval with positive lower end point is a group family. [*Hint:* Let U be uniformly distributed on the interval $(2, 3)$ and let $X = b[g(U)]^\alpha$ where $\alpha, b > 0$ and where g is continuous and $1:1$ from $(2, 3)$ to $(2, 3)$.]

3.10 Find a modification of the transformation group (3.22) which generates a random sample from a population $\{y_1, \ldots, y_N\}$ where the y's, instead of being arbitrary, are restricted to (i) be positive; (ii) satisfy $0 < y_i < 1$.

3.11 Generalize the transformation group of Example 3.6 to the case of s populations $\{y_{ij}, j = 1, \ldots, N_i\}$, $i = 1, \ldots, s$, with a random sample of size n_i being drawn from the i-th.

3.12 Let U be a positive random variable, and let

$$X = bU^{1/c}, \qquad b > 0, \qquad c > 0. \tag{5}$$

(i) Show that (5) defines a group family.

(ii) If U is distributed as $E(0, 1)$, then X is distributed according to the *Weibull* distribution with density

$$(6) \qquad \frac{c}{b}\left(\frac{x}{b}\right)^{c-1} e^{-(x/b)^c}, \qquad x > 0.$$

3.13 If F and F_0 are two continuous, strictly increasing cdf's on the real line, if the cdf of U is F_0 and g is strictly increasing, show that the cdf of $g(U)$ is F if and only if $g = F^{-1}(F_0)$.

3.14 The following two families of distributions are not group families:

(i) The class of binomial distributions $b(p, n)$, with n fixed and $0 < p < 1$.

(ii) The class of Poisson distributions $P(\lambda), 0 < \lambda$.

[*Hint:* (i) How many 1 : 1 transformations are there taking the set of integers $\{0, 1, \ldots, n\}$ into itself?]

3.15 Let $X_1, \ldots, X_r$ have a multivariate normal distribution with $E(X_i) = \xi_i$ and with covariance matrix Σ. If X is the column matrix with elements X_i and B is an $r \times r$ matrix of constants, then BX has a multivariate normal distribution with mean $B\xi$ and covariance matrix $B\Sigma B'$.

Section 4

4.1 Determine the natural parameter space of (4.2) when $s = 1$, $T_1(x) = x$, μ is Lebesgue measure, and $h(x)$ is (i) $e^{-|x|}$ and (ii) $e^{-|x|}/(1 + x^2)$.

4.2 Suppose in (4.2), $s = 2$ and $T_2(x) = T_1(x)$. Explain why it is impossible to estimate η_1. [*Hint:* Compare the model with that obtained by putting $\eta_1' = \eta_1 + c$, $\eta_2' = \eta_2 - c$.]

4.3 Show that the distribution of a sample from the p-variate normal density (3.15) constitutes an s-parameter exponential family. Determine s and identify the functions η_i, T_i, and B of (4.1).

4.4 Find an expression for $E_\theta(T_j)$ in (4.1).

4.5 Verify the relations (i) (4.19) and (ii) (4.23).

4.6 For the binomial distribution (4.25), verify

(i) the moment-generating function (4.27);

(ii) the moments (4.28).

4.7 For the Poisson distribution (4.29), verify the moments (4.32).

4.8 In a binomial sequence of trials with success probability p, let $X + m$ be the number of trials required to achieve m successes.

(i) Show that the distribution of X, the *negative binomial distribution*, is as given in Table 4.1.

(ii) Verify that the negative binomial probabilities add up to 1 by expanding

$$\left(\frac{1}{p} - \frac{q}{p}\right)^{-m} = p^m(1-q)^{-m}. \tag{7}$$

(iii) Show that the distributions of (i) constitute a one-parameter exponential family.

(iv) Show that the moment-generating function of X is

$$M_X(u) = p^m/(1-qe^u)^m. \tag{8}$$

(v) Show that

$$E(X) = mq/p, \qquad \text{var}(X) = mq/p^2. \tag{9}$$

(vi) By expanding $K_X(u)$, show that the first four cumulants of X are $k_1 = mq/p$, $k_2 = mq/p^2$, $k_3 = mq(1+q)/p^3$, and $k_4 = mq(1+4q+q^2)/p^4$.

4.9 In the preceding problem, let $X_i + 1$ be the number of trials required after the $(i-1)$st success has been obtained until the next success occurs. Use the fact that $X = \Sigma_{i=1}^m X_i$ to find an alternative derivation of (9).

4.10 A discrete random variable with probabilities

$$P(X=x) = a(x)\theta^x/C(\theta), \qquad x = 0, 1, \ldots; \quad a(x) \geq 0; \quad \theta > 0 \tag{10}$$

is a *power series distribution*. This is an exponential family (4.1) with $s = 1$, $\eta = \log\theta$, and $T = X$. The moment-generating function is

$$M_X(u) = C(\theta e^u)/C(\theta). \tag{11}$$

4.11 Show that the binomial, negative binomial, and Poisson distributions are special cases of (10) and determine θ and $C(\theta)$.

4.12 The distribution (10) with $a(x) = 1/x$ and $C(\theta) = -\log(1-\theta)$, $x = 1, 2, \ldots$; $0 < \theta < 1$ is the *logarithmic series* distribution. Show that the moment-generating function is $\log(1-\theta e^u)/\log(1-\theta)$ and determine $E(X)$ and $\text{var}(X)$.

4.13 For the multinomial distribution (4.4) verify the moment formulas (4.13).

4.14 As an alternative to using (4.11) and (4.12), obtain the moments (4.13) by representing each X_i as a sum of n indicators, as was done in Example 4.4.

4.15 For the gamma distribution (4.38)

(i) verify the formulas (4.39), (4.40), and (4.41);

(ii) show that (4.40), with the middle term deleted, holds not only for all positive integers r but for all real $r > -\alpha$.

4.16 In Example 4.7, show that

(i) χ_1^2 is the distribution of Y^2 where Y is distributed as $N(0, 1)$;
(ii) χ_n^2 is the distribution of $Y_1^2 + \cdots + Y_n^2$ where the Y_i are independent $N(0, 1)$.

4.17 Determine the values α for which the density (4.38) is (a) a decreasing function of x on $(0, \infty)$, (b) increasing for $x < x_0$ and decreasing for $x > x_0$ $(0 < x_0)$. In case (b), determine the mode of the density.

4.18 A random variable X has the *Pareto distribution* $P(c, k)$ if its cdf is $1 - (k/x)^c$, $x > k > 0$, $c > 0$.

(i) The distributions $P(c, 1)$ constitute a one-parameter exponential family (4.2) with $\eta = -c$, $T = \log X$.
(ii) The statistic T is distributed as $E(\log k, 1/c)$.
(iii) The family $P(c, k)$ $(0 < k, 0 < c)$ is a group family.

4.19 If (X, Y) is distributed according to the bivariate normal distribution (3.16) with $\xi = \eta = 0$:

(i) Show that the moment-generating function of (X, Y) is

$$M_{X,Y}(u_1, u_2) = e^{-[u_1^2\sigma^2 - 2\rho\sigma\tau u_1 u_2 + u_2^2\tau^2]/2}. \tag{12}$$

(ii) Use (12) to show that

$$\mu_{12} = \mu_{21} = 0, \qquad \mu_{11} = \rho\sigma\tau \tag{13}$$

$$\mu_{13} = 3\rho\sigma\tau^3, \qquad \mu_{31} = 3\rho\sigma^3\tau, \qquad \mu_{22} = (1 + 2\rho^2)\sigma^2\tau^2.$$

4.20 (i) If $\underline{X}$ is a random column vector with expectation $\underline{\xi}$, then the covariance matrix of $\underline{X}$ is $\text{cov}(\underline{X}) = E[(\underline{X} - \underline{\xi})(\underline{X}' - \underline{\xi}')]$.
(ii) If the density of $\underline{X}$ is (3.15), then $\underline{\xi} = \underline{a}$ and $\text{cov}(\underline{X}) = \underline{\Sigma}$.

4.21 (i) Let X be distributed with density $p_\theta(x)$ given by (4.1), and let A be any fixed subset of the sample space. Then the distributions of X *truncated* on A, that is, the distributions with density $p_\theta(x)I_A(x)/P_\theta(A)$ again constitute an exponential family.
(ii) Give an example in which the natural parameter space of the original exponential family is a proper subset of the natural parameter space of the truncated family.

4.22 If X_i are independently distributed according to $\Gamma(\alpha_i, b)$, show that ΣX_i is distributed as $\Gamma(\Sigma\alpha_i, b)$. [*Hint: Method 1.* Prove it first for the sum of two gamma variables by a transformation to new variables $Y_1 = X_1 + X_2$, $Y_2 = X_1/X_2$ and then use induction. *Method 2.* Obtain the moment-generating function of ΣX_i and use the fact that a distribution is uniquely determined by its moment-generating function, when the latter exists for at least some $u \neq 0$.]

4.23 When the X_i are independently distributed according to Poisson distributions $P(\lambda_i)$, find the distribution of ΣX_i.

4.24 Let $X_1, \ldots, X_n$ be independently distributed as $\Gamma(\alpha, b)$. Show that the joint distribution is a two-parameter exponential family and identify the functions η_i, T_i and B of (4.1).

4.25 If Y is distributed as $\Gamma(\alpha, b)$, determine the distribution of $c \log Y$ and show that for fixed α and varying b it defines an exponential family.

4.26 For the beta distribution given in Table 4.1, show that $E(X) = a/(a+b)$, $\text{var}(X) = ab/(a+b)^2(a+b+1)$. [*Hint:* Write $E(X)$ and $E(X^2)$ as integrals of $x^{c-1}(1-x)^{d-1}$ and use the constant of $B(c, d)$ of Table 4.1.]

Section 5

5.1 Extend Example 5.1 to the case that $X_1, \ldots, X_r$ are independently distributed with Poisson distributions $P(\lambda_i)$ where $\lambda_i = a_i\lambda$ ($a_i > 0$, known).

5.2 Let $X_1, \ldots, X_n$ be iid according to a distribution with cdf F and probability density f. Show that the conditional distribution given $X_{(i)} = a$ of the $i-1$ values to the left of a and the $n-i$ values to the right of a is that of $i-1$ variables distributed independently according to the probability density $f(x)/F(a)$ and $n-i$ variables distributed independently with density $f(x)/[1-F(a)]$, respectively, with the two sets being (conditionally) independent of each other.

5.3 Let f be a positive integrable function defined over $(0, \infty)$, and let $p_\theta(x)$ be the probability density over $(0, \theta)$ defined by $p_\theta(x) = c(\theta)f(x)$ if $0 < x < \theta$, $= 0$ otherwise. If $X_1, \ldots, X_n$ are iid with density p_θ, show that $X_{(n)}$ is sufficient for θ.

5.4 Let f be a positive integrable function defined over $(-\infty, \infty)$ and let $p_{\xi,\eta}(x)$ be the probability density defined by $p_{\xi,\eta}(x) = c(\xi, \eta)f(x)$ if $\xi < x < \eta$, $= 0$ otherwise. If $X_1, \ldots, X_n$ are iid with density $p_{\xi,\eta}$, show that $(X_{(1)}, X_{(n)})$ is sufficient for (ξ, n).

5.5 Show that each of the statistics $T_1 - T_4$ of Example 5.7 is sufficient.

5.6 Prove Corollary 5.1.

5.7 Let $X_1, \ldots, X_m$; $Y_1, \ldots, Y_n$ be independently distributed according to $N(\xi, \sigma^2)$ and $N(\eta, \tau^2)$, respectively. Find minimal sufficient statistics for the following three cases:

(i) ξ, η, σ, τ are arbitrary: $-\infty < \xi, \eta < \infty$, $0 < \sigma, \tau$.
(ii) $\sigma = \tau$ and ξ, η, σ are arbitrary.
(iii) $\xi = \eta$ and ξ, σ, τ are arbitrary.

5.8 Let $X_1, \ldots, X_n$ be iid according to $N(\sigma, \sigma^2)$, $0 < \sigma$. Find a minimal set of sufficient statistics.

5.9 (i) If $(x_1, \ldots, x_n)$ and $(y_1, \ldots, y_n)$, have the same elementary symmetric functions $\Sigma x_i = \Sigma y_i$, $\Sigma_{i \neq j} x_i x_j = \Sigma_{i \neq j} y_i y_j, \ldots, x_1 \cdots x_n = y_1 \cdots y_n$, then the y's are a permutation of the x's.
(ii) In the notation of Example 5.6, show that U is equivalent to V.

[*Hint:* (i) Compare the coefficients and the roots of the polynomials $P(x) = \Pi(x - u_i)$ and $Q(x) = \Pi(x - v_i)$.]

5.10 Show that the order statistics are minimal sufficient for the location family (5.7) when f is the density of

(i) the double exponential distribution $D(0, 1)$.
(ii) the Cauchy distribution $C(0, 1)$.

5.11 Prove the following generalization of Theorem 5.3 to families without common support.

Theorem. *Let* $\mathcal{P}$ *be a finite family with densities* p_i, $i = 0, \ldots, k$ *and for any* x *let* S(x) *be the set of pairs of subscripts* (i, j) *for which* $p_i(x) + p_j(x) > 0$. *Then the statistic*

$$(14) \qquad T(X) = \left\{ \frac{p_j(X)}{p_i(X)}, \quad i < j \;\; \text{and} \;\; (i,j) \in S(X) \right\}$$

is minimal sufficient. Here $p_j(x)/p_i(x) = \infty$ *if* $p_i(x) = 0$ *and* $p_j(x) > 0$.

5.12 The following example shows that it is not enough to replace $p_i(X)$ by $p_0(X)$ in (14). Let $k = 2$ and $p_0 = U(-1, 0)$, $p_1 = U(0, 1)$ and $p_2(x) = 2x$, $0 < x < 1$.

5.13 Let $k = 2$ and $P_i = U(i, i + 1)$, $i = 0, 1$.

(i) Show that a minimal sufficient statistic for $\mathcal{P} = \{P_0, P_1\}$ is $T(X) = i$ if $i < X < i + 1$, $i = 0, 1$.
(ii) Let X_1, X_2 be iid according to a distribution from $\mathcal{P}$. Show that each of the two statistics $T_1 = T(X_1)$ and $T_2 = T(X_2)$ is sufficient for (X_1, X_2).
(iii) Show that $T(X_1)$ and $T(X_2)$ are equivalent.

5.14 In Lemma 5.1, show that the assumption of common support can be replaced by the weaker assumption that every $\mathcal{P}_0$-null set is also a $\mathcal{P}$-null set so that (a.e. $\mathcal{P}_0$) is equivalent to (a.e. $\mathcal{P}$).

5.15 Let $X_1, \ldots, X_n$ be iid according to a distribution from $\mathcal{P} = \{U(0, \theta), \theta > 0\}$, and let $\mathcal{P}_0$ be the subfamily of $\mathcal{P}$ for which θ is rational. Show that every $\mathcal{P}_0$-null set in the sample space is also a $\mathcal{P}$-null set.

5.16 Let $X_1, \ldots, X_n$ be iid according to a distribution from a family $\mathcal{P}$. Show that T is minimal sufficient in the following cases.

(i) $\mathcal{P} = \{U(0, \theta), \theta > 0\}$; $T = X_{(n)}$
(ii) $\mathcal{P} = \{U(\theta_1, \theta_2), -\infty < \theta_1 < \theta_2 < \infty\}$; $T = (X_{(1)}, X_{(n)})$
(iii) $\mathcal{P} = \{U(\theta - \frac{1}{2}, \theta + \frac{1}{2}), -\infty < \theta < \infty\}$; $T = (X_{(1)}, X_{(n)})$

5.17 Solve the preceding problem for the following cases.

(i) $\mathcal{P} = \{E(\theta, 1), -\infty < \theta < \infty\}$; $T = X_{(1)}$
(ii) $\mathcal{P} = \{E(0, b), 0 < b\}$; $T = \Sigma X_i$
(iii) $\mathcal{P} = \{E(a, b), -\infty < a < \infty, 0 < b\}$; $T = (X_{(1)}, \Sigma[X_i - X_{(1)}])$

5.18 The statistics $X_{(1)}$ and $\Sigma[X_i - X_{(1)}]$ of Problem 5.17(iii) are independently distributed as $E(a, b/n)$ and $\frac{1}{2}b\chi^2_{2n-2}$ respectively.

[*Hint:* If $a = 0$, $b = 1$, the variables $Y_i = (n - i + 1)[X_{(i)} - X_{(i-1)}]$, $i = 2, \dots, n$, are iid as $E(0, 1)$.]

5.19 Show that the sufficient statistics of (i) Problem 5.3 and (ii) Problem 5.4 are minimal sufficient.

5.20 Let (X_i, Y_i), $i = 1, \dots, n$ be iid according to the uniform distribution over a set R in the x, y-plane and let $\mathcal{P}$ be the family of distributions obtained by letting R range over a class $\mathcal{R}$ of sets R. Determine a minimal sufficient statistic for the following cases:

(i) $\mathcal{R}$ is the set of all rectangles $a_1 < x < a_2$, $b_1 < y < b_2$, $-\infty < a_1 < a_2 < \infty$, $-\infty < b_1 < b_2 < \infty$.
(ii) $\mathcal{R}'$ is the subset of $\mathcal{R}$ for which $a_2 - a_1 = b_2 - b_1$.
(iii) $\mathcal{R}''$ is the subset of $\mathcal{R}'$ for which $a_2 - a_1 = b_2 - b_1 = 1$.

5.21 Solve the preceding problem if

(i) $\mathcal{R}$ is the set of all triangles with sides parallel to the x-axis, the y-axis, and the line $y = x$, respectively.
(ii) $\mathcal{R}'$ is the subset of $\mathcal{R}$ in which the sides parallel to the x- and y-axes are equal.

5.22 Formulate a general result of which Problems 5.20(i) and 5.21(i) are special cases.

5.23 If Y is distributed as $E(\eta, 1)$, the distribution of $X = e^{-Y}$ is $U(0, e^{-\eta})$.

5.24 If a minimal sufficient statistic exists, a necessary condition for a sufficient statistic to be complete is for it to be minimal. [*Hint:* Suppose that $T = h(U)$ is minimal sufficient and U is complete. To show that U is equivalent to T, note that otherwise there exists ψ such that $\psi(U) \neq \eta[h(U)]$ with positive probability where $\eta(t) = E[\psi(U)|t]$.]

5.25 Show that the minimal sufficient statistics $T = (X_{(1)}, X_{(n)})$ of Problem 5.16(ii) are complete. [*Hint:* Use the approach of Example 5.14.]

5.26 For each of the following problems determine whether the minimal sufficient statistic is complete.

(i) Problem 5.7(i)–(iii).
(ii) Problem 5.20(i)–(iii).
(iii) Problem 5.21(i) and (ii).

5.27 (i) If $\mathcal{P}_0$, $\mathcal{P}_1$ are two families of distributions such that every null set of $\mathcal{P}_0$ is also a null set of $\mathcal{P}_1$, then a sufficient statistic T that is complete for $\mathcal{P}_0$ is also complete for $\mathcal{P}_1$.
(ii) Let $\mathcal{P}_0$ be the class of binomial distributions $b(p, n)$, $0 < p < 1$, $n =$ fixed, and let $\mathcal{P}_1 = \mathcal{P}_0 \cup \{Q\}$ where Q is the Poisson distribution with expectation 1. Then $\mathcal{P}_0$ is complete but $\mathcal{P}_1$ is not.

5.28 Let $X_1, \ldots, X_n$ be iid each with density $f(x)$ (with respect to Lebesgue measure), which is unknown. Show that the order statistics are complete. [*Hint:* Use Problem 5.27(i) with $\mathcal{P}_0$ the class of distributions of Example 5.8(iv). Alternatively, let $\mathcal{P}_0$ be the exponential family with density $C(\theta_1, \ldots, \theta_n)e^{-\theta_1 \Sigma x_i - \theta_2 \Sigma x_i^2 - \cdots - \theta_n \Sigma x_i^n - \Sigma x_i^{2n}}$.]

5.29 Use Basu's theorem to prove independence of the following pairs of statistics.

(i) $\bar{X}$ and $\Sigma(X_i - \bar{X})^2$ where the X's are iid as $N(\xi, \sigma^2)$.

(ii) $X_{(1)}$ and $\Sigma[X_i - X_{(1)}]$ in Problem 5.18.

5.30 (i) Under the assumptions of Problem 5.18 the ratios $Z_i = [X_{(n)} - X_{(i)}]/[X_{(n)} - X_{(n-1)}]$, $i = 1, \ldots, n-2$ are independent of $\{X_{(1)}, \Sigma[X_i - X_{(1)}]\}$.

(ii) Under the assumptions of Problems 5.16(ii) and 5.25, the ratios $Z_i = [X_{(i)} - X_{(1)}]/[X_{(n)} - X_{(1)}]$, $i = 2, \ldots, n-1$ are independent of $(X_{(1)}, X_{(n)})$.

5.31 Under the assumptions of Theorem 5.2, let A be any fixed set in the sample space, P_θ^* the distribution P_θ truncated on A, and $\mathcal{P}^* = \{P_\theta^*, \theta \in \Omega\}$. Then

(i) if T is sufficient for $\mathcal{P}$, it is sufficient for $\mathcal{P}^*$.

(ii) if, in addition, T is complete for $\mathcal{P}$, it is also complete for $\mathcal{P}^*$.

[For generalizations of this result which is due to Tukey (1949) and Smith (1957) see Smith (1957). The analogous problem for observations that are *censored* rather than truncated is discussed by Bhattacharyya, Johnson, and Mehrotra (1977).]

5.32 (i) If $X_1, \ldots, X_n$ are iid as $B(a, b)$, then $[\Pi X_i, \Pi(1 - X_i)]$ is minimal sufficient for (a, b).

(ii) Determine the minimal sufficient statistic when $a = b$.

Section 6

6.1 Verify the convexity of the functions (i)–(vi) of Example 6.1.

6.2 Show that x^p is concave over $(0, \infty)$ if $0 < p < 1$.

6.3 Give an example showing that a convex function need not be continuous on a closed interval.

6.4 If ϕ is convex on (a, b) and ψ is convex and nondecreasing on the range of ϕ, show that the function $\psi[\phi(x)]$ is convex on (a, b).

6.5 Prove or disprove by counterexample each of the following statements. If ϕ is convex on (a, b) then so is

(i) $e^{\phi(x)}$ and (ii) $\log \phi(x)$ if $\phi > 0$.

6.6 Show that if equality holds in (6.1) for some $0 < \gamma < 1$, then ϕ is linear on $[x, y]$.

6.7 Prove Jensen's inequality for the case that X takes on the values $x_1, \dots, x_n$ with probabilities $\gamma_1, \dots, \gamma_n$ ($\Sigma\gamma_i = 1$) directly from (6.1) by induction over n.

6.8 Let U be uniformly distributed on (0, 1), and let F be a cdf on the real line.

(i) If F is continuous and strictly increasing, show that $F^{-1}(U)$ has cdf F.

(ii) For arbitrary F, show that $F^{-1}(U)$ continues to have cdf F.

[*Hint:* Take F^{-1} to be any nondecreasing function such that $F^{-1}[F(x)] = x$ for all x for which there exists no $x' \neq x$ with $F(x') = F(x)$.]

6.9 Show that the k-dimensional sphere $\Sigma_{i=1}^k x_i^2 \leqslant c$ is convex.

6.10 Show that $f(a) = \sqrt{|x - a|} + \sqrt{|y - a|}$ is minimized by $a = x$ and $a = y$.

6.11 (i) Show that $\phi(\underline{x}) = e^{\Sigma x_i}$ is convex by showing that its Hessian matrix is positive semidefinite.

(ii) Show that the result of Problem 6.4 remains valid if ϕ is a convex function defined over an open convex set in E_k.

(iii) Use (ii) to obtain an alternative proof of the result of part (i).

6.12 Use the convexity of the function ϕ of Problem 6.11 to show that the natural parameter space of the exponential family (4.2) is convex.

6.13 Show that if f is defined and bounded over $(-\infty, \infty)$ or $(0, \infty)$, then f cannot be convex (unless it is constant).

6.14 Show that $\phi(x, y) = -\sqrt{xy}$ is convex over $x > 0$, $y > 0$.

6.15 If f, g are real valued functions such that f^2, g^2 are measurable with respect to the σ-finite measure μ, prove the *Schwarz inequality*

$$(15) \qquad \left(\int fg\, d\mu\right)^2 \leqslant \int f^2 d\mu \cdot \int g^2 d\mu.$$

[*Hint:* Write $\int fg\, d\mu = E_Q(f/g)$, where Q is the probability measure with $dQ = g^2 d\mu / \int g^2 d\mu$, and apply Jensen's inequality with $\varphi(x) = x^2$].

6.16 Show that the loss functions (6.23) are continuously differentiable.

6.17 Prove the statements made in Example 6.6(i) and (ii).

6.18 Let ρ be a real-valued function satisfying

$$(16) \qquad 0 \leqslant \rho(t) \leqslant M < \infty \quad \text{and} \quad \rho(t) \to M \quad \text{as} \quad t \to \pm\infty,$$

and let X be a random variable with a continuous probability density f. Then $\phi(a) = E[\rho(X - a)]$ attains its minimum. [*Hint:* Show that (a) $\phi(a) \to M$ as $a \to \pm\infty$ and (b) ϕ is continuous. Here (b) follows from the fact (see, for example, TSH, Appendix, Section 2) that if f_n, $n = 1, 2, \dots$ and f are probability densities such that $f_n(x) \to f(x)$ a.e., then $\int \psi f_n \to \int \psi f$ for any bounded ψ.]

6.19 Let ϕ be a strictly convex function defined over an interval I (finite or infinite). If there exists a value a_0 in I minimizing $\phi(a)$, then a_0 is unique.

8. REFERENCES

The concept of sufficiency is due to Fisher (1920),* who in his fundamental paper of 1922 also introduced the term and stated the factorization criterion. The criterion was rediscovered by Neyman (1935) and was proved for general dominated families by Halmos and Savage (1949). The theory of minimal sufficiency was initiated by Lehmann and Scheffé (1950) and Dynkin (1951). For further generalizations, see Bahadur (1954) and Landers and Rogge (1972). Theorem 6.4 with squared error loss is due to Rao (1945) and Blackwell (1947). It was extended to the pth power of the error ($p \geqslant 1$) by Barankin (1950) and to arbitrary convex loss functions by Hodges and Lehmann (1950).

Group families were introduced by Fisher (1934) to illustrate his use of ancillary statistics.** One-parameter exponential families as the only (regular) families of distributions for which there exists a one-dimensional sufficient statistic were also introduced by Fisher (1934). His result was generalized to more than one dimension by Darmois (1935), Koopman (1936), and Pitman (1936).† A more recent discussion of this theorem with references to the literature is given, for example, by Hipp (1974). A comprehensive treatment of exponential families is provided by Barndorff-Nielsen (1978); statistical aspects are emphasized in Johansen (1979).

Completeness was introduced by Lehmann and Scheffé (1950). Theorem 5.5 is due to Basu (1955, 1958); some converse results are discussed by Lehmann (1981).

Andersen, E. B.
(1970). "Sufficiency and exponential families for discrete sample spaces." *J. Am. Stat. Assoc.* **65**, 1248–1255.

Bahadur, R. R.
(1954). "Sufficiency and statistical decision functions." *Ann. Math. Statist.* **25**, 423–462.

Bahadur, R. R.
(1955). "Statistics and subfields." *Ann. Math. Statist.* **26**, 490–497.

Bahadur, R. R.
(1957). "On unbiased estimates of uniformly minimum variance." *Sankhya* **18**, 211–224.

Barankin, E. W.
(1949). "Locally best unbiased estimates." *Ann. Math. Statist.* **20**, 477–501.

Barankin, E. W.
(1950). "Extension of a theorem of Blackwell." *Ann. Math. Statist.* **21**, 280–284.

Barankin, E. W. and Maitra, A. P.
(1963). "Generalization of the Fisher–Darmois–Koopman–Pitman theorem on sufficient statistics." *Sankhya A*, **25**, 217–244.

*For some related history, see Stigler (1973).
**For an introduction to ancillarity, see Buehler (1982).
†Their contributions are compared by Barankin and Maitra (1963).

Barndorff–Nielsen, O.
(1978). *Information and Exponential Families in Statistical Theory*. Wiley, New York.

Barndorff–Nielsen, O.; Blaesild, P.; Jensen, J. L.; and Jørgensen, B.
(1982). "Exponential transformation models." *Proc. Roy. Soc. London A.*, **379**, 41–65.

Barndorff–Nielsen, O.; Hoffmann–Jørgensen, J., and Pedersen, K.
(1976). "On the minimal sufficiency of the likelihood function." *Scand. J. Stat.* **3**, 37–38.

Barndorff–Nielsen, O. and Pedersen, K.
(1968). "Sufficient data reduction and exponential families." *Math. Scand.* **22**, 197–202.

Basu, D.
(1955). "On statistics independent of a complete sufficient statistic." *Sankhya* **15**, 377–380.

Basu, D.
(1958). "On statistics independent of sufficient statistics." *Sankhya* **20**, 223–226.

Beran, R.
(1979). "Exponential models for directional data." *Ann. Statist.* **7**, 1162–1178.

Bhattacharyya, G. K.; Johnson, R. A.; and Mehrotra, K. G.
(1977). "On the completeness of minimal sufficient statistics with censored observations." *Ann. Statist.* **5**, 547–553.

Billingsley, P.
(1979). *Probability and Measure*. Wiley, New York.

Blackwell, D.
(1947). "Conditional expectation and unbiased sequential estimation." *Ann. Math. Statist.* **18**, 105–110.

Blackwell, D. and Ryll-Nardzewski, C.
(1963). "Non-existence of everywhere proper conditional distributions." *Ann. Math. Statist.* **34**, 223–225.

Blyth, C. R.
(1970). "On the inference and decision models of statistics (with discussion)." *Ann. Math. Statist.* **41**, 1034–1058.

Blyth, C. R.
(1980). "Expected absolute error of the usual estimator of the binomial parameter." *Amer. Statist.* **34**, 155–157.

Borges, R. and Pfanzagl, J.
(1965). "One-parameter exponential families generated by transformation groups." *Ann. Math. Statist.* **36**, 261–271.

Buehler, R. J.
(1982). "Some ancillary statistics and their properties." *J. Amer. Statist. Assoc.* **77**, 581–589.

Cramér, H.
(1946). *Mathematical Methods of Statistics*. Princeton University Press, Princeton.

Darmois, G.
(1935). "Sur les lois de probabilité à estimation exhaustive." *C. R. Acad. Sci. Paris* **260**, 1265–1266.

Denny, J. L.
(1964). "A real-valued continuous function on R^n almost everywhere 1-1." *Fund. Math.* **55**, 95–99.

Denny, J. L.
(1969). "Note on a theorem of Dynkin on the dimension of sufficient statistics." *Ann. Math. Statist.* **40**, 1474–1476.

Dynkin, E. B.
(1951). "Necessary and sufficient statistics for a family of probability distributions." English translation in *Select. Transl. Math. Statist. Prob.* **1** (1961) 23–41.

Edgeworth, F. Y.
(1883). "The law of error." *Philos. Mag.* (Fifth Series) **16**, 300–309.

Efron, B.
(1975). "Defining the curvature of a statistical problem (with applications to second order efficiency) *Ann. Statist.* **3**, 1189–1242.

Efron, B.
(1978). "The geometry of exponential families." *Ann. Statist.* **6**, 367–376.

Eisenhart, C.
(1964). "The meaning of 'least' in least squares." *J. Wash. Acad. Sci.* **54**, 24–33.

Ferguson, T. S.
(1962, 1963). "Location and scale parameters in exponential families of distributions." *Ann. Math. Statist.* **33**, 986–1001. Correction **34**, 1603.

Ferguson, T. S.
(1967). *Mathematical Statistics: A Decision Theoretic Approach*. Academic Press, New York.

Fisher, R. A.
(1920). "A mathematical examination of the methods of determining the accuracy of an observation by the mean error, and by the mean square error." *Monthly Notices R. Astron. Soc.* **80**, 758–770.

Fisher, R. A.
(1922). "On the mathematical foundations of theoretical statistics." *Philos. Trans. Roy. Soc. London Sec. A* **222**, 309–368.

Fisher, R. A.
(1934). "Two new properties of mathematical likelihood." *Proc. R. Soc. A* **144**, 285–307.

Gauss, W. F.
(1821). in Gauss's Work (1803–1826) on the Theory of least Squares. Trans. H. F. Trotter. Statist. Techniques Res. Group. Tech. Rep. No. 5. Princeton University. Princeton. (Published Translations of these papers are available in French and German.)

Halmos, P. R.
(1950). *Measure Theory*. Van Nostrand, New York.

Halmos, P. R. and Savage, L. J.
(1949). "Application of the Radon–Nikodym theorem to the theory of sufficient statistics." *Ann. Math. Statist.* **20**, 225–241.

Hardy, G. H., Littlewood, J. E., and Polya, G.
(1934). *Inequalities*. Cambridge University Press, Cambridge.

Harter, H. L.
(1974–1976). "The method of least squares and some alternatives." *Int. Statist. Rev.* **42**, 147–174, 235–268, 282; **43**, 1–44, 125–190, 269–278; **44**, 113–159.

Hipp, C.
(1974). "Sufficient statistics and exponential families." *Ann. Statist.* **2**, 1283–1292.

Hodges, J. L., Jr., and Lehmann, E. L.
(1950). "Some problems in minimax point estimation." *Ann. Math. Statist.* **21**, 182–197.

Huber, P. J.
(1964). "Robust estimation of a location parameter." *Ann. Math. Statist.* **35**, 73–101.

Johansen, S.
(1979). Introduction to the Theory of Regular Exponential Families. Lecture Notes, Vol. 3. The Institute of Mathematical Statistics, University of Copenhagen, Copenhagen.

Johnson, N. L.
(1957–58). "A note on the mean deviation of the binomial distribution." *Biometrika* **44**, 532–533 and **45**, 587.

Johnson, N. L. and Kotz, S.
(1969–1972). *Distributions in Statistics* (4 vols.). Wiley, New York.

Johnson, R. A.; Ladalla, J.; and Liu, S. T.
(1979). "Differential relations, in the original parameters, which determine the first two moments of the multi-parameter exponential family." *Ann. Statist.* **7**, 232–235.

Kendall, M. G. and Stuart, A.
(1977). *The Advanced Theory of Statistics*, 4th ed. Vol. 1. Hafner, New York.

Koopman, B. O.
(1936). "On distributions admitting a sufficient statistic." *Trans. Amer. Math. Soc.* **39**, 399–409.

Kruskal, W.
(1968). "When are Gauss–Markov and least squares estimators identical? A coordinate-free approach." *Ann. Math. Statist.* **39**, 70–75.

Landers, D. and Rogge, L.
(1972). "Minimal sufficient σ-fields and minimal sufficient statistics. Two counterexamples." *Ann. Math. Statist.* **43**, 2045–2049.

Laplace, P. S. de
(1820). *Théorie analytique des probabilités*. 3rd ed. Paris.

Lehmann, E. L.
(1981). "An interpretation of completeness and Basu's theorem." *J. Amer. Statist. Assoc.* **76**, 335–340.

Lehmann, E. L. and Scheffé, H.
(1950, 1955, 1956). "Completeness, similar regions, and unbiased estimation." *Sankhya* **10**, 305–340; **15**, 219–236; correction **17**, 250.

Lindley, D. V.
(1965). *Introduction to Probability and Statistics from a Bayesian Viewpoint. Part* 2. *Inference*. Cambridge University Press, Cambridge.

Neyman, J.
(1935). "Sur un teorema concernente le cosidette statistiche sufficienti." *Giorn. Ist. Ital. Att.* **6**, 320–334.

Pfanzagl, J.
(1972). "Transformation groups and sufficient statistics." *Ann. Math. Statist.* **43**, 553–568.

Pitcher, T. S.
(1957). "Sets of measures not admitting necessary and sufficient statistics or subfields." *Ann. Math. Statist.* **28**, 267–268.

Pitman, E. J. G.
(1936). "Sufficient statistics and intrinsic accuracy." *Proc. Camb. Phil. Soc.* **32**, 567–579.

Plackett, R. L.
(1958). "The principle of the arithmetic mean." *Biometrika* **45**, 130–135.

Plackett, R. L.
(1972). "The discovery of the method of least squares." *Biometrika* **59**, 239–251.

Rao, C. R.

(1945). "Information and the accuracy attainable in the estimation of statistical parameters." *Bull. Calc. Math. Soc.* **37**, 81–91.

Roberts, A. W. and Varberg, D. E.

(1973). *Convex Functions.* Academic Press, New York.

Rudin, W.

(1966). Real and Complex Analysis. McGraw-Hill, New York.

Savage, L. J.

(1954, 1972). *The Foundations of Statistics.* Wiley, Rev. ed. Dover Publications, New York.

Simpson, T.

(1755). "A letter to the Right Honorable George Earl of Macclesfield, President of the Royal Society, on the advantage of taking the mean of a number of observations, in practical astronomy." *Phil. Trans. R. Soc. London* **49** (Pt. I), 82–93.

Smith, W. L.

(1957). "A note on truncation and sufficient statistics." *Ann. Math. Statist.* **28**, 247–252.

Stigler, S. M.

(1973). "Laplace, Fisher, and the discovery of the concept of sufficiency." *Biometrika* **60**, 439–445.

Stigler, S. M.

(1980). "An Edgeworth Curiosum." *Ann. Statist.* **8**, 931–934.

Stigler, S. M.

(1981). "Gauss and the invention of least squares." *Ann. Statist.* **9**, 465–474.

Tukey, J. W.

(1949). "Sufficiency, truncation, and selection." *Ann. Math. Statist.* **20**, 309–311.

Tukey, J. W.

(1977). *Exploratory Data Analysis.* Addison-Wesley, Reading, Mass.

Watson, G. S.

(1967). "Linear least squares regression." *Ann. Math. Statist.* **38**, 1679–1699.

CHAPTER 2

Unbiasedness

1. UMVU ESTIMATORS

It was pointed out in Section 1.1 that estimators with uniformly minimum risk typically do not exist, and restricting attention to estimators showing some degree of impartiality was suggested as one way out of this difficulty. As a first such restriction, we shall in the present chapter study the condition of *unbiasedness*, defined in (1.1.11) by

$$(1) \qquad E_\theta[\delta(X)] = g(\theta) \qquad \text{for all} \quad \theta \in \Omega,$$

where δ is the estimator, g the estimand, and Ω the parameter space.

When used repeatedly, an unbiased estimator in the long run will estimate the right value "on the average." This is an attractive feature, but insistence on unbiasedness can lead to problems. To begin with, unbiased estimators of g may not exist.

Example 1.1. Let X be distributed according to the binomial distribution $b(p, n)$ and suppose that $g(p) = 1/p$. Then unbiasedness of an estimator δ requires

$$(2) \qquad \sum_{k=0}^{n} \delta(k)\binom{n}{k} p^k q^{n-k} = g(p) \qquad \text{for all} \quad 0 < p < 1.$$

That no such δ exists can be seen, for example, from the fact that as $p \to 0$, the left side tends to $\delta(0)$ and the right side to ∞. Yet, estimators of $1/p$ exist which (for n not too small) are close to $1/p$ with high probability. For example, since X/n tends to be close to p, n/X (with some adjustment when $X = 0$) will tend to be close to $1/p$.

If there exists an unbiased estimator of g, the estimand g will be called *U-estimable*. (Some authors call such an estimand "estimable," but this conveys the false impression that any g not possessing this property cannot be accurately estimated.) Even when g is U-estimable there is no guarantee that any of its unbiased estimators are desirable in other ways, and one may still prefer to use instead an estimator that does have some bias. On the other hand, a large bias is usually considered a drawback and special methods of bias reduction have been developed for such cases (Quenouille, 1956). Thus, unbiasedness is an attractive condition, but after a best unbiased estimator has been found, its performance should be investigated and the possibility not ruled out that a slightly biased estimator with much smaller risk might exist (see, for example, Sections 4.5 and 4.6).

The motive for introducing unbiasedness was the hope that within the class of unbiased estimators there would exist an estimator with uniformly minimum risk. A natural approach to investigating this possibility is to minimize the risk for some particular value θ_0 and then see whether the result is independent of θ_0. To this end, the following obvious characterization of the totality of unbiased estimators is useful.

Lemma 1.1. *If δ_0 is any unbiased estimator of $g(\theta)$, the totality of unbiased estimators is given by $\delta = \delta_0 - U$ where U is any unbiased estimator of zero, that is, it satisfies*

$$E_\theta(U) = 0 \qquad \text{for all} \quad \theta \in \Omega.$$

To illustrate this approach, suppose the loss function is squared error. The risk of an unbiased estimator δ is then just the variance of δ. Restricting attention to estimators δ_0, δ, and U with finite variance, we have, if δ_0 is unbiased,

$$\text{var}(\delta) = \text{var}(\delta_0 - U) = E(\delta_0 - U)^2 - [g(\theta)]^2$$

so that the variance of δ is minimized by minimizing $E(\delta_0 - U)^2$.

Example 1.2. Let X take on the values $-1, 0, 1, \ldots$ with probabilities (Problem 1.1)

$$(3) \qquad P(X = -1) = p, \qquad P(X = k) = q^2 p^k, \qquad k = 0, 1, \ldots$$

where $0 < p < 1$, $q = 1 - p$, and consider the problems of estimating (a) p and (b) q^2. Simple unbiased estimators of p and q^2 are, respectively,

$$\delta_0 = \begin{cases} 1 & \text{if} \quad X = -1 \\ 0 & \text{otherwise} \end{cases} \quad \text{and} \quad \delta_1 = \begin{cases} 1 & \text{if} \quad X = 0 \\ 0 & \text{otherwise.} \end{cases}$$

It is easily checked that U is an unbiased estimator of zero if and only if [Problem 1.1(ii)]

$$(4) \qquad U(k) = -kU(-1) \qquad \text{for} \quad k = 0, 1, \ldots$$

or equivalently if $U(k) = ak$ for all $k = -1, 0, 1, \ldots$ and some a. The problem of determining the unbiased estimator which minimizes the variance at p_0 thus reduces to that of determining the value of a which minimizes

$$\Sigma P(X = k)[\delta_i(k) - ak]^2. \tag{5}$$

The minimizing values are (Problem 1.2)

$$a_0^* = -p_0 \Big/ \left[p_0 + q_0^2 \sum_{k=1}^{\infty} k^2 p_0^k \right] \quad \text{and} \quad a_1^* = 0$$

in cases (a) and (b), respectively. Since a_1^* does not depend on p_0, the estimator $\delta_1^* = \delta_1 - a_1^* X = \delta_1$ minimizes the variance among all unbiased estimators not only when $p = p_0$ but for all values of p; that is, δ_1 is the *uniformly minimum variance unbiased* (UMVU) estimator of q^2. On the other hand, $\delta_0^* = \delta_0 - a_0^* X$ does depend on p_0, and is therefore only *locally minimum variance unbiased* (LMVU) at p_0. In this case a UMVU estimator of p does not exist.

The existence, uniqueness, and characterization of LMVU estimators have been investigated by Barankin (1949) and Stein (1950). Interpreting $E(\delta_0 - U)^2$ as the distance between δ_0 and U, the minimizing U^* can be interpreted as the projection of δ_0 onto the linear space $\mathfrak{U}$ formed by the unbiased estimators U of zero. The desired results then follow from the projection theorem of linear space theory (see, for example, Bahadur, 1957, and Luenberger, 1969).

Example 1.2 shows that the restriction to unbiased estimators may not be enough to ensure the existence of a UMVU estimator. The following theorem characterizes the situations in which a UMVU estimator exists. In the statement of the theorem, attention will be restricted to estimators with finite variance, since otherwise the problem of minimizing the variance does not arise. The class of estimators δ with $E_\theta \delta^2 < \infty$ for all θ will be denoted by Δ.

Theorem 1.1. *Let* X *have distribution* P_θ, $\theta \in \Omega$, *let* δ *be an estimator in* Δ *and let* $\mathfrak{U}$ *denote the set of all unbiased estimators of zero which are in* Δ. *Then a necessary and sufficient condition for* δ *to be a UMVU estimator of its expectation* $g(\theta)$ *is that*

$$E_\theta(\delta U) = 0 \quad \textit{for all} \quad U \in \mathfrak{U} \quad \textit{and all} \quad \theta \in \Omega. \tag{6}$$

(*Note*: Since $E_\theta(U) = 0$ for all $U \in \mathfrak{U}$, it follows that $E_\theta(\delta U) = \text{cov}_\theta(\delta, U)$, so that (6) is equivalent to the condition that δ is uncorrelated with every $U \in \mathfrak{U}$.)

Proof. (i) *Necessity.* Suppose δ is UMVU for estimating its expectation $g(\theta)$. Fix $U \in \mathfrak{U}$, $\theta \in \Omega$ and for arbitrary real λ let $\delta' = \delta + \lambda U$. Then δ' is also an unbiased estimator of $g(\theta)$, so that

$$\operatorname{var}_\theta(\delta + \lambda U) \geqslant \operatorname{var}_\theta(\delta) \qquad \text{for all} \quad \lambda.$$

Expanding the left side, we see that

$$\lambda^2 \operatorname{var}_\theta U + 2\lambda \operatorname{cov}_\theta(\delta, U) \geqslant 0 \qquad \text{for all} \quad \lambda.$$

This quadratic in λ has the two real roots $\lambda = 0$ and $\lambda = -2\operatorname{cov}_\theta(\delta, U)/\operatorname{var}_\theta(U)$ and hence takes on negative values unless $\operatorname{cov}_\theta(\delta, U) = 0$.

(ii) *Sufficiency.* Suppose $E_\theta(\delta U) = 0$ for all $U \in \mathcal{U}$. To show that δ is UMVU, let δ' be any unbiased estimator of $E_\theta(\delta)$. If $\operatorname{var}_\theta \delta' = \infty$ there is nothing to prove, so assume $\operatorname{var}_\theta \delta' < \infty$. Then $\delta - \delta' \in \mathcal{U}$ (Problem 1.7) so that

$$E_\theta[\delta(\delta - \delta')] = 0$$

and hence $E_\theta(\delta^2) = E_\theta(\delta\delta')$. Since δ and δ' have the same expectation,

$$\operatorname{var}_\theta \delta = \operatorname{cov}_\theta(\delta, \delta'),$$

and from the covariance inequality (8.1) below, we conclude that $\operatorname{var}_\theta(\delta) \leqslant \operatorname{var}_\theta(\delta')$.

The proof of Theorem 1.1 shows that condition (6), if required only for $\theta = \theta_0$, is necessary and sufficient for an estimator δ with $E_{\theta_0}(\delta^2) < \infty$ to be LMVU at θ_0. This result also follows from the characterization of the LMVU estimator as $\delta = \delta_0 - U^*$ where δ_0 is any unbiased estimator of g and U^* is the projection of δ_0 onto $\mathcal{U}$. Interpreting the equation $E_{\theta_0}(\delta U) = 0$ as orthogonality of δ_0 and U, the projection U^* has the property that $\delta = \delta_0 - U^*$ is orthogonal to $\mathcal{U}$, that is, $E_{\theta_0}(\delta U) = 0$ for all $U \in \mathcal{U}$. If the estimator is to be UMVU, this relation must hold for all θ.

Example 1.2. (*Continued*). As an application of Theorem 1.1, let us determine the totality of UMVU estimators in Example 1.2. In view of (4) and (6), a necessary and sufficient condition for δ to be UMVU for its expectation is

$$E_p(\delta X) = 0 \qquad \text{for all} \quad p, \tag{7}$$

that is, for δX to be in $\mathcal{U}$ and hence to satisfy (4). This condition reduces to

$$k\delta(k) = k\delta(-1) \qquad \text{for} \quad k = 0, 1, 2, \ldots,$$

which is satisfied provided

$$\delta(k) = \delta(-1) \qquad \text{for} \quad k = 1, 2, \ldots \tag{8}$$

with $\delta(0)$ being arbitrary. If we put $\delta(-1) = a$, $\delta(0) = b$, the expectation of such a δ is $g(p) = bq^2 + a(1 - q^2)$ and $g(p)$ is therefore seen to possess a UMVU estimator with finite variance if and only if it is of the form $a + cq^2$.

It is interesting to note, although we shall not prove it here, that Theorem 1.1 typically, but not always, holds not only for squared error but for general convex loss functions. This result follows from a theorem of Bahadur (1957). For details, see Padmanabhan (1970) and Linnik and Ruhin (1971).

Constants are always UMVU estimators of their expectations since the variance of a constant is zero. (If δ is a constant, (6), of course, is trivially satisfied.) Deleting the constants from consideration, three possibilities remain concerning the set of UMVU estimators.

CASE 1. No nonconstant U-estimable function has a UMVU estimator.

Example 1.3. Let $X_1, \ldots, X_n$ be a sample from a discrete distribution which assigns probability $1/3$ to each of the points $\theta - 1, \theta, \theta + 1$, and let θ range over the integers. Then no nonconstant function of θ has a UMVU estimator (Problem 1.8). A continuous version of this example is provided by a sample from the uniform distribution $U(\theta - \frac{1}{2}, \theta + \frac{1}{2})$; see Lehmann and Scheffé (1950, 1955, 1956). (For additional examples, see Section 2.3.)

CASE 2. Some, but not all, nonconstant U-estimable functions have UMVU estimators. Example 1.2 provides an instance of this possibility.

CASE 3. Every U-estimable function has a UMVU estimator.

A condition for this to be the case is suggested by the (Rao–Blackwell) Theorem 1.6.4. If T is a sufficient statistic for the family $\mathcal{P} = \{P_\theta, \theta \in \Omega\}$ and $g(\theta)$ is U-estimable, then any unbiased estimator δ of $g(\theta)$ which is not a function of T is improved by its conditional expectation given T, say $\eta(T)$. Furthermore, $\eta(T)$ is again an unbiased estimator of $g(\theta)$ since by (1.5.1) $E_\theta[\eta(T)] = E_\theta[\delta(X)]$. Suppose, now, that there exists only one unbiased estimator of $g(\theta)$ which is a function of T. Then that $\eta(T)$ is UMVU for $g(\theta)$. A class of situations in which this supposition holds for all U-estimable $g(\theta)$ is given by the following lemma.

Lemma 1.2. *Let* X *be distributed according to a distribution from* $\mathcal{P} = \{P_\theta, \theta \in \Omega\}$, *and let* T *be a complete sufficient statistic for* $\mathcal{P}$. *Then every* U*-estimable function* $g(\theta)$ *has one and only one unbiased estimator that is a function of* T. (*Here uniqueness, of course, means that any two such functions differ a.e.* $\mathcal{P}$.)

Proof. That such an unbiased estimator exists was established just preceding the statement of Lemma 1.2. If δ_1 and δ_2 are two unbiased estimators of $g(\theta)$, their difference $f(T) = \delta_1(T) - \delta_2(T)$ satisfies

$$E_\theta f(T) = 0 \qquad \text{for all} \quad \theta \in \Omega,$$

and hence by the completeness of T, $\delta_1(T) = \delta_2(T)$ a.e. $\mathcal{P}$, as was to be proved.

So far, attention has been restricted to squared error loss. However, the Rao–Blackwell theorem applies to any convex loss function, and the preceding argument therefore establishes the following result.

Theorem 1.2. *Let* X *be distributed according to a distribution in* $\mathcal{P} = \{P_\theta, \theta \in \Omega\}$, *and suppose that* T *is a complete sufficient statistic for* $\mathcal{P}$.

(i) *For every* U-*estimable function* $g(\theta)$ *there exists an unbiased estimator that uniformly minimizes the risk for any loss function* $L(\theta, d)$ *which is convex in its second argument; therefore, this estimator in particular is UMVU.*

(ii) *The UMVU estimator of* (*i*) *is the unique unbiased estimator which is a function of* T; *it is the unique unbiased estimator with minimum risk, provided its risk is finite and* L *is strictly convex in* d.

It is interesting to note that under mild conditions the existence of a complete sufficient statistic is not only sufficient but also necessary for Case 3. This result, which is due to Bahadur (1957), will not be proved here.

Corollary 1.1. *If* $\mathcal{P}$ *is an exponential family of full rank given by* (1.4.1), *then the conclusions of Theorem* 1.2 *hold with* $\theta = (\theta_1, \ldots, \theta_s)$ *and* $T = (T_1, \ldots, T_s)$.

Proof. This follows immediately from Theorem 1.5.6.

Theorem 1.2 and its corollary provide best unbiased estimators for large classes of problems, some of which will be discussed in the next three sections. For the sake of simplicity these estimators will be referred to as being UMVU, but it should be kept in mind that their optimality is not tied to squared error as loss but in fact, they minimize the risk for any convex loss function.

Sometimes we happen to know an unbiased estimator δ of $g(\theta)$ which is a function of a complete sufficient statistic. The theorem then states that δ is UMVU. Suppose, for example, that $X_1, \ldots, X_n$ are iid according to $N(\xi, \sigma^2)$ and that the estimand is σ^2. The standard unbiased estimator of σ^2 is then $\delta = \Sigma(X_i - \bar{X})^2/(n-1)$. Since this is a function of the complete sufficient statistic $T = (\Sigma X_i, \Sigma(X_i - \bar{X})^2)$, δ is UMVU. Barring such fortunate accidents, two systematic methods are available for deriving UMVU estimators through Theorem 1.2.

Method One: Solving for δ

If T is a complete sufficient statistic, the UMVU estimator of any U-estimable function $g(\theta)$ is uniquely determined by the set of equations

$$E_\theta \delta(T) = g(\theta) \qquad \text{for all} \quad \theta \in \Omega. \tag{9}$$

Example 1.4. Suppose that T has the binomial distribution $b(p, n)$ and that $g(p) = pq$. Then (9) becomes

$$\sum_{t=0}^{n} \binom{n}{t} \delta(t) p^t q^{n-t} = pq \qquad \text{for all} \quad 0 < p < 1. \tag{10}$$

If $\rho = p/q$ so that $p = \rho/(1+\rho)$ and $q = 1/(1+\rho)$, (10) can be rewritten as

$$\sum_{t=0}^{n} \binom{n}{t} \delta(t) \rho^t = \rho(1+\rho)^{n-2} = \sum_{t=1}^{n-1} \binom{n-2}{t-1} \rho^t, \qquad (0 < \rho < \infty).$$

A comparison of the coefficients on the left and right sides leads to

$$\delta(t) = \frac{t(n-t)}{n(n-1)}.$$

Method Two: Conditioning

If $\delta(X)$ is any unbiased estimator of $g(\theta)$, it follows from Theorem 1.2 that the UMVU estimator can be obtained as the conditional expectation of $\delta(X)$ given T. For this derivation, it does not matter which unbiased estimator δ is being conditioned; one can thus choose δ so as to make the calculation of $\delta'(T) = E[\delta(X)|T]$ as easy as possible.

Example 1.5. Suppose that $X_1, \ldots, X_n$ are iid according to the uniform distribution $U(0, \theta)$ and that $g(\theta) = \theta/2$. Then $T = X_{(n)}$, the largest of the X's, is a complete sufficient statistic. Since $E(X_1) = \theta/2$, the UMVU estimator of $\theta/2$ is $E[X_1 | X_{(n)} = t]$. If $X_{(n)} = t$, then $X_1 = t$ with probability $1/n$, and X_1 is uniformly distributed on $(0, t)$ with the remaining probability $(n-1)/n$ (see Problem 1.5.2). Hence

$$E[X_1|t] = \frac{1}{n} \cdot t + \frac{n-1}{n} \cdot \frac{t}{2} = \frac{n+1}{n} \cdot \frac{t}{2}.$$

Thus $[(n+1)/n] \cdot T/2$ and $[(n+1)/n] \cdot T$ are the UMVU estimators of $\theta/2$ and θ, respectively.

The existence of UMVU estimators under the assumptions of Theorem 1.2 was proved there for convex loss functions. That the situation tends to be very different without that assumption is seen from the following result of Basu (1955).

Theorem 1.3. *Let the loss function* $L(\theta, d)$ *for estimating* $g(\theta)$ *be bounded, say* $L(\theta, d) \leqslant M$, *and assume that* $L[\theta, g(\theta)] = 0$ *for all* θ, *that is, the loss is zero when the estimated value coincides with the true value. Suppose that* g *is* U*-estimable and let* θ_0 *be an arbitrary value of* θ. *Then there exists a sequence of unbiased estimators* δ_n *for which* $R(\theta_0, \delta_n) \to 0$.

Proof. Since $g(\theta)$ is U-estimable, there exists an unbiased estimator $\delta(X)$. For any $0 < \pi < 1$, let

$$\delta'_\pi(x) = \begin{cases} g(\theta_0) & \text{with probability } 1 - \pi \\ \dfrac{1}{\pi}[\delta(x) - g(\theta_0)] + g(\theta_0) & \text{with probability } \pi. \end{cases}$$

Then δ'_π is unbiased for all π and all θ, since

$$E_\theta(\delta'_\pi) = (1 - \pi)g(\theta_0) + \frac{\pi}{\pi}[g(\theta) - g(\theta_0)] + \pi g(\theta_0) = g(\theta).$$

The risk $R(\theta_0, \delta'_\pi)$ at θ_0 is $(1 - \pi) \cdot 0$ plus π times the expected loss of $(1/\pi)[\delta(X) - g(\theta_0)] + g(\theta_0)$, so that

$$R(\theta_0, \delta'_\pi) \leqslant \pi M.$$

As $\pi \to 0$, it is seen that $R(\theta_0, \delta'_\pi) \to 0$.

This result implies that for bounded loss functions no uniformly minimum-risk-unbiased or even locally minimum-risk-unbiased estimator exists except in trivial cases, since at each θ_0 the risk can be made arbitrarily small even by unbiased estimators. [Basu (1955) proved this fact for a more general class of nonconvex loss functions.] The proof lends support to the speculation of Section 1.6 that the difficulty with nonconvex loss functions stems from the possibility of arbitrarily large errors since as $\pi \to 0$, the error $|\delta_\pi(x) - g(\theta_0)| \to \infty$. It is the leverage of these large but relatively inexpensive errors which nullifies the restraining effect of unbiasedness.

This argument applies not only to the limiting case of unbounded errors but also, although to a correspondingly lesser degree, to the case of finite large errors. In the latter situation, convex loss functions receive support from a large-sample consideration. To fix ideas, suppose the observations consist of n iid variables $X_1, \ldots, X_n$. As n increases the error in estimating a given value $g(\theta)$ will decrease and tend to zero as $n \to \infty$. (See Chapter 5 for a precise statement.) Thus, essentially only the local behavior of the loss function near the true value $g(\theta)$ is relevant. If the loss function is smooth, its Taylor expansion about $d = g(\theta)$ gives

$$L(\theta, d) = a(\theta) + b(\theta)[d - g(\theta)] + c(\theta)[d - g(\theta)]^2 + R,$$

where the remainder R becomes negligible as the error $|d - g(\theta)|$ becomes sufficiently small. If the loss is zero when $d = g(\theta)$, then a must be zero, so that $b(\theta)[d - g(\theta)]$ becomes the dominating term for small errors. The condition $L(\theta, d) \geqslant 0$ for all θ, then, implies $b(\theta) = 0$ and hence

$$L(\theta, d) = c(\theta)[d - g(\theta)]^2 + R.$$

Minimizing the risk for large n thus becomes essentially equivalent to minimizing $E[\delta(X) - g(\theta)]^2$, which justifies not only a convex loss function but even squared error. Not only the loss function but also other important aspects of the behavior of estimators and the comparison of different estimators greatly simplifies for large samples, as will be discussed in Chapter 5.

The difficulty which bounded loss functions present for the theory of unbiased estimation is not encountered by a different unbiasedness concept, that of median unbiasedness mentioned in Section 1.1. For estimating $g(\theta)$ in a multiparameter exponential family, it turns out that uniformly minimum risk median unbiased estimators exist for any loss function L for which $L(\theta, d)$ is a nondecreasing function of d as d moves in either direction away from $g(\theta)$. A detailed version of this result can be found in Pfanzagl (1979). We shall not discuss the theory of median unbiased estimation here since the methods required belong to the theory of confidence intervals rather than that of point estimation (see TSH, Section 3.5).

2. THE NORMAL AND EXPONENTIAL ONE- AND TWO-SAMPLE PROBLEM

The problem of estimating an unknown quantity θ from n measurements of θ was considered in Example 1.1.1 as the prototype of an estimation problem. It was formalized by assuming that the n measurements are iid random variables $X_1, \ldots, X_n$ with common distribution belonging to the location family

$$P(X_i \leqslant x) = F(x - \theta). \tag{1}$$

The problem takes different forms according to the assumptions made about F. Some possibilities are

(i) F is completely specified.

(ii) F is specified except for an unknown scale parameter. In this case (1) will be replaced by a location-scale family. It will then be convenient to denote the location parameter by ξ rather than θ (to reserve θ for the totality of unknown parameters) and hence to write the family as

$$P(X_i \leqslant x) = F\left(\frac{x-\xi}{\sigma}\right). \tag{2}$$

Here it will be of interest to estimate both ξ and σ.

(iii) The distribution of the X's is only approximately given by (1) or (2) with specified F. What is meant by "approximately" will be discussed in Section 5.6.

(iv) F is known to be symmetric about 0 (so that the X's are symmetrically distributed about θ or ξ) but is otherwise unknown.

(v) F is unknown except that it has finite variance; the estimand is $\xi = E(X_i)$.

In all of these models, F is assumed to be continuous.

Problem (v) will be considered in Section 2.4 and Problem (iv) in Chapter 5. A treatment of Problems (i) and (ii) for arbitrary known F is given in Chapter 3 from the point of view of equivariance. In the present section we shall be concerned with unbiased estimation of θ or (ξ, σ) in Problems (i) and (ii) and some of their generalizations for some special distributions, particularly for the case that F is normal or exponential.

Example 2.1. The normal one-sample problem. Let $X_1, \ldots, X_n$ be distributed with joint density

$$\frac{1}{(\sqrt{2\pi}\,\sigma)^n} \exp\left[-\frac{1}{2\sigma^2}\Sigma(x_i - \xi)^2\right], \tag{3}$$

and assume to begin with that only one of the parameters is unknown. If σ is known, it follows from Theorem 1.5.6 that the sample mean $\bar{X}$ is a complete sufficient statistic, and since $E(\bar{X}) = \xi$, $\bar{X}$ is the UMVU estimator of ξ. More generally, if $g(\xi)$ is any U-estimable function of ξ, there exists a unique unbiased estimator $\delta(\bar{X})$ based on $\bar{X}$ and it is UMVU. If, in particular, $g(\xi)$ is a polynomial of degree r, $\delta(\bar{X})$ will also be a polynomial of that degree, which can be determined inductively for $r = 2, 3, \ldots$ (Problem 2.1).

If ξ is known, (3) is a one-parameter exponential family with $S^2 = \Sigma(X_i - \xi)^2$ being a complete sufficient statistic. Since $Y = S^2/\sigma^2$ is distributed as χ_n^2 independently of σ^2, it follows that

$$E\left(\frac{S^r}{\sigma^r}\right) = \frac{1}{K_{n,r}},$$

where $K_{n,r}$ is a constant, and hence that

$$K_{n,r}S^r \tag{4}$$

is UMVU for σ^r. Recall from Example 1.4.7 with $a = n/2$, $b = 2$ and with $r/2$ in place of r that

$$E\left(\frac{S^r}{\sigma^r}\right) = E\left[\left(\chi_n^2\right)^{r/2}\right] = \frac{\Gamma[(n+r)/2]}{\Gamma(n/2)}2^{r/2}$$

so that

$$K_{n,r} = \frac{\Gamma(n/2)}{2^{r/2}\Gamma[(n+r)/2]}. \tag{5}$$

As a check, note that for $r = 2$, $K_{n,r} = 1/n$, and hence $E(S^2) = n\sigma^2$.

Formula (5) is established in Example 1.4.7 only for $r > 0$. It is, however, easy to see (Problem 1.4.15) that it holds whenever

$$n > -r, \tag{6}$$

but that the $(r/2)$th moment of χ_n^2 does not exist when $n \leqslant -r$.

We are now in a position to consider the more realistic case in which both parameters are unknown. Then, by Example 1.5.11, $\bar{X}$ and $S^2 = \Sigma(X_i - \bar{X})^2$ jointly are complete sufficient statistics for (ξ, σ^2). This shows that $\bar{X}$ continues to be UMVU for ξ. Since $\text{var}(\bar{X}) = \sigma^2/n$, estimation of σ^2 is, of course, also of great importance. Now S^2/σ^2 is distributed as χ_{n-1}^2 and it follows from (4) with n replaced by $n - 1$ and the new definition of S^2 that

$$K_{n-1,r}S^r \tag{7}$$

is UMVU for σ^r provided $n > -r + 1$, and thus in particular $S^2/(n-1)$ is UMVU for σ^2.

Sometimes it is of interest to measure ξ in σ-units and hence to estimate $g(\xi, \sigma) = \xi/\sigma$. Now $\bar{X}$ is UMVU for ξ and $K_{n-1,-1}/S$ for $1/\sigma$. Since $\bar{X}$ and S are independent, it follows that $K_{n-1,-1}\bar{X}/S$ is unbiased for ξ/σ and hence UMVU, provided $n - 1 > 1$, that is, $n > 2$.

Another class of problems within the framework of the normal one-sample problem relates to the probability

$$p = P(X_1 \leqslant u). \tag{8}$$

Example 2.2. Estimating a probability or a critical value. Suppose that the observations X_i denote the performances of past candidates on an entrance examination and that we wish to estimate the cutoff value u for which the probability of a passing performance, $X \geqslant u$, has a preassigned probability $1 - p$. This is the problem of estimating u in (8) for a given value of p. Solving the equation

$$p = P(X_1 \leqslant u) = \Phi\left(\frac{u - \xi}{\sigma}\right) \tag{9}$$

(where Φ denotes the cdf of the standard normal distribution) for u shows that

$$u = g(\xi, \sigma) = \xi + \sigma\Phi^{-1}(p).$$

It follows that the UMVU estimator of u is

$$\bar{X} + K_{n-1,1}S\Phi^{-1}(p). \tag{10}$$

Consider next the problem of estimating p for a given value of u. Suppose, for example, that a manufactured item is acceptable if some quality characteristic is $\leqslant u$ and that we wish to estimate the probability of an item being acceptable, its *reliability*, given by (9).

To illustrate a method which is applicable to many problems of this type, consider first the simpler case that $\sigma = 1$. An unbiased estimator δ of p is the indicator of the event $X_1 \leqslant u$. Since $\bar{X}$ is a complete sufficient statistic, the UMVU estimator of $p = P(X_1 \leqslant u) = \Phi(u - \xi)$ is therefore

$$E[\delta|\bar{X}] = P[X_1 \leqslant u|\bar{X}].$$

To evaluate this probability, use the fact that $X_1 - \bar{X}$ is independent of $\bar{X}$. This follows from Basu's theorem (Theorem 1.5.5) since $X_1 - \bar{X}$ is ancillary.* Hence

$$P[X_1 \leqslant u|\bar{x}] = P[X_1 - \bar{X} \leqslant u - \bar{x}|\bar{x}] = P[X_1 - \bar{X} \leqslant u - \bar{x}],$$

and the computation of a conditional probability has been replaced by that of an unconditional one. Now $X_1 - \bar{X}$ is distributed as $N(0, (n-1)/n)$, so that

$$P[X_1 - \bar{X} \leqslant u - \bar{x}] = \Phi\left[\sqrt{\frac{n}{n-1}}(u - \bar{x})\right], \tag{11}$$

which is the UMVU estimator of p.

Closely related to the problem of estimating p, which is the cdf

$$F(u) = P[X_1 \leqslant u] = \Phi(u - \xi)$$

of X_1 evaluated at u, is that of estimating the probability density at u: $g(\xi) = \phi(u - \xi)$. We shall now show that the UMVU estimator of the probability density

*Such applications of Basu's theorem can be simplified when invariance is present. The theory and some interesting illustrations are discussed by Eaton and Morris (1970).

$g(\xi) = p_\xi^{X_1}(u)$ of X_1 evaluated at u is the conditional density of X_1 given $\overline{X}$ evaluated at u, $\delta(\overline{X}) = p^{X_1|\overline{X}}(u)$. Since this is a function of $\overline{X}$, it is only necessary to check that δ is unbiased. This can be shown by differentiating the UMVU estimator of the cdf after justifying the required interchange of differentiation and integration, or as follows. Note that the joint density of X_1 and $\overline{X}$ is $p^{X_1|\overline{X}}(u)p_\xi^{\overline{X}}(x)$ and that the marginal density is therefore

$$p_\xi^{X_1}(u) = \int_{-\infty}^{\infty} p^{X_1|\overline{X}}(u)p_\xi^{\overline{X}}(x)\,dx.$$

This equation states just that $\delta(\overline{X})$ is an unbiased estimator of $g(\xi)$. Differentiating the earlier equation

$$P[X_1 \leqslant u|\bar{x}] = \Phi\left[\sqrt{\frac{n}{n-1}}(u - \bar{x})\right]$$

we see that

$$p^{X_1|\overline{X}}(u) = \sqrt{\frac{n}{n-1}}\,\phi\left[\sqrt{\frac{n}{n-1}}(u - \bar{x})\right],$$

which is therefore the UMVU estimator of $p_\xi^{X_1}(u)$.

Suppose now that both ξ and σ are unknown. Then, exactly as in the case $\sigma = 1$, the UMVU estimator of $P[X_1 \leqslant u] = \Phi((u - \xi)/\sigma)$ and of the density $p^{X_1}(u) = (1/\sigma)\phi((u - \xi)/\sigma)$ is given, respectively, by $P[X_1 \leqslant u|\overline{X}, S]$ and the conditional density of X_1 given $\overline{X}, S$ evaluated at u, where $S^2 = \Sigma(X_i - \overline{X})^2$. To replace the conditional distribution with an unconditional one, note that $(X_1 - \overline{X})/S$ is ancillary and therefore, by Basu's theorem, independent of $(\overline{X}, S)$. It follows, as in the earlier case, that

$$P[X_1 \leqslant u|\bar{x}, s] = P\left[\frac{X_1 - \overline{X}}{S} \leqslant \frac{u - \bar{x}}{s}\right] \tag{12}$$

and that

$$p^{X_1|\bar{x}, s}(u) = \frac{1}{s}f\left(\frac{u - \bar{x}}{s}\right) \tag{13}$$

where f is the density of $(X_1 - \overline{X})/S$. A straightforward calculation (Problem 2.7) gives

(14)

$$f(z) = \frac{\Gamma\left(\frac{n-1}{2}\right)}{\Gamma\left(\frac{1}{2}\right)\Gamma\left(\frac{n-2}{2}\right)}\sqrt{\frac{n}{n-1}}\left(1 - \frac{nz^2}{n-1}\right)^{(n/2)-2} \quad \text{if} \quad 0 < |z| < \sqrt{\frac{n-1}{n}}$$

and zero elsewhere. The estimator (13) is obtained by substitution of (14), and the estimator (12) by integrating the conditional density.

We shall next consider two extensions of the normal one-sample model. The first extension is concerned with the two-sample problem, in which there are two independent groups of observations, each with a model of this type, but corresponding to different conditions or representing measurements of two different quantities so that the parameters of the two models are not the same. The second extension deals with the multivariate situation of n p-tuples of observations $(X_{1\nu}, \ldots, X_{p\nu})$, $\nu = 1, \ldots, n$, with $(X_{1\nu}, \ldots, X_{p\nu})$ representing measurements of p different characteristics of the νth subject.

Example 2.3. The normal two-sample problem. Let $X_1, \ldots, X_m$ and $Y_1, \ldots, Y_n$ be independently distributed according to normal distributions $N(\xi, \sigma^2)$ and $N(\eta, \tau^2)$, respectively.

(i) Suppose that ξ, η, σ, τ are completely unknown. Then the joint density

$$\frac{1}{(\sqrt{2\pi})^{m+n}\sigma^m\tau^n} \exp\left[-\frac{1}{2\sigma^2}\Sigma(x_i - \xi)^2 - \frac{1}{2\tau^2}\Sigma(y_j - \eta)^2\right] \tag{15}$$

constitutes an exponential family for which the four statistics

$$\bar{X}, \quad \bar{Y}, \quad S_X^2 = \Sigma(X_i - \bar{X})^2, \quad S_Y^2 = \Sigma(Y_j - \bar{Y})^2$$

are sufficient and complete. The UMVU estimators of ξ and σ^r are therefore $\bar{X}$ and $K_{n-1,r}S_X^r$, as in Example 2.1, and those of η and τ^r are given by the corresponding formulas. In the present model, interest tends to focus on comparing parameters from the two distributions. The UMVU estimator of $\eta - \xi$ is $\bar{Y} - \bar{X}$ and that of τ^r/σ^r is the product of the UMVU estimators of τ^r and $1/\sigma^r$.

(ii) Sometimes it is possible to assume that $\sigma = \tau$. Then $\bar{X}$, $\bar{Y}$, and $S^2 = \Sigma(X_i - \bar{X})^2 + \Sigma(Y_j - \bar{Y})^2$ are complete sufficient statistics [Problem 1.5.26(i)] and the natural unbiased estimators of ξ, η, σ^r, $\eta - \xi$, and $(\eta - \xi)/\sigma$ are all UMVU (Problem 2.8).

(iii) As a third possibility, suppose that $\eta = \xi$ but that σ and τ are not known to be equal, and that it is desired to estimate the common mean ξ. This might arise, for example, when two independent sets of measurements of the same quantity are available. The statistics $T = (\bar{X}, \bar{Y}, S_X^2, S_Y^2)$ are then minimal sufficient (Problem 1.5.7), but they are no longer complete since $E(\bar{Y} - \bar{X}) = 0$.

If $\sigma^2/\tau^2 = \gamma$ is known, the best unbiased linear combination of $\bar{X}$ and $\bar{Y}$ is

$$\delta_\gamma = \alpha\bar{X} + (1 - \alpha)\bar{Y}, \qquad \text{where} \quad \alpha = \frac{\tau^2}{n} \Big/ \left(\frac{\sigma^2}{m} + \frac{\tau^2}{n}\right)$$

(Problem 2.9). Since, in this case, $T' = (\Sigma X_i^2 + \gamma\Sigma Y_j^2, \Sigma X_i + \gamma\Sigma Y_j)$ is a complete sufficient statistic (Problem 2.9) and δ_γ is a function of T', δ_γ is UMVU. When

σ^2/τ^2 is unknown, a UMVU estimator of ξ does not exist (Problem 2.10), but one can first estimate α, and then estimate ξ by $\hat{\xi} = \hat{\alpha}\bar{X} + (1 - \hat{\alpha})\bar{Y}$. It is easy to see that $\hat{\xi}$ is unbiased provided $\hat{\alpha}$ is a function of only S_X^2 and S_Y^2 (Problem 2.10), for example, if σ^2 and τ^2 in α are replaced by $S_X^2/(m-1)$ and $S_Y^2/(n-1)$. The problem of finding a good estimator of ξ has been considered by various authors, among them Brown and Cohen (1974), Cohen and Sackrowitz (1974), Graybill and Deal (1959), Hogg (1960), Rao (1980), Rubin and Weisberg (1975), Seshadri (1963), and Zacks (1966).

It is interesting to note that the nonexistence of a UMVU estimator holds not only for ξ but for any U-estimable function of ξ. This fact, for which no easy proof is available, was established by Unni (1978) using the results of Kagan and Palamadov (1968).

In cases (i) and (ii), the difference $\eta - \xi$ provides one comparison between the distributions of the X's and Y's. An alternative measure of the superiority (if large values of the variables are desirable) of the Y's over the X's is the probability $p = P(X < Y)$. The UMVU estimator of p can be obtained as in Example 2.2 as $P(X_1 < Y_1 | \bar{X}, \bar{Y}, S_X^2, S_Y^2)$ and $P(X_1 < Y_1 | \bar{X}, \bar{Y}, S^2)$ in cases (i) and (ii), respectively (Problem 2.11). In case (iii), the problem disappears since then $p = 1/2$.

Example 2.4. The multivariate normal one-sample problem. Suppose that $(X_i, Y_i, \ldots)$, $i = 1, \ldots, n$ are observations of p characteristics on a random sample of n subjects from a large population, so that the n p-vectors can be assumed to be iid. We shall consider the case that their common distribution is a p-variate normal distribution (Example 1.3.2) and begin with the case $p = 2$.

The joint probability density of the (X_i, Y_i) is then

$$\left(\frac{1}{2\pi\sigma\tau\sqrt{1-\rho^2}}\right)^n \exp\left\{-\frac{1}{2(1-\rho^2)}\left[\frac{1}{\sigma^2}\Sigma(x_i - \xi)^2\right.\right.$$

$$\left.\left. -\frac{2\rho}{\sigma\tau}\Sigma(x_i - \xi)(y_i - \eta) + \frac{1}{\tau^2}\Sigma(y_i - \eta)^2\right]\right\} \tag{16}$$

where $E(X_i) = \xi$, $E(Y_i) = \eta$, $\text{var}(X_i) = \sigma^2$, $\text{var}(Y_i) = \tau^2$, and $\text{cov}(X_i, Y_i) = \rho\sigma\tau$, so that ρ is the correlation coefficient between X_i and Y_i. The bivariate family (16) constitutes a five-parameter exponential family of full rank, and the set of sufficient statistics $T = (\bar{X}, \bar{Y}, S_X^2, S_Y^2, S_{XY})$ where

$$S_{XY} = \Sigma(X_i - \bar{X})(Y_i - \bar{Y}) \tag{17}$$

is therefore complete. Since the marginal distributions of the X_i and Y_i are $N(\xi, \sigma^2)$ and $N(\eta, \tau^2)$, the UMVU estimators of ξ and σ^2 are $\bar{X}$ and $S_X^2/(n-1)$, and those of η and τ^2 are given by the corresponding formulas. The statistic $S_{XY}/(n-1)$ is an unbiased estimator of $\rho\sigma\tau$ (Problem 2.12) and is therefore the UMVU estimator of $\text{cov}(X_i, Y_i)$.

For the correlation coefficient ρ, the natural estimator is the sample correlation coefficient

$$R = S_{XY}\Big/\sqrt{S_X^2 S_Y^2}. \tag{18}$$

However, R is not unbiased, since it can be shown [see, for example, Kendall and Stuart, Vol. 1 (1969, p. 390)] that

$$E(R) = \rho\left[1 - \frac{(1-\rho^2)}{2n} + 0\left(\frac{1}{n^2}\right)\right]. \tag{19}$$

There exists a function $G(R)$ of R which is unbiased and hence UMVU but it is rather complicated [see Olkin and Pratt (1958), who give a table of $G(R)$].

These results extend easily to the general multivariate case. Let us change notation and denote by $(X_{1\nu}, \ldots, X_{p\nu})$, $\nu = 1, \ldots, n$ a sample from a nonsingular p-variate normal distribution with means $E(X_{i\nu}) = \xi_i$ and covariances $\operatorname{cov}(X_{i\nu}, X_{j\nu}) = \sigma_{ij}$. Then the density of the X's is

$$\frac{|\Theta|^{n/2}}{(2\pi)^{pn/2}} \exp\left(-\frac{1}{2}\Sigma\Sigma\theta_{jk}S'_{jk}\right) \tag{20}$$

where

$$S'_{jk} = \sum_{\nu=1}^{n} (X_{j\nu} - \xi_j)(X_{k\nu} - \xi_k) \tag{21}$$

and where $\Theta = (\theta_{jk})$ is the inverse of the covariance matrix (σ_{jk}). This is a full-rank exponential family, for which the $p + \binom{p}{2} = \frac{1}{2}p(p+1)$ statistics $X_{i\cdot} = \Sigma X_{i\nu}/n$ $(i = 1, \ldots, p)$ and $S_{jk} = \Sigma(X_{j\nu} - X_{j\cdot})(X_{k\nu} - X_{k\cdot})$ are complete.

Since the marginal distributions of the $X_{j\nu}$ and the pair $(X_{j\nu}, X_{k\nu})$ are univariate and bivariate normal, respectively, it follows from Example 2.1 and the earlier part of the present example, that $X_{i\cdot}$ is UMVU for ξ_i and $S_{jk}/(n-1)$ for σ_{jk}. Also, the UMVU estimators of the correlation coefficients $\rho_{jk} = \sigma_{jk}/\sqrt{\sigma_{jj}\sigma_{kk}}$ are just those obtained from the bivariate distribution of the $(X_{j\nu}, X_{k\nu})$. The UMVU estimator of the square of the multiple correlation coefficient of one of the p coordinates with the other $p-1$ was obtained by Olkin and Pratt (1958). The problem of estimating a multivariate normal probability density or probability has been treated by Ghurye and Olkin (1969).

Results quite analogous to those found in Examples 2.1–2.3 obtain when the normal density (3) is replaced by the exponential density

$$\frac{1}{b^n} \exp\left[-\frac{1}{b}\Sigma(x_i - a)\right], \qquad x_i > a. \tag{22}$$

Despite its name, this two-parameter family does not constitute an exponential family since its support changes with a. However, for fixed a, it constitutes a one-parameter exponential family with parameter $1/b$.

Example 2.5. The exponential one-sample problem. Suppose, first, that b is known. Then $X_{(1)}$ is sufficient for a and complete (Example 1.5.12). The distribution of $n[X_{(1)} - a]/b$ is the standard exponential distribution $E(0, 1)$ and the UMVU estimator of a is $X_{(1)} - (b/n)$ (Problem 2.14). On the other hand, when a is known, the distribution (22) constitutes a one-parameter exponential family with complete sufficient statistic $\Sigma(X_i - a)$. Since $2\Sigma(X_i - a)/b$ is distributed as χ^2_{2n}, it is seen that $\Sigma(X_i - a)/n$ is the UMVU estimator for b (Problem 2.14).

When both parameters are unknown, $X_{(1)}$ and $\Sigma[X_i - X_{(1)}]$ are jointly sufficient and complete (Example 1.5.14). Since they are independently distributed, $n[X_{(1)} - a]/b$ as $E(0, 1)$ and $2\Sigma[X_i - X_{(1)}]/b$ as $\chi^2_{2(n-1)}$ (Problem 1.5.18), it follows that (Problem 2.15)

$$(23) \qquad \frac{1}{n-1}\Sigma[X_i - X_{(1)}] \quad \text{and} \quad X_{(1)} - \frac{1}{n(n-1)}\Sigma[X_i - X_{(1)}]$$

are UMVU for b and a, respectively.

It is also easy to obtain the UMVU estimators of a/b and of the critical value u for which $P(X_1 \leqslant u)$ has a given value p. If, instead, u is given, the UMVU estimator of $P(X_1 \leqslant u)$ can be found in analogy with the normal case (Problems 2.16 and 2.17). Finally, the two-sample problems corresponding to Example 2.3(i) and (ii) can be handled very similarly to the normal case (Problems 2.18–2.20).

3. DISCRETE DISTRIBUTIONS

The distributions considered in the preceding section were all continuous. We shall now treat the corresponding problems for some of the basic discrete distributions.

Example 3.1. The binomial distribution. In the simplest instance of a one-sample problem with qualitative rather than quantitative "measurements," the observations are dichotomous: cure or no cure, satisfactory or defective, yes or no. The two outcomes will be referred to generically as success or failure.

The results of n independent such observations with common success probability p are conveniently represented by random variables X_i which are 1 or 0 as the ith case or "trial" is a success or failure. Then $P(X_i = 1) = p$, and the joint distribution of the X's is given by

$$(1) \qquad P(X_1 = x_1, \dots, X_n = x_n) = p^{\Sigma x_i} q^{n - \Sigma x_i} \qquad (q = 1 - p).$$

This is a one-parameter exponential family, and $T = \Sigma X_i$—the total number of successes—is a complete sufficient statistic. Since $E(X_i) = E(\bar{X}) = p$ and $\bar{X} = T/n$, it follows that T/n is the UMVU estimator of p. Similarly, $\Sigma(X_i - \bar{X})^2/(n-1) = T(n - T)/n(n-1)$ is the UMVU estimator of $\text{var}(X_i) = pq$ (Problem 3.1; see also Example 1.4).

The distribution of T is the binomial distribution $b(p, n)$, and it was pointed out in Example 1.1 that $1/p$ is not U-estimable on the basis of T, and hence not in the present situation. In fact, it follows from equation (1.2) that a function $g(p)$ can be U-estimable only if it is a polynomial of degree $\leqslant n$.

To see that every such polynomial is actually U-estimable, it is enough to show that p^m is U-estimable for every $m \leqslant n$. This can be established, and the UMVU estimator determined, by Method 1 of Section 1 (Problem 3.2). An alternative approach utilizes Method 2. The quantity p^m is the probability

$$p^m = P(X_1 = \cdots = X_m = 1)$$

and its UMVU estimator is therefore given by

$$\delta(t) = P[X_1 = \cdots = X_m = 1 | T = t].$$

This probability is 0 if $t < m$. For $t \geqslant m$, $\delta(t)$ is the probability of obtaining m successes in the first m trials and $t - m$ successes in the remaining $n - m$ trials, divided by $P(T = t)$, and hence it is

$$p^m \binom{n-m}{t-m} p^{t-m} q^{n-t} \Big/ \binom{n}{t} p^t q^{n-t},$$

or

$$\delta(T) = \frac{T(T-1)\cdots(T-m+1)}{n(n-1)\cdots(n-m+1)}. \tag{2}$$

Since this expression is zero when $T = 0, \ldots, m-1$, it is seen that $\delta(T)$, given by (2) for all $T = 0, 1, \ldots, n$, is the UMVU estimator of p^m. This proves that $g(p)$ is U-estimable on the basis of n binomial trials if and only if it is a polynomial of degree $\leqslant n$.

Consider now the estimation of $1/p$, for which no unbiased estimator exists. This problem arises, for example, when estimating the size of certain animal populations. Suppose that a lake contains an unknown number N of some species of fish. A random sample of size k is caught, tagged, and released again. Somewhat later, a random sample of size n is obtained and the number X of tagged fish in the sample is noted. If, for the sake of simplicity, we assume that each caught fish is immediately returned to the lake (or alternatively that N is very large compared to n), the n fish in this sample constitute n binomial trials with probability $p = k/N$ of success (i.e., obtaining a tagged fish). The population size N is therefore equal to k/p. We shall now discuss a sampling scheme under which $1/p$, and hence k/p, is U-estimable.

Example 3.2. Inverse binomial sampling. Reliable estimation of $1/p$ is clearly difficult when p is close to zero, where a small change of p will cause a large change in $1/p$. To obtain control of $1/p$ for all p, it would therefore seem necessary to take more observations the smaller p is. A sampling scheme achieving this is inverse sampling, which continues until a specified number of successes, say m, have been obtained. Let $Y + m$ denote the required number of trials. Then Y has the *negative* *binomial* distribution given by (Problem 1.4.8)

(3) $$P(Y=y)=\binom{m+y-1}{m-1}p^m q^y; \qquad y=0,1,\ldots$$

and

(4) $$E(Y)=mq/p; \qquad \text{var}(Y)=mq/p^2.$$

It is seen from (4) that $\delta(Y)=(Y+m)/m$, the reciprocal of the proportion of successes, is an unbiased estimator of $1/p$.

The full data in the present situation are not Y but also include the positions in which the m successes occur. However, Y is a sufficient statistic (Problem 3.6), and it is complete since (3) is an exponential family. As a function of Y, $\delta(Y)$ is thus the unique unbiased estimator of $1/p$; based on the full data, it is UMVU.

It is interesting to note that $1/q$ is not U-estimable with the present sampling scheme, for suppose $\delta(Y)$ is an unbiased estimator so that

$$p^m\sum_{y=0}^{\infty}\delta(y)\binom{m+y-1}{m-1}q^y=1/q \qquad \text{for all} \quad 0<q<1.$$

The left side is a power series which converges for all $0<q<1$, and hence converges and is continuous for all $|q|<1$. As $q\to 0$, the left side therefore tends to $\delta(0)$ while the right side tends to infinity. Thus the assumed δ does not exist. (For the estimation of p^r, see Problem 3.4.)

The situations described in Examples 3.1 and 3.2 are special cases of *sequential binomial sampling* in which the number of trials is allowed to depend on the observations. The outcome of such sampling can be represented as a random walk in the plane. The walk starts at (0, 0) and moves a unit to the right or up as the first trial is a success or failure. From the resulting point (1, 0) or (0, 1) it again moves a unit to the right or up, and continues in this way until the sampling plan tells it to stop. A stopping rule is thus defined by a set B of points, a *boundary*, at which sampling stops. We require B to satisfy

(5) $$\sum_{(x,y)\in B}P(x,y)=1$$

since otherwise there is positive probability that sampling will go on indefinitely. A stopping rule that satisfies (5) is called *closed*.

Any particular sample path ending in (x, y) has probability $p^x q^y$, and the probability of a path ending in any particular point (x, y) is therefore

(6) $$P(x,y)=N(x,y)p^x q^y,$$

where $N(x, y)$ denotes the number of paths along which the random walk can reach the point (x, y). As illustrations, consider the plans of Examples 3.1 and 3.2.

(i) In Example 3.1, B is the set of points (x, y) satisfying $x + y = n$; $x = 0, \ldots, n$, and for any $(x, y) \in B$ we have $N(x, y) = \binom{n}{x}$.

(ii) In Example 3.2, B is the set of points (x, y) with $x = m$; $y = 0, 1, \ldots$, and for any such point

$$N(x, y) = \binom{m + y - 1}{y}.$$

The observations in sequential binomial sampling are represented by the sample path, and it follows from (6) and the factorization criterion that the coordinates (X, Y) of the stopping point in which the path terminates constitute a sufficient statistic. This can also be seen from the definition of sufficiency, since the conditional probability of any given sample path given that it ends in (x, y) is

$$\frac{p^x q^y}{N(x, y) p^x q^y} = \frac{1}{N(x, y)},$$

which is independent of p.

Example 3.3. Estimation of p. For any closed sequential binomial sampling scheme, an unbiased estimator of p depending only on the sufficient statistic (X, Y) can be found in the following way. A simple unbiased estimator is $\delta = 1$ if the first trial is a success and $\delta = 0$ otherwise. Application of the Rao–Blackwell theorem then leads to $\delta'(X, Y) = E[\delta|(X, Y)] = P[1^{\text{st}} \text{ trial} = \text{success}|(X, Y)]$ as an unbiased estimator depending only on (X, Y). If the point $(1, 0)$ is a stopping point, then $\delta' = \delta$ and nothing is gained. In all other cases, δ' will have a smaller variance than δ. An easy calculation [Problem 3.8(i)] shows that

$$\delta'(x, y) = N'(x, y)/N(x, y) \tag{7}$$

where $N'(x, y)$ is the number of paths possible under the sampling scheme which pass through $(1, 0)$ and terminate in (x, y).

More generally, if (a, b) is any *accessible point*, that is, if it is possible under the given sampling plan to reach (a, b), the quantity $p^a q^b$ is U-estimable, and an unbiased estimator depending only on (X, Y) is given by (7), where $N'(x, y)$ now stands for the number of paths passing through (a, b) and terminating in (x, y) [Problem 3.8(ii)].

The estimator (7) will be UMVU for any sampling plan for which the sufficient statistic (X, Y) is complete. To describe conditions under which this is the case, let us call an accessible point that is not in B a *continuation point*. A sampling plan is called *simple* if the set of continuation points C_t on each line segment $x + y = t$ is an interval or the empty set. A plan is called finite if the number of accessible points is finite.

Example 3.4. (i) Let a, b, and m be three positive integers with $a < b < m$. Continue observation until either a successes or failures have been obtained. If this does not happen during the first m trials, continue until either b successes or failures have been obtained. This sampling plan is simple and finite.

(ii) Continue until *both* at least a successes and a failures have been obtained. This plan is neither simple nor finite, but it is closed (Problem 3.10).

Theorem 3.1. *A necessary and sufficient condition for a finite sampling plan to be complete is that it is simple.*

We shall here only prove sufficiency. [For a proof of necessity see Girschick, Mosteller, and Savage (1946).] If the restriction to finite plans is dropped, simplicity is no longer sufficient (Problem 3.9). Another necessary condition in that case is stated in Problem 3.13. This condition, together with simplicity, is also sufficient. (For a proof see Lehmann and Stein, 1950.)

For the following proof it may be helpful to consider a diagram of plan (i) of Example 3.4.

Proof of sufficiency. Suppose there exists a nonzero function $\delta(X, Y)$ whose expectation is zero for all p $(0 < p < 1)$. Let t_0 be the smallest value of t for which there exists a boundary point (x_0, y_0) on $x + y = t_0$ such that $\delta(x_0, y_0) \neq 0$. Since the continuation points on $x + y = t_0$ (if any) form an interval, they all lie on the same side of (x_0, y_0). Suppose, without loss of generality, that (x_0, y_0) lies to the left and above C_{t_0}, and let (x_1, y_1) be that boundary point on $x + y = t_0$ above C_{t_0} and with $\delta(x, y) \neq 0$ which has the smallest x-coordinate. Then all boundary points with $\delta(x, y) \neq 0$ satisfy $t \geqslant t_0$ and $x \geqslant x_1$. It follows that for all $0 < p < 1$

$$E[\delta(X, Y)] = N(x_1, y_1)\delta(x_1, y_1)p^{x_1}q^{t_0 - x_1} + p^{x_1 + 1}R(p) = 0$$

where $R(p)$ is a polynomial in p. Dividing by p^{x_1} and letting $p \to 0$, we see that $\delta(x_1, y_1) = 0$, which is a contradiction.

Fixed binomial sampling satisfies the conditions of the theorem, but, there (and for inverse binomial sampling), completeness follows already from the fact that it leads to a full-rank exponential family (1.4.1) with $s = 1$. An example in which this is not the case is curtailed binomial sampling, in which sampling is continued as long as $X < a$, $Y < b$, and

$X + Y < n$ $(a, b < n)$ and is stopped as soon as one of the three boundaries is reached (Problem 3.11). Double sampling and curtailed double sampling provide further applications of the theory. (See Girshick, Mosteller, and Savage, 1946.)

The discrete distributions considered so far were all generated by binomial trials. A large class of examples is obtained by considering one-parameter exponential families (1.4.2) in which $T(x)$ is integer-valued. Without loss of generality, we shall take $T(x)$ to be x and the distribution of X to be given by

$$P(X = x) = e^{\eta x - B(\eta)} a(x). \tag{8}$$

Putting $\theta = e^{\eta}$, we can write (8) as

$$P(X = x) = a(x)\theta^x / C(\theta), \qquad x = 0, 1, \ldots; \qquad \theta > 0. \tag{9}$$

For any function $a(x)$ for which $\Sigma a(x)\theta^x < \infty$ for some $\theta > 0$, this is a family of *power series distributions* (Problems 1.4.10–12). The binomial distribution $b(p, n)$ is obtained from (9) by putting $a(x) = \binom{n}{x}$ for $x = 0, 1, \ldots, n$, and $a(x) = 0$ otherwise; $\theta = p/q$ and $C(\theta) = (\theta + 1)^n$. The negative binomial distribution with $a(x) = \binom{m+x-1}{m-1}$, $\theta = q$, and $C(\theta) = (1 - \theta)^{-m}$ is another example. The family (9) is clearly complete. If $a(x) > 0$ for all $x = 0, 1, \ldots$, then θ^r is U-estimable for any positive integer r, and its unique unbiased estimator is obtained by solving the equations

$$\sum_{x=0}^{\infty} \delta(x) a(x) \theta^x = \theta^r \cdot C(\theta) \qquad \text{for all} \quad \theta \in \Omega.$$

Since $\Sigma a(x)\theta^x = C(\theta)$, comparison of the coefficients of θ^x yields

$$\delta(x) = \begin{cases} 0 & \text{if} \quad x = 0, \ldots, r-1 \\ a(x-r)/a(x) & \text{for} \quad x \geqslant r. \end{cases} \tag{10}$$

Suppose, next, that $X_1, \ldots, X_n$ are iid according to a power series family (9). Then $X_1 + \cdots + X_n$ is sufficient for θ, and its distribution is given by the following lemma.

Lemma 3.1. *The distribution of* $T = X_1 + \cdots + X_n$ *is the power series family*

$$P(T = t) = \frac{A(t, n)\theta^t}{[C(\theta)]^n}, \tag{11}$$

where $A(t, n)$ *is the coefficient of* θ^t *in the power series expansion of* $[C(\theta)]^n$.

Proof. By definition,

$$P(T=t)=\theta^t \sum_t \frac{a(x_1)\cdots a(x_n)}{[C(\theta)]^n}$$

where Σ_t indicates that the summation extends over all n-tuples of integers $(x_1,\dots,x_n)$ with $x_1+\cdots+x_n=t$. If

$$B(t,n)=\sum_t a(x_1)\cdots a(x_n), \tag{12}$$

the distribution of T is given by (11) with $B(t,n)$ in place of $A(t,n)$. On the other hand,

$$[C(\theta)]^n=\left[\sum_{x=0}^{\infty} a(x)\theta^x\right]^n,$$

and for any $t=0,1,\dots,$ the coefficient of θ^t in the expansion of the right side as a power series in θ is just $B(t,n)$. Thus, $B(t,n)=A(t,n)$, and this completes the proof.

It follows from the lemma that T is complete and from (10) that the UMVU estimator of θ^r on the basis of a sample of n is

$$\delta(t)=\begin{cases} 0 & \text{if } t=0,\dots,r-1 \\ \dfrac{A(t-r,n)}{A(t,n)} & \text{if } t\geqslant r. \end{cases} \tag{13}$$

Consider, next, the problem of estimating the probability distribution of X from a sample $X_1,\dots,X_n$. The estimand can be written as

$$g(\theta)=P_\theta(X_1=x)$$

and the UMVU estimator is therefore given by

$$\begin{aligned}\delta(t)&=P[X_1=x|X_1+\cdots+X_n=t]\\ &=\frac{P(X_1=x)P(X_2+\cdots+X_n=t-x)}{P(T=t)}.\end{aligned}$$

In the present case, this reduces to

$$\delta(t)=\frac{a(x)A(t-x,n-1)}{A(t,n)}, \qquad n>1, \qquad 0\leqslant x\leqslant t. \tag{14}$$

Example 3.5. Poisson distribution. This distribution arises as a limiting case of the binomial distribution for large n and small p, and more generally as the number of events occurring in a fixed time period when the events are generated by a Poisson process. The distribution $P(\theta)$ of a Poisson variable with expectation θ is given by (9) with

$$a(x) = \frac{1}{x!}, \qquad C(\theta) = e^{\theta}. \tag{15}$$

Thus $[C(\theta)]^n = e^{n\theta}$ and

$$A(t, n) = \frac{n^t}{t!}. \tag{16}$$

The UMVU estimator of θ^r is therefore, by (13), equal to

$$\delta(t) = \frac{t(t-1)\cdots(t-r+1)}{n^r} \tag{17}$$

for all $t \geqslant r$. Since the right side is zero for $t = 0, \ldots, r-1$, formula (17) holds for all r.

The UMVU estimator of $P_\theta(X = x)$ is given by (14), which, by (16), becomes

$$\delta(t) = \binom{t}{x}\left(\frac{1}{n}\right)^x\left(\frac{n-1}{n}\right)^{t-x}, \qquad x = 0, 1, \ldots, t.$$

For varying x, this is the binomial distribution $b(1/n, t)$.

In some situations, Poisson variables are observed only when they are positive. One is then dealing with a sample from a truncated Poisson distribution distributed according to

$$P(X = x) = \frac{1}{e^{\theta} - 1}\frac{\theta^x}{x!}, \qquad x = 1, 2, \ldots. \tag{18}$$

This is a power series distribution with

$$a(x) = \frac{1}{x!} \quad \text{if} \quad x \geqslant 1, \qquad a(0) = 0$$

and

$$C(\theta) = e^{\theta} - 1.$$

For any values of t and n, the UMVU estimator $\delta(t)$ of θ, for example, can now be obtained from (13). (Problem 3.19; for further discussion see Tate and Goen, 1958.)

To conclude this section, let us consider some multiparameter situations.

Example 3.6. Multinomial distribution. Let $(X_0, X_1, \ldots, X_s)$ have the multinomial distribution (1.4.4). As was seen in Example 1.4.1, this is an s-parameter exponential family, with $(X_1, \ldots, X_s)$ or $(X_0, X_1, \ldots, X_s)$ constituting a complete sufficient statistic. [Recall that $X_0 = n - (X_1 + \cdots + X_s)$.] Since $E(X_i) = np_i$, it follows that X_i/n is the UMVU estimator of p_i. To obtain the UMVU estimator of $p_i p_j$, note that one unbiased estimator is $\delta = 1$ if the first trial results in outcome i and the second trial in outcome j, and $\delta = 0$ otherwise. The UMVU estimator of $p_i p_j$ is therefore

$$E(\delta | X_0, \ldots, X_s) = \frac{(n-2)! X_i X_j}{X_0! \cdots X_s!} \Big/ \frac{n!}{X_0! \cdots X_s!} = \frac{X_i X_j}{n(n-1)}.$$

In the application of multinomial models, the probabilities $p_0, \ldots, p_s$ are frequently subject to additional restrictions, so that the number of independent parameters is less than s. In general, such a restricted family will not constitute a full-rank exponential family. There are, however, important exceptions. Simple examples are provided by certain contingency tables.

Example 3.7. Two-way contingency tables. A number n of subjects are drawn at random from a population sufficiently large that the drawings can be considered to be independent. Each subject is classified according to two characteristics: A, with possible outcomes $A_1, \ldots, A_I$, and B, with possible outcomes $B_1, \ldots, B_J$. [For example, students might be classified as being male or female ($I = 2$) and according to their average performance (A, B, C, D or F; $J = 5$)]. The probability that a subject has properties (A_i, B_j) will be denoted by p_{ij} and the number of such subjects in the sample by n_{ij}. The joint distribution of the IJ variables n_{ij} is an unrestricted multinomial distribution with $s = IJ - 1$, and the results of the sample can be represented in the following $I \times J$ table. From Example 3.6, it follows that the UMVU estimator of p_{ij} is n_{ij}/n.

A special case of Table 3.1 arises when A and B are independent, that is, when $p_{ij} = p_{i+}p_{+j}$ where $p_{i+} = p_{i1} + \cdots + p_{iJ}$ and $p_{+j} = p_{1j} + \cdots + p_{Ij}$. The joint

Table 3.1. $I \times J$ **Contingency Table**

	$B_1 \cdots B_J$	Total
A_1	$n_{11} \cdots n_{1J}$	n_{1+}
$\vdots$		$\vdots$
A_I	$n_{I1} \cdots n_{IJ}$	n_{I+}
Total	$n_{+1} \cdots n_{+J}$	n

probability of the IJ cell counts then reduces to

$$\frac{n!}{\prod_{i,j} n_{ij}!} \prod_i p_{i+}^{n_{i+}} \prod_j p_{+j}^{n_{+j}}.$$

This is an $(I + J - 2)$-parameter exponential family with the complete sufficient statistics (n_{i+}, n_{+j}); $i = 1,\ldots, I$, $j = 1,\ldots, J$ or, equivalently, $i = 1,\ldots, I - 1$; $j = 1,\ldots, J - 1$. In fact, $(n_{1+},\ldots, n_{I+})$ and $(n_{+1},\ldots, n_{+J})$ are independent, with multinomial distributions $M(p_{1+},\ldots, p_{I+}; n)$ and $M(p_{+1},\ldots, p_{+J}; n)$, respectively (Problem 3.25), and the UMVU estimators of p_{i+}, p_{+j} and $p_{ij} = p_{i+}p_{+j}$ are, therefore, n_{i+}/n, n_{+j}/n and $n_{i+}n_{+j}/n^2$, respectively.

When studying the relationship between two characteristics A and B, one may find A and B to be dependent although no mechanism appears to exist through which either factor could influence the other. An explanation is sometimes found in the dependence of both factors on a common third factor, C, a phenomenon known as *spurious correlation*. The following example describes a model for this situation.

Example 3.8. Conditional independence in a three-way table. In the situation of Example 3.7, suppose that each subject is also classified according to a third factor C as $C_1,\ldots,$ or C_K. [The third factor for the students of Example 3.7 might be their major (History, Physics, etc.).] Consider this situation under the assumption that conditionally given C_k $(k = 1,\ldots, K)$, the characteristics A and B are independent, so that

$$p_{ijk} = p_{++k}\, p_{i+|k}\, p_{+j|k} \tag{19}$$

where $p_{i+|k}$, $p_{+j|k}$, and $p_{ij|k}$ denote the probability of the subject having properties A_i, B_j, or (A_i, B_j), respectively, given that it has property C_k.

After some simplification, the joint probability of the IJK cell counts n_{ijk} is seen to be proportional to (Problem 3.26)

$$\prod_{i,j,k} \left(p_{++k}\, p_{i+|k}\, p_{+j|k}\right)^{n_{ijk}} = \prod_k \left[p_{++k}^{n_{++k}} \prod_i p_{i+|k}^{n_{i+k}} \prod_j p_{+j|k}^{n_{+jk}} \right]. \tag{20}$$

This is an exponential family with

$$(K - 1) + K(I + J - 2) = K(I + J - 1) - 1$$

parameters and the complete sufficient statistics $T = \{(n_{++k}, n_{i+k}, n_{+jk}),\ i = 1,\ldots, I;\ j = 1,\ldots, J;\ k = 1,\ldots, K\}$. Since the expectation of any cell count is n times the probability of that cell, the UMVU estimators of p_{++k}, p_{i+k}, and p_{+jk} are n_{++k}/n, n_{i+k}/n, and n_{+jk}/n, respectively.

Consider, now, the estimation of the probability p_{ijk}. The unbiased estimator $\delta_0 = n_{ijk}/n$, which is UMVU in the unrestricted model, is not a function of T and hence no longer UMVU. The relationship (19) suggests the estimator $\delta_1 =$

$(n_{++k}/n) \cdot (n_{i+k}/n_{++k}) \cdot (n_{+jk}/n_{++k})$, which is a function of T. It is easy to see (Problem 3.28) that δ_1 is unbiased and hence is UMVU. (For additional results concerning the estimation of the parameters of this model, see Cohen, 1981.)

4. NONPARAMETRIC FAMILIES

Section 2.2 was concerned with continuous parametric families of distributions such as the normal, uniform, or exponential distributions, and Section 2.3 with discrete parametric families such as the binomial and Poisson distributions. We now turn to nonparametric families in which no specific form is assumed for the distribution.

We begin with the one-sample problem in which $X_1, \ldots, X_n$ are iid with distribution $F \in \mathcal{F}$. About the family $\mathcal{F}$, we shall make only rather general assumptions, for example, that it is the family of distributions F which have a density, or are continuous, or have first moments, and so on. The estimand $g(F)$ might for example be $E(X_i) = \int x\, dF(x)$, or $\operatorname{var} X_i$, or $P(X_i \leqq a) = F(a)$.

It was seen in problem 1.5.28 that for the family $\mathcal{F}_0$ of all probability densities, the order statistics $X_{(1)} < \cdots < X_{(n)}$ constitute a complete sufficient statistic, and the hint given there shows that this result remains valid if $\mathcal{F}_0$ is further restricted by requiring the existence of some moments.* (For alternative proofs see TSH; Fraser, 1953; and Bell, Blackwell, and Breiman, 1960, where the result is also shown to be valid for the family of all continuous distributions.)

An estimator $\delta(X_1, \ldots, X_n)$ is a function of the order statistics if and only if it is symmetric in its n arguments. For families $\mathcal{F}$ for which the order statistics are complete, there can therefore exist at most one symmetric unbiased estimator of any estimand, and this is UMVU. Thus, to find the UMVU estimator of any U-estimable $g(F)$, it suffices to find a symmetric unbiased estimator.

Example 4.1. Let $g(F) = P(X \leqslant a) = F(a)$, a known.

The natural estimator is the number of X's which are $\leqslant a$, divided by n. The number of such X's is the outcome of n binomial trials with success probability $F(a)$, so that this estimator is unbiased for $F(a)$. Since it is also symmetric, it is the UMVU estimator. This can be paraphrased by saying that the empirical cumulative distribution function is the UMVU estimator of the unknown true cumulative distribution function.

Note. In the normal case of Section 2.2 it was possible to find unbiased estimators not only of $P(X \leqslant u)$ but also of the probability density $p_X(u)$ of X. No

*The corresponding problem in which the values of some moments (or expectations of other functions) are given is treated by Hoeffding (1977) and N. Fisher (1982).

unbiased estimator of the density exists for the family $\mathcal{F}_0$. For proofs, see Rosenblatt (1956) and Bickel and Lehmann (1969), and for further discussion of the problem of estimating a nonparametric density see Rosenblatt (1971).

Example 4.2. Let us now restrict $\mathcal{F}_0$ by adding the condition $E|X| < \infty$, and let $g(F) = \int xf(x)\,dx$. Since $\bar{X}$ is symmetric and unbiased for $g(F)$, $\bar{X}$ is UMVU. An alternative proof of this result is obtained by noting that X_1 is unbiased for $g(F)$. The UMVU estimator is found by conditioning on the order statistics: $E[X_1|X_{(1)},\dots, X_{(n)}]$. But, given the order statistics, X_1 assumes each value with probability $1/n$. Hence, the above conditional expectation is equal to $(1/n)\Sigma X_{(i)} = \bar{X}$.

In Section 2.2, it was shown that $\bar{X}$ is UMVU for estimating $E(X_i) = \xi$ in the family of normal distributions $N(\xi, \sigma^2)$; now it is seen to be UMVU in the family of all distributions that have a probability density and finite expectation. Which of these results is stronger? The uniformity makes the nonparametric result appear much stronger. This is counteracted, however, by the fact that the condition of unbiasedness is much more restrictive in that case. Thus, the number of competitors which the UMVU estimator "beats" for such a wide class of distributions is quite small (see Problem 4.1). It is interesting in this connection to note that, for a family intermediate between the two considered here, the family of all symmetric distributions having a probability density, $\bar{X}$ is *not* UMVU (Problem 4.4). The problem of estimating the center of symmetry in that case will be considered in Section 5.4.

Example 4.3. Let $g(F) = \text{var}\, X$. Then $[\Sigma(X_i - \bar{X})^2]/(n-1)$ is symmetric and unbiased, and hence is UMVU.

Example 4.4. Let $g(F) = \xi^2$, where $\xi = EX$. Now, $\sigma^2 = E(X^2) - \xi^2$ and a symmetric unbiased estimator of $E(X^2)$ is $\Sigma X_i^2/n$. Hence the UMVU estimator of ξ^2 is $\Sigma X_i^2/n - \Sigma(X_i - \bar{X})^2/(n-1)$.

An alternative derivation of this result is obtained by noting that X_1X_2 is unbiased for ξ^2. The UMVU estimator of ξ^2 can thus be found by conditioning: $E[X_1X_2|X_{(1)},\dots, X_{(n)}]$. But, given the order statistics, the pair $\{X_1, X_2\}$ assumes the value of each pair $\{X_{(i)}, X_{(j)}\}$, $i \neq j$, with probability $1/n(n-1)$. Hence the above conditional expected value is

$$\frac{1}{n(n-1)}\sum_{i\neq j} X_iX_j,$$

which is equivalent to the earlier result.

Consider, now, quite generally a function $g(F)$ which is U-estimable in $\mathcal{F}_0$. Then there exists an integer $m \leqq n$ and a function $\delta(X_1,\dots, X_m)$, which is unbiased for $g(F)$. We can assume without loss of generality that δ is symmetric in its m arguments; otherwise, it can by symmetrized. Then the estimator

$$\frac{1}{\binom{n}{m}}\sum_{(i_1,\dots,i_m)} \delta(X_{i_1},\dots, X_{i_m}) \tag{1}$$

is UMVU for $g(F)$; here, the sum is all over m-tuples $(i_1,\dots, i_m)$ from the integers $1, 2, \ldots, n$ with $i_1 < \cdots < i_m$. That this estimator is UMVU follows from the facts that it is symmetric and that each of the $\binom{n}{m}$ summands has expectation $g(F)$.

The class of statistics (1) called *U-statistics* were studied by Hoeffding (1948) who, in particular, proved their asymptotic normality.

Two problems suggest themselves:

(i) What kind of functionals $g(F)$ are U-estimable?

(ii) If a functional is U-estimable, what is the smallest number of observations for which it is U-estimable? We shall call this smallest number the *degree* of $g(F)$.

(For the case that F assigns positive probability only to the two values 0 and 1, these questions are answered in the preceding section.)

Example 4.5. Let $g(F)$be the variance σ^2 of F. Then $g(F)$ is U-estimable in the subset $\mathfrak{F}_0'$ of $\mathfrak{F}_0$ with $E_F X^2 < \infty$ with $n = 2$ observations, since $\Sigma(X_i - \bar{X})^2/(n-1) = \frac{1}{2}(X_2 - X_1)^2$ is unbiased for σ^2. Hence, the degree of σ^2 is $\leqq 2$. Furthermore, since in the normal case there is no unbiased estimator of σ^2 based on only one observation (Problem 2.4), there is no such estimator within the class $\mathfrak{F}_0'$. It follows that the degree of σ^2 is 2.

We shall now give another proof that the degree of σ^2 in this example is greater than 1 to illustrate a method that is of more general applicability for problems of this type.

Let g be any estimand that is of degree 1 in $\mathfrak{F}_0'$. Then there exists δ such that

$$\int \delta(x)\, dF(x) = g(F), \qquad \text{for all} \quad F \in \mathfrak{F}_0'.$$

Fix two arbitrary distributions F_1 and F_2, and let $F = \alpha F_1 + (1-\alpha)F_2$, $0 \leqslant \alpha \leqslant 1$. Then

$$(2) \quad g[\alpha F_1 + (1-\alpha)F_2] = \alpha \int \delta(x)\, dF_1(x) + (1-\alpha)\int \delta(x)\, dF_2(x).$$

As a function of α, the right-hand side is linear in α. Thus the only g's that can be of degree 1 are those for which the left-hand side is linear in α.

Now, consider

$$g(F) = \sigma_F^2 = E(X^2) - [EX]^2.$$

In this case,

$$(3) \quad \sigma^2_{\alpha F_1 + (1-\alpha)F_2} = \alpha E(X_1^2) + (1-\alpha)E(X_2^2) - [\alpha EX_1 + (1-\alpha)EX_2]^2,$$

where X_i is distributed according to F_i. The coefficient of α^2 on the right-hand side is seen to be $-[E(X_2) - E(X_1)]^2$. Since this is not zero for all $F_1, F_2 \in \mathcal{F}_0'$, the right-hand side is not linear in α, and it follows that σ^2 is not of degree 1.

Generalizing (2), we see that if $g(F)$ is of degree m, then

$$(4)\quad \begin{cases} g[\alpha F_1 + (1-\alpha)F_2] \\ \qquad = \int \cdots \int \delta(x_1,\ldots,x_m)\, d[\alpha F_1(x_1) + (1-\alpha)F_2(x_1)] \cdots \\ \text{is a polynomial of degree at most } m, \end{cases}$$

which is thus a necessary condition for g to be estimable with m observations. Conditions for (4) to be also sufficient are given by Bickel and Lehmann (1969).

Condition (4) may also be useful for proving that there exists no value of n for which a functional $g(F)$ is U-estimable.

Example 4.6. Let $g(F) = \sigma$. Then $g[\alpha F_1 + (1-\alpha)F_2]$ is the square root of the right-hand side of (3). Since this quadratic in α is not a perfect square for all $F_1, F_2 \in \mathcal{F}_0'$, it follows that its square root is not a polynomial. Hence σ is not U-estimable for any fixed number n of observations.

Let us now turn from the one-sample to the two-sample problem. Let $X_1,\ldots, X_m$ and $Y_1,\ldots, Y_n$ be independently distributed according to distributions F and $G \in \mathcal{F}_0$. Then the order statistics $X_{(1)} < \cdots < X_{(m)}$ and $Y_{(1)} < \cdots < Y_{(n)}$ are sufficient and complete (Problem 4.5). A statistic δ is a function of these order statistics if and only if δ is symmetric in the X_i's and separately symmetric in the Y_j's.

Example 4.7. Let $h(F, G) = E(Y) - E(X)$. Then $\bar{Y} - \bar{X}$ is unbiased for $h(F, G)$. Since it is a function of the complete sufficient statistic, it is UMVU.

The concept of degree runs into difficulty in the present case. Smallest values m_0 and n_0 are sought for which a given functional $h(F, G)$ is U-estimable. One possibility is to find the smallest m for which there exists an n such that $h(F, G)$ is U-estimable, and to let m_0 and n_0 be the smallest values so determined. This procedure is not symmetric in m and n. However, it can be shown that if the reverse procedure is used, the same minimum values are obtained. [See Bickel and Lehmann, (1969).]

As a last illustration, let us consider the bivariate nonparametric problem. Let $(X_1, Y_1),\ldots,(X_n, Y_n)$ be iid according to a distribution $F \in \mathcal{F}$, the family of all bivariate distributions having a probability density. In analogy with the order statistics in the univariate case, the set of pairs

$$T = \{[X_{(1)}, Y_{j_1}],\ldots, [X_{(n)}, Y_{j_n}]\},$$

that is, the n pairs (X_i, Y_i), ordered according to the value of their first coordinate, constitute a sufficient statistic. An equivalent statistic is

$$T' = \{[X_{i_1}, Y_{(1)}], \ldots, [X_{i_n}, Y_{(n)}]\},$$

that is, the set of pairs (X_i, Y_i) ordered according to the value of the second coordinate. Here, as elsewhere, the only aspect of T that matters* is the set of points to which T assigns a constant value. In the present case, these are the $n!$ points that can be obtained from the given point $[(X_1, Y_1), \ldots, (X_n, Y_n)]$ by permuting the n pairs. As in the univariate case, the conditional probability of each of these permutations given T or T' is $1/n!$. Also, as in the univariate case, T is complete (Problem 4.10).

An estimator δ is a function of the complete sufficient statistic if and only if δ is invariant under permutation of the n pairs. Hence, any such function is the unique UMVU estimator of its expectation.

Example 4.8. The estimator $\Sigma(X_i - \bar{X})(Y_i - \bar{Y})/(n-1)$ is UMVU for $\operatorname{cov}(X, Y)$ (Problem 4.8).

5. PERFORMANCE OF THE ESTIMATORS

In concentrating on the derivation of UMVU estimators in the preceding three sections, we have neglected an important aspect. Estimators are of little use unless one has an idea of their accuracy. As was pointed out in Section 1.1, the performance of an estimator can be considered from two different points of view. One can describe it in terms of a rather arbitrary measure of accuracy—and here for the sake of convenience the preferred choice for an unbiased estimator is its variance—or, alternatively, one can try to measure the consequences of an estimator by the losses resulting from incorrect estimates; these losses, of course, would depend on the particular circumstances of the concrete situation. We shall begin by considering the variances of some of the UMVU estimators derived in the preceding sections. Later, we shall discuss an asymptotic approach, in which the choice of a measure of accuracy or loss function becomes less important.

Example 5.1. Suppose, as in Example 2.1, that $X_1, \ldots, X_n$ are iid as $N(\xi, \sigma^2)$, with one of the parameters known, and that the estimand is a polynomial in ξ or σ. Then the UMVU estimator is a polynomial in $\bar{X}$ or $S^2 = \Sigma(X_i - \xi)^2$. The variance of any such polynomial can be estimated if one knows the moments $E(\bar{X}^k)$ and $E(S^k)$ for all $k = 1, 2, \ldots$. A formula for $E(S^k)$ is given by (2.4) and (2.5). To determine $E(\bar{X}^k)$, write $\bar{X} = Y + \xi$, where Y is distributed as $N(0, \sigma^2/n)$. Then

*Besides the relevant measurability conditions.

(Problem 5.1)

$$E(\bar{X}^k) = \sum_{r=0}^{k} \binom{k}{r} \xi^{k-r} E(Y^r) \tag{1}$$

with

$$E(Y^r) = \begin{cases} (r-1)(r-3)\cdot \cdots \cdot 3 \cdot 1(\sigma^2/n)^{r/2} & \text{when } r \geqslant 2 \text{ is even} \\ 0 & \text{when } r \text{ is odd.} \end{cases} \tag{2}$$

As an example, consider the UMVU estimator S^2/n of σ^2. We have $E(S^4) = n(n+2)\sigma^4$ and (Problem 5.1)

$$\operatorname{var}\left(\frac{S^2}{n}\right) = \frac{2\sigma^4}{n}. \tag{3}$$

Since σ is unknown, so is this variance; however, it can be estimated, its UMVU estimator being $2S^4/n^2(n+2)$.

The situation is quite similar when both ξ and σ are unknown. As an example, consider the UMVU estimator of ξ^2 (Problem 5.2) which is

$$\delta = \bar{X}^2 - \frac{S^2}{n(n-1)} \tag{4}$$

where now $S^2 = \Sigma(X_i - \bar{X})^2$. Since $\bar{X}$ and S^2 are independent, the variance of δ and its UMVU estimator are again easily obtained (Problem 5.3).

The calculation of exact variances is often too complicated to be practicable. There do, however, exist simple approximations which tend to work well if n is at least moderately large.

Theorem 5.1. *Let* $X_1, \ldots, X_n$ *be iid with* $E(X_1) = \xi$, *var*$(X_1) = \sigma^2$ *and finite fourth moment. Suppose* h *is a function of a real variable whose first four derivatives* $h'(x)$, $h''(x)$, $h'''(x)$, *and* $h^{(iv)}(x)$ *exist for all* $x \in I$ *where* I *is an interval with* $P(X_1 \in I) = 1$, *and such that* $|h^{(iv)}(x)| \leqslant M$ *for all* $x \in I$, *for some* $M < \infty$. *Then*

$$E[h(\bar{X})] = h(\xi) + \frac{\sigma^2}{2n} h''(\xi) + R_n, \tag{5}$$

and if, in addition, the fourth derivative of h^2 *is also bounded,*

$$var[h(\bar{X})] = \frac{\sigma^2}{n}[h'(\xi)]^2 + R_n, \tag{6}$$

where the remainder R_n *in both cases is* $0(1/n^2)$, *that is, there exist* n_0 *and* $A < \infty$ *such that* $R_n(\xi) < A/n^2$ *for* $n > n_0$ *and all* ξ.

Proof. The reason for the possibility of such a result is the strong set of assumptions concerning h, which permit an expansion of $h(\bar{X})$ about $h(\xi)$ with bounded coefficients, and the fact that $E(\bar{X}-\xi)^{2k-1}$ and $E(\bar{X}-\xi)^{2k}$, if they exist, are of order $1/n^k$ for all $k \geqslant 1$ (Problem 5.4). More specifically, we shall use the following facts.

(A) *Moments of* $\bar{X}$

(7)

$$E(\bar{X}-\xi) = 0, \quad E(\bar{X}-\xi)^2 = \frac{\sigma^2}{n}; \qquad E(\bar{X}-\xi)^3, \quad E(\bar{X}-\xi)^4 = 0\left(\frac{1}{n^2}\right)$$

where σ^2 is the variance of the X_i.

(B) *Expansion of* $h(\bar{X})$

If, for all x, the fourth derivative $h^{(iv)}(x)$ of h exists and satisfies $|h^{(iv)}(x)| \leqslant M$ for some $M < \infty$, then

$$h(\bar{x}) = h(\xi) + h'(\xi)(\bar{x}-\xi) + \tfrac{1}{2}h''(\xi)(\bar{x}-\xi)^2 + \tfrac{1}{6}h'''(\xi)(\bar{x}-\xi)^3 + R(\bar{x}, \xi) \tag{8}$$

where

$$|R(\bar{x}, \xi)| \leqslant \frac{M(\bar{x}-\xi)^4}{24}. \tag{9}$$

From (A) and (B) one finds, by taking the expectation of both sides of (8) that

$$E[h(\bar{X})] = h(\xi) + \tfrac{1}{2}h''(\xi)\frac{\sigma^2}{n} + 0\left(\frac{1}{n^2}\right). \tag{10}$$

Here the term in $h'(\xi)$ is missing since $E(\bar{X}-\xi) = 0$, and the order of the remainder term follows from (7) and (9).

To obtain an expansion of var$[h(\bar{X})]$, apply (10) to h^2 in place of h, using the fact that

$$[h^2(\xi)]'' = 2\{h(\xi)h''(\xi) + [h'(\xi)]^2\}. \tag{11}$$

This yields

$$(12) \quad E\left[h^2(\bar{X})\right] = h^2(\xi) + \left[h(\xi)h''(\xi) + (h'(\xi))^2\right]\frac{\sigma^2}{n} + 0\left(\frac{1}{n^2}\right)$$

while it follows from (10) that

$$(13) \quad \left[Eh(\bar{X})\right]^2 = h^2(\xi) + h(\xi)h''(\xi)\frac{\sigma^2}{n} + 0\left(\frac{1}{n^2}\right).$$

Taking the difference proves the validity of (6).

Example 5.2. Consider the UMVU estimator $T(n - T)/n(n - 1)$ of pq in Example 3.1. Note that $\xi = E(X) = p$ and $\sigma^2 = pq$ and that $\bar{X} = T/n$, and write the estimator as

$$\delta(\bar{X}) = \bar{X}(1 - \bar{X})\frac{n}{n-1}.$$

To obtain an approximation of its variance, let us consider first $h(\bar{X}) = \bar{X}(1 - \bar{X})$. Then $h'(p) = 1 - 2p = q - p$ and $\operatorname{var}[h(\bar{X})] = (1/n)pq(q - p)^2 + 0(1/n^2)$. Also,

$$\left(\frac{n}{n-1}\right)^2 = \frac{1}{(1 - 1/n)^2} = 1 + \frac{2}{n} + 0\left(\frac{1}{n^2}\right).$$

Thus

$$\operatorname{var}\delta(\bar{X}) = \left(\frac{n}{n-1}\right)^2 \operatorname{var}h(\bar{X}) = \left[\frac{pq(q-p)^2}{n} + 0\left(\frac{1}{n^2}\right)\right]\left[1 + \frac{2}{n} + 0\left(\frac{1}{n^2}\right)\right]$$

$$= \frac{pq(q-p)^2}{n} + 0\left(\frac{1}{n^2}\right).$$

The exact variance of $\delta(\bar{X})$ given in Problem 3.1(ii) shows that the error is $2p^2q^2/n(n - 1)$ which is indeed of the order $1/n^2$. The maximum absolute error occurs at $p = \frac{1}{2}$ and is $1/8n(n - 1)$. It is a decreasing function of n which, for $n = 10$, equals $1/720$. On the other hand, the relative error will tend to be large, unless p is close to 0 or 1 (Problem 5.5).

In this example, the bounded derivative condition of Theorem 5.1 is satisfied for all polynomials h because $\bar{X}$ is bounded. On the other hand, the condition fails when h is a polynomial of degree $k \geq 4$ and the X's are, for example, normally distributed. However, (6) continues to hold in these circumstances. To see this, carry out the expansion (8) to the $(k - 1)^{\text{st}}$ power. The kth derivative of h is then a constant M, and instead of (9) the remainder will satisfy $R = M(\bar{X} - \delta)^k/k!$. The result then follows from the fact that all moments of the X's of order $\leqslant k$ exist and from Problem 5.4. This argument proves the following variant of Theorem 5.1.

Theorem 5.1a. *In the situation of Theorem* 5.1, *formulas* (5) *and* (6) *remain valid if for some* $k \geqslant 3$ *the function* h *has* k *derivatives, the* kth *derivative is bounded, and the first* k *moments of the* X's *exist.*

To cover the estimator

$$\delta_n(\bar{X}) = \Phi\left(\sqrt{\frac{n}{n-1}}\,(u - \bar{X})\right) \tag{14}$$

of p in Example 2.2, in which the function h depends on n, a slight generalization of Theorem 5.1 is required.

Theorem 5.1b. *Suppose that the assumptions of Theorem* 5.1 *hold, and that* c_n *is a sequence of constants satisfying*

$$c_n = 1 + \frac{a}{n} + 0\left(\frac{1}{n^2}\right). \tag{15}$$

Then the variance of

$$\delta_n(\bar{X}) = h(c_n\bar{X}) \tag{16}$$

satisfies

$$var\left[\delta_n(\bar{X})\right] = \frac{\sigma^2}{n}[h'(\xi)]^2 + 0\left(\frac{1}{n^2}\right). \tag{17}$$

Proof. Expressions (7) are now replaced by

$$E(c_n\bar{X} - \xi) = \frac{a}{n}\xi + 0\left(\frac{1}{n^2}\right),\ E(c_n\bar{X} - \xi)^2 = \frac{\sigma^2}{n} + 0\left(\frac{1}{n^2}\right) \tag{18}$$

$$\text{and}\quad E(c_n\bar{X} - \xi)^3 = 0\left(\frac{1}{n^2}\right).$$

For example,

$$E(c_n\bar{X} - \xi)^2 = E\left[c_n(\bar{X} - \xi) + (c_n - 1)\xi\right]^2 = c_n^2\frac{\sigma^2}{n} + (c_n - 1)^2\xi^2,$$

and the result follows since $c_n^2 = 1 + 0(1/n)$ and $(c_n - 1)^2 = 0(1/n^2)$. Expanding $h(c_n\bar{X})$ about ξ as in (8) and taking expectations, one now finds analogously to (10)

$$E\left[h\left(c_n\bar{X}\right)\right] = h(\xi) + \xi h'(\xi)\frac{a}{n} + \frac{1}{2}h''(\xi)\frac{\sigma^2}{n} + 0\left(\frac{1}{n^2}\right) \tag{19}$$

and the rest of the proof follows as before.

Example 5.3. For the estimator $\delta_n(\bar{X})$ given by (14), we have

$$c_n = \sqrt{\frac{n}{n-1}} = \left(1 - \frac{1}{n}\right)^{-1/2} = 1 + \frac{1}{2n} + 0\left(\frac{1}{n^2}\right),$$

$$\text{and} \quad \delta_n = h\left(c_n\bar{Y}\right) = \Phi\left(-c_n\bar{Y}\right) \qquad \text{where} \quad Y_i = X_i - u.$$

Thus

$$h'(\xi) = -\phi(\xi),\, h''(\xi) = \xi\phi(\xi) \tag{20}$$

and hence from (17)

$$\operatorname{var} \delta_n\left(\bar{X}\right) = \frac{1}{n}\phi^2(u - \xi) + 0\left(\frac{1}{n^2}\right) \tag{21}$$

Since ξ is unknown, it is of interest to note that to terms of the order $1/n$, the maximum variance is $1/2\pi n$.

If the factor $\sqrt{n/(n-1)}$ is neglected and the estimator $\delta(\bar{X}) = \Phi(u - \bar{X})$ is used instead of δ_n, the variance is unchanged (up to the order $1/n$); however, the estimator is now biased. It follows from (10) and (20) that

$$E\delta\left(\bar{X}\right) = p + \frac{\xi - u}{2n}\phi(u - \xi) + 0\left(\frac{1}{n^2}\right)$$

so that the bias is of the order $1/n$.

The assumptions of Theorem 5.1 and its extensions do not hold for such functions as $h(x) = 1/x$ or $\sqrt{x}$ (unless the X_i are bounded away from zero) and the corresponding fact also limits the applicability of the multivariate version of the theorem given in Problem 5.26. There is, however, an alternative approach to the large-sample evaluation, which does not require the rather stringent conditions of Theorem 5.1 and which at the same time is less dependent on the particular measure of accuracy or loss function chosen.

The variance formula (6) implies that

$$\text{var}\left[\sqrt{n}\,h(\bar{X})\right] \to \sigma^2[h'(\xi)]^2 \quad \text{as} \quad n \to \infty. \tag{22}$$

Instead of the variance of $\sqrt{n}\,h(\bar{X})$, let us consider the distribution of the error $h(\bar{X}) - h(\xi)$, or rather the normalized error $\sqrt{n}\,[h(\bar{X}) - h(\xi)]$. The reason for the factor $\sqrt{n}$ is that with high probability for large n, $\bar{X}$ is close to ξ and hence $h(\bar{X})$ to $h(\xi)$. Multiplication by $\sqrt{n}$ provides just the right magnification to keep the difference in focus. This is discussed in detail in Section 5.1, where the following result is proved.

Theorem 5.2. *Let* $X_1, \ldots, X_n$ *be iid with* $E(X_i) = \xi$ *and* $var(X_i) = \sigma^2$, *and suppose that* h *has a derivative* h' *which at* ξ *is continuous and different from zero. Then the distribution of* $\sqrt{n}\,[h(\bar{X}) - h(\xi)]$ *tends to the normal distribution with zero mean and variance* $v^2 = \sigma^2[h'(\xi)]^2$.

Note that v^2 is now the variance of the limit distribution, the so-called asymptotic variance, rather than the limit of the variance as it was in (22). Under the assumptions of Theorem 5.1 these two agree, but in general they need not (Problem 5.15).

When three estimators are compared in terms of their risk, one might be best in terms of absolute error, another for squared error, and the third in terms of a higher power of the error or the probability of falling within a stated distance of the true value. On the other hand, if the limit distribution is normal with the true value as mean, it is completely determined by its variance, and for that reason comparisons in terms of asymptotic variance tend to have a validity which extends to all reasonable loss functions. For details, see Section 5.1.

As was the case with Theorem 5.1, it is easy to generalize Theorem 5.2 to functions of $c_n \bar{X}$.

Theorem 5.2a. *Under the assumptions of Theorem* 5.2, *let* c_n *be a sequence of constants satisfying* (15). *Then* $\sqrt{n}\,[h(c_n\bar{X}) - h(\xi)]$ *has the normal limit distribution with mean zero and variance* $\sigma^2[h'(\xi)]^2$.

Thus, for the estimator $\delta_n(\bar{X})$ of Example 5.3 in particular, one finds that $\sqrt{n}\,[\delta_n(\bar{X}) - p] \to N[0, \phi^2(u - \xi)]$.

The two approximations of the accuracy of an estimator indicated by Theorems 5.1 and 5.2 appear somewhat restrictive in that they apply only to functions of sample means. However, this covers all estimators based on samples from one-parameter exponential families. For on the basis of such a sample $\bar{T} = \Sigma T(X_i)/n$ [in the notation of (1.4.1)] is a sufficient statistic so that attention can be restricted to estimators that are functions of $\bar{T}$ (and possibly n). Extensions of these approximations to functions of higher sample moments are given by Cramér (1946a, Sections 27.7 and 28.4). On the other hand, this type of approximation is not applicable to optimal estimators for distributions whose support depends on the unknown parameter, such as the uniform or exponential distributions. Here the minimal sufficient statistics and the estimators based on them are governed by different asymptotic laws with different convergence rates. (Problems 5.16–5.19 and Section 6.6).

Theorems 5.1 and 5.2 generalize in a straightforward way to functions of several means. The expansion (8) is replaced by the corresponding Taylor's theorem in several variables and (7) by the results of Problem 5.26(i). The generalization of Theorem 5.1 is given in Problem 5.26(ii), that of Theorem 5.2 in the following result.

Theorem 5.3. *Let* $(X_{1\nu}, \ldots, X_{s\nu})$, $\nu = 1, \ldots, n$ *be* n *independent* s*-tuples of random variables with* $E(X_{i\nu}) = \xi_i$ *and* $cov(X_{i\nu}, X_{j\nu}) = \sigma_{ij}$. *Let* $\overline{X}_i = \Sigma X_{i\nu}/n$, *and suppose that* h *is a function of* s *arguments with continuous first partial derivatives. Then the distribution of* $\sqrt{n}\,[h(\overline{X}_1, \ldots, \overline{X}_s) - h(\xi_1, \ldots, \xi_s)]$ *tends to the normal distribution with mean zero and variance*

$$v^2 = \Sigma\Sigma\sigma_{ij}\frac{\partial h}{\partial \xi_i} \cdot \frac{\partial h}{\partial \xi_j}$$

provided $v^2 > 0$.

For a proof see Problem 5.1.23.

As an illustration, consider the asymptotic distribution of $S^2 = \Sigma(Z_\nu - \overline{Z})^2/n$ where the Z's are iid. Without loss of generality, suppose that $E(Z_\nu) = 0$, $E(Z_\nu^2) = \sigma^2$. Since $S^2 = (1/n)\Sigma Z_\nu^2 - \overline{Z}^2$, Theorem 5.3 applies with $X_{1\nu} = Z_\nu^2$, $X_{2\nu} = Z_\nu$, $h(x_1, x_2) = x_1 - x_2^2$, $\xi_2 = 0$, and $\xi_1 = \text{var}(Z_\nu) = \sigma^2$. Thus, $\sqrt{n}\,(S^2 - \sigma^2) \to N(0, v^2)$ where $v^2 = \text{var}(Z_\nu^2) = \sigma^2$. [For an easier derivation of this result, see Section 5.1 (Problem 5.1.20).]

An important aspect of estimation theory is the comparison of different estimators. As competitors of UMVU estimators, we shall now consider the maximum likelihood estimator (MLE). This comparison is of interest both because of the widespread use of the MLE and because of its asymptotic optimality (which will be discussed in Chapter 6). If a distribution is specified by a parameter θ (which need not be real valued), the MLE of θ is that value $\hat{\theta}$ of θ which maximizes the probability or probability density. The MLE of $g(\theta)$ is defined to be $g(\hat{\theta})$.

Example 5.4. Let $X_1, \ldots, X_n$ be iid according to the normal distribution $N(\xi, \sigma^2)$. Then the joint density of the X's is given by (2.3), and it is easily seen that the MLE's of ξ and σ^2 are (Problem 5.28)

$$\hat{\xi} = \bar{X} \quad \text{and} \quad \hat{\sigma}^2 = \frac{1}{n}\Sigma(X_i - \bar{X})^2. \tag{23}$$

Within the framework of this example, one can illustrate the different possible relationships between UMVU and ML estimators.

(i) When the estimand $g(\xi, \sigma)$ is ξ, then $\bar{X}$ is both the MLE and the UMVU estimator, so that in this case the two estimators coincide.

(ii) Let σ be known, say $\sigma = 1$, and let $g(\xi, \sigma)$ be the probability $p = \Phi(u - \xi)$ considered in Examples 2.2 and 5.3. The UMVU estimator is $\Phi[\sqrt{n/(n-1)}(u - \bar{X})]$ while the MLE is $\Phi(u - \bar{X})$. Since the MLE is biased (by completeness there can only be one unbiased function of $\bar{X}$), the comparison should be based on the mean squared error (rather than on the variance)

$$R_\delta(\xi, \sigma) = E[\delta - g(\xi, \sigma)]^2 \tag{24}$$

as risk. Such a comparison was carried out by Zacks and Even (1966) who found that neither estimator is uniformly better than the other. For $n = 4$, for example, the UMVU estimator is better when $|u - \xi| > 1.3$ or equivalently when $p < .1$ or $p > .9$, whereas for the remaining values of ξ or p the MLE has the smaller mean squared error.

This example raises the question whether there are situations in which the MLE is either uniformly better or uniformly worse than its UMVU competitor. The following two simple examples illustrate these possibilities.

(iii) If ξ and σ^2 are both unknown, the UMVU estimator and the MLE of σ^2 are, respectively, $S^2/(n-1)$ and S^2/n where $S^2 = \Sigma(X_i - \bar{X})^2$. Consider the general class of estimators cS^2. An easy calculation (Problem 5.32) shows that

$$E(cS^2 - \sigma^2)^2 = \sigma^4[(n^2 - 1)c^2 - 2(n-1)c + 1]. \tag{25}$$

For any given c, this risk function is proportional to σ^4. The risk functions corresponding to two different values of c, therefore, do not intersect but one lies entirely above the other. The right side of (25) is minimized by $c = 1/(n+1)$. Since the values $c = 1/(n-1)$ and $c = 1/n$, corresponding to the UMVU and ML estimator, respectively, lie on the same side of $1/(n+1)$ with $1/n$ being the closer, it follows that the MLE has uniformly smaller risk than the UMVU estimator but that the MLE in turn is dominated by $S^2/(n+1)$. (For further discussion of this problem, see Section 3.3.)

(iv) Suppose that σ^2 is known and let the estimand be ξ^2. Then the MLE is $\bar{X}^2$ and the UMVU estimator is $\bar{X}^2 - \sigma^2/n$ (Problem 2.1). That the risk of the MLE is uniformly larger follows from the following lemma.

Lemma 5.1. *Let the risk be expected squared error. If δ is an unbiased estimator of* $g(\theta)$ *and if $\delta^* = \delta + b$ where the bias* b *is independent of θ, then δ^* has uniformly larger risk than δ, in fact*

$$R_{\delta^*}(\theta) = R_\delta(\theta) + b^2.$$

The examples of this section are fairly typical and suggest that the difference between the two estimators tends to be small. For samples from exponential families, which constitute the main area of application of UMVU estimation, it has, in fact, been shown under suitable regularity assumptions that the UMVU and ML estimators are asymptotically equivalent as the sample size tends to infinity, so that the UMVU estimator shares the asymptotic optimality of the MLE. (For an exact statement and counterexamples see Portnoy, 1977.)

For small sample sizes, however, both can be unsatisfactory. One unpleasant possible feature of UMVU estimators is illustrated by the estimation of ξ^2 in the normal case [Problem 2.1; Example 5.4(iv)]. The UMVU estimator is $\bar{X}^2 - \sigma^2/n$ when σ is known, and $\bar{X}^2 - S^2/n(n-1)$ when it is unknown. In either case, the estimator can take on negative values although the estimand is known to be non-negative. Except when $\xi = 0$ or n is small, the probability of such values is not large, but when they do occur they cause an embarrassment. This difficulty can be avoided, and at the same time the risk of the estimator improved by replacing it by zero whenever it leads to a negative value; the resulting estimator, of course, will no longer be unbiased.

To conclude this section, an example is provided in which the UMVU estimator fails completely.

Example 5.5. Let X have the Poisson distribution $P(\theta)$ and let $g(\theta) = e^{-a\theta}$, where a is a known constant. The condition of unbiasedness of an estimator δ leads to

$$\Sigma \frac{\delta(x)\theta^x}{x!} = e^{(1-a)\theta} = \Sigma \frac{(1-a)^x\theta^x}{x!}$$

and hence to

$$\delta(X) = (1-a)^X \tag{26}$$

Suppose $a = 3$. Then $g(\theta) = e^{-3\theta}$, and one would expect an estimator which decreases from 1 to 0 as X goes from 0 to infinity. The MLE e^{-3X} meets this expectation. On the other hand, the unique unbiased estimator $\delta(x) = (-2)^X$ oscillates wildly between positive and negative values and appears to bear no relation to the problem at hand. (A possible explanation for this erratic behavior is suggested in Lehmann, 1983.) It is interesting to see that the difficulty disappears if the sample size is increased. If $X_1, \ldots, X_n$ are iid according to $P(\theta)$, then $T = \Sigma X_i$ is a sufficient statistic and has the Poisson distribution $P(n\theta)$. The condition of unbiasedness now becomes

$$\Sigma \delta(t) \frac{(n\theta)^t}{t!} = e^{(n-a)\theta} = \Sigma \frac{(n-a)^t\theta^t}{t!},$$

and the UMVU estimator is

$$\delta(T) = \left(1 - \frac{a}{n}\right)^T. \tag{27}$$

This is quite reasonable as soon as $n > a$.

6. THE INFORMATION INEQUALITY

The principal applications of UMVU estimators are to exponential families, as illustrated in Sections 2.2–2.3. When a UMVU estimator does not exist, the variance $V_L(\theta_0)$ of the LMVU estimator at θ_0 is the smallest variance that an unbiased estimator can achieve at θ_0. This establishes a useful benchmark against which to measure the performance of a given unbiased estimator δ. If the variance of δ is close to $V_L(\theta)$ for all θ, not much further improvement is possible. Unfortunately, the function $V_L(\theta)$ is usually difficult to determine. Instead, in this section, we shall derive some lower bounds which are typically not sharp [i.e., lie below $V_L(\theta)$] but are much simpler to calculate. One of the resulting inequalities for the variance, the *information inequality*, will be used in Chapter 4 as a tool for minimax estimation. However, its most important role is in Chapter 6, where it provides insight and motivation for the theory of asymptotically efficient estimators.

For any estimator δ of $g(\theta)$ and any function $\psi(x, \theta)$ with finite second moment, the *covariance inequality* (8.1) states that

$$\text{var}(\delta) \geq \frac{[\text{cov}(\delta, \psi)]^2}{\text{var}(\psi)}. \tag{1}$$

In general, this inequality is not helpful since the right side also involves δ. However, when $\text{cov}(\delta, \psi)$ depends on δ only through $E_\theta(\delta) = g(\theta)$, (1) does provide a lower bound for the variance of all unbiased estimators of $g(\theta)$. The following result is due to Blyth (1974).

Theorem 6.1. *A necessary and sufficient condition for* $cov(\delta, \psi)$ *to depend on* δ *only through* $g(\theta)$ *is that for all* θ

$$cov(U, \psi) = 0 \quad \textit{for all} \quad U \in \mathcal{U}, \tag{2}$$

where $\mathcal{U}$ *is the class of statistics defined in Theorem* 1.1.

Proof. Sufficiency. Suppose (2) holds for all θ and that δ_1, δ_2 are two estimators with finite variance and common expectation. Then $\delta_2 - \delta_1 \in \mathcal{U}$ and hence $\text{cov}(\delta_2 - \delta_1, \psi) = 0$ so that $\text{cov}(\delta_1, \psi) = \text{cov}(\delta_2, \psi)$.

Necessity. Suppose, conversely, that cov(δ, ψ) depends on δ only through $g(\theta)$. Then, for any $U \in \mathcal{U}$,

$$\text{cov}(\delta + U, \psi) = \text{cov}(\delta, \psi), \tag{3}$$

and hence cov(U, ψ) = 0, as was to be proved.

Example 6.1. Hammersley–Chapman–Robbins inequality. Suppose X is distributed with density $p_\theta = p(x, \theta)$, and for the moment suppose that $p(x, \theta) > 0$ for all x. If θ and $\theta + \Delta$ are two values for which $g(\theta) \neq g(\theta + \Delta)$, then the function

$$\psi(x, \theta) = \frac{p(x, \theta + \Delta)}{p(x, \theta)} - 1 \tag{4}$$

satisfies the conditions of Theorem 6.1 since

$$E_\theta(\psi) = 0 \tag{5}$$

and hence

$$\text{cov}(U, \psi) = E(\psi U) = E_{\theta+\Delta}(U) - E_\theta(U) = 0.$$

In fact,

$$\text{cov}(\delta, \psi) = E_\theta(\delta\psi) = g(\theta + \Delta) - g(\theta)$$

so that (1) becomes

$$\text{var}(\delta) \geqslant [g(\theta + \Delta) - g(\theta)]^2 / E_\theta\left[\frac{p(X, \theta + \Delta)}{p(X, \theta)} - 1\right]^2. \tag{6}$$

Since this inequality holds for all Δ, it also holds when the right side is replaced by its supremum over Δ. The resulting lower bound is due to Hammersley (1950) and Chapman and Robbins (1951).

In this inequality, the assumption of a common support for the distributions P_θ can be somewhat relaxed. If $S(\theta)$ denotes the support of P_θ, (6) will be valid provided $S(\theta + \Delta)$ is contained in $S(\theta)$. In taking the supremum over Δ, attention must then be restricted to the values of Δ for which this condition holds.

When certain regularity conditions are satisfied, a classic inequality is obtained by letting $\Delta \to 0$ in (4). The inequality (6) is unchanged if (4) is replaced by

$$\frac{p_{\theta+\Delta} - p_\theta}{\Delta} \frac{1}{p_\theta},$$

which tends to $((\partial/\partial\theta)p_\theta)/p_\theta$ as $\Delta \to 0$, provided p_θ is differentiable with respect to θ. This suggests as an alternative to (4)

$$\psi(x, \theta) = \frac{\partial}{\partial\theta} p(x, \theta)/p(x, \theta). \tag{7}$$

Since for any $U \in \mathcal{U}$, clearly $(d/d\theta)E_\theta(U) = 0$, ψ will satisfy (2) provided

$$E_\theta(U) = \int U p_\theta \, d\mu$$

can be differentiated with respect to θ under the integral sign for all $U \in \mathcal{U}$. To obtain the resulting lower bound, let $p_\theta' = (\partial/\partial\theta)p_\theta$ so that

$$\operatorname{cov}(\delta, \psi) = \int \delta p_\theta' \, d\mu.$$

If differentiation under the integral sign is permitted in

$$\int \delta p_\theta \, d\mu = g(\theta),$$

it then follows that

$$\operatorname{cov}(\delta, \psi) = g'(\theta) \tag{8}$$

and hence

$$\operatorname{var}(\delta) \geqslant \frac{[g'(\theta)]^2}{\operatorname{var}\left[\frac{\partial}{\partial\theta} \log p(X, \theta)\right]}. \tag{9}$$

The assumptions required for this inequality will be stated more formally in Theorem 6.4. (For an interesting interpretation of the inequality and discussion of the regularity assumptions, see Pitman, 1979, Chap. 5, and for another approach Ibragimov and Has'minskii, 1981, Section 1.7.)

The function ψ defined by (7) is the relative rate at which the density p_θ changes at x. The average of the square of this rate is denoted by

$$I(\theta) = E\left[\frac{\partial}{\partial\theta} \log p(X, \theta)\right]^2 = \int \left(\frac{p_\theta'}{p_\theta}\right)^2 p_\theta \, d\mu. \tag{10}$$

It is plausible that the greater this expectation is at a given value θ_0, the easier it is to distinguish θ_0 from neighboring values θ, and, therefore, the more accurately θ can be estimated at $\theta = \theta_0$. (Under suitable assumptions this surmise turns out to be correct for large samples; see Chapter 6.) The quantity $I(\theta)$ is called the *information* (or the Fisher information) that X contains about the parameter θ.

It is important to realize that $I(\theta)$ depends on the particular parametrization chosen. In fact, if $\theta = h(\xi)$ and h is differentiable, the information that X contains about ξ is

$$I^*(\xi) = I[h(\xi)] \cdot [h'(\xi)]^2. \tag{11}$$

When different parametrizations are considered in a single problem, the notation $I(\theta)$ is inadequate; however, it suffices for most applications.

To obtain alternative expressions for $I(\theta)$ that sometimes are more convenient, let us make the following assumptions:

(12)
- (i) Ω is an open interval (finite, infinite, or semi-infinite).
- (ii) The distributions P_θ have common support, so that without loss of generality the set $A = \{x: p_\theta(x) > 0\}$ is independent of θ.
- (iii) For any x in A and θ in Ω, the derivative $p'_\theta(x) = \partial p_\theta(x)/\partial\theta$ exists and is finite.

Lemma 6.1. (i) *If* (12) *holds, and the derivative with respect to* θ *of the left side of*

$$\int p_\theta(x)\, d\mu(x) = 1 \tag{13}$$

can be obtained by differentiating under the integral sign, then

$$E_\theta\left[\frac{\partial}{\partial\theta} \log p_\theta(X)\right] = 0 \tag{14}$$

and

$$I(\theta) = \text{var}_\theta\left[\frac{\partial}{\partial\theta} \log p_\theta(X)\right]. \tag{15}$$

(ii) *If, in addition, the second derivative with respect to* θ *of* $\log p_\theta(x)$ *exists for all* x *and* θ *and the second derivative with respect to* θ *of the left side of* (13) *can be obtained by differentiating twice under the integral sign, then*

$$I(\theta) = -E_\theta\left[\frac{\partial^2}{\partial\theta^2} \log p_\theta(X)\right]. \tag{16}$$

Proof. (i) Equation (14) is derived by differentiating (13), and (15) follows from (10) and (14).

(ii) We have

$$\frac{\partial^2}{\partial\theta^2}\log p_\theta(x) = \frac{\frac{\partial^2}{\partial\theta^2}p_\theta(x)}{p_\theta(x)} - \left[\frac{\frac{\partial}{\partial\theta}p_\theta(x)}{p_\theta(x)}\right]^2,$$

and the result follows by taking the expectation of both sides.

Let us now calculate $I(\theta)$ for some of the families discussed in Sections 1.3 and 1.4.

Theorem 6.2. *Let* X *be distributed according to the exponential family* (1.4.1), *with* s = 1, *and with*

$$\theta = \mathrm{E(T)}, \tag{17}$$

the so-called mean-value parametrization. *Then*

$$\mathrm{I}(\theta) = \frac{1}{var(\mathrm{T})}. \tag{18}$$

Proof. It follows from the properties of exponential families discussed in Section 1.4 that (12) and the assumptions of Lemma 6.1 hold for the exponential family (1.4.2) with η in place of θ, so that the information about η contained in X is by (15)

$$\mathrm{var}(T). \tag{19}$$

It follows from (1.4.11) that $\theta = h(\eta) = A'(\eta)$ so that

$$h'(\eta) = A''(\eta) = \mathrm{var}(T).$$

Thus,

$$I[h(\eta)][\mathrm{var}(T)]^2 = \mathrm{var}(T),$$

and this completes the proof.

The following table gives $I(\theta)$ for a number of special cases of (19).

Qualitatively, $I(\theta)$ given by (18) behaves as one would expect. Since T is the UMVU estimator of its expectation θ, the variance of T is a measure

Table 6.1. $I(\theta)$ for Some Exponential Families

Distribution	parameter θ	$I(\theta)$
$N(\xi, \sigma^2)$	ξ	$1/\sigma^2$
$N(\xi, \sigma^2)$	σ^2	$1/2\sigma^4$
$b(p, n)$	p	n/pq
$P(\lambda)$	λ	$1/\lambda$
$\Gamma(\alpha, \beta)$	β	α/β^2

of the difficulty of estimating θ. Thus, the reciprocal of the variance measures the ease with which θ can be estimated, and in this sense the information X contains about θ.

However, in light of this argument, it seems rather surprising that the information about η, which is an increasing function of θ (TSH, p. 74), is given by (19). Suppose, for example, that X has the Poisson distribution $P(\lambda)$, so that the information X contains about $\lambda = E(X)$ is $1/\lambda$. Here $\eta = \log \lambda$, and the information X contains about $\log \lambda$ is λ. Thus for large values of λ, it seems that the parameter $\log \lambda$ can be estimated quite accurately, although the converse is true for λ. This conclusion is correct and is explained by the fact that $\log \lambda$ changes very slowly when λ is large. Hence for large λ, even a large error in the estimate of λ will lead to only a small error in $\log \lambda$, whereas the situation is reversed for λ near zero where $\log \lambda$ changes very rapidly. It is interesting to note that there exists a function of λ [namely, $h(\lambda) = \sqrt{\lambda}$] whose behavior is intermediate between that of $h(\lambda) = \lambda$ and $h(\lambda) = \log \lambda$, in that the amount of information X contains about it is constant, independent of λ (Problem 6.6).

As a second class of distributions for which to evaluate $I(\theta)$, consider location families with density

$$f(x - \theta) \qquad (x, \theta \text{ real-valued}) \tag{20}$$

where $f(x) > 0$ for all x. Conditions (12) are satisfied provided the derivative $f'(x)$ of $f(x)$ exists for all values of x. It is seen that $I(\theta)$ is independent of θ and given by (Problem 6.12)

$$I_f = \int_{-\infty}^{\infty} \frac{[f'(x)]^2}{f(x)}\,dx. \tag{21}$$

The following table shows I_f for a number of distributions (defined in Table 1.3.1).

Table 6.2. I_f **for Some Standard Distributions**

Distribution	$N(0,1)$	$L(0,1)$	$C(0,1)$	$DE(0,1)$
I_f	1	1/3	1/2	1

Actually, the double exponential density does not satisfy the stated assumptions since $f'(x)$ does not exist at $x = 0$. However, (21) is valid under the slightly weaker assumption that f is absolutely continuous [see (1.2.34)], which does hold in the double exponential case. For this and the extensions below, see, for example, Huber (1981), Section 4.4. On the other hand, it does not hold when f is the uniform density on (0, 1) since f is then not continuous and hence, a fortiori, not absolutely continuous. It turns out that whenever f is not absolutely continuous it is natural to put I_f equal to ∞. For the uniform distribution, for example, it is easier by an order of magnitude to estimate θ (for large samples; see Chap. 5) than for any of the distributions listed in Table 6.2, and is thus reasonable to assign to I_f the value ∞. This should be contrasted with the fact that $f'(x) = 0$ for all $x \neq 0, 1$, so that formal application of (21) leads to the incorrect value 0.

When (20) is replaced by

$$\frac{1}{b}f\left(\frac{x-\theta}{b}\right) \tag{22}$$

the amount of information about θ becomes (Problem 6.12)

$$\frac{I_f}{b^2} \tag{23}$$

with I_f given by (21).

The information about θ contained in independent observations is, as one would expect, additive. This is stated formally in the following result.

Theorem 6.3. *Let* X *and* Y *be independently distributed with densities* p_θ *and* q_θ *with respect to measures* μ *and* ν *satisfying* (12) *and* (14).

If $I_1(\theta)$, $I_2(\theta)$, *and* $I(\theta)$ *are the information about* θ *contained in* X, Y *and* (X, Y), *respectively, then*

$$I(\theta) = I_1(\theta) + I_2(\theta). \tag{24}$$

Proof. By definition,

$$I(\theta) = E\left[\frac{\partial}{\partial\theta}\log p_\theta(X) + \frac{\partial}{\partial\theta}\log q_\theta(Y)\right]^2,$$

and the result follows from the fact that the cross product is zero by (14).

Corollary 6.1. *If* $X_1, \ldots, X_n$ *are iid, satisfy* (12) *and* (14), *and each has information* $I(\theta)$, *then the information in* $X = (X_1, \ldots, X_n)$ *is* $nI(\theta)$.

Let us now return to the inequality (9). In view of (15), the denominator on the right side can be replaced by $I(\theta)$. The result is the following version of the *information inequality.*

Theorem 6.4. (Information inequality). *Suppose* (12) *and* (14) *hold, and that* $I(\theta) > 0$. *Let* δ *be any statistic with* $E_\theta(\delta^2) < \infty$ *for which the derivative with respect to* θ *of*

$$E_\theta(\delta) = \int \delta p_\theta \, d\mu \tag{25}$$

exists and can be obtained by differentiating under the integral sign. Then

$$var_\theta(\delta) \geqslant \frac{\left[\frac{\partial}{\partial\theta} E_\theta(\delta)\right]^2}{I(\theta)}. \tag{26}$$

This follows from (9) and Lemma 6.1, and is seen directly by differentiating (25) and then applying (1).

If δ is an estimator of $g(\theta)$, with

$$E_\theta(\delta) = g(\theta) + b(\theta)$$

where $b(\theta)$ is the bias of δ, then (26) becomes

$$\text{var}_\theta(\delta) \geqslant \frac{[b'(\theta) + g'(\theta)]^2}{I(\theta)}, \tag{27}$$

which provides a lower bound for the variance of any estimator in terms of its bias and $I(\theta)$.

If $\delta = \delta(X)$ where $X = (X_1, \ldots, X_n)$ and if the X's are iid then by Corollary 6.1

$$\text{var}_\theta(\delta) \geqslant \frac{[b'(\theta) + g'(\theta)]^2}{nI_1(\theta)} \tag{28}$$

where $I_1(\theta)$ is the information about θ contained in X_1. Inequalities (27) and (28) will be of use in Chapter 4.

Unlike $I(\theta)$, which changes under reparametrization, the lower bound (26), and hence the bounds (27) and (28), do not. Let $\theta = h(\xi)$ with h differentiable. Then

$$\frac{\partial}{\partial \xi} E_{h(\xi)}(\delta) = \frac{\partial}{\partial \theta} E_\theta(\delta) \cdot h'(\xi),$$

and the result follows from (11).

The lower bound (26) for $\text{var}_\theta(\delta)$ typically is not sharp. In fact, under suitable regularity conditions, it is attained if and only if $p_\theta(x)$ is an exponential family (1.4.1) with $s = 1$ and $T(x) = \delta(x)$. For a proof of sufficiency, see Problem 6.14. Necessity is proved by Wijsman (1973); see also Fabian and Hannan, (1977). The result, however, is not correct without some regularity conditions, as is shown by Joshi (1976). Some improvements over (26) that are available when it is not attained are briefly mentioned at the end of the next section.

7. THE MULTIPARAMETER CASE AND OTHER EXTENSIONS

In discussing the information inequality, we have so far assumed that θ is real-valued. To extend the inequalities of the preceding section to the multiparameter case, we begin by generalizing the inequality (6.1) to one involving several functions ψ_i ($i = 1, \ldots, r$). This extension also provides a tool for sharpening the inequality (6.26).

Theorem 7.1. *For any unbiased estimator* δ *of* $g(\theta)$ *and any functions* $\psi_i(x, \theta)$ *with finite second moments, we have*

$$\text{var}(\delta) \geq \gamma' C^{-1} \gamma, \tag{1}$$

where $\gamma' = (\gamma_1 \cdots \gamma_r)$ *and* $C = \|C_{ij}\|$ *are defined by*

$$\gamma_i = \text{cov}(\delta, \psi_i), \qquad C_{ij} = \text{cov}(\psi_i, \psi_j). \tag{2}$$

The right side of (1) will depend on δ only through $g(\theta) = E_\theta(\delta)$ provided each of the functions ψ_i satisfies (6.2).

To prove (1), we require, first, the following extension of the covariance inequality (8.1).

Lemma 7.1. *For any random variables* $X_1, \ldots, X_r$ *with finite second moments, the covariance matrix*

$$C = \|cov(X_i, X_j)\| \tag{3}$$

is positive semidefinite; it is positive definite if and only if the X_i *are affinely independent, that is, there do not exist constants* $(a_1, \ldots, a_r)$ *and* b *such that* $\Sigma a_i X_i = b$ *with probability* 1.

The proof follows immediately from the fact that $\text{var}(\Sigma a_i X_i) = a'Ca \geqslant 0$ for all $a' = (a_1, \ldots, a_r)$.

The inequality (1) is closely related to the theory of multiple correlation. Let $(X_1, \ldots, X_r)$ and Y be random variables with finite second moment, and consider the correlation coefficient

$$\text{corr}(\Sigma a_i X_i, Y). \tag{4}$$

The maximum value ρ^* of (4) over all $(a_1, \ldots, a_r)$ is the *multiple correlation coefficient* between Y and the vector $(X_1, \ldots, X_r)$.

Lemma 7.2. *Let* $(X_1, \ldots, X_r)$ *and* Y *have finite second moment, let* $\gamma_i = cov(X_i, Y)$ *and* Σ *be the covariance matrix of the* X*'s. Without loss of generality, suppose* Σ *is positive definite. Then*

$$\rho^{*2} = \frac{\gamma' \Sigma^{-1} \gamma}{var(Y)}. \tag{5}$$

Proof. Since a correlation coefficient is invariant under scale changes, the a's maximizing (5) are not uniquely determined. Without loss of generality, we therefore impose the condition

$$\text{var}(\Sigma a_i X_i) = a' \Sigma a = 1. \tag{6}$$

The problem then becomes that of maximizing

$$a'\gamma \tag{7}$$

subject to (6). Using the method of undetermined multipliers, one maximizes instead

$$a'\gamma - \tfrac{1}{2}\lambda a' \Sigma a \tag{8}$$

with respect to a and then determines λ so as to satisfy (6). Differentiation with respect to the a_i of (8) leads to a system of linear equations with the unique solution (Problem 7.2)

$$a = \frac{1}{\lambda}\Sigma^{-1}\gamma, \tag{9}$$

and the side condition (6) gives

$$\lambda = \pm\sqrt{\gamma'\Sigma^{-1}\gamma}\,. \tag{10}$$

Substituting these values of λ into (9), one finds that

$$a = \frac{\pm\Sigma^{-1}\gamma}{\sqrt{\gamma'\Sigma^{-1}\gamma}}.$$

In view of (6), the correlation coefficient (4) is $a'\gamma/\sqrt{\text{var}\,Y}$ and its maximum value ρ^* is therefore the positive root of (5) (while the negative root gives the minimum value).

Note that always $0 \leqslant \rho^* \leqslant 1$, and that ρ^* is 1 if and only if constants $a_1, \ldots, a_r$ and b exist such that $Y = \Sigma a_i X_i + b$.

Proof of Theorem 7.1. Replace Y by δ and X_i by $\psi_i(X, \theta)$ in (5). Then the fact that $\rho^{*2} \leqslant 1$ yields (1).

As the first and principal application of (1), we shall extend the information inequality (6.26) to the multiparameter case. Let X be distributed with density p_θ, $\theta \in \Omega$, in relation to μ where θ is vector-valued, say $\theta = (\theta_1, \ldots, \theta_s)$. Replace (6.12) by

(i) and (ii) as in (6.12).
(iii) For any x in A, θ in Ω, and $i = 1, \ldots, s$, the derivative $\partial p_\theta(x)/\partial\theta_i$ exists and is finite.

In generalization of (6.10), define the *information matrix* as the $s \times s$ matrix

$$I(\theta) = \|I_{ij}(\theta)\| \tag{12}$$

where

$$I_{ij}(\theta) = E\left[\frac{\partial}{\partial\theta_i}\log p_\theta(X) \cdot \frac{\partial}{\partial\theta_j}\log p_\theta(X)\right]. \tag{13}$$

If (11) holds and the derivative with respect to each θ_i of the left side of (6.13) can be obtained by differentiating under the integral sign one obtains, as in Lemma 6.1,

$$E\left[\frac{\partial}{\partial \theta_i} \log p_\theta(X)\right] = 0 \tag{14}$$

and

$$I_{ij}(\theta) = \text{cov}\left[\frac{\partial}{\partial \theta_i} \log p_\theta(X), \frac{\partial}{\partial \theta_j} \log p_\theta(X)\right]. \tag{15}$$

Being a covariance matrix, $I(\theta)$ is then positive semidefinite and positive definite unless the $(\partial/\partial\theta_i)\log p_\theta(X)$, $i = 1, \dots, s$ are affinely dependent (and hence, by (14), linearly dependent).

If, in addition to satisfying (11) and (14), the density p_θ also has second derivatives $\partial^2 p_\theta(x)/\partial\theta_i \partial\theta_j$ for all i and j, there is in generalization of (6.16) an alternative expression for $I_{ij}(\theta)$ which is often more convenient (Problem 7.3),

$$I_{ij}(\theta) = -E\left[\frac{\partial^2}{\partial \theta_i \partial \theta_j} \log p_\theta(X)\right]. \tag{16}$$

In the multiparameter situation with $\theta = (\theta_1, \dots, \theta_s)$, Theorem 6.3 and Corollary 6.1 continue to hold with only the obvious changes, that is, information matrices for independent observations are additive.

To see how an information matrix changes under reparametrization, suppose that

$$\theta_i = h_i(\xi_1, \dots, \xi_s), \qquad i = 1, \dots, s \tag{17}$$

and let J be the matrix

$$J = \left\| \frac{\partial \theta_j}{\partial \xi_i} \right\|. \tag{18}$$

Let the information matrix for $(\xi_1, \dots, \xi_s)$ be $I^*(\xi) = \|I^*_{ij}(\xi)\|$ where

$$I^*_{ij}(\xi) = E\left[\frac{\partial}{\partial \xi_i} \log p_{\theta(\xi)}(X) \cdot \frac{\partial}{\partial \xi_j} \log p_{\theta(\xi)}(X)\right]. \tag{19}$$

Then it is seen from the chain rule for differentiating a function of several variables that (Problem 7.6)

$$I^*_{ij}(\xi) = \sum_k \sum_l I_{kl}(\theta) \frac{\partial \theta_k}{\partial \xi_i} \frac{\partial \theta_l}{\partial \xi_j} \tag{20}$$

and hence that

$$I^*(\xi) = JIJ'. \tag{21}$$

In generalization of Theorem 6.2, let us now calculate $I(\theta)$ for multiparameter exponential families.

Theorem 7.2. *Let* X *be distributed according to the exponential family* (1.4.1) *with the mean-value parametrization*

$$\theta_i = ET_i(X), \qquad i = 1, \dots, s. \tag{22}$$

Then

$$I(\theta) = C^{-1} \tag{23}$$

where C *is the covariance matrix of* $(T_1, \dots, T_s)$.

Proof. The information matrix $I^*(\eta)$ for the natural parameter η is by (15) and (1.4.12) equal to C. Furthermore, by (1.4.11) the matrix J defined by (18) is equal to C. It follows from (21) that $C = CI(\theta)C$, and this implies (23).

Example 7.1. Multivariate normal. Let $(X_1, \dots, X_p)$ have a multivariate normal distribution with mean 0 and covariance matrix $\Sigma = \|\sigma_{ij}\|$, so that by (1.3.15) the density is proportional to

$$e^{-\Sigma\Sigma\eta_{ij}x_i x_j/2}$$

where $\|\eta_{ij}\| = \Sigma^{-1}$. Since $E(X_i X_j) = \sigma_{ij}$, we find that the information matrix of the σ_{ij} is

$$I(\Sigma) = \Sigma^{-1} \tag{24}$$

Example 7.2. The following table gives $I(\theta)$ for three two-parameter exponential families, where $\psi(\alpha) = \Gamma'(\alpha)/\Gamma(\alpha)$ and $\psi'(\alpha) = d\psi(\alpha)/d\alpha$ are, respectively, the digamma and trigamma function (Problem 7.4).

Table 7.1. Three Information Matrices

$N(\xi, \sigma^2)$	$\Gamma(\alpha, \beta)$	$B(\alpha, \beta)$
$I(\xi, \sigma) = \begin{pmatrix} 1/\sigma^2 & 0 \\ 0 & 2/\sigma^2 \end{pmatrix}$	$I(\alpha, \beta) = \begin{pmatrix} \psi'(\alpha) & 1/\beta \\ 1/\beta & \alpha/\beta^2 \end{pmatrix}$	$I(\alpha, \beta) = \begin{pmatrix} \psi'(\alpha) - \psi'(\alpha + \beta) & -\psi'(\alpha + \beta) \\ -\psi'(\alpha + \beta) & \psi'(\beta) - \psi'(\alpha + \beta) \end{pmatrix}$

Example 7.3. Location-scale families. For the location-scale families with density $(1/\theta_2)f((x-\theta_1)/\theta_2)$, $\theta_2 > 0$, $f(x) > 0$ for all x, the elements of the information matrix are (Problem 7.4)

$$(25) \quad I_{11} = \frac{1}{\theta_2^2}\int\left[\frac{f'(y)}{f(y)}\right]^2 f(y)\,dy, \qquad I_{22} = \frac{1}{\theta_2^2}\int\left[\frac{yf'(y)}{f(y)} + 1\right]^2 f(y)\,dy$$

and

$$(26) \quad I_{12} = \frac{1}{\theta_2^2}\int y\left[\frac{f'(y)}{f(y)}\right]^2 f(y)\,dy.$$

The covariance term I_{12} is zero whenever f is symmetric about the origin.

Let us now state the information inequality for the multiparameter case, in which $\theta = (\theta_1, \dots, \theta_s)$.

Theorem 7.3. *Suppose that* (11) *and* (14) *hold and that* $\mathrm{I}(\theta)$ *is positive definite. Let* δ *be any statistic with* $\mathrm{E}_\theta(\delta^2) < \infty$ *for which the derivative with respect to* θ_i *of* (6.25) *exists for each* i *and can be obtained by differentiating under the integral sign. Then*

$$(27) \quad var_\theta(\delta) \geqslant \alpha' \mathrm{I}^{-1}(\theta)\alpha$$

where α' *is the row matrix with* i^{th} *element*

$$(28) \quad \alpha_i = \frac{\partial}{\partial\theta_i}\mathrm{E}_\theta[\delta(\mathrm{X})].$$

Proof. If the functions ψ_i of Theorem 7.1 are taken to be $\psi_i = (\partial/\partial\theta_i)\log p_\theta(X)$, (27) follows from (1) and (15).

If δ is an estimator of $g(\theta)$ and $b(\theta)$ is its bias, then (28) reduces to

$$(29) \quad \alpha_i = \frac{\partial}{\partial\theta_i}[b(\theta) + g(\theta)].$$

It is interesting to compare the lower bound (27) with the corresponding bound when the θ's other than θ_i are known. By Theorem 6.4, the latter is equal to $[(\partial/\partial\theta_i)E_\theta(\delta)]^2/I_{ii}(\theta)$. This is the bound obtained by replacing the multiple correlation coefficient between δ and $(\psi_1, \ldots, \psi_s)$ by the correlation between δ and ψ_i, so that $\rho^2 \leqslant \rho^{*2}$, and hence

$$I_{ii}^{-1}(\theta) \leqslant \|I^{-1}(\theta)\|_{ii}. \tag{30}$$

The two sides of (30) are equal if

$$I_{ij}(\theta) = 0 \qquad \text{for all} \quad j \neq i, \tag{31}$$

as is seen from the definition of the inverse of a matrix, and in fact (31) is also necessary for equality in (30) (Problem 7.9). As an illustration, consider the three situations of Table 7.1. In the first of these, the information bound for one of the parametrs is independent of whether the other parameter is known but this is not the case in the second and third situations. The implications of these results for estimation will be taken up in Chapter 6.

The information inequalities (6.26) and (27) have been extended in a number of directions, some of which are briefly sketched in the following

(i) When the lower bound is not sharp, it can usually be improved by considering not only the derivatives ψ_i but also higher derivatives

$$\psi_{i_1, \ldots, i_s} = \frac{1}{p_\theta(x)} \frac{\partial^{i_1 + \cdots + i_s} p_\theta(x)}{\partial\theta_1^{i_1} \cdots \partial\theta_s^{i_s}}. \tag{32}$$

It is then easy to generalize (6.26) and (27) to obtain a lower bound based on any given set S of the ψ's. Assume (11) with (iii) replaced by the corresponding assumption for all the derivatives needed for the set S, and suppose that the covariance matrix $K(\theta)$ of the given set of ψ's is positive definite. Then (1) yields the *Bhattacharyya inequality*

$$\operatorname{var}_\theta(\delta) \geqslant \alpha' K^{-1}(\theta)\alpha \tag{33}$$

where α' is the row matrix with elements

$$\frac{\partial^{i_1 + \cdots + i_s}}{\partial\theta_1^{i_1} \cdots \partial\theta_s^{i_s}} E_\theta \delta(X) = \operatorname{cov}\left(\delta, \psi_{i_1, \ldots, i_s}\right). \tag{34}$$

It is also seen that equality holds in (33) if and only if δ is a linear function of the ψ's in S (Problem 7.10). The problem of whether the

Bhattacharyya bounds become sharp as $s \to \infty$ has been investigated for some one-parameter cases by Blight and Rao (1974).

(ii) A different kind of extension avoids the need for regularity conditions by considering differences instead of derivatives. [See Hammersley (1950), Chapman and Robbins (1951), Kiefer (1952), Fraser and Guttman (1952), Fend (1959), Sen and Ghosh (1976), Chatterji (1982).]

(iii) Extensions to arbitrary convex loss functions are given by Kozek (1976) and to the case that g and δ are vector-valued by Rao (1945), Cramér (1946b), and Seth (1949).

(iv) Applications of the inequality to the sequential case in which the number of observations is not a fixed integer but a random variable, say N, determined from the observations is provided by Wolfowitz (1947), Blackwell and Girshick (1947), and Seth (1949). Under suitable regularity conditions (6.28) then continues to hold with n replaced by $E_\theta(N)$. It has, however, been pointed out by Simons (1980) that counterexamples are possible when these regularity assumptions (which place restrictions on the stopping rule) do not hold.

8. PROBLEMS

Section 1

1.1 Verify

(i) that (1.3) defines a probability distribution;
(ii) condition (1.4).

1.2 In Example 1.2, show that a_i^* minimizes (1.5) for $i = 0, 1$, and simplify the expression for a_0^*. [*Hint*: $\Sigma \kappa p^{\kappa-1}$ and $\Sigma \kappa(\kappa - 1) p^{\kappa-2}$ are the first and second derivatives of $\Sigma p^\kappa = 1/q$.]

1.3 Let X take on the values $-1, 0, 1, 2, 3$ with probabilities $P(X = -1) = 2pq$ and $P(X = k) = p^k q^{3-k}$ for $k = 0, 1, 2, 3$.

(i) Check that this is a probability distribution.
(ii) Determine the LMVU estimator at p_0 of (a) p, and (b) pq, and decide for each whether it is UMVU.

1.4 Any two random variables X, Y with finite second moments satisfy the covariance inequality

$$[\text{cov}(X, Y)]^2 \leqslant \text{var}(X) \cdot \text{var}(Y). \tag{1}$$

[*Hint*: This follows from the Schwarz inequality (1.7.15) with $f = X - E(X)$, $g = Y - E(Y)$.]

1.5 An alternative proof of the Schwarz inequality is obtained by noting that

$$\int (f + \lambda g)^2 \, dP = \int f^2 \, dP + 2\lambda \int fg \, dP + \lambda^2 \int g^2 \, dP \geq 0 \qquad \text{for all} \quad \lambda,$$

so that this quadratic in λ has at most one root.

1.6 Suppose X is distributed on (0, 1) with probability density $p_\theta(x) = (1 - \theta) + \theta/2\sqrt{x}$ for all $0 < x < 1$, $0 \leq \theta \leq 1$. Show that there does not exist an LMVU estimator of θ. [*Hint*: Let $\delta(x) = a[x^{-1/2} + b]$ for $c < x < 1$ and $\delta(x) = 0$ for $0 < x < c$. There exist values a, b, and c such that $E_0(\delta) = 0$, $E_1(\delta) = 1$ (and δ is unbiased) and that $E_0(\delta^2)$ is arbitrarily close to zero.] (Stein, 1950.)

1.7 If δ and δ' have finite variance, so does $\delta' - \delta$. [*Hint*: Inequality (1.1).]

1.8 In Example 1.3, (i) determine all unbiased estimators of zero; (ii) show that no nonconstant estimator is UMVU.

1.9 If δ_1 and δ_2 are in Δ and are UMVU estimators of $g_1(\theta)$ and $g_2(\theta)$, respectively, then $a_1\delta_1 + a_2\delta_2$ is also in Δ and is UMVU for estimating $a_1 g_1(\theta) + a_2 g_2(\theta)$, for any real a_1, a_2.

1.10 Completeness of T is not only sufficient but also necessary for every U-estimable $g(\theta)$ to have only one unbiased estimator that is a function of T.

1.11 (i) If $X_1, \ldots, X_n$ are iid (not necessarily normal) with $\text{var}(X_i) = \sigma^2 < \infty$, show that $\delta = \Sigma(X_i - \bar{X})^2/(n - 1)$ is an unbiased estimator of σ^2.

(ii) If the X_i take on the values 1 and 0 with probabilities p and $q = 1 - p$, the estimator δ of (i) depends only on $T = \Sigma X_i$ and hence is UMVU for estimating $\sigma^2 = pq$. Compare this result with that of Example 1.4.

1.12 If T has the binomial distribution $b(p, n)$ with $n > 3$, use Method 1 to find the UMVU estimator of p^3.

1.13 Let $X_1, \ldots, X_n$ be iid according to the Poisson distribution $P(\lambda)$. Use Method 1 to find the UMVU estimator of (i) λ^k for any positive integer k; (ii) $e^{-\lambda}$.

1.14 Let $X_1, \ldots, X_n$ be distributed as in Example 1.5. Use Method 1 to find the UMVU estimator of θ^k for any integer $k > -n$.

1.15 Solve Problem 1.13(ii) by Method 2, using the fact that an unbiased estimator of $e^{-\lambda}$ is $\delta = 1$ if $X_1 = 0$ and $\delta = 0$ otherwise.

1.16 In n binomial trials, let $X_i = 1$ or 0 as the ith trial is a success or failure, and let $T = \Sigma X_i$. Solve Problem 1.12 by Method 2, using the fact that an unbiased estimator of p^3 is $\delta = 1$ if $X_1 = X_2 = X_3 = 1$ and $\delta = 0$ otherwise.

1.17 Let X take on the values 1, 0 with probability p, q, respectively, and assume that $1/4 < p < 3/4$. Consider the problem of estimating p with loss function $L(p, d) = 1$ if $|d - p| \geq 1/4$, and $= 0$ otherwise. Let δ^* be the randomized estimator which is Y_0 or Y_1 when $X = 0$ or 1 where Y_0 and Y_1 are distributed as $U(-\frac{1}{2}, \frac{1}{2})$ and $U(\frac{1}{2}, \frac{3}{2})$, respectively.

(i) Show that δ^* is unbiased.

(ii) Compare the risk function of δ^* with that of X.

Section 2

2.1 If $X_1, \ldots, X_n$ are iid as $N(\xi, \sigma^2)$ with σ^2 known, find the UMVU estimator of (i) ξ^2, (ii) ξ^3, (iii) ξ^4. [*Hint*: To evaluate the expectation of $\bar{X}^k$, write $\bar{X} = Y + \xi$, where Y is $N(0, \sigma^2/n)$ and expand $E(Y + \xi)^k$.]

2.2 Solve the preceding problem when σ is unknown.

2.3 In Example 2.1 with σ known, let $\delta = \Sigma c_i X_i$ be any linear estimator of ξ. If δ is biased, its risk $E(\delta - \xi)^2$ is unbounded. [*Hint*: If $\Sigma c_i = 1 + k$, the risk is $\geqslant k^2\xi^2$.]

2.4 If X is a single observation from $N(\xi, \sigma^2)$, show that no unbiased estimator δ of σ^2 exists when ξ is unknown. [*Hint*: For fixed $\sigma = a$, X is a complete sufficient statistic for ξ, and $E[\delta(X)] = a^2$ for all ξ implies $\delta(x) = a^2$ a.e.]

2.5 Let X_i, $i = 1, \ldots, n$ be independently distributed as $N(\alpha + \beta t_i, \sigma^2)$ where α, β, and σ^2 are unknown, and the t's are known constants that are not all equal. Find the UMVU estimators of α and β.

2.6 In Example 2.2 with $n = 1$, the UMVU estimator of p is the indicator of the event $X_1 \leqslant u$ whether σ is known or unknown.

2.7 Verify equation (2.14).

2.8 Assuming (2.15) with $\sigma = \tau$, determine the UMVU estimators of σ^2 and $(\eta - \xi)/\sigma$.

2.9 Assuming (2.15) with $\eta = \xi$ and $\sigma^2/\tau^2 = \gamma$, show that when γ is known

(i) T', defined in Example 2.3(iii), is a complete sufficient statistic;

(ii) δ_γ is UMVU for ξ.

2.10 In the preceding problem with γ unknown,

(i) a UMVU estimator of ξ does not exist;

(ii) the estimator $\hat{\xi}$ is unbiased under the conditions stated in Example 2.3. [*Hint*: (i) Problem 2.9(ii) and the fact that δ_γ is unbiased for ξ even when $\sigma^2/\tau^2 \neq \gamma$. (ii) Condition on (S_X, S_Y).]

2.11 For the model (2.15) find the UMVU estimator of $P(X_1 < Y_1)$ when (i) $\sigma = \tau$, (ii) σ, τ are arbitrary. [*Hint*: Use the conditional density (2.13) of X_1 given $\bar{X}, S_X^2$ and that of Y_1 given $\bar{Y}, S_Y^2$ to determine the conditional density of $Y_1 - X_1$ given $\bar{X}, \bar{Y}, S_X^2$, and S_Y^2.]

2.12 If $(X_1, Y_1), \ldots, (X_n, Y_n)$ are iid according to any bivariate distribution with finite second moments, show that $S_{XY}/(n - 1)$ given by (2.17) is an unbiased estimator of $\operatorname{cov}(X_i, Y_i)$.

2.13 In a sample of size $N = n + k + l$, some of the observations are missing. Assume that (X_i, Y_i), $i = 1, \ldots, n$ are iid according to the bivariate normal distribution (2.16), and that $U_1, \ldots, U_k$ and $V_1, \ldots, V_l$ are independent $N(\xi, \sigma^2)$ and $N(\eta, \tau^2)$, respectively.

(i) Show that the minimal sufficient statistics are complete when ξ and η are known but not when they are unknown.

(ii) When ξ and η are known, find the UMVU estimators for σ^2, τ^2, and $\rho\sigma\tau$, and suggest reasonable unbiased estimators for these parameters when ξ and η are unknown.

2.14 For the family (2.22) show that the UMVU estimator of a when b is known and the UMVU estimator of b when a is known are as stated in Example 2.5. [*Hint*: Problem 1.5.18.]

2.15 Show that the estimators (2.23) are UMVU. [*Hint*: Problem 1.5.18.]

2.16 For the family (2.22) with $b = 1$, find the UMVU estimator of $P(X_1 \geqslant u)$ and of the density $e^{-(u-a)}$ of X_1 at u. [*Hint*: Obtain the estimator $\delta(X_{(1)})$ of the density by applying Method 2 of Section 1 and then the estimator of the probability by integration. Alternatively, one can first obtain the estimator of the probability as $P(X_1 \geqslant u | X_{(1)})$ using the fact that $X_1 - X_{(1)}$ is ancillary and that given $X_{(1)}$, X_1 is either equal to $X_{(1)}$ or distributed as $E(X_{(1)}, 1)$.]

2.17 Find the UMVU estimator of $P(X_1 \geqslant u)$ for the family (2.22) when both a and b are unknown.

2.18 Let $X_1, \ldots, X_m$ and $Y_1, \ldots, Y_n$ be independently distributed as $E(a, b)$ and $E(a', b')$, respectively.

(i) If a, b, a', b' are completely unknown, $X_{(1)}$, $Y_{(1)}$, $\Sigma[X_i - X_{(1)}]$, and $\Sigma[Y_j - Y_{(1)}]$ jointly are sufficient and complete.

(ii) Find the UMVU estimators of $a' - a$ and b'/b.

2.19 In the preceding problem, suppose that $b' = b$.

(i) Show that $X_{(1)}$, $Y_{(1)}$, and $\Sigma[X_i - X_{(1)}] + \Sigma[Y_j - Y_{(1)}]$ are sufficient and complete.

(ii) Find the UMVU estimators of b and $(a' - a)/b$.

2.20 In Problem 2.18, suppose that $a' = a$.

(i) Show that the complete sufficient statistic of Problem 2.18(i) is still minimal sufficient but no longer complete.

(ii) Show that a UMVU estimator for $a' = a$ does not exist.

(iii) Suggest a reasonable unbiased estimator for $a' = a$.

2.21 Let $X_1, \ldots, X_n$ be iid according to the uniform distribution $U(\xi - b, \xi + b)$. If ξ, b are both unknown, find the UMVU estimators of ξ, b, and ξ/b. [*Hint*: Problem 1.5.25.]

2.22 Let $X_1, \ldots, X_m$; $Y_1, \ldots, Y_n$ be iid as $U(0, \theta)$ and $U(0, \theta')$, respectively. If $n > 1$, determine the UMVU estimator of θ/θ'.

Section 3

3.1 (i) In Example 3.1, show that $\Sigma(X_i - \bar{X})^2 = T(n - T)/n$.

(ii) The variance of $T(n - T)/n(n - 1)$ in Example 3.1 is $(pq/n)[(q - p)^2 + 2pq/(n - 1)]$.

3.2 If T is distributed as $b(p, n)$, find an unbiased estimator $\delta(T)$ of $p^m (m \leqslant n)$ by Method 1, that is, using (1.9). [*Hint*: Example 1.4.]

3.3 (i) Use the method leading to (3.2) to find the UMVU estimator $\pi_k(T)$ of $P[X_1 + \cdots + X_m = k] = \binom{m}{k} p^k q^{m-k} (m \leqslant n)$.
(ii) For fixed t and varying k, show that the $\pi_k(t)$ are the probabilities of a hypergeometric distribution.

3.4 If Y is distributed according to (3.3), use Method 1 of Section 3.1

(i) to show that the UMVU estimator of $p^r (r < m)$ is

$$\delta(y) = \frac{(m - r + y - 1)(m - r + y - 2) \cdots (m - r)}{(m + y - 1)(m + y - 2) \cdots m}, \tag{2}$$

and hence in particular that the UMVU estimator of $1/p$, $1/p^2$ and p are, respectively, $(m + y)/m$, $(m + y)(m + y + 1)/m(m + 1)$, and $(m - 1)/(m + y - 1)$;
(ii) to determine the UMVU estimator of $\text{var}(Y)$;
(iii) to show how to calculate the UMVU estimator δ of $\log p$.

3.5 Consider the scheme in which binomial sampling is continued until at least a successes *and* b failures have been obtained. Show how to calculate a reasonable estimator of $\log(p/q)$. [*Hint*: To obtain an unbiased estimator of $\log p$, modify the UMVU estimator δ of Problem 3.4(iii).]

3.6 If binomial sampling is continued until m successes have been obtained, let X_i $(i = 1, \ldots, m)$ be the number of failures between the $(i - 1)^{\text{st}}$ and ith success.

(i) The X_i are iid according to the *geometric distribution*

$$P(X_i = x) = pq^x, \qquad x = 0, 1, \ldots. \tag{3}$$

(ii) The statistic $Y = \Sigma X_i$ is sufficient for $(X_1, \ldots, X_m)$ and has the distribution (3.3).

3.7 Suppose that binomial sampling is continued until the number of successes equals the number of failures.

(i) This rule is closed if $p = 1/2$ but not otherwise.
(ii) If $p = 1/2$ and N denotes the number of trials required, $E(N) = \infty$.

3.8 Verify equation (3.7) with the appropriate definition of $N'(x, y)$

(i) for the estimation of p;
(ii) for the estimation of $p^a q^b$.

3.9 Consider sequential binomial sampling with the stopping points $(0, 1)$ and $(2, y)$, $y = 0, 1, \ldots,$.

(i) Show that this plan is closed and simple.
(ii) Show that (X, Y) is not complete by finding a nontrivial unbiased estimator of zero.

3.10 In Example 3.4(ii)

(i) Show that the plan is closed but not simple;
(ii) Show that (X, Y) is not complete;
(iii) Evaluate the unbiased estimator (3.7) of p.

3.11 *Curtailed single sampling.* Let $a, b < n$ be three non-negative integers. Continue observation until either a successes, b failures, or n observations have been obtained. Determine the UMVU estimator of p.

3.12 For any sequential binomial sampling plan, the coordinates (X, Y) of the end point of the sample path are minimal sufficient.

3.13 Consider any closed sequential binomial sampling plan with a set B of stopping points, and let B' be the set $B \cup \{(x_0, y_0)\}$ where (x_0, y_0) is a point not in B that has positive probability of being reached under plan B. Show that the sufficient statistic $T = (X, Y)$ is not complete for the sampling plan which has B' as its set of stopping points. [*Hint*: For any point $(x, y) \in B$, let $N(x, y)$ and $N'(x, y)$ denote the number of paths to (x, y) when the set of stopping points is B and B', respectively, and let $N(x_0, y_0) = 0$, $N'(x_0, y_0) = 1$. Then the statistic

$$1 - \frac{N(X,Y)}{N'(X,Y)}$$

has expectation 0 under B' for all values of p.]

3.14 For any sequential binomial sampling plan under which the point $(1, 1)$ is reached with positive probability but is not a stopping point, find an unbiased estimator of pq depending only on (X, Y). Evaluate this estimator for

(i) taking a sample of fixed size $n > 2$;
(ii) inverse binomial sampling.

3.15 Use (3.3) to determine $A(t, n)$ in (3.11) for the negative binomial distribution with $m = n$, and evaluate the estimators (3.13) of q^r, and (3.14).

3.16 Consider n binomial trials with success probability p, and let r, s be two positive integers with $r + s < n$. To the boundary $x + y = n$, add the boundary point (r, s), that is, if the number of successes in the first $r + s$ trials is exactly r, the process is stopped and the remaining $n - (r + s)$ trials are not carried out.

(i) Show that U is an unbiased estimator of zero if and only if

$$U(k, n-k) = 0 \quad \text{for} \quad k = 0, 1, \ldots, r-1 \quad \text{and} \quad k = n-s+1, n-s+2, \ldots, n \tag{4}$$

and

$$U(k, n-k) = c_k U(r, s) \quad \text{for} \quad k = r, \ldots, n-s \tag{5}$$

where the c's are given constants $\neq 0$.

(ii) Show that δ is the UMVU estimator of its expectation if and only if

(6) $$\delta(k, n-k) = \delta(r, s) \quad \text{for} \quad k = r, \ldots, n-s.$$

3.17 Generalize the preceding problem to the case that two points (r_1, s_1) and (r_2, s_2) with $r_i + s_i < n$ are added to the boundary. Assume that these two points are such that all $n+1$ points $x + y = n$ remain boundary points. [*Hint*: Distinguish the three cases that the intervals (r_1, s_1) and (r_2, s_2) are (i) mutually exclusive; (ii) one contained in the other; (iii) overlapping but neither contained in the other.]

3.18 If X has the Poisson distribution $P(\theta)$, show that $1/\theta$ is not U-estimable.

3.19 If $X_1, \ldots, X_n$ are iid according to (3.18), find the UMVU estimator of θ when (i) $n = 1$, (ii) $n = 2$.

3.20 Suppose that X has the Poisson distribution truncated on the left at a, so that it has the conditional distribution of Y given $Y \geqslant a + 1$, where Y is distributed as $P(\theta)$. Show that θ is not U-estimable.

3.21 For the negative binomial distribution truncated at zero, evaluate the estimators (3.13) and (3.14) for $m = 1$, 2, and 3.

3.22 If $X_1, \ldots, X_n$ are iid according to the logarithmic series distribution of Problem 1.4.12, evaluate the estimators (3.13) and (3.14) for $n = 1$, 2, and 3.

3.23 For the multinomial distribution of Example 3.6,

(i) show that $p_0^{r_0} \cdots p_s^{r_s}$ is U-estimable provided $r_0, \ldots, r_s$ are non-negative integers with $\Sigma r_i \leqslant n$;

(ii) find the totality of U-estimable functions;

(iii) determine the UMVU estimator of the estimand of (i).

3.24 In Example 3.7 when $p_{ij} = p_{i+}p_{+j}$, determine the variances of the two unbiased estimators $\delta_0 = n_{ij}/n$ and $\delta_1 = n_{i+}n_{+j}/n^2$ of p_{ij}, and show directly that $\text{var}(\delta_0) > \text{var}(\delta_1)$ for all $n > 1$.

3.25 In Example 3.7, show that independence of A and B implies that $(n_{1+}, \ldots, n_{I+})$ and $(n_{+1}, \ldots, n_{+J})$ are independent with multinomial distributions as stated.

3.26 Verify (3.20).

3.27 Let X, Y, and g be such that $E[g(X, Y)|y]$ is independent of y. Then $E[f(Y)g(X, Y)] = E[f(Y)]E[g(X, Y)]$, and hence $f(Y)$ and $g(X, Y)$ are uncorrelated, for all f.

3.28 In Example 3.8, show that the estimator δ_1 of p_{ijk} is unbiased for the model (3.20). [*Hint*: Problem 3.27.]

Section 4

4.1 Let $X_1, \ldots, X_n$ be iid with distribution F.

(i) Characterize the totality of functions $f(X_1, \ldots, X_n)$ which are unbiased estimators of zero for the class $\mathfrak{F}_0$ of all distributions F having a density.

(ii) Give one example of a nontrivial unbiased estimator of zero when (a) $n = 2$, (b) $n = 3$.

4.2 Let $\mathcal{F}$ be the class of all univariate distribution functions F that have a probability density function f and finite mth moment.

(i) Let $X_1, \ldots, X_n$ be independently distributed with common distribution $F \in \mathcal{F}$. For $n \geqq m$, find the UMVU estimator of ξ^m where $\xi = \xi(F) = EX_i$.

(i) Show that for the case that $P(X_i = 1) = p$, $P(X_i = 0) = q$, $p + q = 1$, the estimator of (i) reduces to (3.2).

4.3 In the preceding problem, show that $1/\text{var}_F X_i$ does not have an unbiased estimator for any n.

4.4 Let $X_1, \ldots, X_n$ be iid with distribution $F \in \mathcal{F}$ where $\mathcal{F}$ is the class of all symmetric distributions with probability density. There exists no UMVU estimator of the center of symmetry θ of F (if unbiasedness is required only for the distributions F for which the expectation of the estimator exists). [*Hint*: The UMVU estimator of θ when F is $U(\theta - \frac{1}{2}, \theta + \frac{1}{2})$, which was obtained in Problem 2.21, is unbiased for all $F \in \mathcal{F}$; so is $\bar{X}$.]

4.5 If $X_1, \ldots, X_m$; $Y_1, \ldots, Y_n$ are independently distributed according to $F, G \in \mathcal{F}_0$, defined in Problem 4.1, the order statistics $X_{(1)} < \cdots < X_{(m)}$; $Y_{(1)} < \cdots < Y_{(n)}$ are sufficient and complete. [*Hint*: For completeness, generalize the second proof suggested in Problem 1.5.28.]

4.6 Under the assumptions of the preceding problem find the UMVU estimator of $P(X_i < Y_j)$.

4.7 Under the assumptions of Problem 4.5, let $\xi = EX_i$ and $\eta = EY_j$. Show that $\xi^2\eta^2$ possesses an unbiased estimator if and only if $m \geqq 2$ and $n \geqq 2$.

4.8 Let $(X_1, Y_1), \ldots, (X_n, Y_n)$ be iid $F \in \mathcal{F}$, where $\mathcal{F}$ is the family of all distributions with probability density and finite second moments. Show that

$$\delta(X, Y) = \frac{1}{n-1}\Sigma(X_i - \bar{X})(Y_i - \bar{Y})$$

is UMVU for $\text{cov}(X, Y)$.

4.9 Under the assumptions of the preceding problem, find the UMVU estimator of

(i) $P(X_i \leqq Y_i)$;

(ii) $P(X_i \leqq X_j \text{ and } Y_i \leqq Y_j)$, $i \neq j$.

4.10 Let $(X_1, Y_1), \ldots, (X_n, Y_n)$ be iid with $F \in \mathcal{F}$, where $\mathcal{F}$ is the family of all bivariate densities. Show that the sufficient statistic T, which generalizes the order statistics to the bivariate case, is complete. [*Hint*: Generalize the second proof suggested in Problem 1.5.28. As exponential family for (X, Y) take the densities proportional to $e^{Q(x,y)}$ where

$$Q(x,y) = (\theta_{01}x + \theta_{10}y) + (\theta_{02}x^2 + \theta_{11}xy + \theta_{20}y^2) + \cdots$$
$$+ (\theta_{0n}x^n + \cdots + \theta_{n0}y^n) - x^{2n} - y^{2n}.]$$

Section 5

5.1 Verify (i) equation (5.1); (ii) equation (5.3).

5.2 In Example 5.1, when both parameters are unknown, show that the UMVU estimator of ξ^2 is given by (5.4).

5.3 (i) Determine the variance of the estimator (5.4).
(ii) Find the UMVU estimator of the variance of part (i).

5.4 If $X_1, \ldots, X_n$ are iid with $E(X_i) = \xi$ and k is a positive integer, show that $E(\bar{X} - \xi)^{2k-1}$ and $E(\bar{X} - \xi)^{2k}$, if they exist, are both $0(1/n^k)$.
[*Hint*: Without loss of generality, let $\xi = 0$ and note that $E(X_{i_1}^{r_1} X_{i_2}^{r_2} \ldots) = 0$ if any of the r's is equal to 1.]

5.5 Discuss the behavior of the relative error in Example 5.2 for fixed n as a function of p.

5.6 Let $X_1, \ldots, X_n$ be iid as $N(\xi, \sigma^2)$, σ^2 known, and let $g(\xi) = \xi^r$, $r = 2, 3, 4$. Determine, up to terms of order $1/n$,

(i) the variance of the UMVU estimator of $g(\xi)$;
(ii) the bias of the MLE of $g(\xi)$.

5.7 Let $X_1, \ldots, X_n$ be iid as $N(\xi, \sigma^2)$, ξ known. For even r, determine the variance of the UMVU estimator (2.4) of σ^r up to terms of order r.

5.8 Solve the preceding problem for the case that ξ is unknown.

5.9 For estimating p^m in Example 3.1, determine, up to order $1/n$,

(i) the variance of the UMVU estimator (3.2);
(ii) the bias of the MLE.

5.10 Solve the preceding problem if p^m is replaced by the estimand of Problem 3.3.

5.11 Let $X_1, \ldots, X_n$ be iid as Poisson $P(\theta)$.

(i) Determine the UMVU estimator of $P(X_i = 0) = e^{-\theta}$.
(ii) Calculate the variance of the estimator of (i) up to terms of order $1/n$.

[*Hint*: Write the estimator in the form (5.16) where $h(\bar{X})$ is the MLE of $e^{-\theta}$.]

5.12 Solve part (ii) of the preceding problem for the estimator (5.27).

5.13 Under the assumptions of Problem 5.4, show that

$$E|\bar{X} - \xi|^{2k-1} = 0(n^{-k+1/2}). \tag{7}$$

[*Hint*: Use the fact that $E|\bar{X}-\xi|^{2k-1} \leqslant [E(\bar{X}-\xi)^{4k-2}]^{1/2}$ together with the result of Problem 5.4.]

5.14 Obtain a variant of Theorem 5.1, which requires existence and boundedness of only h''' instead of $h^{(iv)}$, but where R_n is only $0(n^{-3/2})$.

[*Hint*: Carry the expansion (5.8) only to the second instead of the third derivative, and apply (7).]

5.15 To see that Theorem 5.1 is not necessarily valid without boundedness of the fourth (or some higher) derivative, suppose that the X's are distributed as $N(\xi, \sigma^2)$ and let $h(X) = e^{x^4}$. Then all moments of the X's and all derivatives of h exist.

(i) Show that the expectation of $h(\bar{X})$ does not exist for any n, and hence that $E\{\sqrt{n}[h(\bar{X}) - h(\xi)]\}^2 = \infty$ for all values of n.

(ii) On the other hand, show that $\sqrt{n}[h(\bar{X}) - h(\xi)]$ has a normal limit distribution with finite variance, and determine that variance.

5.16 Determine the variance of the estimator of Problem 1.14.

5.17 Under the assumptions of the preceding problem, find the MLE of θ^k and compare its expected squared error with the variance of the UMVU estimator.

5.18 Let $X_1, \ldots, X_n$ be iid according to $U(0, \theta)$, let $T = \max(X_1, \ldots, X_n)$, and let h be a function satisfying the conditions of Theorem 5.1. Show that

$$(8)\quad E[h(T)] = h(\theta) - \frac{\theta}{n}h'(\theta) + \frac{1}{n^2}[\theta h'(\theta) + \theta^2 h''(\theta)] + 0\left(\frac{1}{n^3}\right)$$

and

$$(9)\quad \operatorname{var}[h(T)] = \frac{\theta^2}{n^2}[h'(\theta)]^2 + 0\left(\frac{1}{n^3}\right).$$

5.19 Apply (8) and (9) to obtain approximate answers to Problems 5.16 and 5.17, and compare the answers with the exact solutions.

5.20 If the X's are as in Theorem 5.1 and if the first five derivatives of h exist and the fifth derivative is bounded, show that

$$E[h(\bar{X})] = h(\xi) + \frac{1}{2}h''\frac{\sigma^2}{n} + \frac{1}{24n^2}[4h'''\mu_3 + 3h^{(iv)}\sigma^4] + 0(n^{-5/2})$$

and if the fifth derivative of h^2 is also bounded

$$\operatorname{var}[h(\bar{X})] = (h'^2)\frac{\sigma^2}{n} + \frac{1}{n^2}[h'h''\mu_3 + (h'h''' + \tfrac{1}{2}h''^2)\sigma^4] + 0(n^{-5/2})$$

where $\mu_3 = E(X-\xi)^3$.

[*Hint*: Use the facts that $E(\bar{X}-\xi)^3 = \mu_3/n^2$ and $E(\bar{X}-\xi)^4 = 3\sigma^4/n^2 + 0(1/n^3)$.]

5.21 Under the assumptions of the preceding problem, carry the calculation of the variance (5.17) to terms of order $1/n^2$, and compare the result with that of the preceding problem.

5.22 Carry the calculation of Problem 5.6 to terms of order $1/n^2$.

5.23 For the estimands of Problem 5.6, calculate the expected squared error of the MLE to terms of order $1/n^2$, and compare it with the variance calculated in Problem 5.22.

5.24 Calculate the variance (5.21) to terms of order $1/n^2$, and compare it with the expected squared error of the MLE carried to the same order.

5.25 Find the variance of the estimator (3.17) up to terms of the order $1/n^2$.

5.26 (i) Under the assumptions of Theorem 5.3, if all fourth moments of the $X_{i\nu}$ are finite, show that $E(\bar{X}_i - \xi_i)(\bar{X}_j - \xi_j) = \sigma_{ij}/n$ and that all third and fourth moments $E(\bar{X}_i - \xi_i)(\bar{X}_j - \xi_j)(\bar{X}_k - \xi_k)$, and so on are of the order $1/n^2$.

(ii) If, in addition, all derivatives of h of total order $\leqslant 4$ exist and those of order 4 are uniformly bounded, then

$$E\left[h(\bar{X}_1,\ldots,\bar{X}_s)\right] = h(\xi_1,\ldots,\xi_s) + \frac{1}{2n}\sum_{i=1}^{s}\sum_{j=1}^{s}\sigma_{ij}\frac{\partial^2 h(\xi_1,\ldots,\xi_s)}{\partial\xi_i\,\partial\xi_j} + R_n \tag{10}$$

and if the derivatives of h^2 of order 4 are also bounded,

$$\operatorname{var}\left[h(\bar{X}_1,\ldots,\bar{X}_s)\right] = \frac{1}{n}\Sigma\Sigma\sigma_{ij}\frac{\partial h}{\partial\xi_i}\frac{\partial h}{\partial\xi_j} + R_n \tag{11}$$

where the remainder R_n in both cases is $0(1/n^2)$.

5.27 Under the assumptions of the illustration of Theorem 5.3 with $E(Z_\nu) = \zeta$, find the limit distribution of $\sqrt{n}(\bar{Z}-\zeta)/S$. (For a simpler derivation of this distribution, see Section 5.1.)

5.28 Verify the MLE's given in (5.23).

5.29 In Example 5.4(ii), show that

(i) the bias of the MLE is 0 when $\xi = u$;

(ii) at $\xi = u$, the MLE has smaller expected squared error than the UMVU estimator.

[*Hint*: (ii) Note that $u - \bar{X}$ is always closer to 0 than $\sqrt{n/(n-1)}\,(u-\bar{X})$.]

5.30 On the basis of a sample from $N(\xi, \sigma^2)$, let $P_n(\xi, \sigma)$ be the probability that the UMVU estimator $\bar{X}^2 - \sigma^2/n$ of ξ^2 (σ known) is negative.

(i) Show that $P_n(\xi, \sigma)$ is a decreasing function of $\sqrt{n}|\xi|/\sigma$.
(ii) Show that $P_n(\xi, \sigma) \to 0$ as $n \to \infty$ for any fixed $\xi \neq 0$ and σ.
(iii) Determine the value of $P_n(0, \sigma)$.

[*Hint*: $P_n(\xi, \sigma) = P[-1 - \sqrt{n}\xi/\sigma < Y < 1 - \sqrt{n}\xi/\sigma]$ where $Y = \sqrt{n}(\bar{X} - \xi)/\sigma$ is distributed as $N(0, 1)$.]

5.31 Use a table of the t-distribution to find the value of $P_n(0, \sigma)$ in the preceding problem for the UMVU estimator of ξ^2 when σ is unknown for representative values of n.

5.32 Verify (5.25).

5.33 A sequence of numbers R_n is said to be $o(1/k_n)$ as $n \to \infty$ if $k_n R_n \to 0$ and to be $0(1/k_n)$ if there exist M and n_0 such that $|k_n R_n| < M$ for all $n > n_0$ or, equivalently, if $k_n R_n$ is bounded.

(i) If $R_n = o(1/k_n)$ then $R_n = 0(1/k_n)$.
(ii) $R_n = 0(1)$ if and only if R_n is bounded.
(iii) $R_n = o(1)$ if and only if $R_n \to 0$.
(iv) If R_n is $0(1/k_n)$ and k'_n/k_n tends to a finite limit, then R_n is $0(1/k'_n)$.

5.34 (i) If R_n and R'_n are both $0(1/k_n)$, so is $R_n + R'_n$.
(ii) If R_n and R'_n are both $o(1/k_n)$, so is $R_n + R'_n$.

5.35 Suppose $k'_n/k_n \to \infty$.

(i) If $R_n = 0(1/k_n)$, $R'_n = 0(1/k'_n)$, then $R_n + R'_n = 0(1/k_n)$.
(ii) If $R_n = o(1/k_n)$, $R'_n = o(1/k'_n)$, then $R_n + R'_n = o(1/k_n)$.

5.36 Under the assumptions of Lemma 5.1,

(i) if b is replaced by any random variable B which is independent of X and not 0 with probability 1, then $R_\delta(\theta) < R_{\delta^*}(\theta)$;
(ii) if squared error is replaced by any loss function of the form $L(\theta, d) = \rho(d - \theta)$ and if δ is risk unbiased with respect to L, then $R_\delta(\theta) \leqslant R_{\delta^*}(\theta)$.

Section 6

6.1 Under the assumptions of Problem 1.3, determine for each p_1 the value $L_V(p_1)$ of the LMVU estimator of p at p_1 and compare the function $L_V(p)$, $0 < p < 1$ with the variance $V_{p_0}(p)$ of the estimator which is LMVU at (a) $p_0 = 1/3$, (b) $p_0 = 1/2$.

6.2 Determine the conditions under which equality holds in (1).

6.3 Verify $I(\theta)$ for the distributions of Table 6.1.

6.4 If X is normal with mean zero and standard deviation σ, determine $I(\sigma)$.

6.5 Find $I(p)$ for the negative binomial distribution.

6.6 If X is distributed as $P(\lambda)$, show that the information it contains about $\sqrt{\lambda}$ is independent of λ.

6.7 Find a function of θ for which the amount of information is independent of θ

(i) for the gamma distribution $\Gamma(\alpha, \beta)$ with α known and with $\theta = \beta$;
(ii) for the binomial distribution $b(p, n)$ with $\theta = p$.

6.8 Show that (6.13) can be differentiated by differentiating under the integral sign when $p_\theta(x)$ is given by (6.20), for each of the distributions of Table 6.2. [*Hint*: Form the difference quotient and apply the dominated convergence theorem.]

6.9 Verify the entires of Table 6.2.

6.10 Evaluate (6.21) when f is the density of Student's t-distribution with ν degrees of freedom. [*Hint*: Use the fact that

$$\int_{-\infty}^{\infty} \frac{dx}{(1+x^2)^k} = \frac{\Gamma(1/2)\Gamma(k-1/2)}{\Gamma(k)}.]$$

6.11 For the distribution with density (6.20), show that $I(\theta)$ is independent of θ.

6.12 Verify formula (i) (6.21); (ii) (6.23).

6.13 (i) For the scale family with density $(1/\theta)f(x/\theta)$, $\theta > 0$, the amount of information a single observation X has about θ is

$$\frac{1}{\theta^2}\int\left[\frac{yf'(y)}{f(y)} + 1\right]^2 f(y)\,dy. \tag{12}$$

(ii) Show that the information X contains about $\xi = \log\theta$ is independent of θ.
(iii) For the logistic distribution $L(0, \theta)$, $I(\theta) = (3 + \pi^2)/3\pi^2\theta^2$.

6.14 If $p_\theta(x)$ is given by (1.4.1) with $s = 1$ and $T(x) = \delta(x)$, show that $\text{var}[\delta(X)]$ attains the lower bound (6.26). [*Hint*: Use (6.18) and (1.4.12).]

6.15 For a given U-estimable function $g(\theta)$ there exists an unbiased estimator δ which for all θ values attains the lower bound (6.1) for some $\psi(x, \theta)$ satisfying (6.2) if and only if $g(\theta)$ has a UMVU estimator δ_0. [*Hint*: By Theorem 1.1, $\psi(x, \theta) = \delta_0(x)$ satisfies (6.2). For any other unbiased δ, $\text{cov}(\delta - \delta_0, \delta_0) = 0$ and hence $\text{var}(\delta_0) = [\text{cov}(\delta, \delta_0)]^2/\text{var}(\delta_0)$, so that $\psi = \delta_0$ provides an attainable bound.] (Blyth, 1974)

6.16 Let $X_1, \ldots, X_n$ be iid according to a density $p(x, \theta)$ which is positive for all x. Then the variance of any unbiased estimator δ of θ satisfies

$$\operatorname{var}_{\theta_0}(\delta) \geqslant \frac{(\theta - \theta_0)^2}{\left\{\int_{-\infty}^{\infty} \frac{[p(x, \theta)]^2}{p(x, \theta_0)}\right\}^n - 1}, \qquad \theta \neq \theta_0.$$

[*Hint*: Direct consequence of (6.6).]

6.17 If $X_1, \ldots, X_n$ are iid as $N(\theta, \sigma^2)$ where σ is known and θ is known to have one of the values $0, \pm 1, \pm 2, \ldots,$ the inequality of the preceding problem shows that any unbiased estimator δ of the restricted parameter θ satisfies

$$\operatorname{var}_{\theta_0}(\delta) \geqslant \frac{\Delta^2}{e^{n\Delta^2/\sigma^2} - 1}, \Delta \neq 0$$

where $\Delta = \theta - \theta_0$, and hence $\sup_{\Delta \neq 0} \operatorname{var}_{\theta_0}(\delta) \geqslant 1/[e^{n/\sigma^2} - 1]$.

6.18 Under the assumptions of the preceding problem, let $\bar{X}^*$ be the integer closest to $\bar{X}$.

(i) The estimator $\bar{X}^*$ is unbiased for the restricted parameter θ.

(ii) There exist positive constants a and b such that for all sufficiently large n, $\operatorname{var}_\theta(\bar{X}^*) \leqslant ae^{-bn}$ for all integers θ.

[*Hint*: (ii) One finds $P(\bar{X}^* = k) = \int_{I_k} \phi(t)\, dt$, where I_k is the interval $((k - \theta - 1/2)\sqrt{n}/\sigma, (k - \theta + 1/2)\sqrt{n}/\sigma)$, and hence

$$\operatorname{var}(\bar{X}^*) \leqslant 4 \sum_{k=1}^{\infty} k \left\{1 - \Phi\left[\frac{\sqrt{n}}{\sigma}\left(k - \frac{1}{2}\right)\right]\right\}.$$

The result follows from the fact that for all $y > 0$, $1 - \Phi(y) \leqslant \phi(y)/y$. See, for example, Feller, 1957, Chap. VII, Section 1.]

Note. The surprising results of Problems 6.16–6.18 showing a lower bound and variance which decrease exponentially are due to Hammersley (1950), who shows that, in fact,

$$\operatorname{var}(\bar{X}^*) \sim \sqrt{\frac{8\sigma^2}{\pi n}}\, e^{-n/8\sigma^2} \qquad \text{as} \qquad \frac{n}{\sigma^2} \to \infty.$$

Further results concerning the estimation of restricted parameters and properties of $\bar{X}^*$ are given in Ghosh (1974), Ghosh and Meeden (1978), Khan (1973), and Kojima, Morimoto, and Takeuchi (1982).

6.19 *Kiefer inequality*.

(i) Let X have density (with respect to μ) $p(x, \theta)$ which is > 0 for all x, and let Λ_1, Λ_2 be two distributions on the real line with finite first moments. Then any unbiased estimator δ of θ satisfies

$$\operatorname{var}(\delta) \geqslant \frac{\left[\int \Delta\, d\Lambda_1(\Delta) - \int \Delta\, d\Lambda_2(\Delta)\right]^2}{\int \psi^2(x,\theta)\, p(x,\theta)\, d\mu(x)}$$

where

$$\psi(x,\theta) = \frac{\int_{\Omega_\theta} p(x,\theta+\Delta)\left[d\Lambda_1(\Delta) - d\Lambda_2(\Delta)\right]}{p(x,\theta)}$$

with $\Omega_\theta = \{\Delta: \theta + \Delta \varepsilon \Omega\}$.

(ii) If Λ_1, Λ_2 assign probability 1 to $\Delta = 0$ and Δ, respectively, the inequality reduces to (6.6) with $g(\theta) = \theta$. [*Hint*: Apply (6.1).] (Kiefer, 1952.)

Section 7

7.1 For any random variables $(X_1, \ldots, X_s)$, show that the matrix $\|E(X_i X_j)\|$ is positive semidefinite.

7.2 Show that the r equations obtained by setting the derivatives of (7.8) with respect to a_i equal to zero have the unique solution (7.9).

7.3 Prove (7.16) under the assumptions of the text.

7.4 Verify (i) the information matrices of Table 7.1; (ii) equations (7.20) and (7.21).

7.5 If $p(x) = (1-\varepsilon)\phi(x-\xi) + (\varepsilon/\tau)\phi[(x-\xi)/\tau]$ where ϕ is the standard normal density, find $I(\varepsilon, \xi, \tau)$.

7.6 Verify the expressions (7.25) and (7.26).

7.7 (i) Let $A = \begin{pmatrix} A_{11} & A_{12} \\ A_{21} & A_{22} \end{pmatrix}$ be a partitioned matrix with A_{22} square and nonsingular, and let

$$B = \begin{pmatrix} I & -A_{12}A_{22}^{-1} \\ 0 & I \end{pmatrix}.$$

Show that

$$|A| = |A_{11} - A_{12}A_{22}^{-1}A_{21}| \cdot |A_{22}|.$$

7.8 (i) Let

$$A = \begin{pmatrix} a & b' \\ b & C \end{pmatrix}$$

where a is a scalar and b a column matrix, and suppose that A is positive definite. Show that $|A| \leqslant a|C|$ with equality holding if and only if $b = 0$.

(ii) More generally, if the matrix A of Problem 7.7(i) is positive definite, show that $|A| \leqslant |A_{11}| \cdot |A_{22}|$ with equality holding if and only if $A_{12} = 0$.

[*Hint*: Transform A_{11} and the positive semidefinite $A_{12}A_{22}^{-1}A_{21}$ simultaneously to diagonal form.]

7.9 Prove that (7.31) is necessary for equality in (7.30).

[*Hint*: Problem 7.8(i).]

7.10 Prove the Bhattacharyya inequality (7.33) and show that the condition of equality is as stated.

9. REFERENCES

The concept of unbiasedness as "lack of systematic error" in the estimator was introduced by Gauss (1821) in his work on the theory of least squares. It has continued as a basic assumption in the developments of this theory since then.

The amount of information that a data set contains about a parameter was introduced by Edgeworth (1908, 1909) and was developed more systematically by Fisher (1922 and later papers).* The first version of the information inequality appears to have been given by Fréchet (1943). Early extensions and rediscoveries are due to Darmois (1945), Rao (1945), and Cramér (1946b). The designation "information inequality," which replaced the earlier "Cramér–Rao inequality," was proposed by Savage (1954).

The first UMVU estimators were obtained by Aitken and Silverstone (1942) in the situation in which the information inequality yields the same result (Problem 6.14). UMVU estimators as unique unbiased functions of a suitable sufficient statistic were derived in special cases by Halmos (1946) and Kolmogorov (1950), and were pointed out as a general fact by Rao (1947). An early use of Method 1 for determining such unbiased estimators is due to Tweedie (1947). The concept of completeness was defined, its implications for unbiased estimation developed, and Theorem 1.1 obtained, in Lehmann and Scheffé (1950, 1955, 1956).

Theorem 1.2 has been used to determine UMVU estimators in many special cases. A large but undoubtedly incomplete list of such applications is included in the references below.

Abbey, J. L. and David, H. T.
(1970). "The construction of uniformly minimum variance unbiased estimators for exponential distributions." *Ann. Math. Statist.*, **41**, 1217–1222.

Ahuja, J. C.
(1972). "Recurrence relation for minimum variance unbiased estimation of certain left-truncated Poisson distributions." *J. R. Statist. Soc. (C)* **21**, 81–86.

*For more details, see Savage (1976, p. 456).

Aitken, A. C. and Silverstone, H.
(1942). "On the estimation of statistical parameters." *Proc. Roy. Soc. Edinb.* (*A*) **61**, 186–194.

Bahadur, R. R.
(1957). "On unbiased estimates of uniformly minimum variance." *Sankhȳa*, **18**, 211–224.

Barankin, E. W.
(1949). "Locally best unbiased estimates." *Ann. Math. Statist.* **20**, 477–501.

Barankin, E. W
(1951). "Extension of a theorem of Blackwell." *Ann. Math. Statist.* **21**, 280–284.

Barton, D. E.
(1961). "Unbiased estimation of a set of probabilities." *Biometrika* **48**, 227–229.

Basu, A. P.
(1964). "Estimates of reliability for some distributions useful in life testing." *Technometrics* **6**, 215–219.

Basu, D.
(1955). "A note on the theory of unbiased estimation." *Ann. Math. Statist.* **26**, 144–145.

Bell, C. B., Blackwell, D., and Breiman, L.
(1960). "On the completeness of order statistics." *Ann. Math. Statist.* **31**, 794–797.

Bhattacharyya, A.
(1946, 1948). "On some analogues to the amount of information and their uses in statistical estimation." *Sankhȳa* **8**, 1–14, 201–208, 277–280.

Bhattacharyya, G. K., Johnson, R. A., and Mehrotra, K. G.
(1977). "On the completeness of minimal sufficient statistics with censored observations." *Ann. Statist.*, **5**, 547–553.

Bickel, P. J. and Lehmann, E. L.
(1969). "Unbiased estimation in convex families." *Ann. Math. Statist.* **40**, 1523–1535.

Blackwell, D.
(1947). "Conditional expectation and unbiased sequential estimation." *Ann Math. Statist.* **18**, 105–110.

Blackwell, D. and Girshick, M. A.
(1947). "A lower bound for the variance of some unbiased sequential estimates." *Ann. Math. Statist.* **18**, 277–280.

Blackwell, D. and Girshick M. A.
(1954). *Theory of Games and Statistical Decisions*. Wiley, New York.

Blight, J. N. and Rao, P. V.
(1974). "The convergence of Bhattacharyya bounds." *Biometrika* **61**, 137–142.

Blyth, C. R.
(1974). "Necessary and sufficient conditions for inequalities of Cramér–Rao type." *Ann. Statist.* **2**, 464 – 473.

Blyth, C. R. and Curme, G. L.
(1960). "Estimation of a parameter in a classical occupancy problem." *Biometrika* **47**, 180–185.

Brown, L. D. and Cohen, A.
(1974). "Point and confidence estimation of a common mean and recovery of interblock information." *Ann. Statist.* **2**, 963–976.

Cacoullos, T. and Charalambides, C.
(1974). "MVUE for truncated discrete distributions." In *Progress in Statist.* Ed. Gani **1**, 133–144.

Cacoullos, T. and Charalambides, C.
(1975). "On minimum variance unbiased estimation for truncated binomial and negative binomial distributions." *Ann. Inst. Statist. Math.* **27**, 235–244.

Chapman, D. G. and Robbins, H.
(1951). "Minimum variance estimation without regularity assumptions." *Ann. Math. Statist.* **22**, 581–586.

Charalambides, C.
(1974). "Minimum variance unbiased estimation for a class of left-truncated discrete distributions." *Sankhya* **36** (**A**) 397–418.

Chatterji, S. D.
(1982). A remark on the Cramér–Rao inequality." In *Statistics and Probability: Essays in Honor of C. R. Rao*, Ed. Kallianpur, Krishnaiah, and Ghosh, North Holland, New York, pp. 193–196.

Cohen, A.
(1981). "Inference for marginal means in contingency tables with conditional independence." *J. Amer. Statist. Assoc.* **76**, 895–902.

Cohen, A. and Sackrowitz, H. B. (1974). "On estimating the mean of two normal distributions." *Ann. Statist.* **2**, 1274–1282.

Cramér, H.
(1946a). *Mathematical Methods of Statistics*, Princeton University Press, Princeton.

Cramér, H.
(1946b). "A contribution to the theory of statistical estimation." *Skand. Akt. Tidskr.* **29**, 85–94.

Crow, E. L.
(1977). "Minimum variance unbiased estimators of the ratio of means of two log normal variates and of two gamma variates." *Commun. Statist.* **A6**, 967–975.

Darmois, G.
(1945). "Sur les lois limites de la dispersion de certaines estimations." *Rev. Inst. Int. Statist.* **13**, 9–15.

Davis, R. C.
(1951). "On minimum variance in non-regular estimation." *Ann. Math. Statist.* **22**, 43–57.

De Groot, M. H.
(1959). "Unbiased binomial sequential estimation." *Ann. Math. Statist.* **30**, 80–101.

Downton, F.
(1973). "The estimation of $P(Y < X)$ in the normal case." *Technometrics* **15**, 551–558.

Eaton, M. L. and Morris, C. N.
(1970). "The application of invariance to unbiased estimation." *Ann. Math. Statist.* **41**, 1708–1716.

Edgeworth, F. Y.
(1908, 1909) "On the probable errors of frequency constants." *J. R. Statist. Soc.* **71**, 381–397, 499–512, 651–678; **72**, 81–90.

Ellison, B. E.
(1964). "Two theorems for inference about the normal distribution with applications in acceptance sampling." *J. Amer. Statist. Assoc.* **59**, 89–95.

Fabian, V. and Hannan, J.
(1977). "On the Cramér–Rao inequality." *Ann Statist.* **5**, 197–205.

Feller, W.
(1957). *An Introduction to Probability Theory and Its Applications*, vol. 1 2nd Ed., Wiley, New York.

Fend, A. V.
(1959). "On the attainment of the Cramér-Rao and Bhattacharyya bounds for the variance of an estimate." *Ann. Math. Statist.* **30**, 381–388.

Fisher, N. I.
(1982). Unbiased estimation for some new parametric families of distributions. *Ann. Statist.*, **10**, 603–615.

Fisher, R. A.
(1922). "On the mathematical foundations of theoretical statistics." *Philos. Trans. R. Soc. London, Ser. A* **222**, 309–368.

Fraser, D. A. S.
(1952). "Sufficient statistics and selection depending on the parameter." *Ann. Math. Statist.* **23**, 417–425.

Fraser, D. A. S.
(1953). "Completeness of order statistics." *Canad. J. Math.* **6**, 42–45.

Fraser, D. A. S. and Guttman, I.
(1952). "Bhattacharyya bounds without regularity assumptions." *Ann. Math. Statist.* **23**, 629–632.

Fréchet, M.
(1943). "Sur l'extension de certaines evaluations statistiques de petits echantillons." *Rev. Int. Statist.* **11**, 182–205.

Gart, J. J.
(1959). "An extension of the Cramér-Rao inequality." *Ann. Math. Statist.* **30**, 367–380.

Gauss, C. F.
(1821). *Theoria combinationis observationum erroribus minimis obnoxiae.* [An English translation can be found in Gauss's work (1803–1826).]

Ghosh, M.
(1974). "Admissibility and minimaxity of some maximum likelihood estimators when the parameter space is restricted to integers." *J. R. Statist. Soc.* (*B*) **37**, 264–271.

Ghosh, M. and Meeden, G.
(1978). "Admissibility of the MLE of the normal integer mean." *Sankhya* (*B*) **40**, 1–10.

Ghurye, S. G. and Olkin, I.
(1969), "Unbiased estimation of some multivariate probability densities." *Ann. Math. Statist.* **40**, 1261–1271.

Girshick, M. A., Mosteller, F. and Savage, L. J.
(1946). "Unbiased estimates for certain binomial sampling problems with applications." *Ann. Math. Statist.* **17**, 13–23.

Glasser, G. J.
(1962). "Minimum variance unbiased estimators for Poisson probabilities." *Technometrics* **4**, 409–418.

Gray, H. L., Watkins, T. A. and Schucany, W. R.
(1973). "On the jackknife statistic and its relation to UMVU estimators in the normal case." *Commun. Statist.* **2**, 285–320.

Graybill, F. A. and Deal, R. B.
(1959). "Combining unbiased estimators." *Biometrics* **15**, 543–550.

Gupta, R. C.
(1977). "Minimum variance unbiased estimation in a modified power series distribution and some of its applications." *Commun. Statist.* **A6**, 977–991.

Guttman, I.
(1958). "A note on a series solution of a problem in estimation." *Biometrika* **45**, 565–567.

Halmos, P. R.
(1946). "The theory of unbiased estimation." *Ann. Math. Statist.* **17**, 34–43.

Hammersley, J. M.
(1950). "On estimating restricted parameters." *J. Roy. Statist. Soc.* (*B*), **12**, 192–240.

Hoeffding, W.
(1948). "A class of statistics with asymptotically normal distribution." *Ann. Math. Statist.* **19**, 293–325.

Hoeffding, W.
(1977). Some incomplete and boundedly complete families of distributions. *Ann. Statist.* **5**, 278–291.

Hoeffding, W.
(1982). "Unbiased range-preserving estimators." In Festschrift for Erich Lehmann, Ed. Bickel, Doksum, and Hodges, Wadsworth, Belmont.

Hogg, R. V.
(1960). "On conditional expectations of location statistics." *J. Amer. Statist. Assoc.* **55**, 714–717.

Huber, P. J.
(1981). *Robust Statistics*. Wiley, New York.

Ibragimov, I. A. and Has'minskii, R. Z.
(1981). *Statistical Estimation: Asymptotic Theory*. Springer, New York.

Joshi, V. M.
(1976). "On the attainment of the Cramér-Rao lower bound." *Ann. Statist.* **4**, 998–1002.

Kagan, A. M. and Palamadov, V. P.
(1968). "New results in the theory of estimation and testing hypotheses for problems with nuisance parameters." Supplement to Y. V. Linnik, "Statistical problems with nuisance parameters." *Amer. Math. Soc. Transl. of Math. Monographs*, **20**.

Kendall, M. G. and Stuart, A.
(1969). *The Advanced Theory of Statistics*. 3rd ed. Vol. 1. Hafner, New York.

Khan, R. A.
(1973). "On some properties of Hammersley's estimator of an integer mean." *Ann. Statist.* **1**, 838–850.

Kiefer, J.
(1952). "On minimum variance estimators," *Ann. Math. Statist.* **23**, 627–629.

Klebanov, L. B.
(1972). "'Universal' loss functions and unbiased estimators." *Soviet Math. Dokl.* **13**, 539–541.

Klotz, J.
(1970). "The geometric density with unknown location parameter." *Ann. Math. Statist.* **41**, 1078–1082.

Kojima, Y., Morimoto, H., and Takeuchi, K.
(1982). "Two best unbiased estimators of normal integral mean." In *Statistics and Probabil-*

ity: Essays in Honor of C. R. Rao, Ed. Kallianpur, Krishnaiah, and Ghosh, North Holland, New York, pp. 429–441.

Kolmogorov, A. N.
(1950). "Unbiased estimates." *Izvestia Akad. Nauk SSSR, Ser. Math.* **14**, 303–326. (Amer. Math. Soc. Transl. No. 98).

Kozek, A.
(1976). "Efficiency and Cramér–Rao type inequalities for convex loss functions." *Inst. Math., Polish Acad. Sci.*, Preprint No. 90.

Laurent, A. G.
(1963). "Conditional distribution of order statistics and distribution of the reduced ith order statistic of the exponential model." *Ann. Math. Statist.* **34**, 652–657.

Lehmann, E. L. (1983). "Estimation with inadequate information." To appear in *J. Amer. Statist. Assoc.*

Lehmann, E. L. and Scheffé, H.
(1950, 1955, 1956). "Completeness, similar regions and unbiased estimation." *Sankhyā* **10**, 305–340; **15**, 219–236. (Correction **17**, 250.)

Lehmann, E. L. and Stein, C. (1950). "Completeness in the sequential case." *Ann. Math. Statist.* **21**, 376–385.

Lieberman, G. J. and Resnikoff, G. J.
(1955). "Sampling plans for inspection by variables." *J. Amer. Statist. Assoc.* **50**, 457–516.

Linnik, Yu V.
(1970). "A note on Rao–Cramér and Bhattacharyya inequalities." *Sankhya* **32(A)**, 449–452.

Linnik, Yu V. and Ruhin, A. L.
(1971). "Convex loss functions in the theory of unbiased estimation." *Soviet Math. Dokl.* **12**, 839–842.

Luenberger, D. G.
(1969). *Optimization by Vector Space Methods*. Wiley, New York.

Morimoto, H. and Sibuya, M.
(1967). "Sufficient statistics and unbiased estimation of restricted selection parameters." *Sankhyā(A)* **27**, 15–40.

Nath, G. B.
(1975). "Unbiased estimates of reliability for the truncated gamma distribution." *Scand. Act. J.* **1975**, 181–186.

Neyman, J. and Scott, E. L.
(1960). "Correction for bias introduced by a transformation of variables." *Ann. Math. Statist.*, **31**, 643–655.

Olkin, I. and Pratt, J. W.
(1958). "Unbiased estimation of certain correlation coefficients." *Ann. Math. Statist.* **29**, 201–211.

Padmanabhan, A. R.
(1970). "Some results on minimum variance unbiased estimation." *Sankhyā* **32**, 107–114.

Park, C. J.
(1973). "The power series distribution with unknown truncation parameter." *Ann. Statist.* **1**, 395–399.

Patel, J. K.
(1973). "Complete sufficient statistics and UMVU estimation." *Commun. Statist.* **2**, 327–336.

Patel, S. R.
(1978). "Minimum variance unbiased estimation of multivariate modified power series distribution." *Metrika* **25**, 155–161.

Patel, S. R.
(1979). "Minimum variance unbiased estimators of left truncated multivariate modified power series distribution." *Metrika* **26**, 87–94.

Patel, S. R. and Shah, S. M.
(1978). "Minimum variance unbiased estimation of left truncated multivariate power series distributions." *Metrika*, **25**, 209–218.

Patil, G. P.
(1963). "Minimum variance unbiased estimation and certain problems of additive number theory." *Ann. Math. Statist.* **34**, 1050–1056.

Patil, G. P.
(1965). "On multivariate generalized power series distribution and its application to the multinomial and negative multinomial." In *Classical and Contagious Discrete Distributions*, Ed. G. P. Patil, Pergamon Press, London, pp. 183–194.

Patil, G. P. and Bildikar, S.
(1966). "On minimum variance unbiased estimation for the logarithmic series distribution." *Sankhȳa* (*A*) **28**, 239–250.

Patil, G. P. and Wani, J. K.
(1966). "Minimum variance unbiased estimation of the distribution function admitting a sufficient statistic." *Ann. Inst. Statist. Math.* **18**, 39–47.

Pfanzagl, J.
(1979). "On optimal median unbiased estimators in the presence of nuisance parameters." *Ann. Statist.* **7**, 187–193.

Pitman, E. J. G.
(1979). *Some Basic Theory for Statistical Inference.* Chapman and Hall, London.

Portnoy, S.
(1977). "Asymptotic efficiency of minimum variance unbiased estimators." *Ann. Statist.* **5**, 522–529.

Pugh, E. L.
(1963). "The best estimate of reliability in the exponential case." *Oper. Res.* **11**, 57–61.

Quenouille, M. H.
(1956). "Notes on bias in estimation." *Biometrika* **43**, 353–360.

Rao, C. R.
(1945). "Information and accuracy attainable in the estimation of statistical parameters." *Bull. Calc. Math. Soc.* **37**, 81–91.

Rao, C. R.
(1947). "Minimum variance and the estimation of several parameters." *Proc. Camb. Phil. Soc.* **43**, 280–283.

Rao, C. R.
(1949). "Sufficient statistics and minimum variance estimates." *Proc. Camb. Phil. Soc.* **45**, 213–218.

Rao, J. N. K.
(1980). "Estimating the common mean of possibly different normal populations: A simulation study." *J. Amer. Statist. Assoc.* **75**, 447–453.

Rosenblatt, M.
(1956). "Remark on some nonparametric estimates of a density function." *Ann. Math. Statist.* **27**, 832–837.

Rosenblatt, M.
(1971). "Curve estimates." *Ann. Math. Statist.* **42**, 1815–1842.

Roy, J. and Mitra, S. K.
(1957). "Unbiased minimum variance estimation in a class of discrete distributions." *Sankhyā* **18**, 371–378.

Rubin, D. B. and Weisberg, S.
(1975). "The variance of a linear combination of independent estimators using estimated weights." *Biometrika* **62**, 708–709.

Rutemiller, H. C.
(1966). "Point estimation of reliability of a system comprised of k elements from the same exponential distribution." *J. Amer. Statist. Assoc.* **61**, 1029–1032.

Rutemiller, H. C.
(1967). "Estimation of the probability of zero failures in m binomial trials." *J. Amer. Statist. Assoc.* **62**, 272–277.

Sathe, Y. S. and Varde, S. D.
(1965). "On minimum variance unbiased estimation of reliability." *Ann. Math. Statist.* **40**, 710–714.

Savage, L. J.
(1954, 1972). *The Foundations of Statistics.* Wiley, New York. Rev. ed., Dover Publications.

Savage, L. J.
(1976). "On rereading R. A. Fisher" (with discussion). *Ann. Statist.* **4**, 441–500.

Schaeffer, R. L.
(1976). "On the computation of certain minimum variance unbiased estimators." *Technometrics* **18**, 497–499.

Scheffé, H.
(1943). "On a measure problem arising in the theory of nonparametric tests." *Ann. Math. Statist.* **14**, 227–233.

Seheult, A. H. and Quesenberry, C. P.
(1971). "On unbiased estimation of density functions." *Ann. Math. Statist.* **42**, 1434–1438.

Sen, P. K. and Ghosh, B. K.
(1976). "Comparison of some bounds in estimation theory." *Ann. Statist.* **4**, 755–765.

Seshadri, V.
(1963). "Constructing uniformly better estimators." *J. Amer. Statist. Assoc.* **58**, 172–175.

Seth, G. R.
(1949). "On the variance of estimates." *Ann. Math. Statist.* **20**, 1–27.

Simons, G.
(1980). "Sequential estimators and the Cramér-Rao lower bound." *J. Statist. Planning and Inference* **4**, 67–74.

Stein, C.
(1950). "Unbiased estimates of minimum variance. *Ann. Math. Statist.* **21**, 406–415.

Tate, R. F.
(1959). "Unbiased estimation: Functions of location and scale parameters." *Ann. Math. Statist.* **30**, 341–366.

Tate, R. F. and Goen, R. L.
(1958). "Minimum variance unbiased estimation for a truncated Poisson distribution." *Ann. Math. Statist.* **29**, 755–765.

Tweedie, M. C. K.
(1947). "Functions of a statistical variate with given means, with special reference to Laplacian distributions." *Proc. Camb. Phil. Soc.* **43**, 41–49.

Tweedie, M. C. K.
(1965). "Further results concerning expectation-inversion technique." In *Classical and Contagious Discrete Distributions*, Ed. G. P. Patil, Pergamon Press, London, 195–218.

Unni, K.
(1978). "The theory of estimation in algebraic and analytic exponential families with applications to variance components models." Unpublished Ph.D. dissertation, Indian Statistical Institute, Calcutta.

Varde, S. D.
(1970). "Estimation of reliability of a two exponential component series system." *Technometrics*, **12**, 867–875.

Varde, S. D. and Sathe, Y. S.
(1969). "Minimum variance unbiased estimation of reliability for the truncated exponential distribution." *Technometrics* **11**, 609–612.

Vincze, I.
(1979). "On the Cramér–Fréchet–Rao inequality in the non-regular case." In *Contributions to Statistics*. J. Hajek Memorial Volume (Jureckova, Ed.) Academia. Prague.

Wani, J. K. and Kabe, D. G.
(1971). "Point estimation of reliability of a system comprised of K elements from the same gamma model." *Technometrics* **13**, 859–864.

Washio, Y., Morimoto, H. and Ikeda, N.
(1956). "Unbiased estimation based on sufficient statistics." *Bull. Math. Statist.* **6**, 69–94.

Watkins, T. A. and Kern, D. M.
(1973). "On UMVU estimators related to the multivariate normal distribution." *Commun. Statist.* **2**, 321–326.

Wijsman, R. A.
(1973). "On the attainment of the Cramér–Rao lower bound." *Ann. Statist.* **1**, 538–542.

Wolfowitz, J.
(1946). "On sequential binomial estimation." *Ann. Math. Statist.* **17**, 489–493.

Wolfowitz, J.
(1947). "The efficiency of sequential estimates and Wald's equation for sequential processes." *Ann. Math. Statist.* **18**, 215–230.

Woodward, W. A. and Gray, H. L.
(1975). "Minimum variance unbiased estimation in the gamma distribution." *Commun. Statist.* **4**, 907–922.

Woodward, W. A. and Kelley, G. D.
(1977). "Minimum variance unbiased estimation of $P(Y < X)$ in the normal case." *Technometrics* **19**, 95–98.

Zacks, S.
(1966). "Unbiased estimation of the common mean of two normal distributions based on small samples of equal size." *J. Amer. Statist. Assoc.* **61**, 467–476.

Zacks, S. and Even, M.
(1966). "The efficiencies in small samples of the maximum likelihood and best unbiased estimates of reliability functions." *J. Amer. Statist. Assoc.* **61**, 1033–1051, 1052–1062.

CHAPTER 3

Equivariance

1. LOCATION PARAMETERS

In Section 1.1 the principle of unbiasedness was introduced as an impartiality restriction to eliminate estimators such as $\delta(X) \equiv g(\theta_0)$, which would give very low risk for some parameter values at the expense of very high risk for others. As was seen in Sections 2.2–2.4, in many important situations there exists within the class of unbiased estimators a member that is uniformly better for any convex loss function than any other unbiased estimator.

In the present chapter we shall use symmetry considerations as the basis for another such impartiality restriction with a somewhat different domain of applicability. Rather than immediately giving a general definition of the resulting condition of equivariance, we shall in the present section formulate this condition for the special case of location problems.

Let $\underline{X} = (X_1, \ldots, X_n)$ have a joint distribution with probability density

$$f(\underline{x} - \xi) = f(x_1 - \xi, \ldots, x_n - \xi) \tag{1}$$

where f is known and ξ is an unknown *location parameter*. Suppose that for the problem of estimating ξ with loss function $L(\xi, d)$, we have found a satisfactory estimator $\delta(X)$.

Suppose, now, that another statistician would like to refer the observations to another origin. For example, if the X's are readings on a thermometer calibrated on the Celsius scale, another worker might prefer to express them on the Kelvin scale as $X_i' = X_i + 273$. In general, the modified observations would be of the form

$$X_i' = X_i + a. \tag{2}$$

If the corresponding transformation is applied to ξ,

$$\xi' = \xi + a, \tag{3}$$

the joint density of $\underline{X}' = (X_1', \ldots, X_n')$ can be written as

$$f(\underline{x}' - \xi') = f(x_1' - \xi', \ldots, x_n' - \xi'). \tag{4}$$

The estimated value d, when expressed in the new coordinates, becomes

$$d' = d + a, \tag{5}$$

and the loss resulting from its use is $L(\xi', d')$.

If $(\underline{X}', \xi', d')$ and $(\underline{X}, \xi, d)$ represent the same physical situation expressed in two different coordinate systems, the loss function, which is concerned only with the real situation and not how it is represented, should satisfy $L(\xi', d') = L(\xi, d)$ and hence

$$L(\xi + a, d + a) = L(\xi, d). \tag{6}$$

A loss function L satisfies (6) for all values of a if and only if it depends only on the difference $d - \xi$, that is, it is of the form

$$L(\xi, d) = \rho(d - \xi). \tag{7}$$

That (7) implies (6) is obvious. The converse follows by putting $a = -\xi$ in (6) and letting $\rho(d - \xi) = L(0, d - \xi)$.

When (7) holds, the problem of estimating ξ on the basis of $\underline{X}$ is said to be *invariant* under the transformations (2)–(5), since the problems of estimating ξ on the basis of $\underline{X}$ and ξ' on the basis of $\underline{X}'$ are then formally identical. Suppose, now, that in the unprimed situation we have decided to use $\delta(\underline{X})$ as an estimator of ξ. Then the natural estimator of $\xi + a$ is $\delta(\underline{X}) + a$. On the other hand, in view of the formal identity of the two problems, we should use $\delta(\underline{X}')$ to estimate $\xi' = \xi + a$. It seems desirable that these two estimators should agree so that the result of the estimation does not depend on the arbitrary decision to express the problem in terms of the primed or unprimed variables, and hence that

$$\delta(X_1 + a, \ldots, X_n + a) = \delta(X_1, \ldots, X_n) + a \quad \text{for all } a. \tag{8}$$

An estimator satisfying (8) will be called *equivariant** under the transformations (2), (3), and (5) or *location equivariant*.

*Some authors call such estimators "invariant." Since this suggests that the estimator remains unchanged under (2), it seems preferable to reserve that term for functions satisfying (10).

All the usual estimators of a location parameter are location equivariant. This is the case, for example, for the mean, the median, or any weighted average of the order statistics (with weights adding up to one). The MLE $\hat{\xi}$ is also equivariant since, if $\hat{\xi}$ maximizes $f(x - \xi)$, $\hat{\xi} + a$ maximizes $f(x - \xi - a)$.

The following theorem states an important set of properties of location equivariant estimators.

Theorem 1.1. *Let* $\underline{X}$ *be distributed with density* (1), *and let* δ *be equivariant for estimating* ξ *with loss function* (7). *Then the bias, risk, and variance of* δ *are all constant* (*i.e., do not depend on* ξ).

Proof. Note that if $\underline{X}$ has density $f(\underline{x})$ (i.e., $\xi = 0$), then $\underline{X} + \xi$ has density (1). Thus the bias can be written as

$$b(\xi) = E_\xi[\delta(\underline{X})] - \xi = E_0[\delta(\underline{X} + \xi)] - \xi = E_0[\delta(\underline{X})],$$

which does not depend on ξ.

The proofs for risk and variance are analogous (Problem 1.1).

Theorem 1.1 has an important consequence. Since the risk of any equivariant estimator is independent of ξ, the problem of uniformly minimizing the risk within this class of estimators is replaced by the much simpler problem of determining the equivariant estimator for which this constant risk is smallest. Such an estimator will typically exist and will then be called a *minimum risk equivariant* (MRE) estimator. (There could be a sequence of estimators whose risks decrease to a value not assumed, but as we shall see, this typically does not happen.) To derive an explicit expression for the MRE estimator, let us begin by finding a representation of the most general location equivariant estimator.

Lemma 1.1. *If* δ_0 *is any equivariant estimator, then a necessary and sufficient condition for* δ *to be equivariant is that*

$$\delta(\underline{x}) = \delta_0(\underline{x}) + u(\underline{x}) \tag{9}$$

where $u(\underline{x})$ *is any function satisfying*

$$u(\underline{x} + a) = u(\underline{x}), \qquad \textit{for all} \quad \underline{x}, a. \tag{10}$$

Proof. Assume first that (9) and (10) hold. Then $\delta(\underline{x} + a) = \delta_0(\underline{x} + a) + u(\underline{x} + a) = \delta_0(\underline{x}) + a + u(\underline{x}) = \delta(\underline{x}) + a$, so that δ is equivariant.

Conversely, if δ is equivariant, let

$$u(\underline{x}) = \delta(\underline{x}) - \delta_0(\underline{x}).$$

Then

$$u(\underline{x} + a) = \delta(\underline{x} + a) - \delta_0(\underline{x} + a)$$
$$= \delta(\underline{x}) + a - \delta_0(\underline{x}) - a = u(\underline{x})$$

so that (9) and (10) hold.

To complete the representation, we need a characterization of the functions u satisfying (10).

Lemma 1.2. *A function* u *satisfies* (10) *if and only if it is a function of the differences* $y_i = x_i - x_n$, $i = 1, \ldots, n - 1$; $n \geqslant 2$; *for* $n = 1$ *if and only if it is a constant.*

The proof is essentially the same as that of (7).

Combining Lemmas 1.1 and 1.2 gives the following characterization of equivariant estimators.

Theorem 1.2. *If* δ_0 *is any equivariant estimator, then a necessary and sufficient condition for* δ *to be equivariant is that there exists a function* v *of* $n - 1$ *arguments for which*

$$\delta(\underline{x}) = \delta_0(\underline{x}) - v(\underline{y}) \quad \textit{for all} \quad \underline{x}. \tag{11}$$

Example 1.1. Consider the case $n = 1$. Then it follows from Theorem 1.2 that the only equivariant estimators are $X + c$ for some constant c.

We are now in a position to determine the equivariant estimator with minimum risk.

Theorem 1.3. *Let* $\underline{X} = (X_1, \ldots, X_n)$ *be distributed according to* (1), *let* $Y_i = X_i - X_n$ $(i = 1, \ldots, n - 1)$ *and* $\underline{Y} = (Y_1, \ldots, Y_{n-1})$. *Suppose that the loss function is given by* (7) *and that there exists an equivariant estimator* δ_0 *of* ξ *with finite risk. Assume that for each* $\underline{y}$ *there exists a number* $v(\underline{y}) = v^*(\underline{y})$ *which minimizes*

$$E_0\{\rho[\delta_0(\underline{X}) - v(\underline{y})] | \underline{y}\}. \tag{12}$$

Then a location equivariant estimator δ *of* ξ *with minimum risk exists and is given by*

$$\delta^*(\underline{X}) = \delta_0(\underline{X}) - v^*(\underline{Y}).$$

Proof. By Theorem 1.2, the MRE estimator is found by determining v so as to minimize

$$R_\xi(\delta) = E_\xi\{\rho[\delta_0(\underline{X}) - v(\underline{Y}) - \xi]\}.$$

Since the risk is independent of ξ, it suffices to minimize

$$R_0(\delta) = E_0\{\rho[\delta_0(\underline{X}) - v(\underline{Y})]\}$$

$$= \int E_0\{\rho[\delta_0(\underline{X}) - v(\underline{y})]|\underline{y}\} \, dP_0(\underline{y}).$$

The integral is minimized by minimizing the integrand, and hence (12), for each $\underline{y}$. Since δ_0 has finite risk $E_0\{\rho[\delta_0(\underline{X})]|\underline{y}\} < \infty$ (a.e. P_0), so that the minimization of (12) is meaningful. The result now follows from the assumptions of the theorem.

Corollary 1.1. *Under the assumptions of Theorem* 1.3, *suppose that* ρ *is convex and not monotone. Then an MRE estimator of* ξ *exists*; *it is unique if* ρ *is strictly convex.*

Proof. Theorems 1.3 and 1.6.8.

Corollary 1.2. *Under the assumptions of Theorem* 1.3

(i) *if* $\rho(d - \xi) = (d - \xi)^2$, *then*

$$v^*(\underline{y}) = E_0[\delta_0(\underline{X})|\underline{y}]; \tag{13}$$

(ii) *if* $\rho(d - \xi) = |d - \xi|$, *then* $v^*(\underline{y})$ *is any median of* $\delta_0(\underline{X})$ *under the conditional distribution of* $\underline{X}$ *given* $\underline{y}$.

Proof. Examples 1.6.4 and 1.6.5.

Example 1.1. (*Continued*). For the case $n = 1$, if X has finite risk, the arguments of Theorem 1.3 and Corollary 1.1 show that the MRE estimator is $X - v^*$ where v^* is any value minimizing

$$E_0[\rho(X - v)]. \tag{14}$$

In particular, the MRE estimator is $X - E_0(X)$ and $X - \text{med}_0(X)$ when the loss is squared error and absolute error, respectively.

Suppose, now, that X is symmetrically distributed about ξ. Then for any ρ which is convex and even, it follows from Corollary 1.6.3 that (14) is minimized by $v = 0$, so that X is MRE. Under the same assumptions, if $n = 2$, the MRE estimator is $(X_1 + X_2)/2$. (Problem 1.3).

Existence of MRE estimators is, of course, not restricted to convex loss functions. As an important class of nonconvex loss functions, consider the case that ρ is bounded.

Corollary 1.3. *Under the assumptions of Example* 1.1 (*continued*), *suppose that* $0 \leq \rho(t) \leq M$ *for all values of* t, *that* $\rho(t) \to M$ *as* $t \to \pm\infty$, *and that the density* f *of* X *is continuous a.e. Then an MRE estimator of* ξ *exists.*

For a proof see Problem 1.7.

Example 1.2. Suppose that

$$\rho(d-\xi) = \begin{matrix} 1 & \text{if } |d-\xi| > k \\ 0 & \text{otherwise.} \end{matrix}$$

Then v will minimize (14) provided it maximizes

$$P_0\{|X - v| \leqslant k\}. \tag{15}$$

Consider the case that the density f is symmetric about 0. If f is unimodal, then $v = 0$ and the MRE estimator of ξ is X. On the other hand, suppose that f is U-shaped, say $f(x)$ is zero for $|x| > c > k$ and is strictly increasing for $0 < x < c$. Then there are two values of v maximizing (15), namely, $v = c - k$ and $v = -c + k$, hence, $X - c + k$ and $X + c - k$ are both MRE.

Example 1.3. Normal. Let $X_1, \ldots, X_n$ be iid according to $N(\xi, \sigma^2)$ where σ is known. If $\delta_0 = \bar{X}$ in Theorem 1.3, it follows from Basu's theorem that δ_0 is independent of $\underline{Y}$ and hence that $v(\underline{y}) = v$ is a constant determined by minimizing (14) with $\bar{X}$ in place of X. Thus $\bar{X}$ is MRE for all convex and even ρ. It is also MRE for many nonconvex loss functions including that of Example 1.2.

This example has an interesting implication concerning a "least favorable" property of the normal distribution.

Theorem 1.4. *Let $\mathcal{F}$ be the class of all univariate distributions* F *that have a density* f *(w.r.t. to Lebesgue measure) and fixed finite variance, say $\sigma^2 = 1$. Let* $X_1, \ldots, X_n$ *be iid with density* $f(x_i - \xi)$, $\xi = E(X_i)$, *and let* $r_n(F)$ *be the risk of the MRE estimator of ξ with squared error loss. Then* $r_n(F)$ *takes on its maximum value over $\mathcal{F}$ when* F *is normal.*

Proof. The MRE estimator in the normal case is $\bar{X}$ with risk $E(\bar{X} - \xi)^2 = 1/n$. Since this is the risk of $\bar{X}$, regardless of F, the MRE estimator for any other F must have risk $\leqslant 1/n$, and this completes the proof.

For $n \geqslant 3$ the normal distribution is, in fact, the only one for which $r_n(F) = 1/n$. Since the MRE estimator is unique, this will follow if the normal distribution can be shown to be the only one whose MRE estimator is $\bar{X}$. From Corollary 1.2, it is seen that the MRE estimator is $\bar{X} - E_0[\bar{X}|\underline{Y}]$ and hence is $\bar{X}$ if and only if $E_0[\bar{X}|\underline{Y}] = 0$. It was proved by Kagan, Linnik, and Rao (1965) that this last equation holds if and only if F is normal.

Example 1.4. Exponential. Let $X_1, \ldots, X_n$ be iid according to the exponential distribution $E(\xi, b)$ with b known. If $\delta_0 = X_{(1)}$ in Theorem 1.3, it again follows from Basu's theorem that δ_0 is independent of $\underline{Y}$ and hence that $v(\underline{y}) = v$ is determined by minimizing

$$E_0\left[\rho\left(X_{(1)} - v\right)\right]. \tag{16}$$

(i) If the loss is squared error, the minimizing value is $v = E_0[X_{(1)}] = b/n$, and hence the MRE estimator is $X_{(1)} - (b/n)$.

(ii) If the loss is absolute error, the minimizing value is $v = b(\log 2)/n$ (Problem 1.4).

(iii) If the loss function is that of Example 1.2, then v is the center of the interval I of length $2k$ which maximizes $P_{\xi=0}[X_{(1)} \varepsilon I]$. Since for $\xi = 0$, the density of $X_{(1)}$ is decreasing on $(0, \infty)$, $v = k$ and the MRE estimator is $X_{(1)} - k$.

Example 1.5. Uniform. Let $X_1, \ldots, X_n$ be iid according to the uniform distribution $U(\xi - \frac{1}{2}b, \xi + \frac{1}{2}b)$, with b known, and suppose the loss function ρ is convex and even. For δ_0, take $[X_{(1)} + X_{(n)}]/2$ where $X_{(1)} < \cdots < X_{(n)}$ denote the ordered X's. To find $v(y)$ minimizing (12), consider the conditional distribution of δ_0 given y. This distribution depends on y only through the differences $X_{(i)} - X_{(1)}$, $i = 2, \ldots, n$. By Basu's theorem, the pair $(X_{(1)}, X_{(n)})$ is independent of the ratios $Z_i = [X_{(i)} - X_{(1)}]/[X_{(n)} - X_{(1)}]$, $i = 2, \ldots, n-1$ (Problem 1.5.30(ii)). Therefore, the conditional distribution of δ_0 given the differences $X_{(i)} - X_{(1)}$, which is equivalent to the conditional distribution of δ_0 given $X_{(n)} - X_{(1)}$ and the Z's, depends only on $X_{(n)} - X_{(1)}$. However, the conditional distribution of δ_0 given $V = X_{(n)} - X_{(1)}$ is symmetric about 0 (when $\xi = 0$; Problem 1.2). It follows, therefore, as in Example 1.1 (continued) that the MRE estimator of ξ is $[X_{(1)} + X_{(n)}]/2$, the midrange.

When loss is squared error, the MRE estimator

$$\delta^*(\underline{X}) = \delta_0(\underline{X}) - E[\delta_0(\underline{X})|\underline{Y}] \tag{17}$$

can be evaluated more explicitly.

Theorem 1.5. *Under the assumptions of Theorem* 1.3, *with* $L(\xi, d) = (d - \xi)^2$, *the estimator* (17) *is given by*

$$\delta^*(\underline{x}) = \frac{\int_{-\infty}^{\infty} u f(x_1 - u, \ldots, x_n - u)\, du}{\int_{-\infty}^{\infty} f(x_1 - u, \ldots, x_n - u)\, du} \tag{18}$$

and in this form is known as the Pitman estimator *of* ξ.

Proof. Let $\delta_0(\underline{X}) = X_n$. To compute $E_0(X_n|\underline{y})$ (which exists by Problem 1.20), make the change of variables

$$y_i = x_i - x_n \quad (i = 1, \ldots, n-1); \quad y_n = x_n.$$

The Jacobian of the transformation is 1. The joint density of the Y's is therefore

$$p_Y(y_1, \ldots, y_n) = f(y_1 + y_n, \ldots, y_{n-1} + y_n, y_n),$$

and the conditional density of Y_n given $\underline{y} = (y_1, \ldots, y_{n-1})$ is

$$\frac{f(y_1 + y_n, \ldots, y_{n-1} + y_n, y_n)}{\int f(y_1 + t, \ldots, y_{n-1} + t, t)\, dt}.$$

It follows that

$$E_0[X_n|\underline{y}] = E_0[Y_n|\underline{y}] = \frac{\int tf(y_1 + t, \dots, y_{n-1} + t, t)\, dt}{\int f(y_1 + t, \dots, y_{n-1} + t, t)\, dt}.$$

This can be reexpressed in terms of the x's as

$$E_0[X_n|\underline{y}] = \frac{\int tf(x_1 - x_n + t, \dots, x_{n-1} - x_n + t, t)\, dt}{\int f(x_1 - x_n + t, \dots, x_{n-1} - x_n + t, t)\, dt}$$

or, finally, by making the change of variables $u = x_n - t$ as

$$E_0[X_n|\underline{y}] = x_n - \frac{\int uf(x_1 - u, \dots, x_n - u)\, du}{\int f(x_1 - u, \dots, x_n - u)\, du}.$$

This completes the proof.

Example 1.5. *(Continued).* As an illustration of (18), let us apply it to the situation of Example 1.5. Then

$$f(x_1 - \xi, \dots, x_n - \xi) = \begin{cases} b^{-n} & \text{if } \xi - \frac{b}{2} \leqslant X_{(1)} \leqslant X_{(n)} \leqslant \xi + \frac{b}{2} \\ 0 & \text{otherwise.} \end{cases}$$

where b is known. The Pitman estimator is therefore given by

$$\delta^*(\underline{x}) = \int_{x_{(n)} - b/2}^{x_{(1)} + b/2} u\, du \Big/ \int_{x_{(n)} - b/2}^{x_{(1)} + b/2} du = \frac{1}{2}[x_{(1)} + x_{(n)}]$$

which agrees with the result of Example 1.5.

For most densities, the integrals in (18) are difficult to evaluate. The following example illustrates the MRE estimator for one more case.

Example 1.6. Double exponential. Let $X_1, \dots, X_n$ be iid with double exponential distribution $DE(\xi, 1)$, so that their joint density is $(1/2^n)\exp(-\Sigma|x_i - \xi|)$. It is enough to evaluate the integrals in (18) over the set where $x_1 < \cdots < x_n$. If $x_k < \xi < x_{k+1}$,

$$\begin{aligned} \Sigma|x_i - \xi| &= \sum_{k+1}^{n} (x_i - \xi) - \sum_{1}^{k} (x_i - \xi) \\ &= \sum_{k+1}^{n} x_i - \sum_{1}^{k} x_i + (2k - n)\xi. \end{aligned}$$

The integration then leads to two sums, both in numerator and denominator of the Pitman estimator. The resulting expression is the desired estimator.

So far, the estimator δ has been assumed to be nonrandomized. Let us now consider the role of randomized estimators for equivariant estimation. Recall from the proof of Corollary 1.6.2 that a randomized estimator can be obtained as a nonrandomized estimator $\delta(\underline{X}, W)$ depending on $\underline{X}$ and an independent random variable W with known distribution. For such an estimator the equivariance condition (8) becomes

$$\delta(\underline{X} + a, W) = \delta(\underline{X}, W) + a \qquad \text{for all} \quad a.$$

There is no change in Theorem 1.1, and Lemma 1.1 remains valid with (10) replaced by

$$u(\underline{x} + a, w) = u(\underline{x}, w) \qquad \text{for all} \quad \underline{x}, w, \quad \text{and} \quad a.$$

The proof of Lemma 1.2 shows that this condition holds if and only if u is a function only of y and w, so that finally in generalization of (11), an estimator $\delta(\underline{X}, W)$ is equivariant if and only if it is of the form

$$\delta(\underline{X}, W) = \delta_0(\underline{X}, W) - v(\underline{Y}, W). \tag{19}$$

Applying the proof of Theorem 1.3 to (19), we see that the risk is minimized by choosing for $v(\underline{y}, w)$ the function minimizing

$$E_0\{\rho[\delta_0(\underline{X}, w) - v(\underline{y}, w)] | \underline{y}, w\}.$$

Since the starting δ_0 can be any equivariant estimator, let it be nonrandomized, that is, not dependent on W. Since $\underline{X}$ and W are independent, it then follows that the minimizing $v(\underline{y}, w)$ will not involve w, so that the MRE estimator (if it exists) will be nonrandomized.

Suppose now that T is a sufficient statistic for ξ. Then $\underline{X}$ can be represented as (T, W) where W has a known distribution (see Section 1.5) and any estimator $\delta(\underline{X})$ can be viewed as a randomized estimator based on T. The above argument then suggests that a MRE estimator can always be chosen to depend on T only. However, the argument does not apply since the family $\{P_\xi^T, -\infty < \xi < \infty\}$ need no longer be a location family. Let us therefore add the assumption that $T = (T_1, \ldots, T_r)$ where $T_i = T_i(\underline{X})$ are real-valued and equivariant, that is, satisfy

$$T_i(\underline{x} + a) = T_i(\underline{x}) + a \qquad \text{for all} \quad \underline{x} \quad \text{and} \quad a. \tag{20}$$

Under this assumption, the distributions of T do constitute a location family. To see this, let $\underline{V} = \underline{X} - \xi$ so that $\underline{V}$ is distributed with density $f(v_1, \ldots, v_n)$. Then $T_i(\underline{X}) = T_i(\underline{V} + \xi) = T_i(\underline{V}) + \xi$, and this defines a location family. The earlier argument therefore applies, and under assumption (20), an MRE estimator can be found which depends only on T. (For a general discussion of the relationship of invariance and sufficiency, see Hall, Wijsman, and Ghosh, 1965; Basu, 1969; Berk, 1972; and Landers and Rogge, 1973.)

In Examples 1.3, 1.4, and 1.5, the sufficient statistics $\bar{X}$, $X_{(1)}$, and $(X_{(1)}, X_{(n)})$, respectively, satisfy (20), and the previous remark provides an alternative derivation for the MRE estimators in these examples.

It is interesting to compare the results of the present section with those on unbiased estimation in Chapter 2. It was found there that when a UMVU estimator exists, it typically minimizes the risk for all convex loss functions but that for bounded loss functions not even a locally minimum risk unbiased estimator can be expected to exist. In contrast:

(i) An MRE estimator typically exists not only for convex loss functions but even when the loss function is not so restricted.

(ii) On the other hand, even for convex loss functions the MRE estimator often varies with the loss function.

(iii) Randomized estimators need not be considered in equivariant estimation since there are always uniformly better nonrandomized ones.

(iv) Finally, unlike UMVU estimators which are frequently inadmissible, the Pitman estimator is admissible under mild assumptions (see Stein, 1959, and Section 4.4).

The principal area of application of UMVU estimation is that of exponential families, and these have little overlap with location families (see Section 1.4). Thus, for location families, UMVU estimators typically do not exist. (For specific results in this direction, see Bondesson, 1975.) It is, however, of interest to consider whether MRE estimators are unbiased.

Lemma 1.3. *Let the loss function be squared error.*

(i) *When* $\delta(\underline{X})$ *is any equivariant estimator with constant bias* b, *then* $\delta(\underline{X}) - \mathrm{b}$ *is equivariant, unbiased, and has smaller risk than* $\delta(\underline{X})$.

(ii) *The unique MRE estimator is unbiased.*

(iii) *If a UMVU estimator exists and is equivariant, it is MRE.*

Proof. Part (i) follows from Lemma 2.5.1; (ii) and (iii) are immediate consequences of (i).

That an MRE estimator need not be unbiased for general loss functions is seen from Example 1.4 with absolute error as loss. Some light is thrown on the possible failure of MRE estimators to be unbiased by considering the following decision theoretic definition of unbiasedness, which depends on the loss function L. An estimator δ of $g(\theta)$ is said to be *risk-unbiased* if it satisfies

$$(21) \qquad E_\theta L[\theta, \delta(\underline{X})] \leqslant E_\theta L[\theta', \delta(\underline{X})] \qquad \text{for all} \quad \theta' \neq \theta.$$

If one interprets $L(\theta, d)$ as measuring how far the estimated value d is from the estimand $g(\theta)$, then (21) states that on the average δ is at least as close to the true value $g(\theta)$ as it is to any false value $g(\theta')$.

Example 1.7. Mean-unbiasedness. If the loss function is squared error, (21) becomes

$$(22) \qquad E_\theta[\delta(X) - g(\theta')]^2 \geqslant E_\theta[\delta(X) - g(\theta)]^2 \qquad \text{for all} \quad \theta' \neq \theta.$$

Suppose that $E_\theta(\delta^2) < \infty$ and that $E_\theta(\delta) \in \Omega_g$ for all θ, where $\Omega_g = \{g(\theta): \theta \in \Omega\}$. [The latter condition is, of course, automatically satisfied when $\Omega = (-\infty, \infty)$ and $g(\theta) = \theta$, as is the case when θ is a location parameter.] Then the left side of (22) is minimized by $g(\theta') = E_\theta \delta(X)$ (Example 1.6.4) and the condition of risk-unbiasedness, therefore, reduces to the usual unbiasedness condition

$$(23) \qquad E_\theta \delta(X) = g(\theta).$$

Example 1.8. Median-unbiasedness. If the loss function is absolute error, (21) becomes

$$(24) \qquad E_\theta|\delta(X) - g(\theta')| \geqslant E_\theta|\delta(X) - g(\theta)| \qquad \text{for all} \quad \theta' \neq \theta.$$

By Example 1.6.5, the left side of (24) is minimized by any median of $\delta(X)$. It follows that (24) reduces to the condition

$$(25) \qquad \text{med}_\theta \delta(X) = g(\theta),$$

that is, $g(\theta)$ is a median of $\delta(X)$, provided $E_\theta|\delta| < \infty$ and Ω_g contains a median of $\delta(X)$ for all θ. An estimator δ satisfying (25) is called *median-unbiased*.

Theorem 1.6. *If δ is MRE for estimating ξ in model* (1) *with loss function* (7), *then it is risk-unbiased.*

Proof. Condition (21) now becomes

$$E_\xi \rho[\delta(\underline{X}) - \xi'] \geqslant E_\xi \rho[\delta(\underline{X}) - \xi] \qquad \text{for all} \quad \xi' \neq \xi,$$

or, if without loss of generality we put $\xi = 0$,

$$E_0 \rho[\delta(\underline{X}) - a] \geqslant E_0 \rho[\delta(\underline{X})] \qquad \text{for all} \quad a.$$

That this holds, is an immediate consequence of the fact that $\delta(\underline{X}) = \delta_0(\underline{X}) - v^*(\underline{Y})$ where $v^*(\underline{y})$ minimizes (12).

2. THE PRINCIPLE OF EQUIVARIANCE

The key to the results of the preceding section is equation (1.4) which expresses the invariance of the probability model under the transformations (1.2) and (1.3). In the present section, we shall consider the general situation in which the probability model remains invariant under a suitable group of transformations.

Let X be a random observable taking on values in a sample space $\mathcal{X}$ according to a probability distribution from the family

$$\mathcal{P} = \{P_\theta, \theta \in \Omega\}. \tag{1}$$

We shall consider 1 : 1 transformations g of the sample space onto itself and suppose that for each θ the distribution of $X' = gX$ is again a member of $\mathcal{P}$, say $P_{\theta'}$, and that as θ traverses Ω, so does θ'. We shall then say that g leaves model (1) invariant.

Let $\mathcal{C}$ be a class of transformations that leaves model (1) invariant, and let $G = G(\mathcal{C})$ be the set of all compositions (defined in Section 1.3) of a finite number of transformations $g_1^{\pm 1} \cdots g_m^{\pm 1}$ with $g_1, \ldots, g_m \in \mathcal{C}$, where each of the exponents can be $+1$ or -1 and where the elements $g_1, \ldots, g_m$ need not be distinct. Then any element $g \in G$ leaves (1) invariant, and G is a group (Problem 2.1), the group *generated* by $\mathcal{C}$. A class of transformations leaving a model invariant can therefore always without loss of generality be assumed to be a group.

Example 2.1. Location family.

(i) Consider the location family (1.1) and the group of transformations $\underline{X}' = \underline{X} + a$, which was already discussed in (1.2) and Example 1.3.1. It is seen from (1.4) that if $\underline{X}$ is distributed according to (1.1) with $\theta = \xi$, then $\underline{X}' = \underline{X} + a$ has the density (1.1) with $\theta' = \xi' = \xi + a$, so that the model (1.1) is preserved under these transformations.

(ii) Suppose now that, in addition, f has the symmetry property

$$f(-\underline{x}) = f(\underline{x}) \tag{2}$$

where $-\underline{x} = (-x_1, \ldots, -x_n)$, and consider the transformation $\underline{x}' = -\underline{x}$. The density of $\underline{X}'$ is

$$f(-x_1' - \xi, \ldots, -x_n' - \xi) = f(x_1' - \xi', \ldots, x_n' - \xi')$$

if $\xi' = -\xi$. Thus, model (1.1) is invariant under the transformations $\underline{x}' = -\underline{x}$,

$\xi' = -\xi$, and hence under the group consisting of this transformation and the identity (Problem 2.2). This is not true, however, if f does not satisfy (2). If, for example, $X_1, \ldots, X_n$ are iid according to the exponential distribution $E(\xi, 1)$, then the variables $-X_1, \ldots, -X_n$ no longer have an exponential distribution.

Let $\{gX, g \in G\}$ be a group of transformations of the sample space which leave the model invariant. If gX has the distribution $P_{\theta'}$, then $\theta' = \bar{g}\theta$ is a function which maps Ω onto Ω, and the transformation $\bar{g}\theta$ is 1 : 1 provided the distributions P_θ, $\theta \in \Omega$ are distinct (Problem 2.3). It is easy to see that the transformations $\bar{g}$ then also form a group that will be denoted by $\bar{G}$ (Problem 2.4). From the definition of $\bar{g}\theta$ it follows that

$$P_\theta(gX \in A) = P_{\bar{g}\theta}(X \in A) \tag{3}$$

where the subscript on the left side indicates the distribution of X, not that of gX. More generally, for a function ψ whose expectation is defined,

$$E_\theta[\psi(gX)] = E_{\bar{g}\theta}[\psi(X)]. \tag{4}$$

We have now generalized the transformations (1.2) and (1.3), and it remains to consider (1.5). This last generalization is most easily introduced by an example.

Example 2.2. Two-sample location family. Let $\underline{X} = (X_1, \ldots, X_m)$ and $\underline{Y} = (Y_1, \ldots, Y_n)$ and suppose that $(\underline{X}, \underline{Y})$ has the joint density

$$f(\underline{x} - \xi, \underline{y} - \eta) = f(x_1 - \xi, \ldots, x_m - \xi, y_1 - \eta, \ldots, y_n - \eta). \tag{5}$$

This model remains invariant under the transformations

$$g(\underline{x}, \underline{y}) = (\underline{x} + a, \underline{y} + b), \qquad \bar{g}(\xi, \eta) = (\xi + a, \eta + b). \tag{6}$$

Consider the problem of estimating

$$\Delta = \eta - \xi. \tag{7}$$

If the transformed variables are denoted by

$$\underline{x}' = \underline{x} + a, \qquad \underline{y}' = \underline{y} + b, \qquad \xi' = \xi + a, \qquad \eta' = \eta + b,$$

then Δ is transformed into $\Delta' = \Delta + (b - a)$. Hence an estimated value d, when expressed in the new coordinates, becomes

$$d' = d + (b - a). \tag{8}$$

For the problem to remain invariant, we require, analogously to (1.6), that the loss function $L(\xi, \eta; d)$ satisfies

$$L[\xi + a, \eta + b; d + (b - a)] = L(\xi, \eta; d). \tag{9}$$

It is easy to see (Problem 2.5) that this is the case if and only if L depends only on the difference $(\eta - \xi) - d$, that is, if

$$L(\xi, \eta; d) = \rho(\Delta - d). \tag{10}$$

Suppose in Example 2.2 that instead of estimating $\eta - \xi$, the problem is that of estimating

$$h(\xi, \eta) = \xi^2 + \eta^2.$$

Under the transformations (6), $h(\xi, \eta)$ is transformed into $(\xi + a)^2 + (\eta + b)^2$. This does not lead to an analog of (8) since the transformed value does not depend on (ξ, η) only through $h(\xi, \eta)$.

Now, consider the general problem of estimating $h(\theta)$ in model (1), which is assumed to be invariant under the transformations $X' = gX$, $\theta' = \bar{g}\theta$, $g \in G$. The additional assumption required is that for any given $\bar{g}$, $h(\bar{g}\theta)$ depends on θ only through $h(\theta)$, that is,

$$h(\theta_1) = h(\theta_2) \quad \text{implies} \quad h(\bar{g}\theta_1) = h(\bar{g}\theta_2). \tag{11}$$

The common value of $h(\bar{g}\theta)$ for all θ's to which h assigns the same value will then be denoted by

$$h(\bar{g}\theta) = g^*h(\theta). \tag{12}$$

If $\mathcal{H}$ is the set of values taken on by $h(\theta)$ as θ ranges over Ω, the transformations g^* are 1 : 1 from $\mathcal{H}$ onto itself. [Problem 2.8(i)]. As $\bar{g}$ ranges over $\bar{G}$, the transformations g^* form a group G^* (Problem 2.6).

The estimated value d of $h(\theta)$ when expressed in the new coordinates becomes

$$d' = g^*d. \tag{13}$$

Since the problems of estimating $h(\theta)$ in terms of (X, θ, d) or $h(\theta')$ in terms of (X', θ', d') represent the same physical situation expressed in a new coordinate system, the loss function should satisfy $L(\theta', d') = L(\theta, d)$ and hence

$$L(\bar{g}\theta, g^*d) = L(\theta, d). \tag{14}$$

In this discussion, it was tacitly assumed that the set $\mathcal{D}$ of possible decisions coincides with $\mathcal{H}$. This need not, however, be the case. In Chapter 2, for example, estimators of a variance were permitted to take on negative values. In the more general case that $\mathcal{H}$ is a subset of $\mathcal{D}$, one can take the condition that (14) holds for all θ as the definition of g^*d. If $L(\theta, d) = L(\theta, d')$ for all θ implies $d = d'$, as is typically the case, g^*d is uniquely defined by the above condition and g^* and 1 : 1 from $\mathcal{D}$ onto itself [Problem 2.8(ii)].

When (11) and (14) hold, we shall say that the problem of estimating $h(\theta)$ on the basis of model (1) remains invariant under the transformation g. When this is the case and δ is the estimator we should like to use to estimate $h(\theta)$, there are two natural ways of estimating $g^*h(\theta)$, the estimand $h(\theta)$ expressed in the new system. Since X is replaced by $X' = gX$, we would want to use $\delta(X') = \delta(gX)$ in the primed problem. On the other hand, if d is the estimated value of $h(\theta)$, then g^*d should be the estimated value of $g^*h(\theta)$, and this requires the transformed estimator to be $g^*\delta(X)$. An estimator is said to be *equivariant* under g if these two agree, that is, if

$$\delta(gX) = g^*\delta(X) \tag{15}$$

Example 2.2. (*Continued*). In Example 2.2, $h(\xi, \eta) = \eta - \xi$ and by (8), $g^*d = d + (b - a)$. It follows that (15) becomes

$$\delta(\underline{x} + a, \underline{y} + b) = \delta(\underline{x}, \underline{y}) + b - a. \tag{16}$$

If $\delta_0(\underline{X})$ and $\delta_0'(\underline{Y})$ are location equivariant estimators of ξ and η, respectively, then $\delta(\underline{X}, \underline{Y}) = \delta_0'(\underline{Y}) - \delta_0(\underline{X})$ is an equivariant estimator of $\eta - \xi$.

The following theorem generalizes Theorem 1.1 to the present situation.

Theorem 2.1. *If δ is an equivariant estimator in a problem which is invariant under a transformation* g, *then the risk function of δ satisfies*

$$\mathrm{R}(\bar{g}\theta, \delta) = \mathrm{R}(\theta, \delta) \qquad \textit{for all} \quad \theta. \tag{17}$$

Proof. By definition

$$R(\bar{g}\theta, \delta) = E_{\bar{g}\theta} L[\bar{g}\theta, \delta(X)].$$

It follows from (4) that the right side is equal to

$$E_\theta L(\bar{g}\theta, \delta(gX)) = E_\theta L[\bar{g}\theta, g^*\delta(X)] = R(\theta, \delta).$$

Looking back on Section 3.1, we see that the crucial fact underlying the success of the invariance approach was the constancy of the risk function of any equivariant estimator. Theorem 2.1 suggests the following simple condition for this property to obtain.

A group G of transformations of a space is said to be *transitive* if for any two points there is a transformation in G taking the first point into the second.

Corollary 2.1. *Under the assumptions of Theorem* 2.1, *if* $\bar{G}$ *is transitive over the parameter space* Ω, *then the risk function of any equivariant estimator is constant, that is, independent of* θ.

When the risk function of every equivariant estimator is constant, the best equivariant estimator (MRE) is obtained by minimizing that constant, so that a uniformly minimum risk equivariant estimator will then typically exist.

Example 2.2. *(concluded).* In this example, $\theta = (\xi, \eta)$ and $\bar{g}\theta = (\xi + a, \eta + b)$. This group of transformations is transitive over Ω since, given any two points (ξ, η) and (ξ', η'), a and b exist such that $\xi + a = \xi'$, $\eta + b = \eta'$. The MRE estimator can now be obtained in exact analogy to Section 3.1 (Problems 1.12, 1.13).

The estimation problem treated in Section 3.1 was greatly simplified by the fact that it was possible to dispense with randomized estimators. The corresponding result holds quite generally when $\bar{G}$ is transitive. If an estimator δ exists which is MRE among all nonrandomized estimators, it is then also MRE when randomization is permitted. To see this, note that a randomized estimator can be represented as $\delta'(X, W)$ where W is independent of X and has a known distribution and that it is equivariant if $\delta'(gX, W) = g^*\delta'(X, W)$. Its risk is again constant and for any $\theta = \theta_0$ is equal to $E[h(W)]$ where

$$h(w) = E_{\theta_0}\{L[\delta'(X, w), \theta_0]\}.$$

This risk is minimized by minimizing $h(w)$ for each w. However, by assumption $\delta'(X, w) = \delta(X)$ minimizes $h(w)$, and hence the MRE estimator can be chosen to be nonrandomized.

The corresponding result need not hold when $\bar{G}$ is not transitive. A counterexample is given in Example 4.2.2.

In Section 1.3, *group families* were introduced as families of distributions generated by subjecting a random variable with a fixed distribution to a group of transformations. Consider now a family of distributions $\mathcal{P} = \{P_\theta, \theta \in \Omega\}$ which remains invariant under a group G for which $\bar{G}$ is transitive over Ω and $g_1 \neq g_2$ implies $\bar{g}_1 \neq \bar{g}_2$. Let θ_0 be any fixed element of Ω. Then $\mathcal{P}$ is exactly the group family of distributions of $\{gX, g \in G\}$ when X has distribution P_{θ_0}.

Conversely, let $\mathcal{P}$ be the group family of the distributions of gX as g varies over G, when X has a fixed distribution P, so that $\mathcal{P} = \{P_g, g \in G\}$. Then g can serve as the parameter θ and G as the parameter space. In this notation, the starting distribution P becomes P_e where e is the identity transformation. Thus a family of distributions remains invariant under a transitive group of transformations of the sample space if and only if it is a group family.

When an estimation problem is invariant under a group of transformations and an MRE estimator exists, this seems the natural estimator to use—of the various principles we shall consider, equivariance, where it applies, is perhaps the most convincing. Yet even this principle can run into difficulties. The following example illustrates the possibility of a problem remaining invariant under two different groups, G_1 and G_2, which lead to two different MRE estimators δ_1 and δ_2.

Example 2.3. Counterexample. Let the pairs (X_1, X_2) and (Y_1, Y_2) be independent, each with a bivariate normal distribution with mean zero. Let their covariance matrices be $\Sigma = [\sigma_{ij}]$ and $\Delta\Sigma = [\Delta\sigma_{ij}]$, $\Delta > 0$, and consider the problem of estimating Δ.

Let G_1 be the group of transformations

$$(18)\qquad \begin{array}{l|l} X_1' = a_1X_1 + a_2X_2 & Y_1' = c(a_1Y_1 + a_2Y_2) \\ X_2' = bX_2 & Y_2' = cbY_2 \end{array}$$

Then (X_1', X_2'), (Y_1', Y_2') will again be independent and each have a bivariate normal distribution with zero mean. If the covariance matrix of (X_1', X_2') is Σ', that of (Y_1', Y_2') is $\Delta'\Sigma'$ where $\Delta' = c^2\Delta$ (Problem 2.9). Thus G_1 leaves the model invariant.

If $h(\Sigma, \Delta) = \Delta$, (11) clearly holds, (12) and (13) become

$$(19)\qquad \Delta' = c^2\Delta, \qquad d' = c^2 d,$$

and a loss function $L(\Delta, d)$ satisfies (14) provided $L(c^2\Delta, c^2 d) = L(\Delta, d)$. This condition holds if and only if L is of the form

$$(20)\qquad L(\Delta, d) = \rho(d/\Delta).$$

[For the necessity of (20), see Problem 2.9.]

An estimator δ of Δ is equivariant under the above transformation if

$$(21)\qquad \delta(\underline{x}', \underline{y}') = c^2\delta(\underline{x}, \underline{y}).$$

We shall now show that (21) holds if and only if

$$(22)\qquad \delta(\underline{x}, \underline{y}) = \frac{ky_2^2}{x_2^2} \quad \text{for some value of } k \text{ a.e.}$$

It is enough to prove this for the reduced sample space in which the matrix $\begin{pmatrix} x_1 x_2 \\ y_1 y_2 \end{pmatrix}$ is nonsingular and in which both x_2 and y_2 are $\neq 0$, since the rest of the sample space has probability zero.

Let G_1' be the subgroup of G_1 consisting of the transformations (18) with $b = c = 1$. The condition of equivariance under these transformations reduces to

$$\delta(\underline{x}', \underline{y}') = \delta(\underline{x}, \underline{y}). \tag{23}$$

This is satisfied whenever δ depends only on x_2, y_2 since $x_2' = x_2, y_2' = y_2$. To see that this condition is also necessary for (23), suppose that δ satisfies (23) and let $(x_1', x_2; y_1', y_2)$ and $(x_1, x_2; y_1, y_2)$ be any two points in the reduced sample space which have the same second coordinates. Then there exist a_1, a_2 such that

$$x_1' = a_1 x_1 + a_2 x_2; y_1' = a_1 y_1 + a_2 y_2,$$

that is, there exists $g \in G_1'$ for which $g(\underline{x}, \underline{y}) = (\underline{x}', \underline{y}')$, and hence δ depends only on x_2, y_2.

Consider now any $\delta'(x_2, y_2)$. To be equivariant under the full group G_1, δ' must satisfy

$$\delta'(bx_2, cby_2) = c^2\delta'(x_2, y_2). \tag{24}$$

For $x_2 = y_2 = 1$ this condition becomes

$$\delta'(b, cb) = c^2\delta'(1, 1)$$

and hence reduces to (22) with $x_2 = b$, $y_2 = bc$ and $k = \delta'(1, 1)$. This shows that (22) is necessary for δ to be equivariant; that it is sufficient is obvious.

The best equivariant estimator under G_1 is thus $k^* Y_2^2/X_2^2$ where k^* is a value which minimizes

$$E_\Delta \rho\left(\frac{kY_2^2}{\Delta X_2^2}\right) = E_1 \rho\left(\frac{kY_2^2}{X_2^2}\right).$$

Such a minimizing value will typically exist. Suppose, for example, that the loss is 1 if $|d - \Delta|/\Delta > \frac{1}{2}$ and zero otherwise. Then k^* is obtained by maximizing

$$P_1\left(\left|k\frac{Y_2^2}{X_2^2} - 1\right| < \frac{1}{2}\right) = P_1\left(\frac{1}{2k} < \frac{Y_2^2}{X_2^2} < \frac{3}{2k}\right).$$

As $k \to 0$ or ∞, this probability tends to zero, and a maximizing value therefore exists and can be determined from the distribution of Y_2^2/X_2^2 when $\Delta = 1$.

Exactly the same argument applies if G_1 is replaced by the transformations G_2

$$\begin{array}{l|l} X_1' = bX_1 & Y_1' = cbY_1 \\ X_2' = a_1X_1 + a_2X_2 & Y_2' = c(a_1Y_1 + a_2Y_2) \end{array}$$

and leads to the MRE estimator $k^*Y_1^2/X_1^2$.

In the location case, it turned out (Theorem 1.6) that an MRE estimator is always risk-unbiased. The extension of this result to the general case requires some assumptions.

Theorem 2.2. *If* $\bar{G}$ *is transitive and* G^* *commutative, then an MRE estimator is risk-unbiased.*

Proof. Let δ be MRE and $\theta, \theta' \in \Omega$. Then, by the transitivity of $\bar{G}$, there exist $\bar{g} \in \bar{G}$ such that $\theta = \bar{g}\theta'$, and hence

$$E_\theta L[\theta', \delta(X)] = E_\theta L[\bar{g}^{-1}\theta, \delta(X)] = E_\theta L[\theta, g^*\delta(X)]$$

Now, if $\delta(X)$ is equivariant, so is $g^*\delta(X)$ (Problem 2.11), and therefore, since δ is MRE,

$$E_\theta L[\theta, g^*\delta(X)] \geqslant E_\theta L[\theta, \delta(X)],$$

which completes the proof.

Transitivity of $\bar{G}$ will usually [but not always, see Example 2.4(i) below] hold when an MRE estimator exists. On the other hand, commutativity of G^* imposes a severe restriction. That the theorem need not be valid if either condition fails is shown by the following example.

Example 2.4. Counterexample. Let X be $N(\xi, \sigma^2)$ with both parameters unknown, let the estimand be ξ and the loss function

$$L(\xi, \sigma; d) = (d - \xi)^2/\sigma^2. \tag{25}$$

(i) The problem remains invariant under the group G_1: $gx = x + c$. It follows from Section 3.1 that X is MRE under G_1. However, X is not risk-unbiased (Problem 2.12). Here $\bar{G}_1$ is the group of transformations

$$\bar{g}(\xi, \sigma) = (\xi + c, \sigma),$$

which is clearly not transitive.

If the loss function is replaced by $(d - \xi)^2$, the problem will remain invariant under G_1; X remains equivariant but is now risk-unbiased by Example 1.7. Transitivity of $\bar{G}$ is thus not necessary for the conclusion of Theorem 2.2.

(ii) When the loss function is given by (25), the problem also remains invariant under the larger group G_2: $ax + c$, $0 < a$. Since X is equivariant under G_2 and MRE under G_1, it is also MRE under G_2. However, as stated in (i), X is not risk-unbiased with respect to (25). Here G_2^* is the group of transformations $g^*d = ad + c$, and this is not commutative (Problem 2.12).

The location problem considered in Section 3.1 provides an important example in which the assumptions of Theorem 2.2 are satisfied, and Theorem 1.6 is the specialization of Theorem 2.2 to that case. The scale problem, which will be considered in Section 3.3, provides another illustration.

We shall not attempt to generalize to the present setting the characterization of equivariant estimators which was obtained for the location case in Theorem 1.2. Some results in this direction, taking account also of the associated measurability problems, can be found in Berk (1967). Instead, we shall consider in the next section some more special extensions of the problem treated in Section 3.1.

3. LOCATION-SCALE FAMILIES

The location model discussed in Section 3.1 provides a good introduction to the ideas of equivariance, but it is rarely realistic. Even when it is reasonable to assume the form of the density f in (1.1) to be known, it is usually desirable to allow the model to contain an unknown scale parameter. The standard normal model according to which $X_1, \dots, X_n$ are iid as $N(\xi, \sigma^2)$ is the most common example of such a location-scale model. As preparation for the analysis of these models, we begin with the case, which is of interest also in its own right, in which the only unknown parameter is a scale parameter.

Let $\underline{X} = (X_1, \dots, X_n)$ have a joint probability density

$$\frac{1}{\tau^n} f\left(\frac{\underline{x}}{\tau}\right) = \frac{1}{\tau^n} f\left(\frac{x_1}{\tau}, \dots, \frac{x_n}{\tau}\right), \qquad \tau > 0 \tag{1}$$

where f is known and τ is an unknown *scale-parameter*. This model remains invariant under the transformations

$$X_i' = bX_i, \qquad \tau' = b\tau \qquad \text{for} \quad b > 0. \tag{2}$$

The estimand of primary interest is $h(\tau) = \tau^r$. Since h is strictly monotone, (2.11) is vacuously satisfied. Transformations (2) induce the transformations

$$h(\tau) \to b^r\tau^r = b^r h(\tau) \quad \text{and} \quad d' = b^r d, \tag{3}$$

and the loss function L is invariant under these transformations provided

$$L(b\tau, b^r d) = L(\tau, d). \tag{4}$$

This is the case if and only if it is of the form (Problem 3.1)

$$L(\tau, d) = \gamma\left(\frac{d}{\tau^r}\right). \tag{5}$$

Examples are

$$L(\tau, d) = \frac{(d - \tau^r)^2}{\tau^{2r}} \quad \text{and} \quad L(\tau, d) = \frac{|d - \tau^r|}{\tau^r} \tag{6}$$

but not squared error.

An estimator δ of τ^r is equivariant under (2), or *scale-equivariant* provided

$$\delta(b\underline{X}) = b^r \delta(\underline{X}). \tag{7}$$

All the usual estimators of τ are scale-equivariant; for example, the standard deviation $\sqrt{\Sigma(X_i - \bar{X})^2/(n-1)}$, the mean deviation $\Sigma|X_i - \bar{X}|/n$, the range, and the maximum likelihood estimator [Problem 3.1(ii)].

Since the group $\bar{G}$ of transformations $\tau' = b\tau$, $b > 0$, is transitive over Ω, the risk of any equivariant estimator is constant by Corollary 2.1, so that one can expect an MRE estimator to exist. To derive it, we first characterize the totality of equivariant estimators.

Theorem 3.1. *Let* $\underline{X}$ *have density* (1) *and let* $\delta_0(\underline{X})$ *be any scale-equivariant estimator of* τ^r. *Then, if*

$$z_i = \frac{x_i}{x_n} \quad (i = 1, \ldots, n-1) \quad \text{and} \quad z_n = \frac{x_n}{|x_n|} \tag{8}$$

and if $\underline{z} = (z_1, \ldots, z_n)$, *a necessary and sufficient condition for* δ *to satisfy* (7) *is that there exists a function* $w(\underline{z})$ *such that*

$$\delta(\underline{x}) = \frac{\delta_0(\underline{x})}{w(\underline{z})}.$$

Proof. Analogously to Lemma 1.1, a necessary and sufficient condition for δ to satisfy (7) is that it is of the form $\delta(\underline{x}) = \delta_0(\underline{x})/u(\underline{x})$ where

(Problem 3.4)

$$u(b\underline{x}) = u(\underline{x}) \qquad \text{for all } \underline{x} \text{ and all } b > 0. \tag{9}$$

It remains to show that (9) holds if and only if u depends on $\underline{x}$ only through $\underline{z}$. Note here that $\underline{z}$ is defined when $x_n \neq 0$ and hence with probability 1. That any function of $\underline{z}$ satisfies (9) is obvious. Conversely, if (9) holds, then

$$u(x_1, \ldots, x_n) = u\left(\frac{x_1}{x_n}, \ldots, \frac{x_{n-1}}{x_n}, \frac{x_n}{|x_n|}\right),$$

and hence u does depend only on $\underline{z}$, as was to be proved.

Example 3.1. Suppose that $n = 1$. Then the most general estimator satisfying (7) is of the form $X^r/w(Z)$ where $Z = X/|X|$ is ± 1 as $X \gtrless 0$ so that

$$\delta(X) = \begin{matrix} AX^r & \text{if} \quad X > 0 \\ BX^r & \text{if} \quad X < 0 \end{matrix},$$

A, B being two arbitrary constants.

Let us now determine the MRE estimator.

Theorem 3.2.

(i) *Let* $\underline{\mathrm{X}}$ *be distributed according to* (1) *and let* $\underline{\mathrm{Z}}$ *be given by* (8). *Suppose that the loss function is given by* (5) *and that there exists an equivariant estimator* δ_0 *of* τ^r *with finite risk. Assume that for each* $\underline{\mathrm{z}}$ *there exists a number* $\mathrm{w}(\underline{\mathrm{z}}) = \mathrm{w}^*(\underline{\mathrm{z}})$ *which minimizes*

$$\mathrm{E}_1\{\gamma[\delta_0(\underline{\mathrm{X}})/\mathrm{w}(\underline{\mathrm{z}})]|\underline{\mathrm{z}}\}. \tag{10}$$

Then an MRE estimator δ^* *of* τ^r *exists and is given by*

$$\delta^*(\underline{\mathrm{X}}) = \frac{\delta_0(\underline{\mathrm{X}})}{\mathrm{w}^*(\underline{\mathrm{Z}})}. \tag{11}$$

The proof parallels that of Theorem 1.3.

Corollary 3.1. *Under the assumptions of Theorem* 3.2, *suppose that* $\rho(\mathrm{v}) = \gamma(\mathrm{e}^{\mathrm{v}})$ *is convex and not monotone. Then an MRE estimator of* τ^r *exists; it is unique if* ρ *is strictly convex.*

Proof. By replacing $\gamma(w)$ by $\rho(\log w)$ [with $\rho(-\infty) = \gamma(0)$] the result essentially reduces to that of Corollary 1.1. This argument requires that $\delta \geqslant 0$, which can be assumed without loss of generality (Problem 3.2).

Example 3.2. Consider the loss function

$$L(\tau, d) = \frac{|d - \tau^r|^p}{\tau^{pr}} = \left|\frac{d}{\tau^r} - 1\right|^p = \gamma\left(\frac{d}{\tau^r}\right) \tag{12}$$

with $\gamma(v) = |v - 1|^p$. Then ρ is strictly convex for $v > 0$ provided $p \geqslant 1$ (Problem 3.5).

Lemma 3.1. *Let* X *be a positive random variable.*

(i) *If* $E(X^2) < \infty$, *then*

$$E\left(\frac{X}{c} - 1\right)^2 = \frac{1}{c^2} E(X - c)^2 \tag{13}$$

is minimized by

$$c = \frac{E(X^2)}{E(X)}. \tag{14}$$

(ii) *If* $E(X) < \infty$, *and* X *has a probability density* f, *then*

$$E\left|\frac{X}{c} - 1\right| = \frac{1}{|c|} E|X - c| \tag{15}$$

is minimized by any c *satisfying*

$$\int_0^c x\, dP(x) = \int_c^\infty x\, dP(x). \tag{16}$$

Proof. Part (i) is proved by expanding $E(X - c)^2$ and minimizing the resulting quadratic in $1/c$. To prove (ii), note that the minimizing c is positive and write the right side of (15) as

$$G(c) = \int_0^c f(x)\, dx - \int_c^\infty f(x)\, dx + \frac{1}{c}\left[\int_c^\infty xf(x)\, dx - \int_0^c xf(x)\, dx\right]$$

The result follows by considering the derivative of $G(c)$ (Problem 3.6).

Any c satisfying (16) will be called a *scale-median* of X.

Corollary 3.2. *Under the assumptions of Theorem* 3.2, *if*

$$\gamma\left(\frac{d}{\tau^r}\right) = \frac{(d - \tau^r)^2}{\tau^{2r}}, \tag{17}$$

then

$$\delta^*(\underline{X}) = \frac{\delta_0(\underline{X}) E_1[\delta_0(\underline{X}) | \underline{Z}]}{E_1[\delta_0^2(\underline{X}) | \underline{Z}]}; \tag{18}$$

if

$$\gamma\left(\frac{d}{\tau^r}\right) = \frac{|d - \tau^r|}{\tau^r}, \tag{19}$$

then $\delta^*(\underline{X})$ *is given by* (11), *with* $w^*(\underline{Z})$ *any scale-median of* $\delta_0(\underline{X})$ *under the conditional distribution of* $\underline{X}$ *given* $\underline{Z}$ *and with* $\tau = 1$.

The proof is left to Problem 3.9.

Example 3.1. (*Continued*). Suppose that $n = 1$, and $X > 0$ with probability 1. Then the arguments of Theorem 3.2 and Corollary 3.2 show that if X^r has finite risk the MRE estimator of τ^r is X^r/w^* where w^* is any value minimizing

$$E_1[\gamma(X^r/w)]. \tag{20}$$

In particular, the MRE estimator is

$$X^r E_1(X^r)/E_1(X^{2r}) \tag{21}$$

when the loss is (17), and is X^r/w^* where w^* is any scale-median of X^r for $\tau = 1$ when the loss is (19).

Example 3.3. Normal. Let $X_1, \ldots, X_n$ be iid according to $N(0, \sigma^2)$ and consider the estimation of σ^2. For $\delta_0 = \Sigma X_i^2$ it follows from Basu's theorem that δ_0 is independent of $\underline{Z}$ and hence that $w^*(\underline{z}) = w^*$ is a constant determined by minimizing (20) with ΣX_i^2 in place of X^r. For the loss function (17) with $r = 2$ the MRE estimator turns out to be $\Sigma X_i^2/(n + 2)$ [Equation (2.5.25) or Problem 3.10].

Example 3.4. Uniform. Let $X_1, \ldots, X_n$ be iid according to the uniform distribution $U(0, \theta)$. Since $X_{(n)}$ is a complete sufficient statistic, it is independent of $\underline{Z}$ by Basu's theorem, and the MRE estimator of θ for the loss function (17) with $r = 1$ is $\delta^*(\underline{X}) = X_{(n)}/w$ where (Problem 3.11)

$$w = \frac{E_1\left(X_{(n)}^2\right)}{E_1\left(X_{(n)}\right)} = \frac{n+1}{n+2}.$$

Hence the MRE estimator of θ is $[(n + 2)/(n + 1)]X_{(n)}$.

If, instead of (17), the loss function is (19) with $r = 1$, the MRE estimator is $\sqrt[n+1]{2}\, X_{(n)}$ (Problem 3.11).

Quite generally, when the loss function is (17), the MRE estimator of τ^r is given by

$$\delta^*(\underline{x}) = \frac{\int_0^\infty v^{n+r-1} f(vx_1, \ldots, vx_n)\, dv}{\int_0^\infty v^{n+2r-1} f(vx_1, \ldots, vx_n)\, dv}, \tag{22}$$

and in this form is known as the Pitman estimator of τ^r. The proof parallels that of Theorem 1.5 (Problem 3.15).

So far, the estimator δ has been assumed to be nonrandomized. Since $\bar{G}$ is transitive over Ω, it follows from the result proved in the preceding section that randomized estimators need not be considered. It is further seen, as for the corresponding result in the location case, that if a sufficient statistic T exists which permits a representation $T = (T_1, \dots, T_r)$ with

$$T_i(b\underline{X}) = bT_i(\underline{X}) \qquad \text{for all} \quad b > 0,$$

then an MRE estimator can be found which depends only on T. Illustrations are provided by Example 3.3 and 3.4, with $T = (\Sigma X_i^2)^{1/2}$ and $T = X_{(n)}$, respectively. When the loss function is (17), it follows from the factorization criterion that the MRE estimator (22) depends only on T.

Since the group $\tau' = b\tau$, $b > 0$ is transitive and the group $d' = \tau^r d$ is commutative, Theorem 2.2 applies and an MRE estimator is always risk-unbiased, although the MRE estimators of Examples 3.3 and 3.4 are not unbiased in the sense of Chapter 2.

Example 3.5. If the loss function is (17), the condition of risk-unbiasedness reduces to

$$E_\tau[\delta^2(\underline{X})] = \tau^r E_\tau[\delta(\underline{X})]. \tag{23}$$

Given any scale-equivariant estimator $\delta_0(\underline{X})$ of τ^r, there exists a value of c for which $c\delta_0(\underline{X})$ satisfies (23), and for this value $c\delta_0(\underline{X})$ has uniformly smaller risk than $\delta_0(\underline{X})$ unless $c = 1$ (Problem 3.18).

If the loss function is (19), the condition of risk-unbiasedness requires that $E_\tau|\delta(\underline{X}) - a|/a$ be minimized by $a = \tau^r$. It follows from Lemma 3.1 that for this loss function, risk-unbiasedness is equivalent to the condition that the estimand τ^r is equal to the scale-median of $\delta(\underline{X})$.

Let us now turn to location-scale families, where the density of $\underline{X} = (X_1, \dots, X_n)$ is given by

$$\frac{1}{\tau^n} f\left(\frac{x_1 - \xi}{\tau}, \dots, \frac{x_n - \xi}{\tau}\right) \tag{24}$$

with both parameters unknown. Consider first the estimation of τ^r with loss function (5). This problem remains invariant under the transformations

$$X_i' = a + bX_i, \qquad \xi' = a + b\xi, \qquad \tau' = b\tau, \qquad (b > 0) \tag{25}$$

and $d' = b^r d$, and an estimator δ is equivariant under this group if

$$\delta(a + b\underline{X}) = b^r\delta(\underline{X}). \tag{26}$$

Consider, first, only a change in location,

$$X_i' = X_i + a, \tag{27}$$

which takes ξ into $\xi' = \xi + a$ but leaves τ unchanged. By (26), δ must then satisfy

$$\delta(\underline{x} + a) = \delta(\underline{x}), \tag{28}$$

that is, remain *invariant*, and, by Lemma 1.2, condition (28) holds if and only if δ is a function only of the differences $y_i = x_i - x_n$. The joint density of the Y's is

$$\begin{aligned}(29)\quad &\frac{1}{\tau^n}\int_{-\infty}^{\infty} f\left(\frac{y_1 + t}{\tau}, \dots, \frac{y_{n-1} + t}{\tau}, \frac{t}{\tau}\right) dt \\ &= \frac{1}{\tau^{n-1}}\int_{-\infty}^{\infty} f\left(\frac{y_1}{\tau} + u, \dots, \frac{y_{n-1}}{\tau} + u, u\right) du\end{aligned}$$

Since this density has the structure (1) of a scale family, Theorem 3.2 applies and provides the estimator that uniformly minimizes the risk among all estimators satisfying (26).

It follows from Theorem 3.2 that such an MRE estimator is given by

$$\delta(\underline{X}) = \frac{\delta_0(\underline{Y})}{w^*(\underline{Z})} \tag{30}$$

where $\delta_0(\underline{Y})$ is any finite risk scale-equivariant estimator of τ^r based on $\underline{Y} = (Y_1, \dots, Y_{n-1})$, where $\underline{Z} = (Z_1, \dots, Z_{n-1})$ with

$$Z_i = \frac{Y_i}{Y_{n-1}} \quad (i = 1, \dots, n-2) \quad \text{and} \quad Z_{n-1} = \frac{Y_{n-1}}{|Y_{n-1}|}, \tag{31}$$

and where $w^*(\underline{z})$ is any number minimizing

$$E_{\tau=1}\{\gamma[\delta_0(\underline{Y})/w(\underline{z})|\underline{z}]\}. \tag{32}$$

Example 3.6. ***(Normal).*** Let $X_1, \dots, X_n$ be iid according to $N(\xi, \sigma^2)$ and consider the estimation of σ^2 with loss function (17), $r = 2$. By Basu's theorem $(\bar{X}, \Sigma(X_i - \bar{X})^2)$ is independent of $\underline{Z}$. If $\delta_0 = \Sigma(X_i - \bar{X})^2$, then δ_0 is equivariant under (25) and independent of $\underline{Z}$. Hence $w^*(\underline{z}) = w^*$ in (30) is a constant determined by minimizing (20) with $\Sigma(X_i - \bar{X})^2$ in place of X^r. Since $\Sigma(X_i - \bar{X})^2$ has the distribution of δ_0 of Example 3.3 with $n - 1$ in place of n, the MRE estimator for the loss function (17) with $r = 2$ is $\Sigma(X_i - \bar{X})^2/(n + 1)$.

Example 3.7. Uniform. Let $X_1, \ldots, X_n$ be iid according to $U(\xi - \frac{1}{2}\tau, \xi + \frac{1}{2}\tau)$, and consider the problem of estimating τ with loss function (17), $r = 1$. By Basu's theorem $(X_{(1)}, X_{(n)})$ is independent of $\underline{Z}$. If δ_0 is the range $R = X_{(n)} - X_{(1)}$, it is equivariant under (25) and independent of $\underline{Z}$. It follows from (21) with $r = 1$ that (Problem 3.19) $\delta^*(\underline{X}) = [(n+2)/n]R$.

Quite generally, for the loss function (17), it follows from (30) that the MRE estimator is given as a ratio of two double integrals by replacing $f(x_1, \ldots, x_n)$ in (22) by the joint density of the Y's when $\tau = 1$, which is given by (29) with $\tau = 1$.

Since the group $\xi' = a + b\xi$, $\tau' = b\tau$ is transitive and the group $d' = b^r d$ is commutative, it follows (as in the pure scale case) that an MRE estimator is always risk-unbiased.

Finally, consider the problem of estimating the location parameter ξ in (24). The transformations (25) relating to the sample space and parameter space remain the same, but the transformations of the decision space now become $d' = a + bd$. A loss function $L(\xi, \tau; d)$ is invariant under these transformations if and only if it is of the form

$$L(\xi, \tau; d) = \rho\left(\frac{d - \xi}{\tau}\right). \tag{33}$$

That any such loss function is invariant is obvious. Conversely, suppose that L is invariant and that $(\xi, \tau; d)$ and $(\xi', \tau'; d')$ are two points with $(d' - \xi')/\tau' = (d - \xi)/\tau$. Putting $b = \tau'/\tau$ and $\xi' - a = b\xi$, one has $d' = a + bd$, $\xi' = a + b\xi$, $\tau' = b\tau$, and hence $L(\xi', \tau'; d') = L(\xi, \tau; d)$, as was to be proved.

Equivariance in the present case becomes

$$\delta(a + b\underline{x}) = a + b\delta(\underline{x}), \qquad b > 0. \tag{34}$$

Since $\overline{G}$ is transitive over the parameter space, the risk of any equivariant estimator is constant so that an MRE estimator can be expected to exist. In some special cases, the MRE estimator reduces to that derived in Section 3.1 with τ known, as follows.

For fixed τ, write

$$g_\tau(x_1, \ldots, x_n) = \frac{1}{\tau^n} f\left(\frac{x_1}{\tau}, \ldots, \frac{x_n}{\tau}\right) \tag{35}$$

so that (24) becomes

$$g_\tau(x_1 - \xi, \ldots, x_n - \xi). \tag{36}$$

Lemma 3.2. *Suppose that for the location family* (36) *and loss function* (33) *there exists an MRE estimator δ^* of ξ with respect to the transformations*

(1.2) *and* (1.3) *and that*

(i) δ^* *is independent of* τ,

and

(ii) δ^* *satisfies* (34).

Then δ^* *minimizes the risk among all estimators satisfying* (34).

Proof. Suppose δ is any other estimator which satisfies (34) and hence a fortiori is equivariant with respect to the transformations (1.2) and (1.3), and that the value τ of the scale parameter is known. It follows from the assumptions about δ^* that for this τ the risk of δ^* does not exceed the risk of δ. Since this is true for all values of τ, the result follows.

Example 3.8. Normal. Let $X_1, \ldots, X_n$ be iid as $N(\xi, \tau^2)$, both parameters being unknown. Then it follows from Example 1.3 that $\delta^* = \bar{X}$ for any loss function $\rho[(d - \xi)/\tau]$ for which ρ satisfies the assumptions of Example 1.3. Since (i) and (ii) of Lemma 3.2 hold for this δ^*, it is the MRE estimator of ξ under the transformations (25).

Example 3.9. Uniform. Let $X_1, \ldots, X_n$ be iid as $U(\xi - \frac{1}{2}\tau, \xi + \frac{1}{2}\tau)$. Then, analogously to Example 3.8, it follows from Example 1.5 that $[X_{(1)} + X_{(n)}]/2$ is MRE for the loss functions of Example 3.8.

Unfortunately, the MRE estimators of Section 3.1 typically do not satisfy the assumptions of Lemma 3.2. This is the case, for instance, with the estimators of Examples 1.4 and 1.6. To derive the MRE estimator without these assumptions, let us first characterize the totality of equivariant estimators.

Theorem 3.3. *Let* δ_0 *be any estimator of* ξ *satisfying* (34) *and* δ_1 *any estimator of* τ *taking on positive values only and satisfying*

$$\delta_1(\mathrm{a} + \mathrm{b}\underline{\mathrm{x}}) = \mathrm{b}\delta_1(\underline{\mathrm{x}}) \quad \textit{for all} \quad \mathrm{b} > 0 \quad \textit{and all} \quad \mathrm{a}. \tag{37}$$

Then δ *satisfies* (34) *if and only if it is of the form*

$$\delta(\underline{\mathrm{x}}) = \delta_0(\underline{\mathrm{x}}) - \mathrm{w}(\underline{\mathrm{z}})\delta_1(\underline{\mathrm{x}}) \tag{38}$$

where $\underline{\mathrm{z}}$ *is given by* (31).

Proof. Analogously to Lemma 1.1, it is seen that δ satisfies (34) if and only if it is of the form

$$\delta(\underline{x}) = \delta_0(\underline{x}) - u(\underline{x})\delta_1(\underline{x}), \tag{39}$$

where

$$u(a + b\underline{x}) = u(\underline{x}) \quad \text{for all} \quad b > 0 \quad \text{and all} \quad a \tag{40}$$

(Problems 3.21). That (39) holds if and only if u depends on $\underline{x}$ only through $\underline{z}$, follows from Lemma 1.2 and Theorem 3.1.

An argument paralleling that of Theorem 1.3 now shows that the MRE estimator of ξ is

$$\delta(\underline{X}) = \delta_0(\underline{X}) - w^*(\underline{Z})\delta_1(\underline{X})$$

where for each $\underline{z}$, $w^*(\underline{z})$ is any number minimizing

$$E_{0,1}\{\rho[\delta_0(\underline{X}) - w^*(\underline{z})\delta_1(\underline{X})]|\underline{z}\}. \tag{41}$$

Here, $E_{0,1}$ indicates that the expectation is evaluated at $\xi = 0$, $\tau = 1$.

If, in particular,

$$\rho\left(\frac{d - \xi}{\tau}\right) = \frac{(d - \xi)^2}{\tau^2}, \tag{42}$$

it is easily seen that $w^*(\underline{z})$ is

$$w^*(\underline{z}) = E_{0,1}[\delta_0(\underline{X})\delta_1(\underline{X})|\underline{z}]/E_{0,1}[\delta_1^2(\underline{X})|\underline{z}]. \tag{43}$$

Example 3.10. Exponential. Let $X_1, \ldots, X_n$ be iid according to the exponential distribution $E(\xi, \tau)$. If $\delta_0(\underline{X}) = X_{(1)}$ and $\delta_1(\underline{X}) = \Sigma[X_i - X_{(1)}]$, it follows from Example 1.5.14 that (δ_0, δ_1) are jointly independent of $\underline{Z}$ and are also independent of each other. Then (Problem 3.20)

$$w^*(\underline{z}) = w^* = E\frac{[\delta_0(\underline{X})\delta_1(\underline{X})]}{E[\delta_1^2(\underline{X})]} = \frac{1}{n^2},$$

and the MRE estimator of ξ is therefore

$$\delta^*(\underline{X}) = X_{(1)} - \frac{1}{n^2}\Sigma[X_i - X_{(1)}].$$

When the best location-equivariant estimate is not also scale-equivariant, its risk is of course smaller than that of the MRE under (34). Some numerical values of the increase that results from the additional requirement are given for a number of situations by Hoaglin (1975).

For the loss function (42) no risk-unbiased estimator δ exists, since this would require that for all ξ, ξ', τ, τ'

$$\frac{1}{\tau^2}E_{\xi,\tau}[\delta(\underline{X}) - \xi]^2 \leqslant \frac{1}{\tau'^2}E_{\xi,\tau}[\delta(\underline{X}) - \xi']^2, \tag{44}$$

which is clearly impossible. Perhaps (44) is too strong and should be required only when $\tau' = \tau$. It then reduces to (1.22) with $\theta = (\xi, \tau)$, $g(\theta) = \xi$, and this weakened form of (44) reduces to the classical unbiasedness condition $E_{\xi,\tau}[\delta(\underline{X})] = \xi$. A UMVU estimator of ξ exists in Example 3.10 (Problem 2.2.15), but it is

$$\delta(\underline{X}) = X_{(1)} - \frac{1}{n(n-1)}\Sigma[X_i - X_{(1)}]$$

rather than $\delta^*(\underline{X})$, and the latter is not unbiased (Problem 3.22).

4. LINEAR MODELS (NORMAL)

Having developed the theory of unbiased estimation in Chapter 2 and of equivariant estimation in the first three sections of the present chapter, we shall now apply these results to some important classes of statistical models. One of the most widely used bodies of statistical techniques, comprising particularly the analysis of variance, regression, and the analysis of covariance, is formalized in terms of linear models, which will be defined and illustrated in the following. The examples, however, are not enough to give an idea of the full richness of the applications. For a more complete treatment see, for example, Scheffé (1959) and Seber (1977).

Consider the problem of investigating the effect of a number of different factors on a response. Typically, each factor can occur in a number of different forms or at a number of different levels. Factor levels can be qualitative or quantitative. Three possibilities arise, corresponding to three broad categories of linear models.

(i) All factor levels qualitative
(ii) All factor levels quantitative
(ii) Some factors of each kind

Example 4.1. ***One-way layout.*** A simple illustration of category (i) is provided by the *one-way layout* in which a single factor occurs at a number of qualitatively different levels. For example, we may wish to study the effect on performance of a number of different textbooks or the effect on weight loss of a number of diets. If X_{ij} denotes the response of the jth subject receiving treatment i, it is often reasonable to assume that the X_{ij} are independently distributed as

$$(1) \qquad X_{ij}: N(\xi_i, \sigma^2), \qquad j = 1, \ldots, n_i; \qquad i = 1, \ldots, s.$$

Estimands that may be of interest are ξ_i and $\xi_i - (1/s)\Sigma_{j=1}^s \xi_j$.

Example 4.2. ***A simple regression model.*** As an example of type (ii), consider the time required to memorize a list of words. If the number of words presented to the ith subject and the time it takes the subject to learn the words are denoted by t_i

and X_i, respectively, one might assume that for the range of t's of interest the X's are independently distributed as

$$(2) \qquad X_i: N(\alpha + \beta t_i + \gamma t_i^2, \sigma^2)$$

where α, β, and γ are the unknown regression coefficients, which are to be estimated.

This would turn into an example of the third type, if there were several groups of subjects. One might, for example, wish to distinguish between women and men or to see how learning ability is influenced by the form of the word list (whether it is handwritten, typed, or printed). The model might then become

$$(3) \qquad X_{ij}: N(\alpha_i + \beta_i t_{ij} + \gamma_i t_{ij}^2, \sigma^2)$$

where X_{ij} is the response of the jth subject in the ith group. Here, the group is a qualitative factor and the length of the list a quantitative one.

The general *linear model*, which covers all three cases, assumes that

$$(4) \qquad X_i \text{ is distributed as } N(\xi_i, \sigma^2),\ i = 1, \dots, n,$$

where the X_i are independent and $(\xi_1, \dots, \xi_n) \in \Pi_\Omega$, an s-dimensional linear subspace of E_n $(s < n)$.

It is convenient to reduce this model to a canonical form by means of an orthogonal transformation

$$(5) \qquad \underline{Y} = \underline{X}C$$

where we shall use $\underline{Y}$ to denote both the vector with components $(Y_1, \dots, Y_n)$ and the row matrix $(Y_1, \dots, Y_n)$. If $\eta_i = E(Y_i)$, the η's and ξ's are related by

$$(6) \qquad \underline{\eta} = \underline{\xi}C$$

where $\underline{\eta} = (\eta_1, \dots, \eta_n)$ and $\underline{\xi} = (\xi_1, \dots, \xi_n)$.

To find the distribution of the Y's note that the joint density of $X_1, \dots, X_n$ is

$$\frac{1}{(\sqrt{2\pi}\,\sigma)^n} \exp\left[-\frac{1}{2\sigma^2}\Sigma(x_i - \xi_i)^2\right],$$

that

$$\Sigma(x_i - \xi_i)^2 = \Sigma(y_i - \eta_i)^2,$$

since C is orthogonal, and that the Jacobian of the transformation is 1. Hence the joint density of $Y_1, \dots, Y_n$ is

$$\frac{1}{(\sqrt{2\pi}\,\sigma)^n} \exp\left[-\frac{1}{2\sigma^2}\Sigma(y_i - \eta_i)^2\right].$$

The Y's are therefore independent normal with $Y_i \sim N(\eta_i, \sigma^2)$, $i = 1, \dots, n$. If $\underline{c}_i'$ denotes the ith column of C, the desired form is obtained by choosing the $\underline{c}_i$ so that the first s columns $\underline{c}_1', \dots, \underline{c}_s'$ span Π_Ω. Then

$$\underline{\xi} \in \Pi_\Omega \Leftrightarrow \underline{\xi} \text{ is orthogonal to the last } n - s \text{ columns of } C.$$

Since $\underline{\eta} = \underline{\xi}C$, it follows that

$$\underline{\xi} \in \Pi_\Omega \Leftrightarrow \eta_{s+1} = \cdots = \eta_n = 0. \tag{7}$$

In terms of the Y's, the model (4) thus becomes

$$Y_i: N(\eta_i, \sigma^2), i = 1, \dots, s \quad \text{and} \quad Y_j: N(0, \sigma^2), \qquad j = s+1, \dots, n. \tag{8}$$

As $(\xi_1, \dots, \xi_n)$ varies over Π_Ω, $(\eta_1, \dots, \eta_s)$ varies unrestrictedly over E_s while $\eta_{s+1} = \cdots = \eta_n = 0$.

In this canonical model, $Y_1, \dots, Y_s$ and $S^2 = \Sigma_{j=s+1}^n Y_j^2$ are complete sufficient statistics for $(\eta_1, \dots, \eta_s, \sigma^2)$.

Theorem 4.1. (i) *The UMVU estimators of* $\Sigma_{i=1}^s \lambda_i \eta_i$ *(where the* λ*'s are known constants) and* σ^2 *are* $\Sigma_{i=1}^s \lambda_i Y_i$ *and* $S^2/(n - s)$. *(Here UMVU is used in the strong sense of Section* 2.1.)

(ii) *Under the transformations*

$$Y_i' = Y_i + a_i \ (i = 1, \dots, s); \qquad Y_j' = Y_j \ (j = s+1, \dots, n)$$

$$\eta_i' = \eta_i + a_i \ (i = 1, \dots, s) \quad \textit{and} \quad d' = d + \sum_{i=1}^{s} a_i \lambda_i$$

and with loss function $L(\eta, d) = \rho(d - \Sigma\lambda_i\eta_i)$ *where* ρ *is convex and even, the UMVU estimator* $\Sigma_{i=1}^s \lambda_i Y_i$ *is also the MRE estimator of* $\Sigma_{i=1}^s \lambda_i \eta_i$.

(iii) *Under the loss function* $(d - \sigma^2)^2/\sigma^4$, *the MRE estimator of* σ^2 *is* $S^2/(n - s + 2)$.

Proof. (i) Since $\Sigma_{i=1}^s \lambda_i Y_i$ and $S^2/(n - s)$ are unbiased and are functions of the complete sufficient statistics, they are UMVU.

(ii) The condition of equivariance is that

$$\delta(Y_1 + c_1, \dots, Y_s + c_s, Y_{s+1}, \dots, Y_n)$$
$$= \delta(Y_1, \dots, Y_s, Y_{s+1}, \dots, Y_n) + \sum_{i=1}^{s} \lambda_i c_i$$

and the result follows from Problem 2.19.

(iii) This follows essentially from Example 3.3 (see Problem 4.3).

It would be more convenient to have the estimator expressed in terms of the original variables $X_1, \dots, X_n$, rather than the transformed variables $Y_1, \dots, Y_n$. For this purpose we introduce the following definition.

Let $\underline{\xi} = (\xi_1, \dots, \xi_n)$ be any vector in Π_Ω. Then the *least squares estimators* (LSE) $(\hat{\xi}_1, \dots, \hat{\xi}_n)$ of $(\xi_1, \dots, \xi_n)$ are those estimators which minimize $\Sigma_{i=1}^n (X_i - \xi_i)^2$ subject to the condition $\underline{\xi} \in \Pi_\Omega$.

Theorem 4.2. *Under the model* (4), *the UMVU estimator of* $\Sigma_{i=1}^n \gamma_i \xi_i$ *is* $\Sigma_{i=1}^n \gamma_i \hat{\xi}_i$.

Proof. By Theorem 4.1 (and the completeness of $Y_1, \dots, Y_s$ and S^2), it suffices to show that $\Sigma_{i=1}^n \gamma_i \hat{\xi}_i$ is a linear function of $Y_1, \dots, Y_s$, and that it is unbiased for $\Sigma_{i=1}^n \gamma_i \xi_i$. Now,

$$\sum_{i=1}^{n} (X_i - \xi_i)^2 = \sum_{i=1}^{n} [Y_i - E(Y_i)]^2 = \sum_{i=1}^{s} (Y_i - \eta_i)^2 + \sum_{j=s+1}^{n} Y_j^2. \tag{9}$$

The right side is minimized by $\hat{\eta}_i = Y_i$ $(i = 1, \dots, s)$, while the left side is minimized by $\hat{\xi}_1, \dots, \hat{\xi}_n$. Hence

$$(Y_1 \cdots Y_s 0 \cdots 0) = (\hat{\xi}_1 \cdots \hat{\xi}_n)C = \underline{\hat{\xi}}C$$

so that

$$\underline{\hat{\xi}} = (Y_1 \cdots Y_s 0 \cdots 0)C^{-1}.$$

It follows that each $\hat{\xi}_i$ and therefore also $\Sigma_{i=1}^n \gamma_i \hat{\xi}_i$ is a linear function of $Y_1, \dots, Y_s$. Furthermore

$$E(\underline{\hat{\xi}}) = E[(Y_1 \cdots Y_s 0 \cdots 0)C^{-1}] = (\eta_1 \cdots \eta_s 0 \cdots 0)C^{-1} = \underline{\xi}.$$

Thus each $\hat{\xi}_i$ is unbiased for ξ_i, and consequently $\Sigma_{i=1}^n \gamma_i \hat{\xi}_i$ is unbiased for $\Sigma_{i=1}^n \gamma_i \xi_i$.

It is interesting to note that each of the two quite different equations

$$\underline{X} = (Y_1 \cdots Y_n)C^{-1} \quad \text{and} \quad \hat{\underline{\xi}} = (Y_1 \cdots Y_s 0 \cdots 0)C^{-1}$$

leads to $\underline{\xi} = (\eta_1, \ldots, \eta_s\, 0 \cdots 0)C^{-1}$ by taking expectations.

Let us next reinterpret the equivariance considerations of Theorem 4.1 in terms of the original variables. It is necessary first to specify the group of transformations leaving the problem invariant. The transformations of Y-space defined in Theorem 4.1(ii), in terms of the X's become $X_i' = X_i + b_i$, $i = 1, \ldots, n$ but the b_i are not arbitrary since the problem remains invariant only if $\underline{\xi}' = \underline{\xi} + \underline{b} \in \Pi_\Omega$; that is, the b_i must satisfy $\underline{b} = (b_1, \ldots, b_n) \in \Pi_\Omega$. Theorem 4.1(ii) thus becomes the following Corollary.

Corollary 4.1. *Under the transformations*

$$\underline{X}' = \underline{X} + \underline{b} \quad \textit{with} \quad \underline{b} \in \Pi_\Omega, \tag{10}$$

$\Sigma_{i=1}^n \gamma_i \hat{\xi}_i$ *is MRE for estimating* $\Sigma_{i=1}^n \gamma_i \xi_i$ *with the loss function* $\rho(d - \Sigma\gamma_i\xi_i)$ *provided* ρ *is convex and even.*

To obtain the UMVU and MRE estimators of σ^2 in terms of the X's it is only necessary to reexpress S^2. From the minimization of the two sides of (9) it is seen that

$$\sum_{i=1}^{n} (X_i - \hat{\xi}_i)^2 = \sum_{j=s+1}^{n} Y_j^2 = S^2. \tag{11}$$

The UMVU and MRE estimators of σ^2 given in Theorem 4.1, in terms of the X's, are therefore $\Sigma(X_i - \hat{\xi}_i)^2/(n - s)$ and $\Sigma(X_i - \hat{\xi}_i)^2/(n - s + 2)$, respectively.

Let us now illustrate these results.

Example 4.1. (*Continued*). Let X_{ij} be independent $N(\xi_i, \sigma^2)$, $j = 1, \ldots, n_i$, $i = 1, \ldots, s$. To find the UMVU or MRE estimator of a linear function of the ξ_i, it is only necessary to find the least squares estimators $\hat{\xi}_i$. Minimizing

$$\sum_{i=1}^{s} \sum_{j=1}^{n_i} (X_{ij} - \xi_i)^2 = \sum_{i=1}^{s} \left[\sum_{j=1}^{n_i} (X_{ij} - X_{i\cdot})^2 + n_i (X_{i\cdot} - \xi_i)^2 \right]$$

we see that

$$\hat{\xi}_i = X_{i\cdot} = \frac{1}{n_i} \sum_{j=1}^{n_i} X_{ij}.$$

From (11) the UMVU estimator of σ^2 in the present case is seen to be

$$\hat{\sigma}^2 = \sum_{i=1}^{s} \sum_{j=1}^{n_i} (X_{ij} - X_{i\cdot})^2/(\Sigma n_i - s).$$

Example 4.3. ***Simple linear regression.*** Let X_i be independent $N(\xi_i, \sigma^2)$, $i = 1,\ldots, n$, with $\xi_i = \alpha + \beta t_i$, t_i known and not all equal. Here Π_Ω is spanned by the vectors $(1,\ldots, 1)$ and $(t_1,\ldots, t_n)$ so that the dimension of Π_Ω is $s = 2$. The least squares estimators of ξ_i are obtained by minimizing $\Sigma_{i=1}^n (X_i - \alpha - \beta t_i)^2$ with respect to α and β. It is easily seen that for any i, j with $t_i \neq t_j$

$$\beta = \frac{\xi_j - \xi_i}{t_j - t_i}, \qquad \alpha = \frac{t_j \xi_i - t_i \xi_j}{t_j - t_i} \tag{12}$$

and that $\hat{\beta}$, $\hat{\alpha}$ are given by the same functions of $\hat{\xi}_i$, $\hat{\xi}_j$ (Problem 4.4). Hence $\hat{\alpha}$ and $\hat{\beta}$ are the best unbiased and equivariant estimators of α and β, respectively.

Note that the representation of α and β in terms of the ξ_i's is not unique. Any two ξ_i and ξ_j values with $t_i \neq t_j$ determine α and β and thus all the ξ's. The reason, of course, is that the vectors $(\xi_1,\ldots, \xi_n)$ lie in a two-dimensional linear subspace of n-space.

Example 4.3 is a special case of the model specified by the equation

$$\underline{\xi} = \underline{\theta} A \tag{13}$$

where $\underline{\theta} = (\theta_1 \cdots \theta_s)$ are s unknown parameters and A is a known $s \times n$ matrix of rank s, the so-called *full-rank model*. In Example 4.3

$$\underline{\theta} = (\alpha\ \beta) \quad \text{and} \quad A = \begin{pmatrix} 1 & \cdots & 1 \\ t_1 & \cdots & t_n \end{pmatrix}.$$

The least squares estimators of the ξ_i in (13) are obtained by minimizing

$$\sum_{i=1}^{n} [X_i - \xi_i(\underline{\theta})]^2$$

with respect to $\underline{\theta}$. The minimizing values $\hat{\theta}_i$ are the LSEs of θ_i, and the LSEs of the ξ_i are given by

$$\hat{\underline{\xi}} = \hat{\underline{\theta}} A. \tag{14}$$

Theorems 4.1 and 4.2 establish that the various optimality results apply to the estimators of the ξ_i and their linear combinations. The following theorem shows that they also apply to the estimators of the θ's and their linear functions.

Theorem 4.3. *Let* $X_i \sim N(\xi_i, \sigma^2)$, $i = 1, \ldots, n$ *be independent, and let* $\underline{\xi}$ *satisfy* (13) *with* A *of rank* s. *Then the least squares estimator* $\hat{\underline{\theta}}$ *of* $\underline{\theta}$ *is a linear function of the* $\hat{\xi}_i$ *and hence has the optimality properties established in Theorems* 4.1 *and* 4.2 *and Corollary* 4.1.

Proof. It need only be shown that $\underline{\theta}$ is a linear function of $\underline{\xi}$; then by (13) and (14) $\hat{\underline{\theta}}$ is the corresponding linear function of $\hat{\underline{\xi}}$.

Assume without loss of generality that the first s columns of A are linearly independent, and form the corresponding nonsingular $s \times s$ submatrix A^*. Then

$$(\xi_1 \cdots \xi_s) = (\theta_1 \cdots \theta_s)A^*,$$

so that

$$(\theta_1 \cdots \theta_s) = (\xi_1 \cdots \xi_s)A^{*-1},$$

and this completes the proof.

Typical examples in which $\underline{\xi}$ is given in terms of (13) are polynomial regressions such as

$$\xi_i = \alpha + \beta t_i + \gamma t_i^2$$

or regression in more than one variable such as

$$\xi_i = \alpha + \beta t_i + \gamma u_i$$

where the t's and u's are given, and α, β, and γ are the unknown parameters. Or there might be several regression lines with a common slope, say

$$\xi_{ij} = \alpha_i + \beta t_{ij} \qquad (j = 1, \ldots, n_i;\ i = 1, \ldots, a),$$

and so on.

The full rank model does not always provide the most convenient parametrization; for reasons of symmetry, it is often preferable to use a model (13) with more parameters than are needed. Before discussing such models more fully, let us illustrate the resulting difficulties on a trivial example. Suppose that $\xi_i = \xi$ for all i and that we put $\xi_i = \lambda + \mu$. Such a model does not define λ and μ uniquely but only their sum. One can then either let this ambiguity remain but restrict attention to clearly defined functions such as $\lambda + \mu$, or alternatively one can remove the ambiguity by placing an additional restriction on λ and μ, such as $\mu - \lambda = 0$, $\mu = 0$, or $\lambda = 0$.

More generally, let us suppose that the model is given by

$$\underline{\xi} = \underline{\theta} A \tag{15}$$

where A is a $t \times n$ matrix of rank $s < t$. To define the θ's uniquely, (15) is supplemented by side conditions

$$\underline{\theta} B = 0 \tag{16}$$

chosen so that the set of equations (15) and (16) has a unique solution $\underline{\theta}$ for every $\underline{\xi} \in \Pi_\Omega$.

Example 4.4. One-way layout. Consider the one-way layout of Example 4.1, with X_{ij}($j = 1, \ldots, n_i$; $i = 1, \ldots, s$) independent normal variables with means ξ_i and variance σ^2. When the principal concern is a comparison of the s treatments or populations, one is interested in the differences of the ξ's and may represent these by means of the differences between the ξ_i and some mean value μ, say $\alpha_i = \xi_i - \mu$. The model then becomes

$$\xi_i = \mu + \alpha_i, \qquad i = 1, \ldots, s \tag{17}$$

which expresses the s ξ's in terms of $s + 1$ parameters. To specify the parameters, an additional restriction is required, for example,

$$\Sigma \alpha_i = 0. \tag{18}$$

Adding the s equations (17) and using (18), one finds

$$\mu = \Sigma \frac{\xi_i}{s} = \bar{\xi} \tag{19}$$

and hence

$$\alpha_i = \xi_i - \bar{\xi}. \tag{20}$$

The quantity α_i measures the effect of the ith treatment. Since $X_{i\cdot}$ is the least squares estimator of ξ_i, the UMVU estimators of μ and the α's are

$$\hat{\mu} = \Sigma \frac{X_{i\cdot}}{s} = \Sigma\Sigma \frac{X_{ij}}{sn_i} \quad \text{and} \quad \hat{\alpha}_i = X_{i\cdot} - \hat{\mu}. \tag{21}$$

When the sample sizes n_i are not all equal, a possible disadvantage of this representation is that the vectors of the coefficients of the X_{ij} in the $\hat{\alpha}_i$ are not orthogonal to the corresponding vector of coefficients of $\hat{\mu}$ (Problem 4.7(i)). As a result, $\hat{\mu}$ is not independent of the $\hat{\alpha}_i$. Also, when the α_i are known to be zero, the estimator of μ is no longer given by (21) (Problem 4.8).

For these reasons, the side condition (18) is sometimes replaced by

$$\Sigma n_i \alpha_i = 0, \tag{22}$$

which leads to

$$\mu = \Sigma \frac{n_i \xi_i}{N} = \tilde{\xi} \quad (N = \Sigma n_i) \tag{23}$$

and hence

$$\alpha_i = \xi_i - \tilde{\xi}. \tag{24}$$

This α_i seems to be a less natural measure of the effect of the ith treatment, but the resulting UMVU estimators $\hat{\alpha}_i$ and $\hat{\mu}$ have the orthogonality property not possessed by the estimators (21) [Problem 4.7(ii)]. The side conditions (18) and (22), of course, agree when the n_i are all equal.

The following theorem shows that the conclusion of Theorem 4.3 continues to hold when the θ's are defined by (15) and (16) instead of (13).

Theorem 4.4. *Let* X_i *be independent* $N(\xi_i, \sigma^2)$, $i = 1, \ldots, n$, *with* $\underline{\xi} \in \Pi_\Omega$, *an* s*-dimensional linear subspace of* E_n. *Suppose that* $(\theta_1, \ldots, \theta_t)$ *are uniquely determined by* (15) *and* (16), *where* A *is of rank* $s < t$ *and* B *of rank* k. *Then* $k = t - s$, *and the optimality results of Theorem* 4.2 *and Corollary* 4.1 *apply to the parameters* $\theta_1, \ldots, \theta_t$ *and their least squares estimators* $\hat{\theta}_1, \ldots, \hat{\theta}_t$.

Proof. Let $\hat{\theta}_1, \ldots, \hat{\theta}_t$ be the LSEs of $\theta_1, \ldots, \theta_t$, that is, the values that minimize

$$\sum_{i=1}^{n} [X_i - \xi_i(\underset{\sim}{\theta})]^2$$

subject to (15) and (16). It must be shown, as in the proof of Theorem 4.3, that the $\hat{\theta}_i$'s are linear functions of $\hat{\xi}_1, \ldots, \hat{\xi}_n$, and that the θ_i's are the same functions of $\xi_1, \ldots, \xi_n$.

Without loss of generality, suppose that the θ's are numbered so that the last k columns of B are linearly independent. Then one can solve for $\theta_{t-k+1}, \ldots, \theta_t$ in terms of $\theta_1, \ldots, \theta_{t-k}$, obtaining the unique solution

$$\theta_j = L_j(\theta_1, \ldots, \theta_{t-k}) \quad \text{for } j = t - k + 1, \ldots, t. \tag{25}$$

Substituting into $\underline{\xi} = \underline{\theta} A$ gives

$$\underline{\xi} = (\theta_1 \cdots \theta_{t-k}) A^*$$

for some matrix A^*, with $(\theta_1, \ldots, \theta_{t-k})$ varying freely in E_{t-k}. Since each $\underline{\xi} \in \Pi_\Omega$ uniquely determines $\underline{\theta}$, in particular the value $\underline{\xi} = \underline{0}$ has the unique solution $\underline{\theta} = \underline{0}$, so that $(\theta_1 \cdots \theta_{t-k})A^* = \underline{0}$ has a unique solution. This implies that A^* has rank $t - k$. On the other hand, since $\underline{\xi}$ ranges over a linear space of dimension s, it follows that $t - k = s$, and hence that $k = t - s$.

The situation is now reduced to that of Theorem 4.3 with $\underline{\xi}$ a linear function of $t - k = s$ freely varying θ's, so the earlier result applies to $\theta_1, \ldots, \theta_{t-k}$. Finally, the remaining parameters $\theta_{t-k+1}, \ldots, \theta_t$ and their LSEs are determined by (25), and this completes the proof.

Example 4.5. Two-way layout. A typical illustration of the above approach is provided by a two-way layout. This arises in the investigation of the effect of two factors on a response. In a medical situation, for example, one of the factors might be the kind of treatment (e.g., surgical, nonsurgical, or no treatment at all), the other the severity of the disease. Let X_{ijk} denote the response of the kth subject to which factor 1 is applied at level i and factor 2 at level j. We assume that the X_{ijk} are independently, normally distributed with means ξ_{ij} and common variance σ^2. To avoid the complications of Example 4.4, we shall suppose that each treatment combination (i, j) is applied to the same number of subjects. If the number of levels of the two factors is a and b, respectively, the model is thus

$$X_{ijk}: N(\xi_{ij}, \sigma^2), \qquad i = 1, \ldots, I; \qquad j = 1, \ldots, J; \qquad k = 1, \ldots, m. \tag{26}$$

This model is frequently parametrized by

$$\xi_{ij} = \mu + \alpha_i + \beta_j + \gamma_{ij} \tag{27}$$

with the side conditions

$$\sum_i \alpha_i = \sum_j \beta_j = \sum_i \gamma_{ij} = \sum_j \gamma_{ij} = 0. \tag{28}$$

It is easily seen that (27) and (28) uniquely determine μ and the α's, β's, and γ's. Using a dot to denote averaging over the indicated subscript, we find by averaging (27) over both i and j, and separately over i and over j that

$$\xi_{\cdot\cdot} = \mu, \qquad \xi_{i\cdot} = \mu + \alpha_i, \qquad \xi_{\cdot j} = \mu + \beta_j$$

and hence that

$$\mu = \xi_{\cdot\cdot}, \qquad \alpha_i = \xi_{i\cdot} - \xi_{\cdot\cdot}, \qquad \beta_j = \xi_{\cdot j} - \xi_{\cdot\cdot} \tag{29}$$

and

$$\gamma_{ij} = \xi_{ij} - \xi_{i\cdot} - \xi_{\cdot j} + \xi_{\cdot\cdot}. \tag{30}$$

Thus α_i is the average effect (averaged over the levels of the second factor) of the first factor at level i, and β_j is the corresponding effect of the second factor at level j. The quantity γ_{ij} can be written as

$$\gamma_{ij} = (\xi_{ij} - \xi_{..}) - [(\xi_{i.} - \xi_{..}) + (\xi_{.j} - \xi_{..})]. \tag{31}$$

It is therefore the difference between the joint effect of the two treatments at levels i and j, respectively, and the sum of the separate effects $\alpha_i + \beta_j$. The quantity γ_{ij} is called the *interaction* of the two factors when they are at levels i and j, respectively.

The UMVU estimators of these various effects follow immediately from Theorem 4.1 and Example 4.1 (continued). This example shows that the UMVU estimator of ξ_{ij} is $X_{ij.}$ and the associated estimators of the various parameters are thus

$$\hat{\mu} = X_{...}, \qquad \hat{\alpha}_i = X_{i..} - X_{...}, \qquad \beta_j = X_{.j.} - X_{...} \tag{32}$$

and

$$\hat{\gamma}_{ij} = X_{ij.} - X_{i..} - X_{.j.} + X_{...}. \tag{33}$$

The UMVU estimator of σ^2 is

$$\frac{1}{IJ(m-1)} \Sigma\Sigma\Sigma (X_{ijk} - X_{ij.})^2. \tag{34}$$

These results for the two-way layout easily generalize to other *factorial experiments*, that is, experiments concerning the joint effect of several factors, provided the numbers of observations at the various combinations of factor levels are equal. Theorems 4.3 and 4.4, of course, apply without this restriction, but then the situation is less simple.

Model (4) assumes that the random variables X_i are independently normally distributed with common unknown variance σ^2 and means ξ_i, which are subject to certain linear restrictions. We shall now consider some models that retain the linear structure but drop the assumption of normality.

(i) A very simple treatment is possible if one is willing to restrict attention to unbiased estimators that are linear functions of the X_i and to squared error loss. Suppose we retain from (4) only the assumptions about the first and second moments of the X_i, namely

$$\begin{aligned} &E(X_i) = \xi_i, \qquad \underline{\xi} \in \Pi_\Omega \\ &\operatorname{var}(X_i) = \sigma^2, \qquad \operatorname{cov}(X_i, X_j) = 0 \qquad \text{for} \quad i \neq j. \end{aligned} \tag{35}$$

Thus, both the normality and independence assumptions are dropped.

Theorem 4.5 (Gauss' theorem on least squares)*. *Under assumptions* (35), $\Sigma_{i=1}^{n}\gamma_i\hat{\xi}_i$ *of Theorem* 4.2 *is UMVU among all linear estimators of* $\Sigma_{i=1}^{n}\gamma_i\xi_i$.

Proof. The estimator is still unbiased, since the expectations of the X_i are the same under (35) as under (4). Let $\Sigma_{i=1}^{n}c_iX_i$ be any other linear unbiased estimator of $\Sigma_{i=1}^{n}\gamma_i\xi_i$. Since $\Sigma_{i=1}^{n}\gamma_i\hat{\xi}_i$ is UMVU in the normal case, and variances of linear functions of the X_i depend only on first and second moments, it follows that $\operatorname{var}\Sigma_{i=1}^{n}\gamma_i\hat{\xi}_i \leqq \operatorname{var}\Sigma_{i=1}^{n}c_iX_i$. Hence $\Sigma_{i=1}^{n}\gamma_i\hat{\xi}_i$ is UMVU among linear unbiased estimators.

Corollary 4.2. *Under the assumptions* (35) *and with squared error loss,* $\Sigma_{i=1}^{n}\gamma_i\hat{\xi}_i$ *is MRE with respect to the transformations* (10) *among all linear equivariant estimators of* $\Sigma_{i=1}^{n}\gamma_i\xi_i$.

Proof. This follows from the argument of Lemma 1.3, since $\Sigma_{i=1}^{n}\gamma_i\hat{\xi}_i$ is UMVU and equivariant.

For estimating σ^2, it is natural to restrict attention to unbiased quadratic (rather than linear) estimators Q of σ^2. Among these, does the estimator $S^2/(n-s)$ which is UMVU in the normal case continue to minimize the variance? Under mild additional restrictions—for example, invariance under the transformations (10) or restrictions to Q's taking on only positive values—it turns out that this is true in some cases (for instance, in Example 4.6 below when the n_i are equal) but not in others. For details, see Hsu (1938) and Rao (1952).

Example 4.6. Let X_{ij} ($j = 1,\dots, n_i$; $i = 1,\dots, s$) be independently distributed with means $E(X_{ij}) = \xi_i$ and common variance and fourth moment

$$\sigma^2 = E(X_{ij} - \xi_i)^2 \quad \text{and} \quad \beta = E(X_{ij} - \xi_i)^4/\sigma^4.$$

Consider estimators of σ^2 of the form $Q = \Sigma\lambda_i S_i^2$ where $S_i^2 = \Sigma(X_{ij} - X_{i\cdot})^2$ and $\Sigma\lambda_i(n_i - 1) = 1$ so that Q is an unbiased estimator of σ^2. Then the variance of Q is minimized (Problem 4.16) when the λ's are proportional to $1/(\alpha_i + 2)$ where $\alpha_i = [(n_i - 1)/n_i](\beta - 3)$. The standard choice of the λ_i (which is to make them equal) is, therefore, best if either the n_i are equal or $\beta = 3$, which is the case when the X_{ij} are normal.

(ii) Let us now return to the model obtained from (4) by dropping the assumption of normality but without restricting attention to linear estimators. More specifically, we shall assume that $X_1,\dots, X_n$ are random varia-

*This theorem, which is frequently referred to as the Gauss–Markov theorem, has been extensively generalized. See, for example, Rao (1976).

bles such that

(36) the variables $X_i - \xi_i$ are iid with a common distribution F which has expectation zero and an otherwise unknown probability density f,

and such that (13) holds with A an $s \times n$ matrix of rank s.

In Section 2.4, we found that for the case $\xi_i = \theta$, the LSE $\bar{X}$ of θ is UMVU in this nonparametric model. To show that the corresponding result does not generally hold when ξ is given by (13), consider the two-way layout of Example 4.5 and the estimation of

$$\alpha_i = \xi_{i\cdot} - \xi_{\cdot\cdot} = \frac{1}{IJ} \sum_{j=1}^{I} \sum_{k=1}^{J} (\xi_{ik} - \xi_{jk}). \tag{37}$$

To avoid calculations, suppose that F is t_2, the t-distribution with 2 degrees of freedom. Then the least squares estimators have infinite variance. On the other hand, let $\tilde{X}_{ij}$ be the median of the observations $X_{ij\nu}$, $\nu = 1, \dots, m$. Then $\tilde{X}_{ik} - \tilde{X}_{jk}$ is an ubiased estimator of $\xi_{ik} - \xi_{jk}$ so that $\delta = (1/ab)\Sigma\Sigma(\tilde{X}_{ik} - \tilde{X}_{jk})$ is an unbiased estimator of α_i. Furthermore, if $m \geqq 3$, the $\tilde{X}_{ij}$ have finite variance and so, therefore, does δ. (A sum of random variables with finite variance has finite variance.) This shows that the least squares estimators of the α_i are not UMVU when F is unknown. The same argument applies to the β's and γ's.

The situation is quite different for the estimation of μ. Let $\mathfrak{U}$ be the class of unbiased estimators of μ in model (27) with F unknown, and let $\mathfrak{U}'$ be the corresponding class of unbiased estimators when the α's, β's, and γ's are all zero. Then clearly $\mathfrak{U} \subset \mathfrak{U}'$; furthermore, it follows from Section 2.4 that $X_{\dots}$ uniformly minimizes the variance within $\mathfrak{U}'$. Since $X_{\dots}$ is a member of $\mathfrak{U}$, it uniformly minimizes the variance within $\mathfrak{U}$ and hence is UMVU for μ in model (27) when F is unknown.

For a more detailed discussion of this problem, see Anderson (1962).

(iii) Instead of assuming the density f in (36) to be unknown, we may be interested in the case in which f is known but not normal. The model then remains invariant under the transformations

$$X'_\nu = X_\nu + \sum_{j=1}^{s} a_{j\nu}\gamma_j, \qquad -\infty < \gamma_1, \dots, \gamma_s < \infty. \tag{38}$$

Since $E(X'_\nu) = \Sigma a_{j\nu}(\theta_j + \gamma_j)$, the induced transformations in the parameter space are given by

$$\theta'_j = \theta_j + \gamma_j \qquad (j = 1, \dots, s). \tag{39}$$

The problem of estimating θ_j remains invariant under the transformations (38), (39), and

$$d' = d + \gamma_j \tag{40}$$

for any loss function of the form $\rho(d - \theta_j)$, and an estimator δ of θ_j is equivariant with respect to these transformations if it satisfies

$$\delta(\underline{X}') = \delta(\underline{X}) + \gamma_j. \tag{41}$$

Since (39) is transitive over Ω, the risk of any equivariant estimator is constant, and an MRE estimator of θ_j can be found by generalizing Theorems 1.2 and 1.3 to the present situation (see Verhagen, 1961).

(iv) Two important extensions to general exponential families and to robust estimation will be taken up in the next section and at the end of Section 5.6, respectively.

5. EXPONENTIAL LINEAR MODELS

The great success of the normal linear models described in the preceding section suggests the desirability of extending these models beyond the normal case. A natural generalization combines a general exponential family with the structure of a linear model by assuming that the parameter vector $(\eta_1, \ldots, \eta_s)$ in the exponential family (1.4.2) is restricted to the intersection of an r-dimensional space $(r < s)$ with the natural parameter space (which may not be a linear space). This provides a large and flexible class of models which (together with a further generalization) will be illustrated in this section with three of its principal areas of application. If the exponential family is of full rank, the restricted model will continue to have a set of sufficient statistics that is complete, so that all U-estimable functions of the η's have UMVU estimators. Unfortunately, these are not always as satisfactory as they are in the normal case. Equivariance tends to play a small role in these models; although they are therefore somewhat out of place in this Chapter, it seems convenient to present them immediately after the linear models.

Random Effects and Mixed Models

In many applications of linear models, the effects of the various factors $A, B, C, \ldots$ which were considered to be unknown constants in Section 3.4 are instead random. One then speaks of a *random effects* model (or Model II); in contrast, the corresponding model of Section 3.4 is a *fixed effects* model (or Model I). If both fixed and random effects occur, the model is said to be *mixed.*

Example 5.1. Suppose that as a measure of quality control an auto manufacturer tests a sample of new cars, observing for each car the mileage achieved on a number of occasions on a gallon of gas. Suppose X_{ij} is the mileage of the ith car on the jth occasion, at time t_{ij}, with all the t_{ij} being selected at random and independently of each other. This would have been modeled in Example 4.1 as

$$X_{ij} = \mu + \alpha_i + U_{ij}$$

where the U_{ij} are independent $N(0, \sigma^2)$. Such a model would be appropriate if these particular cars were the object of study and a replication of the experiment thus consisted of a number of test runs by the same cars. However, the manufacturer is interested in the performance of the thousands of cars to be produced that year and for this reason has drawn a random sample of cars for the test. A replication of the experiment would start by drawing a new sample. The effect of the ith car is therefore a random variable, and the model becomes

$$(1) \qquad X_{ij} = \mu + A_i + U_{ij} \qquad (j = 1, \dots, n_i; i = 1, \dots, s).$$

Here and following, the populations being sampled are assumed to be large enough so that independence and normality of the unobservable random variables A_i and U_{ij} can be assumed as a reasonable approximation. Without loss of generality, one can put $E(A_i) = E(U_{ij}) = 0$ since the means can be absorbed into μ. The variances will be denoted by $\text{var}(A_i) = \sigma_A^2$, $\text{var}(U_{ij}) = \sigma^2$.

The X_{ij} are dependent and their joint distribution, and hence the estimation of σ_A^2 and σ^2, is greatly simplified if the model is assumed to be *balanced*, that is, to satisfy $n_i = n$ for all i. In that case, in analogy with the transformation (4.5), let each set $(X_{i1}, \dots, X_{in})$ be subjected to an orthogonal transformation to $(Y_{i1}, \dots, Y_{in})$ such that $Y_{i1} = \sqrt{n}\, X_{i\cdot}$. An additional orthogonal transformation is made from $(Y_{11}, \dots, Y_{s1})$ to $(Z_{11}, \dots, Z_{s1})$ such that $Z_{11} = \sqrt{s}\, Y_{\cdot 1}$, while for $i > 1$ we put $Z_{ij} = Y_{ij}$. Unlike the X_{ij}, the Y_{ij} and Z_{ij} are all independent (Problem 5.1). They are normal with means

$$E(Z_{11}) = \sqrt{sn}\,\mu; \qquad E(Z_{ij}) = 0 \qquad \text{if} \quad i > 1 \text{ or } j > 1$$

and variances

$$\text{var}(Z_{i1}) = \sigma^2 + n\sigma_A^2; \qquad \text{var}(Z_{ij}) = \sigma^2 \qquad \text{for} \quad j > 1,$$

so that the joint density of the Z's is proportional to

$$(2) \qquad \exp\left\{ -\frac{1}{2(\sigma^2 + n\sigma_A^2)} \left[(Z_{11} - \sqrt{sn}\,\mu)^2 + S_A^2 \right] - \frac{1}{2\sigma^2} S^2 \right\}$$

with

$$S_A^2 = \sum_{i=2}^{s} Z_{i1}^2 = n\Sigma(X_{i\cdot} - X_{\cdot\cdot})^2, \quad S^2 = \sum_{i=1}^{s}\sum_{j=2}^{n} Z_{ij}^2 = \sum_{i=1}^{s}\sum_{j=1}^{n} (X_{ij} - X_{i\cdot})^2.$$

This is a three-parameter exponential family with

$$\eta_1 = \frac{\mu}{\sigma^2 + n\sigma_A^2}, \quad \eta_2 = \frac{1}{\sigma^2 + n\sigma_A^2}, \quad \eta_3 = \frac{1}{\sigma^2}.$$

The variance of X_{ij} is $\text{var}(X_{ij}) = \sigma^2 + \sigma_A^2$, and we are interested in estimating the *variance components* σ_A^2 and σ^2. Since

$$E\left(\frac{S_A^2}{s-1}\right) = \sigma^2 + n\sigma_A^2 \quad \text{and} \quad E\left(\frac{S^2}{s(n-1)}\right) = \sigma^2,$$

it follows that

$$\hat{\sigma}^2 = \frac{S^2}{s(n-1)} \quad \text{and} \quad \hat{\sigma}_A^2 = \frac{1}{n}\left[\frac{S_A^2}{s-1} - \frac{S^2}{s(n-1)}\right] \tag{3}$$

are UMVU estimators of σ^2 and σ_A^2, respectively. The UMVU estimator of the ratio σ_A^2/σ^2 is

$$\frac{1}{n}\left[\frac{K_{f,-2}}{s(n-1)}\frac{\hat{\sigma}^2 + n\hat{\sigma}_A^2}{\hat{\sigma}^2} - 1\right],$$

where $K_{f,-2}$ is given by (2.2.5) with $f = s(n-1)$ (Problem 5.3). Typically, the only linear subspace of the η's of interest here is the trivial one defined by $\sigma_A^2 = 0$, which corresponds to $\eta_2 = \eta_3$ and to the case in which the sn X_{ij} are iid as $N(\mu, \sigma^2)$.

Example 5.2. Balanced two-way layout. In analogy to Example 4.5, consider next the random effects two-way layout

$$X_{ijk} = \mu + A_i + B_j + C_{ij} + U_{ijk} \tag{4}$$

where the unobservable random variables A_i, B_j, C_{ij}, and U_{ijk} are independently normally distributed with zero mean and with variances σ_A^2, σ_B^2, σ_C^2, and σ^2, respectively. We shall restrict attention to the balanced case $i = 1,\ldots, I$; $j = 1,\ldots, J$; $k = 1,\ldots, n$. As in the preceding example, a linear transformation leads to independent normal variables Z_{ijk} with means $E(Z_{111}) = \sqrt{IJn}\,\mu$ and $= 0$ for all other Z's, and with variances

$$
\begin{aligned}
&\operatorname{var}(Z_{111}) = nJ\sigma_A^2 + nI\sigma_B^2 + n\sigma_C^2 + \sigma^2, \\
&\operatorname{var}(Z_{i11}) = nJ\sigma_A^2 + n\sigma_C^2 + \sigma^2, \qquad i > 1 \\
(5) \qquad &\operatorname{var}(Z_{1j1}) = nI\sigma_B^2 + n\sigma_C^2 + \sigma^2, \qquad j > 1 \\
&\operatorname{var}(Z_{ij1}) = n\sigma_C^2 + \sigma^2, \qquad i, j > 1 \\
&\operatorname{var}(Z_{ijk})^2 = \sigma^2, \qquad k > 1.
\end{aligned}
$$

As an example in which such a model might arise, consider a reliability study of blood counts, in which blood samples from each of J patients are divided into nI subsamples of which n are sent to each of I laboratories. The study is not concerned with these particular patients and laboratories, which instead are assumed to be random samples from suitable patient and laboratory populations. From (4) it follows that $\operatorname{var}(X_{ijk}) = \sigma_A^2 + \sigma_B^2 + \sigma_C^2 + \sigma^2$. The terms on the right are the variance components due to laboratories, patients, the interaction between the two, and the subsamples from a patient.

The joint distribution of the Z_{ijk} constitutes a five-parameter exponential family with the complete set of sufficient statistics (Problem 5.9)

$$
\begin{aligned}
&S_A^2 = \sum_{i=2}^{I} Z_{i11}^2 = nJ\sum_{i=1}^{I}(X_{i\cdot\cdot} - X_{\cdot\cdot\cdot})^2, \\
&S_B^2 = \sum_{j=2}^{J} Z_{1j1}^2 = nI\sum_{j=1}^{J}(X_{\cdot j\cdot} - X_{\cdot\cdot\cdot})^2, \\
(6) \qquad &S_C^2 = \sum_{i=2}^{I}\sum_{j=2}^{J} Z_{ij1}^2 = n\sum_{i=1}^{I}\sum_{j=1}^{J}(X_{ij\cdot} - X_{i\cdot\cdot} - X_{\cdot j\cdot} + X_{\cdot\cdot\cdot})^2, \\
&S^2 = \sum_{i=1}^{I}\sum_{j=1}^{J}\sum_{k=2}^{n} Z_{ijk}^2 = \sum_{i=1}^{I}\sum_{j=1}^{J}\sum_{k=1}^{n}(X_{ijk} - X_{ij\cdot})^2, \text{ and} \\
&Z_{111} = \sqrt{IJn}\, X_{\cdots}.
\end{aligned}
$$

From the expectations of these statistics, one finds the UMVU estimators of the variance components σ^2, σ_C^2, σ_A^2, and σ_B^2 to be

$$
\hat\sigma^2 = \frac{S^2}{IJ(n-1)}, \qquad \hat\sigma_C^2 = \frac{1}{n}\left[\frac{S_C^2}{(I-1)(J-1)} - \hat\sigma^2\right],
$$

$$
\hat\sigma_A^2 = \frac{1}{nJ}\left[\frac{S_A^2}{I-1} - n\hat\sigma_C^2 - \hat\sigma^2\right], \qquad \hat\sigma_B^2 = \frac{1}{nI}\left[\frac{S_B^2}{J-1} - n\hat\sigma_C^2 - \hat\sigma^2\right].
$$

A submodel of (4), which is sometimes appropriate is the *additive* model corresponding to the absence of the interaction terms C_{ij} and hence to the assumption $\sigma_C^2 = 0$. If $\eta_1 = \mu/\text{var}(Z_{111})$, $1/\eta_2 = nJ\sigma_A^2 + n\sigma_C^2 + \sigma^2$, $1/\eta_3 = nI\sigma_B^2 + n\sigma_C^2 + \sigma^2$, $1/\eta_4 = n\sigma_C^2 + \sigma^2$, and $1/\eta_5 = \sigma^2$, this assumption is equivalent to $\eta_4 = \eta_5$ and thus restricts the η's to a linear subspace. The submodel constitutes a four-parameter exponential family, with the complete set of sufficient statistics Z_{111}, S_A^2, S_B^2, and $S'^2 = S_C^2 + S^2 = \Sigma\Sigma\Sigma(X_{ijk} - X_{i..} - X_{.j.} + X_{...})^2$. The UMVU estimators of the variance components σ_A^2, σ_B^2, and σ^2 are now easily obtained as before (Problem 5.10).

Another submodel of (4) which is of interest is obtained by setting $\sigma_B^2 = 0$, thus eliminating the B_j-terms from (4). However, this model, which corresponds to the linear subspace $\eta_3 = \eta_4$, does not arise naturally in the situations leading to (4) as illustrated by the laboratory example. These situations are characterized by a *crossed* design in which each of the I A-units (laboratories) is observed in combination with each of the J B-units (patients). On the other hand, the model without the B terms arises naturally in the very commonly occurring *nested* design illustrated in the following example.

Example 5.3. Two nested random factors. For the two factors A and B, suppose that each of the units corresponding to different values of i (i.e., different levels of A) is itself a collection of smaller units from which the values of B are drawn. Thus, the A units might be hospitals, schools, or farms that constitute a random sample from a population of such units from each of which a random sample of patients, students, or trees is drawn. On each of the latter, a number of observations is taken (for example, a number of blood counts, grades, or weights of a sample of apples). The resulting model [with a slight change of notation from (4)] may be written as

$$X_{ijk} = \mu + A_i + B_{ij} + U_{ijk}. \tag{7}$$

Here the A's, B's, and U's are again assumed to be independent normal with zero means and variances σ_A^2, σ_B^2, and σ^2, respectively. In the balanced case ($i = 1,\dots, I$; $j = 1,\dots, J$; $k = 1,\dots, n$), a linear transformation produces independent variables with means $E(Z_{111}) = \sqrt{IJn}\,\mu$ and $= 0$ for all other Z's and variances

$$\text{var}(Z_{i11}) = \sigma^2 + n\sigma_B^2 + Jn\sigma_A^2 \qquad (i = 1,\dots, I),$$

$$\text{var}(Z_{ij1}) = \sigma^2 + n\sigma_B^2 \qquad (j > 1),$$

$$\text{var}(Z_{ijk}) = \sigma^2 \qquad (k > 1).$$

The joint distribution of the Z's constitutes a four parameter exponential family with the complete set of sufficient statistics

$$S_A^2 = \sum_{i=2}^{I} Z_{i11}^2 = Jn\Sigma(X_{i\cdot\cdot} - X_{\cdot\cdot\cdot})^2,$$

$$(8) \qquad S_B^2 = \sum_{j=2}^{J} Z_{1j1}^2 = n\Sigma\Sigma(X_{ij\cdot} - X_{i\cdot\cdot})^2,$$

$$S^2 = \sum_{i=1}^{I}\sum_{j=1}^{J}\sum_{k=2}^{n} Z_{ijk}^2 = \sum_{i=1}^{I}\sum_{j=1}^{J}\sum_{k=1}^{n}(X_{ijk} - X_{ij\cdot})^2,$$

$$Z_{111} = \sqrt{IJn}\, X_{\cdot\cdot\cdot},$$

and the UMVU estimators of the variance components can be obtained as before (Problem 5.12).

The models illustrated in Examples 5.2 and 5.3 extend in a natural way to more than two factors, and in the balanced cases the UMVU estimators of the variance components are easily derived.

The estimation of variance components described above suffers from two serious difficulties.

(i) The UMVU estimators of all the variance components except σ^2 can take on negative values with probabilities as high as .5 and even in excess of that value (Problems 5.5–5.7) (and, correspondingly, their expected squared errors are quite unsatisfactory; see Klotz, Milton, and Zacks, 1969). The interpretation of such negative values either as indications that the associated components are negligible (which is sometimes formalized by estimating them to be zero) or that the model is incorrect is not always convincing because negative values do occur even when the model is correct and the components are positive. An alternative possibility, here and throughout this section, is to fall back on maximum likelihood estimation or to the restricted MLE's obtained by maximizing the likelihood after first reducing the data through location invariance (Thompson, 1962). These methods have, however, no small-sample justification. A promising approach has been suggested by Hartung (1981), who minimizes the bias, subject to the condition of non-negativity. Still another class of estimators will be discussed in Section 4.1.

(ii) Models as simple as those obtained in Examples 5.1–5.3 are not available when the layout is not balanced.

The joint density of the X's can then be obtained by noting that they are linear functions of normal variables and thus have a joint multivariate normal distribution. To obtain it, one only need write down the covariance matrix of the X's and invert it. The result is an exponential family which typically is not complete unless the model is balanced. (This is illustrated for the one-way layout in Problem 5.4.) UMVU estimators cannot be expected in this case (see Pukelsheim, 1981). A characterization of U-estimable functions permitting UMVU estimators is given by Unni (1978). Two general methods for the estimation of variance components have been developed in some detail; these are maximum and restricted maximum likelihood, and the minimum norm quadratic unbiased estimation (Minque) introduced by Rao (1970). Surveys of the area are given by Searle (1971), Harville (1977), and Kleffe (1977). More detailed introductions can be found, for example, in the books by Kendall and Stuart (1968) and Graybill (1976). An extensive bibliography is provided by Sahai (1979).

So far we have restricted attention to situations in which the factors are either all fixed or all random. We conclude the present subsection by discussing one example of a mixed model.

Example 5.4. In Example 5.3 it was assumed that the hospitals, schools, or farms were obtained as a random sample from a population of such units. Let us now suppose that it is only these particular hospitals that are of interest (perhaps it is the set of all hospitals in the city) while the patients continue to be drawn at random from these hospitals. Instead of (7), we shall assume that the observations are given by

$$X_{ijk} = \mu + \alpha_i + B_{ij} + U_{ijk} \qquad (\Sigma\alpha_i = 0). \tag{9}$$

A transformation very similar to the earlier one (Problem 5.14) now leads to independent normal variables W_{ijk} with joint density proportional to

$$\exp\left\{-\frac{1}{2(\sigma^2 + n\sigma_B^2)}\left[\Sigma(w_{i11} - \mu - \alpha_i) + S_B^2\right] - \frac{1}{2\sigma^2}S^2\right\} \tag{10}$$

with S_B^2 and S^2 given by (8), and with $W_{i11} = \sqrt{Jn}\,X_{i\cdot\cdot}$. This is an exponential family with the complete set of sufficient statistics $X_{i\cdot\cdot}$, S_B^2, and S^2. The UMVU estimators of σ_B^2 and σ^2 are the same as in Example 5.3, whereas the UMVU estimator of α_i is $X_{i\cdot\cdot} - X_{\cdots}$ as it would be if the B's were fixed.

Contingency Tables

Suppose, next, that the underlying exponential family is the set of multinomial distributions (1.4.4), which may be written as

$$\exp\left(\sum_{i=0}^{s} x_i \log p_i\right) h(x), \tag{11}$$

and that a linear structure is imposed on the parameters $\eta_i = \log p_i$. Expositions of the resulting theory of *log linear models* can be found in the books by Bishop, Fienberg, and Holland (1975), Haberman (1974), and Plackett (1974). The models have close formal similarities with the corresponding normal models, and a natural linear subspace of the $\log p_i$ often corresponds to a natural restriction on the p's. In particular, since sums of the log p's correspond to products of the p's, a subspace defined by setting suitable interaction terms equal to zero often is equivalent to certain independence properties in the multinomial model.

The exponential family (11) is not of full rank since the p's must add up to 1. A full-rank form is

$$\left[\exp \sum_{i=1}^{s} x_i \log(p_i/p_0)\right] h(x). \tag{12}$$

If we let

$$\eta_i' = \log \frac{p_i}{p_0} = \eta_i - \eta_0, \tag{13}$$

we see that arbitrary linear functions of the η_i' correspond to arbitrary contrasts (i.e., functions of the differences) of the η_i. From Example 2.3.6, it follows that $(X_1, \ldots, X_s)$ or $(X_0, X_1, \ldots, X_s)$ is sufficient and complete for (12) and hence also for (11). In applications, we shall find (11) the more convenient form to use.

If the η's are required to satisfy r independent linear restrictions $\Sigma a_{ij}\eta_j = b_i$ $(i = 1, \ldots, r)$, the resulting distributions will form an exponential family of rank $s - r$, and the associated minimal sufficient statistics T will continue to be complete. Since $E(X_i/n) = p_i$, the probabilities p_i are always U-estimable; their UMVU estimators can be obtained as the conditional expectations of X_i/n given T. If $\hat{p}_i$ is the UMVU estimator of p_i, a natural estimator of η_i is $\hat{\eta}_i = \log \hat{p}_i$, but, of course, this is no longer unbiased. In fact, no unbiased estimator of η_i exists because only polynomials of the p_i can be U-estimable (Problem 2.3.23). When $\hat{p}_i$ is also the MLE of p_i, $\hat{\eta}_i$ is the MLE of η_i. However, the MLE $\hat{\hat{p}}_i$ does not always coincide with the UMVU estimator $\hat{p}_i$. An example of this possibility with $\log p_i = \alpha + \beta t_i$ (t's known; α and β unknown) is given by Haberman (1974, pp. 29 and 64). It is a disadvantage of the $\hat{p}_i$ in this case that, unlike $\hat{\hat{p}}_i$ they do not always satisfy the restrictions of the model, that is, for some values of the X's no α, β exist for which $\log \hat{p}_i = \alpha + \beta t_i$. Typically, if $\hat{p}_i \neq \hat{\hat{p}}_i$, the difference between the two is moderate.

For estimating the η_i, Goodman (1970) has recommended in some cases applying the estimators not to the cell frequencies X_i/n but to $X_i/n + \frac{1}{2}$, in order to decrease the bias of the MLE. This procedure also avoids difficulties that may arise when some of the cell counts are zero.

Example 5.5. ***Two-way contingency table.*** Consider the situation of Example 2.3.7 in which n subjects are classified according to two characteristics A and B with possible outcomes $A_1, \ldots, A_I$ and $B_1, \ldots, B_J$. If n_{ij} is the number of subjects with properties A_i and B_j, the joint distribution of the n_{ij} can be written as

$$\frac{n!}{\Pi_{i,j}(n_{ij})!} \exp \Sigma\Sigma n_{ij}\xi_{ij}, \qquad \xi_{ij} = \log p_{ij}.$$

Write $\xi_{ij} = \mu + \alpha_i + \beta_j + \gamma_{ij}$ as in Example 4.5, with the side conditions (4.28). This implies no restrictions since any IJ numbers ξ_{ij} can be represented in this form. The p_{ij} must, of course, satisfy $\Sigma\Sigma p_{ij} = 1$ and the ξ_{ij} therefore $\Sigma \exp \xi_{ij} = 1$. This equation determines μ as a function of the α's, β's, and γ's which are free, subject only to (4.28). The UMVU estimators of the p_{ij} were seen in Example 2.3.7 to be n_{ij}/n.

In Example 4.5 (normal two-way layout) it is sometimes reasonable to suppose that all the γ_{ij}'s (the interactions) are zero. In the present situation, this corresponds exactly to the assumption that the characteristics A and B are independent, that is, that $p_{ij} = p_{i+}p_{+j}$ (Problem 5.17). The UMVU estimator of p_{ij} is now $n_{i+}n_{+j}/n^2$.

Example 5.6. ***Conditional independence in a three-way table.*** In Example 2.3.8, it was assumed that the subjects are classified according to three characteristics A, B, and C and that conditionally, given outcome C, the two characteristics A and B are independent. If $\xi_{ijk} = \log p_{ijk}$ and ξ_{ijk} is written as

$$\xi_{ijk} = \mu + \alpha_i^A + \alpha_j^B + \alpha_k^C + \alpha_{ij}^{AB} + \alpha_{ik}^{AC} + \alpha_{jk}^{BC} + \alpha_{ijk}^{ABC}$$

with the α's subject to the usual restrictions and with μ determined by the fact that the p_{ijk} add up to 1, it turns out that the conditional independence of A and B given C is equivalent to the vanishing of both the three-way interactions α_{ijk}^{ABC} and the A, B-interactions α_{ij}^{AB} (Problem 5.18). The UMVU estimators of the p_{ijk} in this model were obtained in Example 2.3.8.

Independent Binomial Experiments

The submodels considered in Examples 5.2–5.6 corresponded to natural assumptions about the variances or probabilities in question. However, in general the assumption of linearity in the η's made at the beginning of this section is rather arbitrary and is dictated by mathematical convenience rather than by meaningful structural assumptions. We shall now consider a particularly simple class of problems, in which this linearity assumption is inconsistent with more customary assumptions. Agreement with these assumptions can be obtained by not insisting on a linear structure for the parameters η_i themselves but permitting a linear structure for a suitable function of the η's.

The problems are concerned with a number of independent random variables X_i having the binomial distributions $b(p_i, n_i)$. Suppose the X's have been obtained from some unobservable variables Z_i distributed independently as $N(\zeta_i, \sigma^2)$ by setting

$$X_i = \begin{cases} 0 & \text{if} \quad Z_i \leqslant u \\ 1 & \text{if} \quad Z_i > u. \end{cases} \tag{14}$$

Then

$$p_i = P(Z_i > u) = \Phi\left(\frac{\zeta_i - u}{\sigma}\right) \tag{15}$$

and hence

$$\zeta_i = u + \sigma\Phi^{-1}(p_i). \tag{16}$$

Now consider a two-way layout for the Z's in which the effects are additive, as in Example 4.5. The subspace of the ζ_{ij} $(i = 1,\dots, a; j = 1,\dots, b)$ defining this model is characterized by the fact that the interactions satisfy

$$\gamma_{ij} = \zeta_{ij} - \zeta_{i\cdot} - \zeta_{\cdot j} + \zeta_{\cdot\cdot} = 0 \tag{17}$$

which by (16) implies that

(18)

$$\Phi^{-1}(p_{ij}) - \frac{1}{J}\sum_j \Phi^{-1}(p_{ij}) - \frac{1}{I}\sum_i \Phi^{-1}(p_{ij}) + \frac{1}{IJ}\sum_i\sum_j \Phi^{-1}(p_{ij}) = 0.$$

The "natural" linear subspace of the parameter space for the Z's thus translates into a linear subspace in terms of the parameters $\Phi^{-1}(p_{ij})$ for the X's, and the corresponding fact by (16) is true quite generally for subspaces defined in terms of differences of the ζ's. On the other hand, the joint distribution of the X's is proportional to

$$\exp\left[\Sigma x_i \log \frac{p_i}{q_i}\right] h(x), \tag{19}$$

and the natural parameters of this exponential family are $\eta_i = \log(p_i/q_i)$. The restrictions (18) are not linear in the η's, and the minimal sufficient statistics for the exponential family (19) with the restrictions (18) are not complete.

It is interesting to ask whether there exists a distribution F for the underlying variables Z_i such that a linear structure for the ζ_i will result in a linear structure for $\eta_i = \log(p_i/q_i)$ when the p_i and the ζ_i are linked by the equation

$$q_i = P(Z_i \leqslant u) = F(u - \zeta_i) \tag{20}$$

instead of by (15). Then $\zeta_i = u - F^{-1}(q_i)$ so that linear functions of the ζ_i correspond to linear functions of the $F^{-1}(q_i)$ and hence of $\log(p_i/q_i)$ provided

$$F^{-1}(q_i) = a - b\log\frac{p_i}{q_i}. \tag{21}$$

Suppressing the subscript i and putting $x = a - b\log(p/q)$, we see that (21) is equivalent to

$$q = F(x) = \frac{1}{1 + e^{-(x-a)/b}}, \tag{22}$$

which is the cdf of the logistic distribution $L(a, b)$ whose density is shown in Table 1.3.1.

Inferences based on the assumption of linearity in the $\Phi^{-1}(p_i)$ and the $\log(p_i/q_i) = F^{-1}(q_i)$ with F given by (22) where, without loss of generality, we can take $a = 0$, $b = 1$, are known as *probit* and *logit analysis*, respectively. For more details and many examples, see Cox (1970) and also Bishop, Fienberg, and Holland (1975). As is shown by Cox (p. 28), the two analyses may often be expected to give very similar results, provided the p's are not too close to 0 or 1.

The outcomes of s independent binomial experiments can be represented by a $2 \times s$ contingency table, as in Table 2.3.1, with $I = 2$, $J = s$, and the outcomes A_1 and A_2 corresponding to success and failure, respectively. The column totals $n_{+1}, \ldots, n_{+s}$ are simply the s sample sizes, and are therefore fixed in the present model. In fact, this is the principal difference between the present model and that assumed for a $2 \times J$ table in Example 2.3.7. The case of s independent binomials arises in the situation of that example, if the n subjects, instead of being drawn at random from the population at large, are obtained by drawing n_{+j} subjects from the subpopulation having property B_j for $j = 1, \ldots, s$.

A $2 \times J$ contingency table, with fixed column totals and with the distribution of the cell counts given by independent binomials, occurs not only in its own right through the sampling of $n_{+1}, \ldots, n_{+J}$ subjects from categories $B_1, \ldots, B_J$, respectively, but also in the multinomial situation of Example 5.5 with $I = 2$, as the conditional distribution of the cell counts given the column totals. This relationship leads to an apparent paradox. In the conditional model, the UMVU estimator of the probability $p_j = p_{1j}/(p_{1j} + p_{2j})$ of success, given that the subject is in B_j, is $\delta_j = n_{1j}/n_{+j}$. Since δ_j satisfies

$$E(\delta_j | B_j) = p_j, \tag{23}$$

it appears also to satisfy $E(\delta_j) = p_j$ and hence to be an unbiased estimator of $p_{1j}/(p_{1j} + p_{2j})$ in the original multinomial model. On the other hand, an easy extension of the argument of Example 2.3.1 (see Problem 2.3.23) shows that in this model only polynomials in the p_{ij} can be U-estimable, and the ratio in question clearly is not a polynomial.

The explanation lies in the tacit assumption made in (23) that $n_{+j} > 0$ and in the fact that δ_j is not defined when $n_{+j} = 0$. To ensure at least one observation in B_j, one needs a sampling scheme under which an arbitrarily large number of observations is possible. For such a scheme, the U-estimability of $p_{1j}/(p_{1j} + p_{2j})$ would no longer be surprising.

It is clear from the discussion leading to (18) that the generalization of normal linear models to models linear in the natural parameters η_i of an exponential family is too special and that instead linear spaces in suitable functions of the η_i should be permitted. Because in exponential families the parameters of primary interest often are the expectations $\theta_i = E(T_i)$ [for example in (19) the $p_i = E(X_i)$], generalized linear models are typically defined by restricting the parameters to lie in a space defined by linear conditions on $v(\theta_i)$ [or in some cases $v_i(\theta_i)$] for a suitable link-function v (linking the θ's with the linear space). A theory of such models was developed by Dempster (1971) and Nelder and Wedderburn (1972), who in particular discuss maximum likelihood estimation of the parameters. Further aspects are treated in Wedderburn (1976), in lecture notes by Stuart (1977), and in Pregibon (1980). A generalized linear interactive modeling (GLIM) package has been developed by Baker and Nelder (1978).

6. SAMPLING FROM A FINITE POPULATION

In the location-scale models of Sections 3.1 and 3.3, and the more general linear models of Section 3.4, observations are measurements that are subject to random errors. The parameters to be estimated are the true values of the quantities being measured, or differences and other linear functions of these values, and the variance of the measurement errors. We shall now consider a class of problems in which the measurements are assumed to be without error, but in which the observations are nevertheless random because the subjects (or objects) being observed are drawn at random from a finite population.

Problems of this kind occur whenever one wishes to estimate the average income, days of work lost to illness, reading level, or the proportion of a population supporting some measure or candidate. The elements being sampled need not be human but may be trees, food items, financial records, schools, and so on. We shall consider here only the simplest sampling schemes. For a fuller account of the principal methods of sampling, see, for example, Cochran (1977); a systematic treatment of the more theoretical aspects is given by Cassel, Särndal, and Wretman (1977).

The prototype of the problems to be considered is the estimation of a population average on the basis of a simple random sample from that population. In order to draw a random sample, one needs to be able to identify the members of the population. Telephone subscribers, for example, can conveniently be identified by the page and position on the page, trees by their coordinates, and students in a class by their names or by the row and number of their seat. In general, a list or other identifying description of the members of the population is called a *frame*. To represent the sampling frame suppose that N population elements are labeled $1, \ldots, N$; in addition, a value a_i (the quantity of interest) is associated with the element i. (This notation is somewhat misleading because, in any realization of the model, the a's will simply be N real numbers without identifying subscripts.) For the purpose of estimating $\bar{a} = \sum_{i=1}^{N} a_i/N$, a sample of size n is drawn in order, one element after another, without replacement. It is a *simple random sample* if all $N(N-1)\ldots(N-n+1)$ possible n-tuples are equally likely.

The data resulting from such a sampling process consist of the n labels of the sampled elements and the associated a values, in the order in which they were drawn, say

$$X = \{(I_1, Y_1), \ldots, (I_n, Y_n)\} \tag{1}$$

where the I's denote the labels, and the Y's the associated a values, $Y_k = a_{I_k}$. The unknown aspect of the situation, which as usual we shall denote by θ, is the set of population a values of the N elements,

$$\theta = \{(1, a_1), \ldots, (N, a_N)\}. \tag{2}$$

In the classic approach to sampling, the labels are discarded. Let us for a moment follow this approach, so that what remains of the data is the set of

n observed a values: $Y_1, \ldots, Y_n$. Under simple random sampling, the order statistics $Y_{(1)} \leqslant \cdots \leqslant Y_{(n)}$ are then sufficient. To obtain UMVU estimators of $\bar{a}$ and other functions of the a's, one needs to know whether this sufficient statistic is complete. The answer depends on the parameter space Ω, which we have not yet specified.

It frequently seems reasonable to assume that the set V of possible values is the same for each of the a's and does not depend on the values taken on by the other a's. (This would not be the case, for example, if the a's were the grades obtained by the students in a class which is being graded "on the curve.") The parameter space is then the set Ω of all θ's given by (2) with $(a_1, \ldots, a_N)$ in the Cartesian product

$$V \times V \times \cdots \times V. \tag{3}$$

Here V may, for example, be the set of all real numbers, all positive real numbers, or all positive integers. Or it may just be the set $V = \{0, 1\}$ representing a situation in which there are only two kinds of elements—those who vote yes or no, which are satisfactory or defective, and so on.

Theorem 6.1. *If the parameter space is given by* (3), *the order statistics* $Y_{(1)}, \ldots, Y_{(n)}$ *are complete.*

Proof. Denote by s an unordered sample of n elements and by $Y_{(1)}(s, \theta), \ldots, Y_{(n)}(s, \theta)$ its a values in increasing size. Then the expected value of any estimator δ depending only on the order statistics is

$$E_\theta\{\delta[Y_{(1)}, \ldots, Y_{(n)}]\} = \Sigma P(s)\, \delta[Y_{(1)}(s, \theta), \ldots, Y_{(n)}(s, \theta)], \tag{4}$$

where the summation extends over all $\binom{N}{n}$ possible samples, and where for simple random sampling $P(s) = 1/\binom{N}{n}$ for all s. We need to show that

$$E_\theta\{\delta[Y_{(1)}, \ldots, Y_{(n)}]\} = 0 \qquad \text{for all} \quad \theta \in \Omega \tag{5}$$

implies that $\delta[y_{(1)}, \ldots, y_{(n)}] = 0$ for all $y_{(1)} \leqslant \cdots \leqslant y_{(n)}$.

Let us begin by considering (5) for all parameter points θ for which $(a_1, \ldots, a_N)$ is of the form $(a, \ldots, a)$, $a \in V$. Then (5) reduces to

$$\sum_s P(s)\, \delta(a, \ldots, a) = 0 \qquad \text{for all} \quad a,$$

which implies $\delta(a, \ldots, a) = 0$. Next, suppose that $N - 1$ elements in θ are $= a$, and one is $= b > a$. Now (5) will contain two kinds of terms: those corresponding to samples consisting of n a's and those in which the sample contains b, and (5) becomes

$$p\,\delta(a,\dots,a) + q\,\delta(a,\dots,a,b) = 0$$

where p and q are known numbers $\neq 0$. Since the first term has already been shown to be zero, it follows that $\delta(a,\dots,a,b) = 0$. Continuing inductively, we see that $\delta(a,\dots,a,b,\dots,b) = 0$ for any k a's and $n-k$ b's, $k = 0,\dots,n$.

As the next stage in the induction argument consider θ's of the form $(a,\dots,a,b,c)$ with $a < b < c$, then θ's of the form $(a,\dots,a,b,b,c)$, and so on, showing successively that $\delta(a,\dots,a,b,c)$, $\delta(a,\dots,a,b,b,c),\dots$ are equal to zero. Continuing in this way, we see that $\delta[y_{(1)},\dots,y_{(n)}] = 0$ for all possible $(y_{(1)},\dots,y_{(n)})$, and this proves completeness.

It is interesting to note the following.

(i) No use has been made of the assumption of simple random sampling, so that the result is valid also for other sampling methods for which the probabilities $P(s)$ are known and positive for all s.

(ii) The result need not be true for other parameter spaces Ω (Problem 6.1).

Corollary 6.1. *On the basis of the sample values* $Y_1,\dots,Y_n$, *a UMVU estimator exists for any* U*-estimable function of the* a*'s, and it is the unique unbiased estimator* $\delta(Y_1,\dots,Y_n)$ *that is symmetric in its* n *arguments.*

Proof. The result follows from Theorem 2.1.2 and the fact that a function of $y_1,\dots,y_n$ depends only on $y_{(1)},\dots,y_{(n)}$ if and only if it is symmetric in its n arguments (see Section 2.4).

Example 6.1. If the sampling method is simple random sampling and the estimand is $\bar{a}$, the sample mean $\bar{Y}$ is clearly unbiased since $E(Y_i) = \bar{a}$ for all i (Problem 6.2). Since $\bar{Y}$ is symmetric in $Y_1,\dots,Y_n$, it is UMVU and among unbiased estimators minimizes the risk for any convex loss function. The variance of $\bar{Y}$ is (Problem 6.3)

$$\text{var}(\bar{Y}) = \frac{N-n}{N-1}\cdot\frac{1}{n}\tau^2 \tag{6}$$

where

$$\tau^2 = \frac{1}{N}\Sigma(a_i - a)^2 \tag{7}$$

is the *population variance*. To obtain an unbiased estimator of τ^2, note that (Problem 6.3)

$$E\left[\frac{1}{n-1}\Sigma(Y_i - \bar{Y})^2\right] = \frac{N}{N-1}\tau^2. \tag{8}$$

Thus, $[(N-1)/N(n-1)]\Sigma_{i=1}^{n}(Y_i - \bar{Y})^2$ is unbiased for τ^2 and, because it is symmetric in its n arguments, UMVU.

So far we have ignored the labels. That Theorem 6.1 and Corollary 6.1 no longer hold when the labels are included in the data is seen by the following result.

Theorem 6.2. *Given any sampling scheme of fixed size* n *which assigns to the sample* s *a known probability* P(s) (*which may depend on the labels but not on the* a *values of the sample*), *given any* U-*estimable function* g(θ), *and given any preassigned parameter point* $\theta_0 = \{(1, a_{10}), \ldots, (N, a_{N0})\}$, *there exists an unbiased estimator* δ^* *of* g(θ) *with variance* $var_{\theta_0}(\delta^*) = 0$.

Proof. Let δ be any unbiased estimator of $g(\theta)$, which may depend on both labels and y values, say

$$\delta(s) = \delta[(i_1, y_1), \ldots, (i_n, y_n)],$$

and let

$$\delta_0(s) = \delta\left[\left(i_1, a_{i_1 0}\right), \ldots, \left(i_n, a_{i_n 0}\right)\right].$$

Note that δ_0 depends on the labels whether or not δ does and thus would not be available if the labels had been discarded. Let

$$\delta^*(s) = \delta(s) - \delta_0(s) + g(\theta_0).$$

Since

$$E_\theta(\delta) = g(\theta) \quad \text{and} \quad E_\theta(\delta_0) = g(\theta_0),$$

it is seen that δ^* is unbiased for estimating $g(\theta)$. When $\theta = \theta_0$, $\delta^* = g(\theta_0)$ and is thus a constant. Its variance is therefore zero, as was to be proved.

To see under what circumstances the labels are likely to be helpful and when, instead, it is reasonable to discard them, let us consider an example.

Example 6.2. Suppose the population is a class of several hundred students. A random sample is drawn and each of the sampled students is asked to provide a numerical evaluation of the instructor. (Such a procedure may be more accurate than distributing reaction sheets to the whole class, if for the much smaller sample it is possible to obtain a considerably higher rate of response.) Suppose that the frame is an alphabetically arranged class list and that the label is the number of the student on this list. Typically, one would not expect this label to carry any useful information since the place of a name in the alphabet does not usually shed much light on the student's attitude toward the instructor. (In exceptional circumstances, ethnic differences in the initial letters of names could vitiate this argument.) On the other hand, suppose the students are seated alphabetically. In a large class, the students sitting in front may have the advantage of hearing and seeing better, receiving more attention from the instructor, and being less likely to read the campus newspaper or fall asleep. Their attitude could thus be affected by the place of their name in the alphabet, and thus the labels could carry some information.

We shall discuss two ways of formalizing the idea that the labels can reasonably be discarded if they appear to be unrelated to the associated a values.

(i) *Invariance*. Consider the transformations of the parameter and sample space obtained by an arbitrary permutation of the labels:

$$\bar{g}\theta = \{(j(1), a_1), \ldots, (j(N), a_N)\} \tag{9}$$

$$gX = \{(j(I_1), Y_1), \ldots, (j(I_n), Y_n)\}.$$

The estimand $\bar{a}$ [or, more generally, any function $h(a_1, \ldots, a_N)$ that is symmetric in the a's] is unchanged by these transformations, so that $g^*d = d$ and a loss function $L(\theta, d)$ is invariant if it depends on θ only through the a's (in fact, as a symmetric function of the a's) and not the labels. [For estimating $\bar{a}$, a typical such loss function would be of the form $\rho(d - \bar{a})$.] Since $g^*d = d$, an estimator δ is equivariant if it satisfies the condition

$$\delta(gX) = \delta(X) \qquad \text{for all} \quad g, X. \tag{10}$$

In this case equivariance thus reduces to invariance. Condition (10) holds if and only if the estimator δ depends only on the observed Y values and not on the labels. Combining this result with Corollary 6.1, we see that for any U-estimable function $h(a_1, \ldots, a_N)$ the estimator of Corollary 6.1 uniformly minimizes the risk for any convex loss function that does not depend on the labels among all estimators of h which are both unbiased and invariant.

The appropriateness of the principle of equivariance, which permits restricting consideration to equivariant (in the present case, invariant) estimators, depends on the assumption that the transformations (9) leave the problem invariant. This is clearly not the case when there is a relationship between the labels and the associated a values, for example, when low a values tend to be associated with low labels and high a values with high labels, since permutation of the labels will destroy this relationship. Equivariance considerations therefore justify discarding the labels if, in our judgment, the problem is symmetric in the labels, that is, unchanged under any permutation of the labels.

(ii) *Random labels.* Sometimes it is possible to adopt a slightly different formulation of the model which makes an appeal to equivariance unnecessary. Suppose that the labels have been assigned at random, that is, so that all $N!$ possible assignments are equally likely. Then the observed a values $Y_1, \ldots, Y_n$ are sufficient. To see this, note that given these values, any n labels $(I_1, \ldots, I_n)$ associated with them are equally likely, so that the conditional distribution of X given $(Y_1, \ldots, Y_n)$ is independent of θ. In this model, the estimators of Corollary 6.1 are therefore UMVU without any further restriction.

Of course, the assumption of random labeling is legitimate only if the labels really were assigned at random rather than in some systematic way such as alphabetically or first-come, first-labeled. In the latter cases, rather than incorporating a very shaky assumption into the model, it seems preferable to invoke equivariance when it comes to the analysis of the data with the implied admission that we believe the labels to be unrelated to the a values but without denying that a hidden relationship may exist.

Simple random sampling tends to be inefficient unless the population being sampled is fairly homogeneous with respect to the a's. To see this, suppose that $a_1 = \cdots = a_{N_1} = a$, $a_{N_1+1} = \cdots = a_{N_1+N_2} = b$ $(N_1 + N_2 = N)$. Then (Problem 6.3)

$$\text{var}(\bar{Y}) = \frac{N-n}{N-1} \cdot \frac{\gamma(1-\gamma)}{n}(b-a)^2 \tag{11}$$

where $\gamma = N_1/N$. On the other hand, suppose that the subpopulations Π_i consisting of the a's and b's, respectively, can be identified and that one observation X_i is taken from each of the $\Pi_i (i = 1, 2)$. Then $X_1 = a$, $X_2 = b$ and $(N_1 X_1 + N_2 X_2)/N = \bar{a}$ is an unbiased estimator of $\bar{a}$ with variance zero.

This suggests that rather than taking a simple random sample from a heterogeneous population Π, one should try to divide Π into more homogeneous subpopulations Π_i, called *strata*, and sample each of the strata separately. Human populations are frequently stratified by such factors as age, sex, socioeconomic background, severity of disease or by administrative units such as schools, hospitals, counties, voting districts, and so on.

Suppose that the population Π has been partitioned into s strata $\Pi_1, \ldots, \Pi_s$ of sizes $N_1, \ldots, N_s$ and that independent simple random samples of size n_i are taken from each Π_i $(i = 1, \ldots, s)$. If a_{ij} $(j = 1, \ldots, N_i)$ denote the a values in the ith stratum, the parameter is now $\theta = (\theta_1, \ldots, \theta_s)$ where

$$\theta_i = \{(1, a_{i1}), \ldots, (N_i, a_{iN_i}); i\}$$

and the observations are $X = (X_1, \ldots, X_s)$ where

$$X_i = \{(K_{i1}, Y_{i1}), \ldots, (K_{in_i}, Y_{in_i}); i\}.$$

Here K_{ij} is the label of the jth element drawn from Π_i and Y_{ij} is its a value.

It is now easy to generalize the optimality results for simple random sampling to stratified sampling.

Theorem 6.3. *Let the* Y_{ij} $(j = 1, \ldots, n_i)$, *ordered separately for each* i, *be denoted by* $Y_{i(1)} < \cdots < Y_{i(n_i)}$. *On the basis of the* Y_{ij} *(i.e., without the labels), these ordered sample values are sufficient. They are also complete if the parameter space* Ω_i *for* θ_i *is of the form* $V_i \times \cdots \times V_i$ (N_i *factors) and the overall parameter space is* $\Omega = \Omega_1 \times \cdots \times \Omega_s$. *(Note that the value sets* V_i *may be different for different strata.)*

The proof is left to the reader (Problem 6.5).

It follows from Theorem 6.3 that on the basis of the Y's a UMVU estimator exists for any U-estimable function of the a's and that it is the unique unbiased estimator $\delta(Y_{11}, \ldots, Y_{1n_i}; Y_{21}, \ldots, Y_{2n_2}; \ldots)$ which is symmetric in its first n_1 arguments, symmetric in its second set of n_2 arguments,....

Example 6.3. Let $a_{\cdot\cdot} = \Sigma\Sigma a_{ij}/N$ be the average of the a's for the population Π. If $a_{i\cdot}$ is the average of the a's in Π_i, $Y_{i\cdot}$ is unbiased for estimating $a_{i\cdot}$ and hence

$$\delta = \Sigma \frac{N_i Y_{i\cdot}}{N} \tag{12}$$

is an unbiased estimator of $a_{\cdot\cdot}$. Since δ is symmetric for each of the s subsamples, it is UMVU for $a_{\cdot\cdot}$ on the basis of the Y's. From (6) and the independence of the $Y_{i\cdot}$'s, it is seen that

$$\operatorname{var}(\delta) = \Sigma \frac{N_i^2}{N^2} \cdot \frac{N_i - n_i}{N_i - 1} \cdot \frac{1}{n_i} \tau_i^2, \tag{13}$$

where τ_i^2 is the population variance of Π_i, and from (8) one can read off the UMVU estimator of (13).

Discarding the labels within each stratum (but not the strata labels) can again be justified by invariance considerations if these labels appear to be unrelated to the associated a values. Permutation of the labels within each stratum then leaves the problem invariant, and the condition of equivariance reduces to the invariance condition (10). In the present situation, an estimator again satisfies (10) if and only if it does not depend on the within-strata labels. The estimator (12), and other estimators which are UMVU when these labels are discarded, are therefore also UMVU invariant without this restriction.

A central problem in stratified sampling is the choice of the sample sizes n_i. This is a design question and hence outside the scope of this book. We only mention that a natural choice is *proportional allocation*, in which the sample sizes n_i are proportional to the population sizes N_i. If the τ_i are known, the best possible choice in the sense of minimizing the approximate variance

$$\Sigma\left(N_i^2\tau_i^2/n_iN^2\right) \tag{14}$$

is the *Tschuprow–Neyman* allocation with n_i proportional to $N_i\tau_i$ (Problem 6.8).

Stratified sampling, in addition to providing greater precision for the same total sample size than simple random sampling, often has the advantage of being administratively more convenient, which may mean that a larger sample size is possible on the same budget. Administrative convenience is the principal advantage of a third sampling method, *cluster sampling*, which we shall consider next. The population is divided into K clusters of sizes $M_1, \ldots, M_K$. A simple random sample of k clusters is taken and the a values of all the elements in the sampled clusters are obtained. The clusters might, for example, be families or city blocks. A field worker obtaining information about one member of a family can often at relatively little additional cost obtain the same information for all the members.

An important special case of cluster sampling is *systematic sampling*. Suppose the items on a conveyor belt or the cards in a card catalog are being sampled. The easiest way of drawing a sample in these cases and in many situations in which the sampling is being done in the field, is to take every rth element where r is some positive number. To inject some randomness into the process, the starting point is chosen at random. Here there are r clusters consisting of the items labeled

$$\{1, r+1, 2r+1, \ldots\}, \{2, r+2, 2r+2, \ldots\}, \ldots, \{r, 2r, 3r, \ldots\}$$

of which one is chosen at random, so that $K = r$, $k = 1$. In general let the elements of the ith cluster be $\{a_{i1}, \ldots, a_{iM_i}\}$ and let $u_i = \Sigma_{j=1}^{M_i} a_{ij}$ be the total for the ith cluster. We shall be interested in estimating some function of the u's such as the population average $a_{..} = \Sigma u_i/\Sigma M_i$. Of the a_{ij}, we shall assume that the vector of values $(a_{i1}, \ldots, a_{iM_i})$ belongs to some set W_i (which may but need not be of the form $V \times \cdots \times V$), and that $(a_{11}, \ldots, a_{1M_1}; a_{21}, \ldots, a_{2M_2}; \ldots) \in W_1 \times \cdots \times W_K$. The observations consist of the labels of the clusters included in the sample together with the full set of labels and values of the elements of each such cluster:

$$X = \left\{\left[i_1; \left(1, a_{i_1,1}\right), \left(2, a_{i_1,2}\right), \ldots\right]; \left[i_2; \left(1, a_{i_2,1}\right), \left(2, a_{i_2,2}\right), \ldots\right]; \ldots\right\}.$$

Let us begin the reduction of the statistical problem with invariance considerations. Clearly the problem remains invariant under permutations of the labels within each cluster, and this reduces the observation to

$$X' = \left\{\left[i_1, \left(a_{i_1,1}, \ldots, a_{i_1, M_{i_1}}\right)\right]; \left[i_2, \left(a_{i_2,1}, \ldots, a_{i_2, M_{i_2}}\right)\right]; \ldots\right\}$$

in the sense that an estimator is invariant under these permutations if and only if it depends on X only through X'.

The next group is different from any we have encountered so far. Consider any transformation taking $(a_{i1}, \ldots, a_{iM_i})$ into $(a'_{i1}, \ldots, a'_{iM_i})$ $i = 1, \ldots, K$, where the a'_{ij} are arbitrary except that they must satisfy

$$(a) \qquad \left(a'_{i1}, \ldots, a'_{iM_i}\right) \in W_i$$

and

$$(b) \qquad \sum_{j=1}^{M_i} a'_{ij} = u_i.$$

Note that for some vectors $(a_{i1}, \ldots, a_{iM_i})$ there may be no such transformations except the identity; for others, there may be just the identity and one other, and so on, depending on the nature of W_i.

It is clear that these transformations leave the problem invariant provided both the estimand and the loss function depend on the a's only through the u's. Since the estimand remains unchanged, the same should then be true for δ, which, therefore, should satisfy

$$(15) \qquad \delta(gX') = \delta(X')$$

for all these transformations. It is easy to see (Problem 6.13) that δ satisfies (15) if and only if δ depends on X' only through the observed cluster labels, cluster sizes, and the associated cluster totals, that is, only on

$$(16) \qquad X'' = \left\{\left(i_1, u_{i_1}, M_{i_1}\right), \ldots, \left(i_k, u_{i_k}, M_{i_k}\right)\right\}$$

and the order in which the clusters were drawn.

This differs from the set of observations we would obtain in a simple random sample from the collection

$$\{(1, u_1), \ldots, (K, u_K)\} \tag{17}$$

through the additional observations provided by the cluster sizes. For the estimation of the population average or total, this information may be highly relevant and the choice of estimator must depend on the relationship between M_i and u_i. The situation does, however, reduce to that of simple random sampling from (17) under the additional assumption that the cluster sizes M_i are equal, say $M_i = M$, where M can be assumed to be known. This is the case, either exactly or as a very close approximation, for systematic sampling, and also in certain applications to industrial, commercial, or agricultural sampling—for example, when the clusters are cartons of eggs or other packages or boxes containing a fixed number of items. From the discussion of simple random sampling we know that the average $\bar{Y}$ of the observed u values is then the UMVU invariant estimator of $\bar{u} = \Sigma u_i/K$ and hence that $\bar{Y}/M$ is UMVU invariant for estimating $a_{\ldots}$. The variance of the estimator is easily obtained from (6) with $\tau^2 = \Sigma(u_i - \bar{u})^2/K$.

In stratified sampling, it is desirable to have the strata as homogeneous as possible: the more homogeneous a stratum, the smaller the sample size it requires. The situation is just the reverse in cluster sampling where the whole cluster will be observed in any case. The more homogeneous a cluster, the less benefit is derived from these observations: "If you have seen one, you have seen them all." Thus, it is desirable to have the clusters as heterogeneous as possible. For example, families for some purposes constitute good clusters by being both administratively convenient and heterogeneous with respect to age and variables related to age. The advantages of stratified sampling apply not only to the sampling of single elements but equally to the sampling of clusters. *Stratified cluster sampling* consists of drawing a simple random sample of clusters from each stratum and combining the estimates of the strata averages or totals in the obvious way. The resulting estimator is again UMVU invariant provided the cluster sizes are constant within each stratum, although they may differ from one stratum to the next. (For a more detailed discussion of stratified cluster sampling, see, for example, Kish, 1965.)

To conclude this section, we shall briefly indicate two ways in which the equivariance considerations in the present section differ from those in the rest of the chapter.

(i) In all of the present applications, the transformations leave the estimand unchanged rather than transforming it into a different value, and the condition of equivariance then reduces to the invariance condition: $\delta(gX) = \delta(X)$. Correspondingly, the group $\bar{G}$ is not transitive over the parameter space and a UMRE estimator cannot be expected to exist. To obtain an optimal estimator, one has to invoke unbiasedness in addition to invariance. (For an alternative optimality property, see Section 4.4.)

(ii) Instead of starting with transformations of the sample space which would then induce transformations of the parameter space, we inverted the order and began by transforming θ, thereby inducing transformations of X. This does not involve a new approach but was simply more convenient than the usual order. To see how to present the transformations in the usual order, let us consider the sample space as the totality of possible samples s together with the labels and values of their elements. Suppose, for example, that the transformations are permutations of the labels. Since the same elements appear in many different samples, one must ensure that the transformations g of the samples are consistent, that is, that the transform of an element is independent of the particular sample in which it appears. If a transformation has this property, it will define a permutation of all the labels in the population and hence a transformation $\bar{g}$ of θ. Starting with g or $\bar{g}$ thus leads to the same result; the latter is more convenient because it provides the required consistency property automatically.

7. PROBLEMS

Section 1

1.1 Prove the parts of Theorem 1.1 relating to (i) risk, (ii) variance.

1.2 In model (1.1), suppose that $n = 2$ and that f satisfies

$$f(-x_1, -x_2) = f(x_2, x_1). \tag{1}$$

Then the distribution of $(X_1 + X_2)/2$ given $X_2 - X_1 = y$ is symmetric about 0. Note that (1) holds if X_1, X_2 are iid according to a distribution which is symmetric about 0.

1.3 If X_1, X_2 are distributed according to (1.1) with $n = 2$ and f satisfying the assumptions of Problem 1.2, and if ρ is convex and even, then the MRE estimator of ξ is $(X_1 + X_2)/2$.

1.4 Under the assumptions of Example 1.4, show that (i) $E[X_{(1)}] = b/n$; (ii) $\text{med}[X_{(1)}] = b \log 2/n$.

1.5 If T is a sufficient statistic for the family (1.1), the estimator (1.18) is a function of T only. [*Hint*: Use the factorization theorem.]

1.6 Let X_i ($i = 1, 2, 3$) be independently distributed with density $f(x_i - \xi)$ and let $\delta = X_1$ if $X_3 > 0$ and $= X_2$ if $X_3 \leqslant 0$. Then the estimator δ of ξ has constant risk for any invariant loss function, but δ is not location-equivariant.

1.7 Prove Corollary 1.3. [*Hint:* Show that (i) $\phi(v) = E_0\rho(X - v) \to M$ as $v \to \pm\infty$ and (ii) that ϕ is continuous; (ii) follows from the fact (see TSH, Cor. App. Section 2) that if f_n, $n = 1, 2, \dots$ and f are probability densities such that $f_n(x) \to f(x)$ a.e., then $\int \psi f_n \to \int \psi f$ for any bounded ψ.]

1.8 Let $X_1, \dots, X_n$ be distributed as in Example 1.5 and let the loss function be that of Example 1.2. Determine the totality of MRE estimators and show that the midrange is one of them.

1.9 Consider the loss function

$$\rho(t) = \begin{cases} -At & \text{if } t < 0 \\ Bt & \text{if } t \geqslant 0, \end{cases} \qquad (A, B \geqslant 0). \tag{2}$$

If X is a random variable with density f and distribution function F, then $E\rho(X - v)$ is minimized for any v satisfying $F(v) = B/(A + B)$.

1.10 In Example 1.3 find the MRE estimator of ξ when the loss function is given by (2).

1.11 An estimator $\delta(X)$ of $g(\theta)$ is risk-unbiased with respect to the loss function (2) if $F_\theta[g(\theta)] = B/(A + B)$, where F_θ is the cdf of $\delta(X)$ under θ.

1.12 Suppose $X_1, \dots, X_m$; $Y_1, \dots, Y_n$ have joint density

$$f(x_1 - \xi, \dots, x_m - \xi; y_1 - \eta, \dots, y_n - \eta) \tag{3}$$

and consider the problem of estimating $\Delta = \eta - \xi$. Explain why it is desirable for the loss function $L(\xi, \eta; d)$ to be of the form $\rho(d - \Delta)$ and for an estimator δ of Δ to satisfy

$$\delta(\underline{x} + a, \underline{y} + b) = \delta(\underline{x}, \underline{y}) + (b - a). \tag{4}$$

1.13 Under the assumptions of the preceding problem, prove the equivalents of Theorems 1.1–1.4, and Corollaries 1.1–1.3 for estimators satisfying (4).

1.14 In Problem 1.12, determine the totality of estimators satisfying (4) when $m = n = 1$.

1.15 In Problem 1.12, suppose the X's and Y's are independently normally distributed with known variances σ^2 and τ^2. Find conditions on ρ under which the MRE estimator is $\bar{Y} - \bar{X}$.

1.16 In Problem 1.12, suppose the X's and Y's are independently distributed as $E(\xi, 1)$ and $E(\eta, 1)$, respectively, and that $m = n$. Find conditions on ρ under which the MRE estimator of Δ is $Y_{(1)} - X_{(1)}$.

1.17 In Problem 1.12, suppose that $\underline{X}$ and $\underline{Y}$ are independent and that the loss function is squared error. If $\hat{\xi}$ and $\hat{\eta}$ are the MRE estimators of ξ and η, respectively, the MRE estimator of Δ is $\hat{\eta} - \hat{\xi}$.

1.18 Suppose the X's and Y's are distributed as in Problem 1.16 but with $m \neq n$. Determine the MRE estimator of Δ when the loss is squared error.

1.19 For any density f of $\underline{X} = (X_1, \ldots, X_n)$ the probability of the set $A = \{\underline{x}: 0 < \int_{-\infty}^{\infty} f(\underline{x} - u)\, du < \infty\}$ is 1. [*Hint:* With probability 1 the integral in question is equal to the marginal density of $\underline{Y} = (Y_1, \ldots, Y_{n-1})$ where $Y_i = X_i - X_n$, and $P[0 < g(\underline{Y}) < \infty] = 1$ holds for any probability density g.]

1.20 Under the assumptions of Theorem 1.3, if there exists an equivariant estimator δ_0 of ξ with finite expected squared error, then

(i) $E_0(|X_n| \,|\, \underline{Y}) < \infty$ with probability 1;

(ii) the set $B = \{\underline{x}: \int |u| f(\underline{x} - u)\, du < \infty\}$ has probability 1.

[*Hint:* (i) $E|\delta_0| < \infty$ implies $E(|\delta_0| \,|\, \underline{Y}) < \infty$ with probability 1 and hence $E[|\delta_0 - v(\underline{Y})| \,|\, \underline{Y}] < \infty$ with probability 1 for any $v(\underline{Y})$. (ii) $P(B) = 1$ if and only if $E(|X_n| \,|\, \underline{Y}) < \infty$ with probability 1.]

1.21 Let δ_0 be location-equivariant and let $\mathcal{U}$ be the class of all functions u satisfying (1.10) and such that $u(X)$ is an unbiased estimator of zero. Then δ_0 is MRE if and only if

$$\operatorname{cov}[\delta_0, u(X)] = 0 \qquad \text{for all} \quad u \in \mathcal{U}.*$$

(Note the analogy with Theorem 2.1.1.)

Section 2

2.1 Show that the class $G(\mathcal{C})$ is a group.

2.2 In Example 2.1(ii), show that the transformation $\underline{x}' = -\underline{x}$ together with the identity transformation form a group.

2.3 Let $\{gX, g \in G\}$ be a group of transformations that leave the model (2.1) invariant. If the distributions P_θ, $\theta \in \Omega$ are distinct, then the induced transformations $\bar{g}$ are $1:1$ transformations of Ω. [*Hint:* To show that $\bar{g}\theta_1 = \bar{g}\theta_2$ implies $\theta_1 = \theta_2$, use the fact that $P_{\theta_1}(A) = P_{\theta_2}(A)$ for all A implies $\theta_1 = \theta_2$.]

2.4 Under the assumptions of the preceding problem

(i) the transformations $\bar{g}$ satisfy

$$\overline{g_2 g_1} = \bar{g}_2 \cdot \bar{g}_1 \quad \text{and} \quad (\bar{g})^{-1} = \left(\overline{g^{-1}}\right); \tag{5}$$

(ii) the transformations $\bar{g}$ corresponding to $g \in G$ form a group.

2.5 A loss function satisfies (2.9) if and only if it is of the form (2.10).

*Communicated to the author by P. Bickel.

2.6 (i) The transformations g^* defined by (2.12) satisfy

$$(g_2 g_1)^* = g_2^* \cdot g_1^* \quad \text{and} \quad (g^*)^{-1} = (g^{-1})^*. \tag{6}$$

(ii) If G is a group leaving (2.1) invariant and $G^* = \{g^*, g \in G\}$, then G^* is a group.

2.7 Let X be distributed as $N(\xi, \sigma^2)$, $-\infty < \xi < \infty$, $0 < \sigma$, and let $h(\xi, \sigma) = \sigma^2$. The problem is invariant under the transformations $x' = ax + c$; $0 < a$, $-\infty < c < \infty$. Show that the only equivariant estimator is $\delta(X) \equiv 0$.

2.8 (i) If (2.11) holds, the transformations g^* defined by (2.12) are 1 : 1 from $\mathcal{H}$ onto itself.

(ii) If $L(\theta, d) = L(\theta, d')$ for all θ implies $d = d'$, then g^* defined by (2.14) is unique, and is a 1 : 1 transformation from $\mathcal{D}$ onto itself.

2.9 (i) If g is the transformation (2.18), determine $\bar{g}$.

(ii) In Example 2.3, show that (2.20) is not only sufficient for (2.14) but also necessary.

2.10 (i) In Example 2.3, determine the smallest group G containing both G_1 and G_2.

(ii) Show that the only estimator that is invariant under G is $\delta(\underline{X}, \underline{Y}) \equiv 0$.

2.11 If $\delta(X)$ is an equivariant estimator of $h(\theta)$ under a group G, then so is $g^*\delta(X)$ with g^* defined by (2.12) and (2.13), provided G^* is commutative.

2.12 (i) In Example 2.4(i), X is not risk-unbiased.

(ii) The group of transformations $ax + c$ of the real line ($0 < a$, $-\infty < c < \infty$) is not commutative.

2.13 Let θ be real-valued and h strictly increasing, so that (2.11) is vacuously satisfied. If $L(\theta, d)$ is the loss resulting from estimating θ by d, suppose that the loss resulting from estimating $\theta' = h(\theta)$ by $d' = h(d)$ is

$$M(\theta', d') = L[\theta, h^{-1}(d')]. \tag{7}$$

(i) If the problem of estimating θ with loss function L is invariant under G, then so is the problem of estimating $h(\theta)$ with loss function M.

(ii) If δ is equivariant under G for estimating θ with loss function L, then $h[\delta(X)]$ is equivariant for estimating $h(\theta)$ with loss function M.

(iii) If δ is MRE for θ with L, then $h[\delta(X)]$ is MRE for $h(\theta)$ with M.

2.14 If $\delta(\underline{X})$ is MRE for estimating ξ in Example 2.1(i) with loss function $\rho(d - \xi)$, state an optimum property of $e^{\delta(\underline{X})}$ as an estimator of e^{ξ}.

2.15 Let X_{ij}, $j = 1, \dots, n_i$; $i = 1, \dots, s$, and W be distributed according to a density of the form

$$\left[\prod_{i=1}^{s} f_i(\underline{x}_i - \xi_i)\right] h(w) \tag{8}$$

where $\underline{x}_i - \xi_i = (x_{i1} - \xi_i, \dots, x_{in_i} - \xi_i)$, and consider the problem of estimating

(9)
$$\theta = \Sigma c_i \xi_i$$

with loss function

(10)
$$L(\xi_1, \dots, \xi_s; d) = \rho(d - \theta).$$

(i) This problem remains invariant under the transformations

(11)
$$X'_{ij} = X_{ij} + a_i, \qquad \xi'_i = \xi_i + a_i, \qquad \theta' = \theta + \Sigma a_i c_i,$$
$$d' = d + \Sigma a_i c_i.$$

(ii) An estimator δ of θ is equivariant under these transformations if

(12)
$$\delta(\underline{x}_1 + a_1, \dots, \underline{x}_s + a_s, w) = \delta(\underline{x}_1, \dots, \underline{x}_s, w) + \Sigma a_i c_i.$$

2.16 Generalize Theorem 1.1 to the situation of Problem 2.15.

2.17 If δ_0 is any equivariant estimator of θ in Problem 2.15, and if $\underline{y}_i = (x_{i1} - x_{in_i}, x_{i2} - x_{in_i}, \dots, x_{in_i - 1} - x_{in_i})$, show that the most general equivariant estimator of θ is of the form

(13)
$$\delta(\underline{x}_1, \dots, \underline{x}_s, w) = \delta_0(\underline{x}_1, \dots, \underline{x}_s, w) - v(\underline{y}_1, \dots, \underline{y}_s, w).$$

2.18 (i) Generalize Theorem 1.3 and Corollary 1.2 to the situation of Problems 2.15 and 2.17.

(ii) Show that the MRE estimators of (i) can be chosen to be independent of W.

2.19 Suppose that the variables X_{ij} in Problem 2.15 are independently distributed as $N(\xi_i, \sigma^2)$, σ known.

(i) The MRE estimator of θ is then $\Sigma c_i \bar{X}_i - v^*$, where $\bar{X}_i = (X_{i1} + \cdots + X_{in_i})/n_i$, and where v^* minimizes (1.14) with $X = \Sigma c_i \bar{X}_i$.

(ii) If ρ is convex and even, the MRE estimator of θ is $\Sigma c_i \bar{X}_i$.

(iii) The results of (i) and (ii) remain valid when σ is unknown and the distribution of W depends on σ (but not the ξ's).

Section 3

3.1 (i) A loss function L satisfies (3.4) if and only if it satisfies (3.5) for some γ.

(ii) The sample standard deviation, the mean deviation, the range, and the MLE of τ all satisfy (3.7) with $r = 1$.

3.2 If $\delta(\underline{X})$ is scale-invariant, so is $\delta^*(\underline{X})$ defined to be $\delta(\underline{X})$ if $\delta(\underline{X}) \geqslant 0$ and $= 0$ otherwise, and the risk of δ^* is no larger than that of δ for any loss function (3.5) for which $\gamma(v)$ is nonincreasing for $v \leqslant 0$.

3.3 The bias of any equivariant estimator of τ^r in (3.1) is proportional to τ^r.

3.4 A necessary and sufficient condition for δ to satisfy (3.7) is that it is of the form $\delta = \delta_0/u$ with δ_0 and u satisfying (3.7) and (3.9), respectively.

3.5 The function ρ of Corollary 3.1 with γ defined in Example 3.2 is strictly convex for $p \geqslant 1$.

3.6 Prove Lemma 3.1(ii).

3.7 Under the assumptions of Lemma 3.1(ii), the set of scale medians of X is an interval. If $f(x) > 0$ for all $x > 0$, the scale median of X is unique.

3.8 Determine the scale median of X when the distribution of X is (i) $U(0, \theta)$; (ii) $E(0, b)$.

3.9 Prove Corollary 3.2.

3.10 (i) Let Y have the gamma distribution $\Gamma(\alpha, 1)$. Then the value of w minimizing $E[(Y/w) - 1]^2$ is $w = \alpha + 1$.
(ii) Let Y have a χ^2-distribution with f degrees of freedom. Then the minimizing value is $w = f + 2$. [*Hint:* (ii) Example 1.4.7.]

3.11 Verify the values of w given in Example 3.4.

3.12 In Example 3.3, find the MRE estimator of $\text{var}(X_1)$ when the loss function is (i) (3.17), (ii) (3.19) with $r = 2$.

3.13 Let $X_1, \ldots, X_n$ be iid according to the exponential distribution $E(0, \tau)$. Determine the MRE estimator of τ for the loss functions (i) (3.17), (ii) (3.19) with $r = 1$.

3.14 In the preceding problem find the MRE estimator of $\text{var}(X_1)$ when the loss function is (3.17) with $r = 2$.

3.15 Prove formula (3.22).

3.16 Let $X_1, \ldots, X_n$ be iid each with density $(2/\tau)[1 - (x/\tau)]$, $0 < x < \tau$. Determine the MRE estimator (3.22) of τ^r when (i) $n = 2$, (ii) $n = 3$, (iii) $n = 4$.

3.17 In the preceding problem, find $\text{var}(X_1)$ and its MRE estimator for $n = 2, 3, 4$ when the loss function is (3.17) with $r = 2$.

3.18 (i) If δ_0 satisfies (3.7) and $c\delta_0$ satisfies (3.23), then $c\delta_0$ cannot be unbiased in the sense of satisfying $E(c\delta_0) \equiv \tau^r$.
(ii) Prove the statement made in Example 3.5.

3.19 Verify the estimator δ^* of Example 3.7.

3.20 If $X_1, \ldots, X_n$ are iid according to $E(\xi, \tau)$, determine the MRE estimator of τ for the loss functions (i) (3.17), (ii) (3.19) with $r = 1$ and the MRE estimator of ξ for the loss function (3.42).

3.21 Show that δ satisfies (3.34) if and only if it satisfies (3.39) and (3.40).

3.22 Determine the bias of the estimator $\delta^*(\underline{X})$ of Example 3.10.

3.23 Let $X_1, \ldots, X_m$; $Y_1, \ldots, Y_n$ have joint density

$$\frac{1}{\sigma^m \tau^n} f\left(\frac{x_1}{\sigma}, \ldots, \frac{x_m}{\sigma}; \frac{y_1}{\tau}, \ldots, \frac{y_n}{\tau}\right), \tag{14}$$

and consider the problem of estimating $\theta = (\tau/\sigma)^r$ with loss function $L(\sigma, \tau; d) = \gamma(d/\theta)$. This problem remains invariant under the transformations $X_i' = aX_i$, $Y_j' = bY_j$, $\sigma' = a\sigma$, $\tau' = b\tau$, and $d' = (b/a)^r d$, $(a, b > 0)$, and an estimator δ is equivariant under these transformations if

$$\delta(a\underline{x}, b\underline{y}) = (b/a)^r \delta(\underline{x}, \underline{y}).$$

Generalize Theorems 3.1 and 3.2, Corollaries 3.1 and 3.2 and formula (3.22) to the present situation.

3.24 Under the assumptions of the preceding problem and with loss function $(d - \theta)^2/\theta^2$, determine the MRE estimator of θ in the following situations:

(i) $m = n = 1$ and X, Y are independently distributed as $\Gamma(\alpha, \sigma^2)$ and $\Gamma(\beta, \tau^2)$, respectively, (α, β known).

(ii) $X_1, \ldots, X_m$; $Y_1, \ldots, Y_n$ are independently distributed as $N(0, \sigma^2)$, $N(0, \tau^2)$, respectively.

(iii) $X_1, \ldots, X_m$; $Y_1, \ldots, Y_n$ are independently distributed as $U(0, \sigma)$, $U(0, \tau)$, respectively.

3.25 Generalize the results of Problem 3.23 to the case that the joint density of $\underline{X}, \underline{Y}$ is

$$\frac{1}{\sigma^m \tau^n} f\left(\frac{x_1 - \xi}{\sigma}, \ldots, \frac{x_m - \xi}{\sigma}; \frac{y_1 - \eta}{\tau}, \ldots, \frac{y_n - \eta}{\tau}\right). \tag{15}$$

3.26 Obtain the MRE estimator of $\theta = (\tau/\sigma)^r$ with the loss function of Problem 3.24 when (15) specializes to

$$\frac{1}{\sigma^m \tau^n} \prod_i f\left(\frac{x_i - \xi}{\sigma}\right) \prod_j f\left(\frac{y_j - \eta}{\tau}\right) \tag{16}$$

and f is (i) normal, (ii) exponential, or (iii) uniform.

3.27 In the model (16) with $\tau = \sigma$, discuss the equivariant estimation of $\Delta = \eta - \xi$ with loss function $(d - \Delta)^2/\sigma^2$ and obtain explicit results for the three distributions of Problem 3.26.

3.28 Suppose in (16) an MRE estimator δ^* of $\Delta = \eta - \xi$ under the transformations $X_i' = a + bX_i$, $Y_j' = a + bY_j$, $b > 0$, exists when the ratio $\tau/\sigma = c$ is known and that δ^* is independent of c. Show that δ^* is MRE also when σ and τ are completely unknown despite the fact that the induced group of transformations of the parameter space is not transitive.

Section 4

4.1 (i) Suppose X_i: $N(\xi_i, \sigma^2)$ with $\xi_i = \alpha + \beta t_i$. If the first column of the matrix C leading to the canonical form (4.7) is $(1/\sqrt{n}, \ldots, 1/\sqrt{n})'$, find the second column of C.

(ii) If X_i: $N(\xi_i, \sigma^2)$ with $\xi_i = \alpha + \beta t_i + \gamma t_i^2$, and the first two columns of C are those of (i), find the third column under the simplifying assumptions $\Sigma t_i = 0$, $\Sigma t_i^2 = 1$. [*Note:* The orthogonal polynomials that are progressively built up in this way are frequently used to simplify regression analysis.]

4.2 Write out explicit expressions for the transformations (4.10) when Π_Ω is given by

(i) $\xi_i = \alpha + \beta t_i$.
(ii) $\xi_i = \alpha + \beta t_i + \gamma t_i^2$.

4.3 Use Problem 3.10 to prove (iii) of Theorem 4.1.

4.4 (i) In Example 4.3, determine $\hat{\alpha}$, $\hat{\beta}$, and hence $\hat{\xi}_i$ by minimizing $\Sigma(X_i - \alpha - \beta t_i)^2$.

(ii) Verify the expressions (4.12) for α and β, and the corresponding expressions for $\hat{\alpha}$ and $\hat{\beta}$.

4.5 In Example 4.2, find the UMVU estimators of α, β, γ, and σ^2 when $\Sigma t_i = 0$, $\Sigma t_i^2 = 1$.

4.6 Let X_{ij} be independent $N(\xi_{ij}, \sigma^2)$ with $\xi_{ij} = \alpha_i + \beta t_{ij}$. Find the UMVU estimators of the α_i and β.

4.7 (i) In Example 4.4, show that the vectors of the coefficients in the $\hat{\alpha}_i$ are not orthogonal to the vector of the coefficients of $\hat{\mu}$.

(ii) The conclusion of (i) is reversed if $\hat{\alpha}_i$, $\hat{\mu}$ are replaced by $\hat{\hat{\alpha}}_i$, $\hat{\hat{\mu}}$.

4.8 In Example 4.4 find the UMVU estimator of μ when the α_i are known to be zero and compare it with $\hat{\mu}$.

4.9 The coefficient vectors of the X_{ijk} given by (4.32) for $\hat{\mu}$, $\hat{\alpha}_i$, $\hat{\beta}_j$ are orthogonal to the coefficient vectors for the $\hat{\gamma}_{ij}$ given by (4.33).

4.10 In the model defined by (4.26) and (4.27) determine the UMVU estimators of α_i, β_j, and σ^2 under the assumption that the γ_{ij} are known to be zero.

4.11 (i) In Example 4.5 show that

$$\Sigma\Sigma\Sigma\left(X_{ijk} - \mu - \alpha_i - \beta_j - \gamma_{ij}\right)^2 = S^2 + S_\mu^2 + S_\alpha^2 + S_\beta^2 + S_\gamma^2$$

where $S^2 = \Sigma\Sigma\Sigma(X_{ijk} - X_{ij\cdot})^2$, $S_\mu^2 = IJm(X_{\cdots} - \mu)^2$, $S_\alpha^2 = Jm\Sigma(X_{i\cdot\cdot} - X_{\cdots} - \alpha_i)^2$, and S_β^2, S_γ^2 are defined analogously.

(ii) Use the decomposition of (i) to show that the least squares estimators of μ, $\alpha_i, \ldots$ are given by (4.32) and (4.33).

(iii) Show that the *error sum of squares* S^2 is equal to $\Sigma\Sigma\Sigma(X_{ijk} - \hat{\xi}_{ij})^2$ and hence in the canonical form to $\Sigma_{j=s+1}^n Y_j^2$.

4.12 (i) Show how the decomposition in Problem 4.11(i) must be modified when it is known that the γ_{ij} are zero.

(ii) Use the decomposition of (i) to solve Problem 4.10.

4.13 Let $X_{ijk}(i = 1,\ldots, I; j = 1,\ldots, J;\ k = 1,\ldots, K)$ be $N(\xi_{ijk}, \sigma^2)$ with

$$\xi_{ijk} = \mu + \alpha_i + \beta_j + \gamma_k$$

where $\Sigma\alpha_i = \Sigma\beta_j = \Sigma\gamma_k = 0$. Express μ, α_i, β_j, and γ_k in terms of the ξ's and find their UMVU estimators. Viewed as a special case of (4.4), what is the value of s?

4.14 Extend the results of the preceding problem to the model

$$\xi_{ijk} = \mu + \alpha_i + \beta_j + \gamma_k + \delta_{ij} + \varepsilon_{ik} + \lambda_{jk}$$

where

$$\sum_i \delta_{ij} = \sum_j \delta_{ij} = \sum_i \varepsilon_{ik} = \sum_k \varepsilon_{ik} = \sum_j \lambda_{jk} = \sum_k \lambda_{jk} = 0.$$

4.15 In the preceding problem, if it is known that the λ's are zero, determine whether the UMVU estimators of the remaining parameters remain unchanged.

4.16 (i) Under the assumptions of Example 4.6, find the variance of $\Sigma\lambda_i S_i^2$.

(ii) Show that the variance of (i) is minimized by the values stated in the example.

4.17 In the linear model (4.4), a function $\Sigma c_i\xi_i$ with $\Sigma c_i = 0$ is called a *contrast*. Show that a linear function $\Sigma d_i\xi_i$ is a contrast if and only if it is translation invariant, that is, it satisfies $\Sigma d_i(\xi_i + a) = \Sigma d_i\xi_i$ for all a, and hence if and only if it is a function of the differences $\xi_i - \xi_j$.

4.18 Determine which of the following are contrasts:

(i) The regression coefficients α, β, or γ of (4.2).

(ii) The parameters μ, α_i, β_j, or γ_{ij} of (4.27).

(iii) The parameters μ or α_i of (4.23) and (4.24).

Section 5

5.1 In Example 5.1

(i) Show that the joint density of the Z_{ij} is given by (5.2).

(ii) Obtain the joint multivariate normal density of the X_{ij} directly by evaluating their covariance matrix and then inverting it.

[*Hint:* The covariance matrix of $X_{11}, \ldots, X_{1n}; \ldots; X_{s1}, \ldots, X_{sn}$ has the form

$$\Sigma = \begin{pmatrix} \Sigma_1 & 0 & \cdots & 0 \\ 0 & \Sigma_2 & \cdots & 0 \\ \vdots & & \ddots & \vdots \\ 0 & 0 & \cdots & \Sigma_s \end{pmatrix}$$

where each Σ_i is an $n \times n$ matrix with a value a_i for all diagonal elements and a value b_i for all off-diagonal elements. For the inversion of Σ_i, see the next problem.]

5.2 Let $A = (a_{ij})$ be a nonsingular $n \times n$ matrix with $a_{ii} = a$ and $a_{ij} = b$ for all $i \neq j$. Determine the elements of A^{-1}. [*Hint:* Assume that $A^{-1} = (c_{ij})$ with $c_{ii} = c$, $c_{ij} = d$ for all $i \neq j$, calculate c and d as the solutions of the two linear equations $\Sigma a_{1j}c_{j1} = 1, \Sigma a_{1j}c_{j2} = 0$, and check the product AC.]

5.3 Verify the UMVU estimator of σ_A^2/σ^2 given in Example 5.1.

5.4 Obtain the joint density of the X_{ij} in Example 5.1 in the unbalanced case in which $j = 1, \ldots, n_i$, with the n_i not all equal, and determine a minimal set of sufficient statistics (which depends on the number of distinct values of n_i).

5.5 In the balanced one-way layout of Example 5.1, determine $\lim P(\hat{\sigma}_A^2 < 0)$ as $n \to \infty$ for $\sigma_A^2/\sigma^2 = 0, 0.2, 0.5, 1$, and $s = 3, 4, 5, 6$. [*Hint:* The limit of the probability can be expressed as a probability for a χ_{s-1}^2 variable, for which good tables exist.]

5.6 In the preceding problem, use tables of the incomplete beta function to obtain some values of $P(\hat{\sigma}_A^2 < 0)$ for finite n. [*Hint:* Utilize the relationship between the F- and beta-distributions.]

5.7 The following problem shows that in Examples 5.1–5.3 every unbiased estimator of the variance components (except σ^2) takes on negative values. (For some related results, see Pukelsheim, 1981.)

Let X have distribution $P \in \mathcal{P}$ and suppose that T is a complete sufficient statistic for $\mathcal{P}$. If $g(P)$ is any U-estimable function defined over $\mathcal{P}$ and its UMVU estimator $\eta(T)$ takes on negative values with probability > 0, then so does every unbiased estimator of $g(P)$. [*Hint:* For any unbiased estimator δ, recall that $E(\delta|T) = \eta(T)$.]

5.8 Modify the car illustration of Example 5.1 so that it illustrates (5.4).

5.9 In Example 5.2, define a linear transformation of the X_{ijk} leading to the joint distribution of the Z_{ijk} stated in connection with (5.5), and verify the complete sufficient statistics (5.6).

5.10 In Example 5.2 obtain the UMVU estimators of the variance components σ_A^2, σ_B^2, σ^2 when $\sigma_C^2 = 0$, and compare them to those obtained without this assumption.

5.11 For the X_{ijk} given by (5.7), determine a transformation taking them to variables Z_{ijk} with the distribution stated in Example 5.3.

5.12 In Example 5.3, obtain the UMVU estimators of the variance components σ_A^2, σ_B^2, and σ^2.

5.13 In Example 5.3, obtain the UMVU estimators of σ_A^2 and σ^2 when $\sigma_B^2 = 0$ so that the B-terms in (5.7) drop out, and compare them with those of Problem 5.12.

5.14 In Example 5.4

(i) Give a transformation taking the variables X_{ijk} into the W_{ijk} with density (5.10).

(ii) Obtain the UMVU estimators of μ, α_i, σ_B^2 and σ^2.

5.15 A general class of models containing linear models of Type I, II, and mixed models as special cases assumes that the $1 \times n$ observation vector $\underline{X}$ is normally distributed with mean $\underline{\theta}A$ as in (4.13) and with covariance matrix $\Sigma_{i=1}^m \gamma_i V_i$ where the γ's are the components of variance and the V_i's are known symmetric positive semidefinite $n \times n$ matrices. Show that the following models are of this type and in each case specify the γ's and V's: (i) (5.1); (ii) (5.4); (iii) (5.4) without the terms C_{ij}; (iv) (5.7); (v) (5.9).

5.16 Consider a nested three-way layout with

$$X_{ijkl} = \mu + a_i + b_{ij} + c_{ijk} + U_{ijkl}$$

$(i = 1,\ldots, I;\ j = 1,\ldots, J;\ k = 1,\ldots, K;\ l = 1,\ldots, n)$ in the versions

(i) $a_i = \alpha_i,\ b_{ij} = \beta_{ij},\ c_{ijk} = \gamma_{ijk}$;
(ii) $a_i = \alpha_i,\ b_{ij} = \beta_{ij},\ c_{ijk} = C_{ijk}$;
(iii) $a_i = \alpha_i;\ b_{ij} = B_{ij},\ c_{ijk} = C_{ijk}$;
(iv) $a_i = A_i,\ b_{ij} = B_{ij},\ c_{ijk} = C_{ijk}$

where the α's, β's, and γ's are unknown constants defined uniquely by the usual conventions, and the A's, B's, C's, and U's are unobservable random variables, independently normally distributed with means zero and with variances σ_A^2, σ_B^2, σ_C^2, and σ^2.

In each case, transform the X_{ijkl} to independent variables Z_{ijkl} and obtain the UMVU estimators of the unknown parameters.

5.17 In Example 5.5 show that $\gamma_{ij} = 0$ for all i, j is equivalent to $p_{ij} = p_{i+}p_{+j}$. [*Hint:* $\gamma_{ij} = \xi_{ij} - \xi_{i\cdot} - \xi_{\cdot j} + \xi_{\cdot\cdot} = 0$ implies $p_{ij} = a_i b_j$ and hence $p_{i+} = ca_i$, $p_{+j} = b_j/c$ for suitable a_i, b_j, and $c > 0$.]

5.18 In Example 5.6, show that the conditional independence of A, B given C is equivalent to $\alpha_{ijk}^{ABC} = \alpha_{ij}^{AB} = 0$ for all i, j, and k.

5.19 In Example 5.5 show that the conditional distribution of the vectors $(n_{i1},\ldots, n_{iJ})$ given the values of $n_{i+}(i = 1,\ldots, I)$, is that of I independent vectors with multinomial distribution $M(p_{1|i},\ldots, p_{J|i};\ n_{i+})$ where $p_{j|i} = p_{ij}/p_{i+}$.

5.20 Show that the distribution of the preceding problem also arises in Example 5.5 when the n subjects, rather than being drawn from the population at large, are randomly drawn: n_{1+} from Category $A_1,\ldots, n_{I+}$ from Category A_I.

5.21 A city has been divided into I major districts and the ith district into J_i subdistricts, all of which have populations of roughly equal size. From the police records for a given year, a random sample of n robberies is obtained. Write the joint multinomial distribution of the numbers n_{ij} of robberies in subdistrict (i, j) for this nested two-way layout as $e^{\Sigma\Sigma n_{ij}\xi_{ij}}$ with $\xi_{ij} = \mu + \alpha_i + \beta_{ij}$ where $\Sigma_i \alpha_i = \Sigma_j \beta_{ij} = 0$, and show that the assumption $\beta_{ij} = 0$ for all i, j is equivalent to the assumption that $p_{ij} = p_{i+}/J_i$ for all i, j.

5.22 Instead of a sample of fixed size n in the preceding problem, suppose the observations consist of all robberies taking place within a given time period, so that n is the value taken on by a random variable N. Suppose that N has a Poisson distribution with unknown expectation λ and that the conditional distribution of the n_{ij} given $N = n$ is the distribution assumed for the n_{ij} in the preceding problem. Find the UMVU estimator of λp_{ij} and show that no unbiased estimator of p_{ij} exists. [*Hint:* See the following problem.]

5.23 Let N be an integer-valued random variable with distribution $P_\theta(N = n) = P_\theta(n)$, $n = 0, 1, \ldots,$ for which N is complete. Given $N = n$, let X have the binomial distribution $b(p, n)$ for $n > 0$, with p unknown, and let $X = 0$ when $n = 0$. For the observations (N, X)

(i) Show that (N, X) is complete.
(ii) Determine the UMVU estimator of $pE_\theta(N)$.
(iii) Show that no unbiased estimator of any function $g(p)$ exists if $P_\theta(0) > 0$ for some θ.
(iv) Determine the UMVU estimator of p if $P_\theta(0) = 0$ for all θ.

Section 6

6.1 (i) Consider a population $\{a_1, \ldots, a_N\}$ with the parameter space defined by the restriction $a_1 + \cdots + a_N = A$ (known). A simple random sample of size n is drawn in order to estimate τ^2. Assuming the labels to have been discarded, show that $Y_{(1)}, \ldots, Y_{(n)}$ are not complete.
(ii) Show that Theorem 6.1 need not remain valid when the parameter space is of the form $V_1 \times V_2 \times \cdots \times V_N$. [*Hint:* Let $N = 2$, $n = 1$, $V_1 = \{1, 2\}$, $V_2 = \{3, 4\}$.]

6.2 If $Y_1, \ldots, Y_n$ are the sample values obtained in a simple random sample of size n from the finite population (6.2) then (i) $E(Y_i) = \bar{a}$, (ii) $\text{var}(Y_i) = \tau^2$, (iii) $\text{cov}(Y_i, Y_j) = -\tau^2/(N-1)$.

6.3 Verify equations (i) (6.6), (ii) (6.8), (iii) (6.11).

6.4 In simple random sampling, with the labels discarded, a necessary condition for $h(a_1, \ldots, a_N)$ to be U-estimable is that h is symmetric in its N arguments.

6.5 Prove Theorem 6.3.

6.6 The approximate variance (6.14) for stratified sampling with $n_i = nN_i/N$ (proportional allocation) is never greater than the corresponding approximate variance τ^2/n for simple random sampling with the same total sample size.

6.7 Let V_p be the exact variance (6.13) and V_r the corresponding variance for simple random sampling given by (6.6) with $n = \Sigma n_i$, $N = \Sigma N_i$, $n_i/n = N_i/N$ and $\tau^2 = \Sigma\Sigma(a_{ij} - a_{..})^2/N$.

(i) Show that

$$V_r - V_p = \frac{N-n}{n(N-1)N}\left[\Sigma N_i(a_{i\cdot} - a_{\cdot\cdot})^2 - \frac{1}{N}\Sigma\frac{N-N_i}{N_i-1}N_i\tau_i^2\right].$$

(ii) Give an example in which $V_r < V_p$.

6.8 The approximate variance (6.14) for stratified sampling with a total sample size $n = n_1 + \cdots + n_s$ is minimized when n_i is proportional to $N_i\tau_i$.

6.9 Suppose that an auxiliary variable is available for each element of the population (6.2) so that

$$\theta = \{(1, a_1, b_1), \dots, (N, a_N, b_N)\}.$$

If $Y_1, \dots, Y_n$ and $Z_1, \dots, Z_n$ denote the values of a and b observed in a simple random sample of size n, and $\bar{Y}$ and $\bar{Z}$ denote their averages, then

$$\operatorname{cov}(\bar{Y}, \bar{Z}) = E(\bar{Y} - \bar{a})(\bar{Z} - \bar{b}) = \frac{N-n}{nN(N-1)}\Sigma(a_i - \bar{a})(b_i - \bar{b}).$$

6.10 Under the assumptions of Problem 6.9, if $B = b_1 + \cdots + b_N$ is known, an alternative unbiased estimator of $\bar{a}$ is

$$(17) \qquad \left(\frac{1}{n}\sum_{i=1}^{n}\frac{Y_i}{Z_i}\right)\bar{b} + \frac{n(N-1)}{(n-1)N}\left[\bar{Y} - \left(\frac{1}{n}\sum_{i=1}^{n}\frac{Y_i}{Z_i}\right)\bar{Z}\right].$$

[*Hint:* Use the facts that $E(Y_1/Z_1) = (1/N)\Sigma(a_i/b_i)$ and that by the preceding problem

$$E\left[\frac{1}{n-1}\Sigma\frac{Y_i}{Z_i}(Z_i - \bar{Z})\right] = \frac{1}{N-1}\Sigma\frac{a_i}{b_i}(b_i - \bar{b})\Bigg].$$

6.11 In connection with cluster sampling, consider a set W of vectors $(a_1, \dots, a_M)$ and the totality G of transformations taking $(a_1, \dots, a_M)$ into $(a_1', \dots, a_M')$ such that $(a_1', \dots, a_M') \in W$ and $\Sigma a_i' = \Sigma a_i$. Give examples of W such that for any real number a_1 there exist $a_2, \dots, a_M$ with $(a_1, \dots, a_M) \in W$ and such that

(i) G consists of the identity transformation only.
(ii) G consists of the identity and one other element.
(iii) G is transitive over W.

6.12 For cluster sampling with unequal cluster sizes M_i, (17) provides an alternative estimator of $\bar{a}$, with M_i in place of b_i. Show that (17) reduces to $\bar{Y}$ if $b_1 = \cdots = b_N$ and hence when the M_i are equal.

6.13 Show that (6.15) holds if and only if δ depends only on X'', defined by (6.16).

8. REFERENCES

The theory of equivariant estimation of location and scale parameters is due to Pitmann (1939), and the first general discussions of equivariant estimation were provided by Peisakoff (1950) and Kiefer (1957). The concept of risk-unbiasedness (but not the term) and its relationship to equivariance were given in Lehmann (1951).

The linear models of Section 3.4 and Theorem 4.5 are due to Gauss. The history of both is discussed in Seal (1967); see also Stigler (1981). The generalization to exponential linear models was introduced by Dempster (1971) and Nelder and Wedderburn (1972).

Estimation in finite populations has, until recently, been developed outside the mainstream of statistics. The book by Cassel, Särndal, and Wretman (1977) constitutes an important effort at a systematic presentation of this topic within the framework of theoretical statistics. The first steps in this direction were taken by Neyman (1934) and by Blackwell and Girshick (1954). The need to consider the labels as part of the data was first emphasized by Godambe (1955). Theorem 6.1 is due to Watson (1964) and Royall (1968), and Theorem 6.2 to Basu (1971).

Anderson, T. W.
(1962). "Least squares and best unbiased estimates." *Ann. Math. Statist.* **33**, 266–272.

Baker, R. J. and Lane, P. W.
(1978). "Fitting generalized linear models with the GLIM system." *Comp. Statist.* **3**, 31–36.

Baker, R. J. and Nelder, J. A.
(1978). *The GLIM System Release 3: Generalized Linear Interactive Modeling.* Numerical Algorithms Group, Oxford.

Basu, D.
(1969). "On sufficiency and invariance." In "Essays in Probability and Statistics." Ed. Bose. University of North Carolina Press, Chapel Hill.

Basu, D.
(1971). "An essay on the logical foundations of survey sampling, Part I." In *Foundations of Statistical Inference.* Ed. Godambe and Sprott. Holt, Rinehart and Winston, Toronto, pp. 203–242.

Berk, R.
(1967). "A special group structure and equivariant estimation." *Ann. Math. Statist.* **38**, 1436–1445.

Berk, R.
(1972). "A note on sufficiency and invariance." *Ann. Math. Statist.* **43**, 647–650.

Bishop, Y. M. M., Fienberg, S. E., and Holland, P. W.
(1975). *Discrete Multivariate Analysis.* MIT Press, Cambridge, Mass.

Blackwell, D. and Girshick, M. A.
(1954). *Theory of Games and Statistical Decisions.* Wiley, New York.

Bondar, J. V. and Milnes, P.
(1981). "Amenability: A survey for statistical applications of Hunt–Stein and related conditions on groups." *Zeitschr. Wahrsch. Verw. Geb.* **57**, 103–128

Bondesson, L.
(1975). "Uniformly minimum variance estimation in location parameter families." *Ann. Statist.* **3**, 637–660.

Brown, L. D.
(1966). "On the admissibility of invariant estimators of one or more location parameters." *Ann. Math. Statist.* **37**, 1087–1136.

Brown, L. D.
(1968). "Inadmissibility of the usual estimators of scale parameters in problems with unknown location and scale parameters." *Ann. Math. Statist.* **39**, 29–48.

Brown, L. D. and Fox, M.
(1974). "Admissibility in statistical problems involving a location or scale parameter." *Ann. Statist.* **2**, 807–814.

Cassel, C., Särndal, C., and Wretman, J. H.
(1977). *Foundations of Inference in Survey Sampling.* Wiley, New York.

Cochran, W. G.
(1977). *Sampling Techniques*, 3rd ed. Wiley, New York.

Cohen, A. and Strawderman, W. E.
(1973). "Admissible confidence interval and point estimation for translation or scale parameters." *Ann. Statist.* **1**, 545–550.

Cox, D. R.
(1970). *The Analysis of Binary Data.* Methuen, London.

Dawid, A. P.
(1977). "Invariant distributions and analysis of variance models." *Biometrika* **64**, 291–297.

Dempster, A. P.
(1971). "An overview of multivariate data analysis." *J. Multiv. Analysis* **1**, 316–346.

Eaton, M. L.
(1972). *Multivariate Statistical Analysis.* Institute of Mathematical Statistics, Copenhagen.

Eaton, M. L. and Morris, C. N.
(1970). "The application of invariance to unbiased estimation." *Ann. Math. Statist.* **41**, 1708–1716.

Eisenhart, C.
(1947). "The assumptions underlying the analysis of variance." *Biometrics* **3**, 1–21.

Farrell, R. H.
(1964). "Estimators of a location parameter in the absolutely continuous case." *Ann. Math. Statist.* **35**, 949–998.

Godambe, V. P.
(1955). "A unified theory of sampling from finite populations." *J. R. Statist. Soc. (B)* **17**, 269–278.

Goodman, L. A.
(1953). "A simple method for improving some estimators." *Ann. Math. Statist.* **24**, 114–117.

Goodman, L. A.
(1960). "A note on the estimation of variance." *Sankhya* **22**, 221–228.

Goodman, L. A.
(1970). "The multivariate analysis of qualitative data: Interactions among multiple classifications." *J. Amer. Statist. Assoc.* **65**, 226–256.

Graybill, F. A.
(1976). *Theory and Application of the Linear Model.* Duxbury Press, North Scituate, R.I.

Haberman, S. J.
(1973). "Loglinear models for frequency data: Sufficient statistics and likelihood equations." *Ann. Statist.* **1**, 617–632.

Haberman, S. J.
(1974). *The Analysis of Frequency Data.* University of Chicago Press, Chicago.

Hall, W. J., Wijsman, R. A., and Ghosh, J. R.
(1965). "The relationship between sufficiency and invariance with applications in sequential analysis." *Ann. Math. Statist.* **36**, 575–614.

Hartung, J.
(1981). "Nonnegative minimum biased invariant estimation in variance component models." *Ann. Statist.* **9**, 278–292.

Harville, D. N.
(1977). "Maximum likelihood approaches to variance component estimation and to related problems." *J. Amer. Statist. Assoc.* **72**, 320–340.

Hoaglin, C. D.
(1975). "The small-sample variance of the Pitman location estimators." *J. Amer. Statist. Assoc.* **70**, 880–888.

Hora, R. B. and Buehler, R. J.
(1966). "Fiducial theory and invariant estimation." *Ann. Math. Statist.* **37**, 643–656.

Hsu, P. L.
(1938). "On the best unbiased quadratic estimate of the variance." *Statist. Research Mem.* **2**, 91–104.

Huber, P. J.
(1973). "Robust regression: Asymptotics, Conjectures, and Monte Carlo." *Ann. Statist.* **1**, 799–821.

Kagan, A. M., Linnik, Yu V., and Rao, C. R.
(1965). "On a characterization of the normal law based on a property of the sample average." *Sankhya, Ser. A* **27**, 405–406.

Kagan, A. M., Linnik, Yu V., and Rao, C. R.
(1973). *Characterization Problems in Mathematical Statistics.* Wiley, New York.

Kendall, M. G. and Stuart, A.
(1968). *The Advanced Theory of Statistics*, 3rd ed. Vol. 3 Hafner, New York.

Kiefer, J.
(1957). "Invariance, minimax sequential estimation, and continuous time processes." *Ann. Math. Statist.* **28**, 573–601.

Kish, L.
(1965). *Survey Sampling.* Wiley, New York.

Kleffe, J.
(1977). "Optimal estimation of variance components—A survey." *Sankhya, Series B* **39**, 211–244.

Klotz, J. H., Milton, R. C., and Zacks, S.
(1969). "Mean square efficiency of estimators of variance components." *J. Amer. Statist. Assoc.* **64**, 1383–1402.

Kudo, H.
(1955). "On minimal invariant estimators of the transformation parameter." *Nat. Sci. Rep. Ochanomizu Univ.* **6**, 31–73.

Landers, D. and Rogge, L.
(1973). "On sufficiency and invariance." *Ann. Statist.* **1**, 543–544.

Lehmann, E. L.
(1951). "A general concept of unbiasedness." *Ann. Math. Statist.* **22**, 587–592.

Lindley, D. V. and Smith, A. F. M.
(1972). "Bayes estimates for the linear model" (with discussion). *J. Roy. Statist. Soc.* (*B*) **34**, 1–41.

Nelder, J. A. and Wedderburn, R. W. M.
(1972). "Generalized linear models." *J. R. Statist. Soc.* (*A*) **135**, 370–384.

Neyman, J.
(1934). "On the two different aspects of the representative method: The method of stratified sampling and the method of purposive selection." *J. R. Statist. Soc.* **97**, 558–625.

Peisakoff, M.
(1950). "Transformation parameters." Unpublished Ph.D. thesis. Princeton University, Princeton, N.J.

Perng, S. K.
(1970). "Inadmissibility of various 'good' statistical procedures which are translation invariant." *Ann. Math. Statist.* **41**, 1311–1321.

Pitman, E. J. G.
(1939). "The estimation of the location and scale parameters of a continuous population of any given form." *Biometrika* **30**, 391–421.

Plackett, R. L.
(1974). *The Analysis of Categorical Data*. Hafner, New York.

Pregibon, D.
(1980). "Goodness of link tests for generalized linear models." *Appl. Statist.* **29**, 15–24.

Pukelsheim, F.
(1981). "On the existence of unbiased nonnegative estimates of variance covariance components." *Ann. Statist.* **9**, 293–299.

Rao, C. R.
(1952). "Some theorems on minimum variance estimation." *Sankhya* **12**, 27–42.

Rao, C. R.
(1970). "Estimation of heteroscedastic variances in linear models." *J. Amer. Statist. Assoc.* **65**, 161–172.

Rao, C. R.
(1976). "Estimation of parameters in a linear model." *Ann. Statist.* **4**, 1023–1037.

Royall, R. M.
(1968). "An old approach to finite population sampling theory." *J. Amer. Statist. Assoc.* **63**, 1269–1279.

Sacks, J. and Ylvisaker, D.
(1978). "Linear estimation for approximately linear models." *Ann. Statist.* **6**, 1122–1137.

Sahai, H.
(1979). "A bibliography on variance components." *Int. Statist. Rev.* **47**, 177–222.

Scheffé, H.
(1959). *The Analysis of Variance*. Wiley, New York.

Seal, H. L.
(1967). "The historical development of the Gauss linear model," *Biometrika* **54**, 1–24.

Searle, S. R.
(1971). "Topics in variance component estimation." *Biometrics* **27**, 1–76.

Seber, G. A. F.
(1977). *Linear Regression Analysis*. Wiley, New York.

Staudte, R. G., Jr.
(1971). "A characterization of invariant loss functions." *Ann. Math. Statist.* **42**, 1322–1327.

Stein, C.
(1959). "The admissibility of Pitman's estimator of a single location parameter." *Ann. Math. Statist.* **30**, 970–979.

Stigler, S. M.
(1981). "Gauss and the invention of least squares." *Ann. Statist.* **9**, 465–474.

Stone, C. J.
(1974). "Asymptotic properties of estimators of a location parameter." *Ann. Statist.* **2**, 1127–1137.

Stuart, M.
(1977). Unpublished lecture notes.

Thompson, W. A., Jr.
(1962). "The problem of negative estimates of variance components." *Ann. Math. Statist.* **33**, 273–289.

Unni, K.
(1978). "The theory of estimation in algebraic and analytical exponential families with applications to variance components models." Ph.D. thesis, Indian Statistical Institute, Calcutta, India.

Verhagen, A. M. W.
(1961). "The estimation of regression and error-scale parameters when the joint distribution of the errors is of any continuous form and known apart from a scale parameter." *Biometrika* **48**, 125–132.

Watson, G. S.
(1964). "Estimation in finite populations." Unpublished report.

Wedderburn, R. W. M.
(1974). "Quasi-likelihood functions, generalized linear models, and the Gauss–Newton method." *Biometrika* **61**, 439–447.

Wedderburn, R. W. M.
(1976). "On the existence and uniqueness of the maximum likelihood estimates for certain generalized linear models." *Biometrika* **63**, 27–32.

CHAPTER 4

Global Properties

1. BAYES ESTIMATION

So far we have been concerned with finding estimators which minimize the risk $R(\theta, \delta)$ at every value of θ. This was possible only by restricting the class of estimators to be considered by an impartiality requirement such as unbiasedness or equivariance. We shall now drop such restrictions, admitting all estimators into competition, but shall then have to be satisfied with a weaker optimum property than uniformly minimum risk. We shall look for estimators that make the risk function $R(\theta, \delta)$ small in some overall sense. Two such optimality properties will be considered: minimizing the (weighted) average risk for some suitable non-negative weight function and minimizing the maximum risk. The present section is concerned with the first of these approaches, the problem of minimizing

$$\int R(\theta, \delta)\, d\Lambda(\theta) \tag{1}$$

where we shall assume that the weights represented by Λ add up to 1, that is,

$$\int d\Lambda(\theta) = 1, \tag{2}$$

so that Λ is a probability distribution. An estimator δ minimizing (1) is called a *Bayes estimator* with respect to Λ.

The problem of determining such Bayes estimators arises in a number of different contexts.

(i) As Mathematical Tools

Bayes estimators play a central role in Wald's decision theory. It is one of the main results of this theory that in any given statistical problem attention can be restricted to Bayes solutions and suitable limits of Bayes solutions;

given any other procedure δ there exists a procedure δ' in this class such that $R(\theta, \delta') \leq R(\theta, \delta)$ for all values of θ. In view of this result, it is not surprising that Bayes estimators provide a tool for solving minimax problems, as will be seen in the next section.

(ii) As a Way of Utilizing Past Experience

It is frequently reasonable to treat the parameter θ of a statistical problem as the realization of a random variable Θ with known distribution rather than as an unknown constant. Suppose, for example, that we wish to estimate the probability of a penny showing heads when spun on a flat surface. So far we would have considered n spins of the penny as a set of n binomial trials with an unknown probability p of showing heads. Suppose, however, that we have had considerable experience with spinning pennies, experience perhaps which has provided us with approximate values of p for a large number of similar pennies. If we believe this experience to be relevant to the present penny, it might be reasonable to represent this past knowledge as a probability distribution for p, the approximate shape of which is suggested by the earlier data.

This is not as unlike the modeling we have done in the earlier sections as it may seem at first sight. When assuming that the random variables representing the outcomes of our experiments have normal, Poisson, exponential distributions, and so on, we also draw on past experience. Furthermore, we also realize that these models are in no sense exact but at best represent reasonable approximations. There is the difference that in the earlier models we have assumed only the shape of the distribution to be known but not the values of the parameters, while now we take the prior distribution to be completely specified. However, this is a difference in degree rather than in kind and may be quite reasonable if the past experience is sufficiently extensive.

A difficulty, of course, is the assumption that past experience is relevant to the present case. Perhaps the mint has recently changed its manufacturing process, and the present coin, though it looks like the earlier ones, has totally different spinning properties. Similar kinds of judgment are required also for the models considered earlier. In addition, the conclusions derived from statistical procedures are typically applied not only to the present situation or population, but also to those in the future, and extrastatistical judgment is again required in deciding how far such extrapolation is justified.

The choice of the prior distribution Λ is typically made like that of the distributions P_θ by combining experience with convenience. When we make the assumption that the amount of rainfall has a gamma distribution, we don't do so because we really believe this to be the case but because the gamma family is a two-parameter family which seems to fit such data reasonably well and which is mathematically very convenient. Analogously, we can obtain a prior distribution by starting with a flexible family that is mathematically easy to handle and selecting a member from this family which approximates our past experience. Such an approach, in which the model incorporates a prior distribution for θ to reflect past experience, is useful in fields in which a large amount of past experience is available. It can be brought to bear, for example, in many applications in education, business, and medicine.

There is one important difference between the modeling of the distributions P_θ and that of Λ. Typically, we have a number of observations from P_θ and can use these to check the assumption of the form of the distribution. Such a check of Λ is not possible on the basis of one experiment because the value of θ under study represents only a single observation from this distribution. This requires special safeguards in the case of a Bayesian analysis, to which we shall return.

Another difference concerns the meaning of a replication of the experiment. In the models preceding this section the replication would consist of drawing another set of observations from P_θ with the same value of θ. In the model of the present section, we would replicate the experiment by first drawing another value, θ', of Θ from Λ and then a set of observations from $P_{\theta'}$.

This points to another weakness of the Bayes approach. The sampling of the θ-values (choice of penny, for example) tends to be even more haphazard and less well controlled than the choice of subjects for an experiment of a study, which assumes these subjects to be a random sample from the population of interest.

(iii) As a Description of a State of Mind

A formally similar approach is adopted by the so-called Bayesian school which interprets Λ as expressing the subjective feeling about the likelihood of different θ-values. In the presence of a large amount of previous experience, the chosen Λ would often be close to that made under (ii), but the subjective approach can be applied even when little or no prior knowledge is available. In the latter case, for example, the prior distribution Λ then models the state of ignorance about θ. The subjective Bayesian uses the observations X to modify prior beliefs. After $X = x$ has been observed, the belief about θ is expressed by the posterior (i.e., conditional) distribution of Θ given x.

Detailed discussions of this approach, which we shall not pursue here, can be found, for example, in books by Berger (1980b), Box and Tiao (1973), de Finetti (1970), Lindley (1965), Novick and Jackson (1974), and Savage (1954).

The determination of a Bayes estimator is in principle quite simple. First, consider the situation before any observations are taken. Then Θ has distribution Λ and the Bayes estimator of $g(\Theta)$ is any number d minimizing $EL(\Theta, d)$. Once the data have been obtained and are given by the observed value x of X, the prior distribution Λ of Θ is replaced by the posterior, that is, conditional, distribution of Θ given x and the Bayes estimator is any number $\delta(x)$ minimizing the posterior risk $E\{L[\Theta, \delta(x)]\|x\}$. The following is a precise statement of this result where, as usual, measurability considerations, are ignored.

Theorem 1.1. *Let Θ have distribution Λ and, given $\Theta = \theta$, let X have distribution P_θ. Suppose, in addition, the following assumptions hold for the problem of estimating $g(\Theta)$ with non-negative loss function $L(\theta, d)$.*

(a) There exists an estimator δ_0 with finite risk.

(b) For almost all x, there exists a value $\delta_\Lambda(x)$ minimizing

$$E\{L[\Theta, \delta(x)]|X = x\} \tag{3}$$

Then $\delta_\Lambda(X)$ is a Bayes estimator.

Proof. Let δ be any estimator with finite risk. Then (3) is finite a.e. since L is non-negative. Hence

$$E\{L[\Theta, \delta(x)]|X = x\} \geqslant E\{L[\Theta, \delta_\Lambda(x)]|X = x\} \qquad \text{a.e.,}$$

and the result follows by taking the expectation of both sides.

[For a discussion of some measurability aspects and more detail when $L(\theta, d) = \rho(d - \theta)$, see DeGroot and Rao, 1963.]

Corollary 1.1. *Suppose the assumptions of Theorem 1.1 hold.*

(i) *If $L(\theta, d) = [d - g(\theta)]^2$, then*

$$\delta_\Lambda(x) = E[g(\Theta)|x] \tag{4}$$

and more generally, if

$$L(\theta, d) = w(\theta)[d - g(\theta)]^2, \tag{5}$$

then

$$\delta_\Lambda(x) = \frac{\int w(\theta)g(\theta)\, d\Lambda(\theta|x)}{\int w(\theta) d\Lambda(\theta|x)} = \frac{E[w(\Theta)g(\Theta)|x]}{E[w(\Theta)|x]}. \tag{6}$$

(ii) *If* $L(\theta, d) = |d - g(\theta)|$, *then* $\delta_\Lambda(x)$ *is any median of the conditional distribution of* θ *given* x.

(iii) *If*

$$L(\theta, d) = \begin{cases} 0 & \text{when } |d - \theta| \leq c, \\ 1 & \text{when } |d - \theta| > c, \end{cases} \tag{7}$$

then $\delta_\Lambda(x)$ *is the midpoint of the interval* I *of length* 2c *which maximizes* $P[\Theta \in I|x]$.

Proof of (i). By Theorem 1.1, the Bayes estimator is obtained by minimizing

$$E\{[g(\Theta) - \delta(x)]^2|x\}. \tag{8}$$

By assumption (*a*) of Theorem 1.1 there exists $\delta_0(x)$ for which (8) is finite for almost all values of x, and it then follows from Example 1.6.4 that (8) is minimized by (4).

The proofs of the other parts are completely analogous.

It is frequently important to know whether a Bayes solution is unique. The following are sufficient conditions for this to be the case.

Corollary 1.2. *If the loss function* $L(\theta, d)$ *is squared error, or more generally if it is strictly convex in* d, *a Bayes solution* δ_Λ *is unique (a.e.* $\mathcal{P}$*), where* $\mathcal{P}$ *is the class of distributions* P_θ, *provided*

(*i*) *its average risk with respect to* Λ *is finite,*

and (*ii*) *if* Q *is the marginal distribution of* X *given by*

$$Q(A) = \int P_\theta(X \in A)\, d\Lambda(\theta),$$

then a.e. Q *implies a.e.* $\mathcal{P}$.

Proof. For squared error, it follows from Corollary 1.1 that any Bayes estimator $\delta_\Lambda(x)$ with finite risk must satisfy (4) except on a set N of x-values with $Q(N) = 0$. For strictly convex loss functions in general, the result follows by the same argument from Problem 1.6.19.

As an example of a case in which condition (ii) does not hold, let X have the binomial distribution $b(p, n), 0 \leq p \leq 1$, and suppose that Λ assigns probability 1/2 to each of the values $p = 0$ and $p = 1$. Then any estimator $\delta(X)$ of p with $\delta(0) = 0$ and $\delta(n) = 1$ is Bayes.

On the other hand, condition (ii) is satisfied when the parameter space is an open set which is the support of Λ and if the probability $P_\theta(X \in A)$ is continuous in θ for any A. To see this, note that $Q(N) = 0$ implies $P_\theta(N) = 0$ (a.e. Λ) by (1.2.23). If there exists θ_0 with $P_{\theta_0}(N) > 0$ there exists a neighborhood ω of θ_0 in which $P_\theta(N) > 0$. By the support assumption, $P_\Lambda(\omega) > 0$ and this contradicts the assumption that $P_\theta(N) = 0$ (a.e. Λ).

Three different aspects of the performance of a Bayes estimator, or of any other estimator δ, may be of interest in the present model. These are (a) the Bayes risk (1); (b) the risk function $R(\theta, \delta)$ of Section 1.1, which is now the conditional risk of $\delta(X)$ given θ; (c) the posterior risk given x which is defined by (3).

For the determination of the Bayes estimator the relevant criterion is, of course, (a). However, consideration of (b), the conditional risk given θ, as a function of θ provides an important safeguard against an inappropriate choice of Λ. Finally, consideration of (c) is of interest primarily to the Bayesian. From the Bayesian point of view the posterior distribution of Θ given x summarizes the investigator's belief about θ in the light of the observation, and hence the posterior risk is the only measure of risk or accuracy that is of interest.

The possibility of evaluating the risk function (b) of δ_Λ suggests still another use of Bayes estimators.

(iv) As a General Method for Generating Reasonable Estimators

Postulating some plausible distributions Λ provides a method for generating interesting estimators which can then be studied in the conventional way. A difficulty with this approach is, of course, the choice of Λ. One possibility is to select a mathematically convenient and flexible parametric family of priors, and from it select a plausible member to obtain an estimator for consideration. A modification that is sometimes available is the *empirical Bayes method*, in which the parameters of the prior are themselves estimated from the data. This approach will be illustrated in Section 4.6.

A third possibility leading to a particular choice of Λ corresponds to the case of (iii), in which the state of mind can be described as "ignorance." One would then select for Λ a *noninformative* prior which tries (in the spirit of invariance) to treat all parameter values equitably. Such an approach was developed by Jeffreys (1939, 1948, 1961), who, on the basis of invariance considerations, suggests as noninformative prior for θ a density that is proportional to $\sqrt{|I(\theta)|}$, where $|I(\theta)|$ is the determinant of the information matrix. An excellent modern account of this approach with many applications is given by Box and Tiao (1973). For further developments in this direction see Jaynes (1979).

An interesting discussion of different Bayesian attitudes is presented by Good (1965), and an account of criticisms of the Bayesian approach is found in Rothenberg (1977); see also Berger (1980b, pp. 84–85).

Example 1.1. Binomial. Suppose that X has the binomial distribution $b(p, n)$. A two-parameter family of prior distributions for p which is flexible and for which the calculation of the conditional distribution is particularly simple is the family of beta distributions $B(a, b)$. These densities can take on a variety of shapes (see Problem 1.1) and we note for later reference that the expectation and variance of a random variable p with density $B(a, b)$ are (Problem 1.4.26)

$$E(p) = \frac{a}{a+b} \quad \text{and} \quad \operatorname{var}(p) = \frac{ab}{(a+b)^2(a+b+1)}. \tag{9}$$

To determine the Bayes estimator of a given estimand $g(p)$, let us first obtain the conditional distribution (posterior distribution) of p given x. The joint density of X and p is

$$\binom{n}{x}\frac{\Gamma(a+b)}{\Gamma(a)\Gamma(b)}p^{x+a-1}q^{n-x+b-1}.$$

The conditional density of p given x is obtained by dividing by the marginal of x, which is a function of x alone. Thus, the conditional density of p given x has the form

$$C(a, b, x)p^{x+a-1}q^{n-x+b-1}. \tag{10}$$

This is recognized to be again a beta distribution, with parameters

$$a' = a + x, \qquad b' = b + n - x. \tag{11}$$

Let us now determine the Bayes estimator of p when the loss function is squared error. By (4) this is

$$\delta_\Lambda(x) = E(p|x) = \frac{a'}{a'+b'} = \frac{a+x}{a+b+n}. \tag{12}$$

It is interesting to compare this Bayes estimator with the usual estimator X/n. Before any observations are taken, the estimator from the Bayesian approach is the expectation of the prior: $a/(a+b)$. Once X has been observed, the standard non-Bayesian (for example, UMVU) estimator is X/n. The estimator $\delta_\Lambda(X) = (a + X)/(a+b+n)$ lies between these two. In fact,

$$\frac{a+X}{a+b+n} = \left(\frac{a+b}{a+b+n}\right)\frac{a}{a+b} + \left(\frac{n}{a+b+n}\right)\frac{X}{n}, \tag{13}$$

is a weighted average of $a/(a+b)$, the estimator of p before any observations are taken, and X/n, the estimator without consideration of a prior.

The estimator (13) can be considered as a modification of the standard estimator X/n in the light of the prior information about p expressed by (9) or as a modification of the prior estimator $a/(a+b)$ in the light of the observation X. From this point of view, it is interesting to notice what happens as a and $b \to \infty$, with the ratio b/a being kept fixed. Then the estimator (12) tends in probability to $a/(a+b)$, that is, the prior information is so overwhelming that it essentially determines the estimator. The explanation is, of course, that in this case the beta distribution $B(a, b)$ concentrates all its mass essentially at $a/(a+b)$ [the variance in (9) tends toward 0], so that the value of p is taken to be essentially known and is not influenced by X. ("Don't confuse me with the facts!")

On the other hand, if a and b are fixed, but $n \to \infty$, it is seen from (12) that δ_Λ essentially coincides with X/n. This is the case in which the information provided by X overwhelms the initial information contained in the prior distribution.

The UMVU estimator X/n corresponds to the case $a = b = 0$. However, $B(0,0)$ is no longer a probability distribution since $\int_0^1 (1/pq)\, dp = \infty$. Even with such an improper prior distribution, it is possible formally to calculate a posterior distribution given x, and it is interesting to note that the posterior distribution is a proper distribution provided $0 < x < n$, that is, provided there has been at least one success and one failure. Putting $a = b = 0$ in (12), we see that the Bayes estimator corresponding to this improper prior distribution is just X/n. It is interesting to ask whether X/n is also the Bayes estimator corresponding to a proper prior distribution. This will be answered in Theorem 1.2.

Example 1.2. Sequential Binomial Sampling. Consider a sequence of binomial trials with a stopping rule as in Section 2.3. Let X, Y, and N denote, respectively, the number of successes, the number of failures, and the total number of trials at the moment sampling stops. The probability of any sample path is then $p^x q^y$ and we shall suppose that p again has the prior distribution $B(a, b)$. What now is the posterior distribution of p given X and Y (or equivalently X and $N = X + Y$)? The calculation in Example 1.1 shows that as in the fixed sample size case it is the beta distribution with parameters a' and b' given by (11), so that in particular the Bayes estimator of p is given by (12) *regardless* of the stopping rule.

This illustrates a quite general feature of Bayesian inference: the posterior distribution does not depend on the sampling rule but only on the likelihood of the observed results.

Example 1.3. Normal Mean. Let $X_1, \ldots, X_n$ be iid $N(\theta, \sigma^2)$, with σ known, and let the estimand be θ. A convenient prior distribution for Θ is a normal distribution, say $N(\mu, b^2)$. The joint density of Θ and $X = (X_1, \ldots, X_n)$ is then proportional to

$$f(x, \theta) = \exp\left[-\frac{1}{2\sigma^2}\sum_{i=1}^{n}(X_i - \theta)^2\right]\exp\left[-\frac{1}{2b^2}(\theta - \mu)^2\right].$$

To obtain the posterior distribution of $\Theta|x$, the joint density is divided by the marginal density of X, so that the posterior distribution has the form $C(x)f(x, \theta)$. If $C(x)$ is used generically to denote any function of x not involving θ, the posterior density of $\Theta|x$ is

$$C(x)e^{-1/2\theta^2[n/\sigma^2+1/b^2]+\theta[n\bar{x}/\sigma^2+\mu/b^2]}$$

$$= C(x)\exp\left\{\left[-\frac{1}{2}\left(\frac{n}{\sigma^2}+\frac{1}{b^2}\right)\right]\left[\theta^2 - 2\theta\frac{n\bar{x}/\sigma^2+\mu/b^2}{n/\sigma^2+1/b^2}\right]\right\}.$$

This is recognized to be the normal density with mean

$$E(\Theta|x) = \frac{n\bar{x}/\sigma^2+\mu/b^2}{n/\sigma^2+1/b^2} \tag{14}$$

and variance

$$\operatorname{var}(\Theta|x) = \frac{1}{n/\sigma^2+1/b^2}. \tag{15}$$

When the loss is squared error, the Bayes estimator of θ is given by (14) and can be rewritten as

$$\delta_\Lambda(X) = \left(\frac{n/\sigma^2}{n/\sigma^2+1/b^2}\right)\bar{X} + \left(\frac{1/b^2}{n/\sigma^2+1/b^2}\right)\mu, \tag{16}$$

and by Corollary 1.6.3, this result remains true for any loss function $\rho(d-\theta)$ for which ρ is convex and even. This shows δ_Λ to be a weighted average of the standard estimator $\bar{X}$, and the mean μ of the prior distribution, which is the Bayes estimator before any observations are taken. As $n \to \infty$ with μ and b fixed, $\delta_\Lambda(X)$ becomes essentially the estimator $\bar{X}$, and $\delta_\Lambda(X) \to \theta$ in probability [Problem 1.7(i)]. As $b \to 0$, $\delta_\Lambda(X) \to \mu$ in probability [Problem 1.7(ii)], as is to be expected when the prior becomes more and more concentrated about μ. As $b \to \infty$, $\delta_\Lambda(X)$ essentially coincides with $\bar{X}$ [Problem 1.7(iii)], which again is intuitively reasonable. These results are analogous to those in the binomial case.

It was seen above that $\bar{X}$ is the limit of the Bayes estimators as $b \to \infty$. As $b \to \infty$, the prior density tends to Lebesgue measure. Since the Fisher information $I(\theta)$ of a location parameter is constant, this is actually the Jeffrey's prior mentioned under (iv) earlier in the section. It is easy to check that the posterior distribution calculated from this improper prior is a proper distribution as soon as an observation has been taken. This is not surprising; since X is normally distributed about θ with variance 1, even a single observation provides a good idea of the position of θ.

As in the binomial case, the question arises whether $\bar{X}$ is the Bayes solution also with respect to a proper prior Λ. This question is answered for both cases by the following theorem.

Theorem 1.2. *Let Θ have a distribution Λ, and let P_θ denote the conditional distribution of X given θ. Consider the estimation of $g(\theta)$ when the loss* *function is squared error. Then no unbiased estimator* $\delta(X)$ *can be a Bayes solution unless*

$$E[\delta(X) - g(\Theta)]^2 = 0, \tag{17}$$

where the expectation is taken with respect to variation in both X *and* Θ.

Proof. Suppose $\delta(X)$ is a Bayes estimator, and is unbiased for estimating $g(\theta)$. Since $\delta(X)$ is Bayes and the loss is squared error,

$$\delta(X) = E[g(\Theta)|X],$$

with probability 1. Since $\delta(X)$ is unbiased,

$$E[\delta(X)|\theta] = g(\theta) \qquad \text{for all} \quad \theta.$$

Conditioning on X and using (1.5.2) leads to

$$E[g(\Theta)\,\delta(X)] = E\{\delta(X)E[g(\Theta)|X]\} = E[\delta^2(X)].$$

Conditioning instead on Θ, we find

$$E[g(\Theta)\,\delta(X)] = E\{g(\Theta)E[\delta(X)|\Theta]\} = E[g^2(\Theta)].$$

It follows that

$$E[\delta(X) - g(\Theta)]^2 = E[\delta^2(X)] + E[g^2(\Theta)] - 2E[\delta(X)g(\Theta)] = 0$$

as was to be proved.

Let us now apply this result to the normal and binomial cases.

(i) Normal

The estimator under consideration is $\delta(X) = \bar{X}$, so that the risk (given θ) is

$$E(\bar{X} - \theta)^2 = \frac{\sigma^2}{n}.$$

Then for any prior distribution on Θ,

$$E[\bar{X} - \Theta]^2 = \frac{\sigma^2}{n} \neq 0.$$

Thus $\bar{X}$ is not a Bayes estimator.

(ii) Binomial

Consider $\delta(X) = X/n$. For fixed p, its risk function is

$$E\left(\frac{X}{n} - p\right)^2 = \frac{pq}{n},$$

and the left side of (17) is therefore

$$\frac{1}{n}\int_0^1 pq\, d\Lambda(p).$$

The integral is zero if and only if Λ assigns probability 1 to the set $\{0, 1\}$. For such a distribution, Λ,

$$\delta_\Lambda(0) = 0 \quad \text{and} \quad \delta_\Lambda(n) = 1,$$

and any estimator satisfying this condition is a Bayes estimator for such a Λ. Hence in particular X/n is a Bayes estimator. Of course, if Λ is true, then the values $X = 1, 2, \ldots, n-1$ are never observed. Thus X/n is Bayes only in a rather trivial sense.

Theorem 1.2 has been extended to the general case of risk-unbiasedness by Bickel and Blackwell (1967).

The beta and normal prior distributions in the binomial and normal cases are the so-called *conjugate* families of prior distributions. These are frequently defined as distributions with densities proportional to the density of P_θ. It has been pointed out by Diaconis and Ylvisaker (1979) that this definition is ambiguous; they show that in the above examples and more generally in the case of exponential families, conjugate priors can be characterized by the fact that the resulting Bayes estimators are linear in X. They also extend the weighted-average representation (13) of the Bayes estimator to general exponential families.

As another example of the use of conjugate priors, consider the estimation of a normal variance.

Example 1.4. Normal Variance. Let $X_1, \ldots, X_n$ be iid according to $N(0, \sigma^2)$, so that the joint density of the X's is $C\tau^r e^{-\tau\Sigma x_i^2}$, where $\tau = 1/2\sigma^2$ and $r = n/2$. As conjugate prior for τ, we take the gamma density $\Gamma(g, 1/\alpha)$ noting that by (1.4.40)

$$E(\tau) = \frac{g}{\alpha}, \qquad E(\tau^2) = \frac{g(g+1)}{\alpha^2}, \tag{18}$$

$$E\left(\frac{1}{\tau}\right) = \frac{\alpha}{g-1}, \qquad E\left(\frac{1}{\tau^2}\right) = \frac{\alpha^2}{(g-1)(g-2)}.$$

Writing $y = \Sigma x_i^2$, we see that the posterior density of τ given the x's is

$$C(y)\tau^{r+g-1}e^{-\tau(\alpha+y)},$$

which is $\Gamma[r+g, 1/(\alpha+y)]$. If the loss is squared error, the Bayes estimator of $2\sigma^2 = 1/\tau$ is the posterior expectation of $1/\tau$, which by (18) is $(\alpha+y)/(r+g-1)$. The Bayes estimator of $\sigma^2 = 1/2\tau$ is therefore

$$\frac{\alpha+Y}{n+2g-2}. \tag{19}$$

In the present situation, we might instead prefer to work with the scale invariant loss function

$$\frac{(d-\sigma^2)^2}{\sigma^4}, \tag{20}$$

which leads to the Bayes estimator (Problem 1.13)

$$\frac{E(1/\sigma^2)}{E(1/\sigma^4)} = \frac{E(\tau)}{2E(\tau^2)}, \tag{21}$$

and hence by (18) after some simplification to

$$\frac{\alpha+Y}{n+2g+2}. \tag{22}$$

Since the Fisher information for σ is proportional to $1/\sigma^2$ (Table 2.6.1), the Jeffreys prior density in the present case is proportional to the improper density $1/\sigma$, which induces for τ the density $(1/\tau)\,d\tau$. This corresponds to the limiting case $\alpha = 0$, $g = 0$, and hence by (19) and (22) to the Bayes estimators $Y/(n-2)$ and $Y/(n+2)$ for squared error and loss function (20), respectively. The first of these has uniformly larger risk than the second which is MRE.

Let us finally consider two examples involving more than one parameter.

Example 1.5. Normal. Let $X_1, \ldots, X_n$ be iid as $N(\theta, \sigma^2)$ and consider the Bayes estimation of θ and σ^2 when the prior assigns to $\tau = 1/2\sigma^2$ the distribution $\Gamma(g, 1/\alpha)$ as in Example 1.4 and takes θ to be independent of τ with (for the sake of simplicity) the uniform improper prior $d\theta$ corresponding to $b = \infty$ in Example 1.3. Then the joint posterior density of (θ, τ) is proportional to

$$\tau^{r+g-1}e^{-\tau[\alpha+z+n(\bar{x}-\theta)^2]} \tag{23}$$

where $Z = \Sigma(x_i - \bar{x})^2$ and $r = n/2$. By integrating out θ, it is seen that the posterior distribution of τ is $\Gamma[r+g-\frac{1}{2}, 1/(\alpha+Z)]$ (Problem 1.14). In particular for $\alpha = g = 0$, the Bayes estimator of $\sigma^2 = 1/2\tau$ is $Z/(n-3)$ and $Z/(n+1)$ for squared error and loss function (20), respectively. To see that the Bayes estimator of θ is $\bar{X}$ regardless of the values of α and g, it is enough to notice that the posterior density of θ is symmetric about $\bar{X}$ (Problem 1.16).

A problem for which the theories of Chapters 2 and 3 does not lead to a satisfactory solution is that of components of variance. The following example treats the simplest case from the present point of view.

Example 1.6. One-way Layout with Random Effects. In the model (3.5.12), suppose for the sake of simplicity that μ and Z_{11} have been eliminated either by invariance or by assigning to μ the uniform prior on $(-\infty, \infty)$. In either case, this restricts the problem to the remaining Z's with joint density proportional to

$$(24) \qquad \frac{1}{\sigma^{s(n-1)}\left(\sigma^2 + n\sigma_A^2\right)^{(s-1)/2}}\exp\left[-\frac{1}{2\left(\sigma^2 + n\sigma_A^2\right)}\sum_{i=2}^{s} z_{i1}^2 - \frac{1}{2\sigma^2}\sum_{i=1}^{s}\sum_{j=2}^{n} z_{ij}^2\right].$$

The most natural noninformative prior postulates σ and σ_A to be independent with improper densities $1/\sigma$ and $1/\sigma_A$, respectively. Unfortunately, however, in this case the posterior distribution of (σ, σ_A) continues to be improper, so that the calculation of a posterior expectation is meaningless (Problem 1.18).

Instead, let us consider the Jeffreys prior Λ which has the improper density $(1/\sigma)(1/\tau)$ but with $\tau^2 = \sigma^2 + n\sigma_A^2$ so that the density is zero for $\tau < \sigma$. [For a discussion of the appropriateness of this and related priors see Hill (1965), Stone and Springer (1965), Tiao and Tan (1965), and Box and Tiao (1973).] The posterior distribution is then proper (Problem 1.19). The resulting Bayes estimator δ_Λ of σ_A^2 is obtained by Klotz, Milton, and Zacks (1969), who compare it with the more traditional estimators discussed in Example 3.5.1. Since the risk of δ_Λ is quite unsatisfactory, Portnoy (1971) replaces squared error by the scale-invariant loss function $(d - \sigma_A^2)^2/(\sigma^2 + n\sigma_A^2)^2$, and shows the resulting estimator to be

$$(25) \qquad \delta'_\Lambda = \frac{1}{2n}\left[\frac{S_A^2}{a} - \frac{S^2}{c-a-1} + \frac{c-1}{ca(c-a-1)}\cdot\frac{S_A^2 + S^2}{F(R)}\right]$$

where $c = \frac{1}{2}(sn+1)$, $a = \frac{1}{2}(s+3)$, $R = S^2/(S_A^2 + S^2)$, and

$$F(R) = \int_0^1 \frac{v^a}{[R + v(1-R)]^{c+1}}\,dv.$$

Portnoy's risk calculations suggest that δ'_Λ is a satisfactory estimator of σ_A^2 for his loss function or equivalently for squared error loss. The estimation of σ^2 is analogous.

A general Bayesian treatment of linear models is given by Lindley and Smith (1972); sampling from a finite population is discussed from a Bayesian point of view by Ericson (1969), (see also Godambe, 1982); a Bayesian approach to contingency tables is developed by Lindley (1964),

Good (1965), and Bloch and Watson (1967), (see also Bishop, Fienberg, and Holland, 1975.)

2. MINIMAX ESTIMATION

In the preceding section we considered the problem of minimizing a weighted average risk $\int R(\theta, \delta)\, d\Lambda(\theta)$ for a given weight function Λ. We shall now turn to the problem of finding the estimator, δ, which minimizes the maximum risk: $\sup_\theta R(\theta, \delta)$; such an estimator is said to be *minimax*. Unfortunately, unlike what happened in UMVU, equivariant, and Bayes estimation, we shall not be able to determine minimax estimators for large classes of problems. As we shall see, explicit minimax solutions are not easy to find and each problem must be treated on its own merits (but see Section 4.4).

As we mentioned at the beginning of the preceding section, a search for estimators with small risk can be restricted to Bayes estimators and suitable limits of such estimators. But for what prior distribution Λ is the Bayes solution δ_Λ likely to be minimax? A minimax procedure, by minimizing the maximum risk, tries to do as well as possible in the worst case. One might, therefore, expect that the minimax estimator would be Bayes for the worst possible distribution. To make this concept precise, let us denote the average risk (Bayes risk) of the Bayes solution δ_Λ by

$$r_\Lambda = \int R(\theta, \delta_\Lambda)\, d\Lambda(\theta). \tag{1}$$

A prior distribution Λ is said to be *least favorable* if $r_\Lambda \geqq r_{\Lambda'}$ for all prior distributions Λ'. From a Bayesian point of view, this is the prior distribution which causes the statistician the greatest unavoidable average loss.

The following theorem provides a simple condition for a Bayes estimator δ_Λ to be minimax.

Theorem 2.1. *Suppose that Λ is a distribution of Θ such that*

$$\int \mathrm{R}(\theta, \delta_\Lambda)\, \mathrm{d}\Lambda(\theta) = sup_\theta \mathrm{R}(\theta, \delta_\Lambda). \tag{2}$$

Then

(i) *δ_Λ is minimax.*

(ii) *If δ_Λ is the unique Bayes solution with respect to Λ, it is the unique minimax procedure.*

(iii) *Λ is least favorable.*

Proof. (i) Let δ be any other procedure. Then

$$\sup_\theta R(\theta, \delta) \geqq \int R(\theta, \delta)\, d\Lambda(\theta)$$

$$\geqq \int R(\theta, \delta_\Lambda)\, d\Lambda(\theta) = \sup_\theta R(\theta, \delta_\Lambda).$$

(ii) This follows by replacing $\geqq$ by $>$ in the second equality of the proof of (i).

(iii) Let Λ' be some other distribution of Θ. Then

$$r_{\Lambda'} = \int R(\theta, \delta_{\Lambda'})\, d\Lambda'(\theta) \leqq \int R(\theta, \delta_\Lambda)\, d\Lambda'(\theta)$$

$$\leqq \sup_\theta R(\theta, \delta_\Lambda) = r_\Lambda.$$

Condition (2) states that the average of $R(\theta, \delta_\Lambda)$ is equal to its maximum. This will be the case when the risk function is constant or, more generally, when Λ assigns probability 1 to the set on which the risk function takes on its maximum value. A more formal statement is provided by the following two corollaries.

Corollary 2.1. *If a Bayes solution δ_Λ has constant risk, then it is minimax.*

Proof. If δ_Λ has constant risk, (2) clearly holds.

Corollary 2.2. *Let ω_Λ be the set of parameter points at which the risk function of δ_Λ takes on its maximum, that is,*

$$\omega_\Lambda = \{\theta\colon \mathrm{R}(\theta, \delta_\Lambda) = sup_{\theta'}\mathrm{R}(\theta', \delta_\Lambda)\}. \tag{3}$$

Then δ_Λ is minimax if

$$\Lambda(\omega_\Lambda) = 1. \tag{4}$$

This can be rephrased by saying that a sufficient condition for δ_Λ to be minimax is that there exists a set ω such that

$$\Lambda(\omega) = 1 \tag{5}$$

and

$$R(\theta, \delta_\Lambda) \text{ attains its maximum at all points of } \omega. \tag{6}$$

Example 2.1. Binomial. Suppose that X has the binomial distribution $b(p, n)$ and that we wish to estimate p with squared error loss. To see whether X/n is minimax, note that its risk function pq/n has a unique maximum at $p = 1/2$. To apply Corollary 2.2, we should choose as prior for p the distribution Λ, which assigns probability 1 to $p = 1/2$. The corresponding Bayes estimator is $\delta(X) \equiv 1/2$, and not X/n. Thus if X/n is minimax, the approach suggested by Corollary 2.2 does not work in the present case. It is in fact easy to see that X/n is not minimax (Problem 2.1).

To determine a minimax estimator by the method of Theorem 2.1, let us utilize the result of Example 1.1 and try a beta distribution for Λ. If Λ is $B(a, b)$, the Bayes estimator is given by (1.12) and its risk function is

$$\frac{1}{(a+b+n)^2}\left[npq + (aq - bp)^2\right]. \tag{7}$$

Corollary 2.1 suggests seeing whether there exist values a and b for which the risk function (7) is constant. Setting the coefficients of p^2 and p in (7) equal to zero shows that (7) is constant if and only if

$$(a+b)^2 = n \quad \text{and} \quad 2a(a+b) = n. \tag{8}$$

Since a and b are positive, $a + b = \sqrt{n}$ and hence

$$a = b = \tfrac{1}{2}\sqrt{n}. \tag{9}$$

It follows that the estimator

$$\delta = \frac{X + \frac{1}{2}\sqrt{n}}{n + \sqrt{n}} = \frac{X}{n} \cdot \frac{\sqrt{n}}{1 + \sqrt{n}} + \frac{1}{2(1 + \sqrt{n})} \tag{10}$$

is constant risk Bayes and hence minimax. Because of the uniqueness of the Bayes estimator (1.4), it is seen that (10) is the unique minimax estimator of p.

Of course, the estimator (10) is biased (Problem 2.2) because X/n is the only unbiased estimator that is a function of X. A comparison of its risk, which is

$$r_n = E(\delta - p)^2 = 1/4(1 + \sqrt{n})^2 \tag{11}$$

with the risk function

$$R_n(p) = pq/n \tag{12}$$

of X/n shows that (Problem 2.3) $r_n < R_n(p)$ in an interval $I_n = (\frac{1}{2} - c_n < p < \frac{1}{2} + c_n)$ and $r_n > R_n(p)$ outside I_n. For small values of n, c_n is close to $1/2$ so that the minimax estimator is better (and in fact substantially better) for most of the range of p. However, as $n \to \infty$, $c_n \to 0$ and I_n shrinks toward the point $1/2$. Furthermore $\sup_p R_n(p)/r_n = R_n(\frac{1}{2})/r_n \to 1$ so that even at $p = 1/2$, where the comparison is least favorable to X/n, the improvement achieved by the minimax estimator is negligible. Thus, for large and even moderate n, X/n is much the better of the two estimators. In the limit as $n \to \infty$ (although not for any finite n) X/n dominates the minimax estimator. Problems for which such a *subminimax* sequence does not exist are discussed by Ghosh (1964).

The present example illustrates an asymmetry between parts (ii) and (iii) of Theorem 2.1. Part (ii) asserts the uniqueness of the minimax estimator while no such claim is made in part (iii) for the least favorable Λ. In the present case, it follows from (1.4) that for any Λ, the Bayes estimator of p is

$$\delta_\Lambda(x) = \frac{\int_0^1 p^{x+1}(1-p)^{n-x}\,d\Lambda(p)}{\int_0^1 p^x(1-p)^{n-x}\,d\Lambda(p)}. \tag{13}$$

Expansion of $(1-p)^{n-x}$ in powers of p shows that $\delta_\Lambda(x)$ depends on Λ only through the first $n+1$ moments of Λ. This shows in particular that the least favorable distribution is not unique in the present case. Any prior distribution with the same first $n+1$ moments gives the same Bayes solution and hence by Theorem 2.1 is least favorable (Problem 2.5).

Viewed as a loss function, squared error may be unrealistic when estimating p since in many situations an error of fixed size seems much more serious for values of p close to 0 or 1 than for values near $1/2$. To take account of this difficulty let

$$L(p,d) = \frac{(d-p)^2}{pq}. \tag{14}$$

With this loss function, X/n becomes a constant risk estimator and is seen to be a Bayes estimator with respect to the uniform distribution on $(0,1)$ and hence a minimax estimator. It is interesting to note that with (14) the risk function of the estimator (10) is unbounded. This indicates how strongly the minimax property depends on the loss function.

When the loss function is convex in d, as was the case in Example 2.1, it follows from Corollary 1.6.2 that attention may be restricted to nonrandomized estimators. The next example shows that this is no longer true when the convexity assumption is dropped.

Example 2.2. Randomized Minimax Estimator. In the preceding example, suppose that the loss is zero when $|d-p| \leqslant \alpha$ and is one otherwise, where $\alpha < 1/2(n+1)$. Since any nonrandomized $\delta(X)$ can take on at most $n+1$ distinct values, the maximum risk of any such δ is then equal to 1. To exhibit a randomized estimator with smaller maximum risk, consider the extreme case in which the estimator of p does not depend on the data at all but is a random variable U, which is uniformly distributed on $(0,1)$. The resulting risk function is

$$R(p,U) = 1 - P(|U-p| \leqslant \alpha) \tag{15}$$

and it is easily seen that the maximum of (15) is $1-\alpha < 1$ (Problem 2.6).

The loss function in this example was chosen to make the calculations easy, but the possibility of reducing the maximum risk through randomization exists also for other nonconvex loss functions. In particular, for the problem of Example 2.1 with loss function $|d - p|^r$ $(0 < r < 1)$, it can be proved that no nonrandomized estimator can be minimax (Hodges and Lehmann (1950)).

Example 2.3. Difference of Two Binomials. Consider the case of two independent variables X and Y with distributions $b(p_1, m)$ and $b(p_2, n)$, respectively, and the problem of estimating $p_2 - p_1$ with squared error loss. We shall now obtain the minimax estimator when $m = n$; no solution is known when $m \neq n$.

The derivation of the estimator in Example 1.1 suggests that in the present case, too, the minimax estimator might be a linear estimator $aX + bY + k$ with constant risk. However, it is easy to see (Problem 2.10) that such an estimator does not exist. Still hoping for a linear estimator, we shall therefore try to apply Corollary 2.2. Before doing so, let us simplify the hoped-for solution by an invariance consideration.

The problem remains invariant under the transformation

$$(X', Y') = (Y, X), \qquad (p_1', p_2') = (p_2, p_1), \qquad d' = -d, \tag{16}$$

and an estimator $\delta(X, Y)$ is equivariant under this transformation provided $\delta(Y, X) = -\delta(X, Y)$ and hence if

$$(a + b)(x + y) + 2k = 0 \qquad \text{for all} \quad x, y.$$

This leads to the condition $a + b = k = 0$ and therefore to an estimator of the form

$$\delta(X, Y) = c(Y - X). \tag{17}$$

As will be seen in Section 4.4, if a problem remains invariant under a finite group G and if a minimax estimator exists, then there exists an equivariant minimax estimator. In our search for a linear minimax estimator, we may therefore restrict attention to estimators of the form (17).

Application of Corollary 2.2 requires determination of the set ω of pairs (p_1, p_2) for which the risk of (17) takes on its maximum. The risk of (17) is

$$\begin{aligned} R_c(p_1, p_2) &= E[c(Y - X) - (p_2 - p_1)]^2 \\ &= c^2 n(p_1 q_1 + p_2 q_2) + (cn - 1)^2 (p_2 - p_1)^2. \end{aligned}$$

Taking partial derivatives with respect to p_1 and p_2 and setting the resulting expressions equal to 0 leads to the two equations

$$\begin{aligned} \left[2(cn - 1)^2 - 2c^2 n\right] p_1 - 2(cn - 1)^2 p_2 &= -c^2 n \\ -2(cn - 1)^2 p_1 + \left[2(cn - 1)^2 - 2c^2 n\right] p_2 &= -c^2 n. \end{aligned} \tag{18}$$

Typically, these equations have a unique solution, say (p_1^0, p_2^0), which is the point of maximum risk. Application of Corollary 2.2 would then have Λ assign probability 1 to the point (p_1^0, p_2^0) and the associated Bayes estimator would be $\delta(X, Y) \equiv p_2^0 - p_1^0$, whose risk does not have a maximum at (p_1^0, p_2^0).

This impasse does not occur if the two equations (18) are linearly dependent. This will be the case only if

$$c^2 n = 2(cn - 1)^2$$

and hence if

$$c = \frac{\sqrt{2n}}{n[\sqrt{2n} \pm 1]}. \tag{19}$$

Now a Bayes estimator (1.4) does not take on values outside the convex hull of the range of the estimand, which in the present case is $(-1, 1)$. This rules out the minus sign in the denominator of c. Substituting (19) with the plus sign into (18) reduces these two equations to the single equation

$$p_1 + p_2 = 1. \tag{20}$$

The hoped-for minimax estimator is thus

$$\delta(X, Y) = \frac{\sqrt{2n}}{n(\sqrt{2n} + 1)}(Y - X). \tag{21}$$

We have shown (and it is easily verified directly, see Problem 2.11) that in the (p_1, p_2)-plane the risk of this estimator takes on its maximum value at all points of the line segment (20), with $0 < p_1 < 1$, which therefore is the conjectured ω of Corollary 2.2.

It remains to be shown that (21) is the Bayes estimator of a prior distribution Λ, which assigns probability 1 to the set (20).

Let us now confine attention to this subset and note that $p_1 + p_2 = 1$ implies $p_2 - p_1 = 2p_2 - 1$. The following lemma reduces the problem of estimating $2p_2 - 1$ to that of estimating p_2.

Lemma 2.1. *Let δ be a Bayes (respectively, UMVU, minimax, admissible) estimator of $g(\theta)$ for squared error loss. Then, $a\delta + b$ is Bayes (respectively, UMVU, minimax, admissible) for $ag(\theta) + b$.*

Proof. This follows immediately from the fact that

$$R(ag(\theta) + b, a\delta + b) = a^2 R(g(\theta), \delta).$$

For estimating p_2, we have in the present case n binomial trials with parameter $p = p_2$ and n binomial trials with parameter $p = p_1 = 1 - p_2$. If we interchange the meanings of "success" and "failure" in the latter n trials, we have $2n$ binomial trials with success probability p_2, resulting in $Y + (n - X)$ successes. According to Example 2.1, the estimator

$$\frac{Y + n - X}{2n} \cdot \frac{\sqrt{2n}}{1 + \sqrt{2n}} + \frac{1}{2(1 + \sqrt{2n})}$$

is unique Bayes for p_2. Applying Lemma 2.1 and collecting terms, we see that the estimator (21) is unique Bayes for estimating $p_2 - p_1 = 2p_2 - 1$ on ω. It now follows from the properties of this estimator and Corollary 2.1 that δ is minimax for estimating $p_2 - p_1$. It is interesting that $\delta(X, Y)$ is not the difference of the minimax estimators for p_2 and p_1. This is unlike the behavior of UMVU estimators.

That $\delta(X, Y)$ is the unique Bayes (and hence minimax) estimator for $p_2 - p_1$, even when attention is not restricted to ω, follows from the remark after Corollary 1.2. It is only necessary to observe that the subsets of the sample space which have positive probability are the same whether (p_1, p_2) is in ω or not.

The comparison of the minimax estimator (21) with the UMVU estimator $(Y - X)/n$ gives results similar to those in the case of a single p. In particular, the UMVU estimator is again much better for large $m = n$ (Problem 2.12).

Assumption (2) implies that a least favorable distribution exists. When such a distribution does not exist, Theorem 2.1 is not applicable. Consider, for example, the problem of estimating the mean θ of a normal distribution with known variance. Since all possible values of θ play a completely symmetrical role, in the sense that none is easier to estimate than any other, it is natural to conjecture that the least favorable distribution is "uniform" on the real line, that is, that the least favorable distribution is Lebesgue measure. This is the Jeffreys prior and in this case is not a proper distribution.

There are two ways in which the approach of Theorem 2.1 can be generalized to include such improper priors.

(*a*) As was seen in Section 4.1 it may turn out that the posterior distribution given x is a proper distribution. One can then compute the expectation $E[g(\Theta)|x]$ for this distribution and hope that it is the desired estimator. This approach is discussed, for example, by Sacks (1963) and Berger and Srinivasan (1978).

(*b*) Alternatively, one can approximate the improper prior distribution with a sequence of proper distributions, for example, Lebesgue measure by the uniform distributions on $(-N, N)$, $N = 1, 2, \ldots$, and generalize the concept of least favorable distribution to that of least favorable sequence. We shall here follow the second approach.

Let $\{\Lambda_n\}$ be a sequence of prior distributions, and δ_n the Bayes estimator corresponding to Λ_n. Suppose that its Bayes risk is

$$r_n = \int R(\theta, \delta_n)\, d\Lambda_n(\theta) \tag{22}$$

and that

$$r_n \to r. \tag{23}$$

Then the sequence $\{\Lambda_n\}$ is said to be *least favorable* if for every prior distribution Λ we have

$$r_\Lambda \leqq r. \tag{24}$$

Theorem 2.2. *Suppose that* $\{\Lambda_n\}$ *is a sequence of prior distributions with Bayes risks* r_n *satisfying* (23) *and that* δ *is an estimator for which*

$$\sup_\theta R(\theta, \delta) = r. \tag{25}$$

Then

(i) δ *is minimax*

and

(ii) *the sequence* $\{\Lambda_n\}$ *is least favorable.*

Proof. (i) Suppose δ' is any other estimator. Then,

$$\sup_\theta R(\theta, \delta') \geqq \int R(\theta, \delta')\, d\Lambda_n(\theta) \geqq r_n,$$

and this holds for every n. Hence,

$$\sup_\theta R(\theta, \delta') \geqq \sup_\theta R(\theta, \delta),$$

and δ is minimax.

(ii) If Λ is any distribution, then

$$r_\Lambda = \int R(\theta, \delta_\Lambda)\, d\Lambda(\theta) \leqq \int R(\theta, \delta)\, d\Lambda(\theta) \leqq \sup_\theta R(\theta, \delta) = r.$$

This completes the proof.

This theorem is less satisfactory than Theorem 2.1 in two respects. First, even if the Bayes estimators δ_n are unique, it is not possible to conclude that δ is the unique minimax estimator. The reason for this is that the second inequality in the second line of the proof of (i), which is strict when δ_n is unique Bayes, becomes weak under the limit operation.

The other difficulty is that, in order to check condition (25), it is necessary to evaluate r and hence the Bayes risks r_n. This evaluation is helped by the following lemma.

Lemma 2.2. *If* δ_Λ *is the Bayes estimator of* $g(\theta)$ *with respect to* Λ, *and if*

$$r_\Lambda = E[\delta_\Lambda(X) - g(\Theta)]^2 \tag{26}$$

is its Bayes risk, then

$$r_\Lambda = \int var[g(\Theta)|x]\, dP(x). \tag{27}$$

In particular, if the posterior variance of $g(\Theta)|x$ *is independent of* x, *then*

$$r_\Lambda = var[g(\Theta)|x]. \tag{28}$$

Proof. The right side of (26) is equal to

$$\int \{E[g(\Theta) - \delta_\Lambda(x)]^2 | x\}\, dP(x)$$

and the result follows from (1.4).

Example 2.4. Normal Mean. Let $X = (X_1, \ldots, X_n)$, with the X_i iid according to $N(\theta, \sigma^2)$. Let the estimand be θ, the loss squared error, and suppose at first that σ^2 is known. We shall prove that $\bar{X}$ is minimax by finding a sequence of Bayes estimators δ_n satisfying (23) with $r = \sigma^2/n$.

As prior distribution for θ let us try the normal distribution $N(\mu, b^2)$. Then it follows from Example 1.3 that the Bayes estimator is

$$\delta_\Lambda(x) = \frac{n\bar{x}/\sigma^2 + \mu/b^2}{n/\sigma^2 + 1/b^2}. \tag{29}$$

The posterior variance is given by (1.15) and is independent of x, so that

$$r_\Lambda = \frac{1}{n/\sigma^2 + 1/b^2}. \tag{30}$$

As $b \to \infty$, $r_\Lambda \to \sigma^2/n$, and this completes the proof of the fact that $\bar{X}$ is minimax.

Suppose now that σ^2 is unknown. It follows from the result just proved that the maximum risk of every estimator will be infinite unless σ^2 is bounded. We shall therefore assume that

$$\sigma^2 \leqslant M. \tag{31}$$

Under this restriction, the maximum risk of $\bar{X}$ is

$$\sup E(\bar{X} - \theta)^2 = \frac{M}{n}. \tag{32}$$

That $\bar{X}$ is minimax subject to (31), then, is an immediate consequence of Lemma 2.3 below.

It is interesting to note that although the boundedness condition (31) was required for the minimax problem to be meaningful, the minimax estimator does not, in fact, depend on the value of M.

The following lemma will be helpful for the next example.

Lemma 2.3. *Let* X *be a random quantity with distribution* F, *and let* g(F) *be a functional defined over a set* $\mathcal{F}_1$ *of distributions* F. *Suppose that* δ *is a minimax estimator of* g(F) *when* F *is restricted to some subset* $\mathcal{F}_0$ *of* $\mathcal{F}_1$. *Then if*

$$\sup_{F \in F_0} R(F, \delta) = \sup_{F \in F_1} R(F, \delta), \tag{33}$$

δ *is minimax also when* F *is permitted to vary over* $\mathcal{F}_1$.

Proof. If an estimator δ' existed with smaller sup risk over F_1 than δ, it would also have smaller sup risk over $\mathcal{F}_0$ and thus contradict the minimax property of δ over $\mathcal{F}_0$.

Example 2.5. Nonparametric Mean. Let $X_1, \ldots, X_n$ be iid with distribution F and finite expectation θ, and consider the problem of estimating θ with squared error loss. If the maximum risk of every estimator of θ is infinite, the minimax problem is meaningless. To rule this out, we shall consider two possible restrictions on F:

(i) Bounded variance,

$$\text{var}_F(X_i) \leqslant M < \infty; \tag{34}$$

(ii) bounded range,

$$-\infty < a < X_i < b < \infty. \tag{35}$$

Under (i), it is easy to see that $\bar{X}$ is minimax by applying Lemma 2.3 with $\mathcal{F}_1$ the family of all distributions F satisfying (34), and $\mathcal{F}_0$ the family of normal distributions satisfying (34). Then $\bar{X}$ is minimax for $\mathcal{F}_0$ by Example 2.4. Since (33) holds with $\delta = \bar{X}$, it follows that $\bar{X}$ is minimax for $\mathcal{F}_1$. We shall see in the next section that it is in fact the unique minimax estimator of θ.

To find a minimax estimator of θ under (ii), suppose without loss of generality that $a = 0$, $b = 1$, and let $\mathcal{F}_1$ denote the class of distributions F with $F(1) - F(0) = 1$. It seems plausible in the present case that a least favorable distribution over $\mathcal{F}_1$ would concentrate on those distributions $F \in \mathcal{F}_1$ which are as spread out as possible, that is, which put all their mass on the points 0 and 1. But these are just the binomial distributions with $n = 1$. If this conjecture is correct, the minimax estimator of θ should reduce to (10) when all the X_i are 0 or 1, with X in (10) given by $X = \Sigma X_i$. This suggests the estimator

$$\delta(X_1, \ldots, X_n) = \frac{\sqrt{n}}{1 + \sqrt{n}} \bar{X} + \frac{1}{2(1 + \sqrt{n})}, \tag{36}$$

and we shall now prove that (36) is indeed a minimax estimator of θ.

Let $\mathcal{F}_0$ denote the set of distributions F according to which

$$P(X_i = 0) = q, \qquad P(X_i = 1) = p, \qquad 0 < p < 1.$$

Then it was seen in Example 2.1 that (36) is the minimax estimator of $p = E(X_i)$ as F varies over $\mathcal{F}_0$. To prove that (36) is minimax with respect to $\mathcal{F}_1$ it is by Lemma 2.3 enough to prove that the risk function of the estimator (36) takes on its maximum over $\mathcal{F}_0$.

Let $R(F, \delta)$ denote the risk of (36). Then

$$R(F, \delta) = E\left[\frac{\sqrt{n}}{1 + \sqrt{n}}\bar{X} + \frac{1}{2(1 + \sqrt{n})} - \theta\right]^2.$$

By adding and subtracting $[\sqrt{n}/(1 + \sqrt{n})]\theta$ inside the square brackets, this is seen to simplify to

$$R(F, \delta) = \frac{1}{(1 + \sqrt{n})^2}\left[\operatorname{var}_F(X) + \left(\tfrac{1}{2} - \theta\right)^2\right]. \tag{37}$$

Now

$$\operatorname{var}_F(X) = E(X - \theta)^2 = E(X^2) - \theta^2 \leqq E(X) - \theta^2$$

since $0 \leqq X \leqq 1$ implies $X^2 \leqq X$. Thus

$$\operatorname{var}_F(X) \leqq \theta - \theta^2. \tag{38}$$

Substitution of (38) into (37) shows after some simplification that

$$R(F, \delta) \leqslant \frac{1}{4(1 + \sqrt{n})^2}. \tag{39}$$

Since the right side of (39) is the (constant) risk of δ over $\mathcal{F}_0$, the minimax property of δ follows.

Let us next return to the situation, considered at the beginning of Section 3.6 of estimating the mean $\bar{a}$ of a population $\{a_1, \dots, a_N\}$ from a simple random sample $Y_1, \dots, Y_n$ drawn from this population. To make the minimax estimation of $\bar{a}$ meaningful, restrictions on the a's are needed. In analogy to (34) and (35), we shall consider the following cases:

(i) bounded population variance,

$$\frac{1}{N}\Sigma(a_i - \bar{a})^2 \leqslant M, \tag{40}$$

and (ii) bounded range,

$$0 \leqslant a_i \leqslant 1, \tag{41}$$

to which the more general case $a \leqslant a_i \leqslant b$ can always be reduced. The loss function will be squared error, and for the time being we shall ignore the labels. It will be seen in Section 4.4 that the minimax results remain valid when the labels are included in the data.

Example 2.6. Simple Random Sampling. We begin with case (ii) and consider first the special case in which all the values of a are either 1 or 0, say D equal to 1, $N - D$ equal to 0. The total number X of 1s in the sample is then a sufficient statistic and has the hypergeometric distribution

$$P(X = x) = \binom{D}{x}\binom{N-D}{n-x}\Big/\binom{N}{n} \tag{42}$$

where $\max[0, n - (N - D)] \leqslant x \leqslant \min(n, D)$ (Problem 2.20) and where D can take on the values $0, 1, \ldots, N$. The estimand is $\bar{a} = D/N$, and, following the method of Example 1.1, one finds that $\alpha X/n + \beta$ with

$$\alpha = \frac{1}{1 + \sqrt{\dfrac{N-n}{n(N-1)}}}, \qquad \beta = \frac{1}{2}(1 - \alpha) \tag{43}$$

is a linear estimator with constant risk (Problem 2.21). That (43) is minimax then is a consequence of the fact that it is the Bayes estimator of D/N with respect to the prior distribution (Problem 2.21)

$$P(D = d) = \int_0^1 \binom{N}{d} p^d q^{N-d} \frac{\Gamma(a+b)}{\Gamma(a)\Gamma(b)} p^{a-1} q^{b-1}\, dp, \tag{44}$$

where

$$a = b = \frac{\beta}{\alpha/n - 1/N}. \tag{45}$$

It is easily checked that as $N \to \infty$, (43) $\to$ (10) and (45) $\to \frac{1}{2}\sqrt{n}$, as one would expect since the hypergeometric distribution then tends toward the binomial.

The special case just treated plays the same role as a tool for the problem of estimating $\bar{a}$ subject to (41) that the binomial case played in Example 2.5. To show that

$$\delta = \alpha\bar{Y} + \beta \tag{46}$$

is minimax, it is only necessary to check that

$$E(\delta - \bar{a})^2 = \alpha^2 \operatorname{var}(\bar{Y}) + [\beta + (\alpha - 1)\bar{a}]^2 \tag{47}$$

takes on its maximum when all the values of a are 0 or 1, and this is seen as in Example 2.5 (Problem 2.23). Unfortunately, δ shares the poor risk properties of the binomial risk estimator for all but very small n.

The minimax estimator of $\bar{a}$ subject to (40), as might be expected from Example 2.5, is $\bar{Y}$. For a proof of this result, which will not be given here, see Bickel and Lehmann (1981).

As was seen in Examples 1.1 and 1.3, minimax estimators can be quite unsatisfactory over a large part of the parameter space. This is perhaps not surprising since, as Bayes estimator with respect to a least favorable prior, a minimax estimator takes the most pessimistic view possible. This is illustrated by Example 1.1 in which the least favorable prior, $B(a_n, b_n)$ with $a_n = b_n = \sqrt{n}/2$, concentrates nearly its entire attention on the neighborhood of $p = 1/2$ for which accurate estimation of p is most difficult. On the other hand, a Bayes estimator corresponding to a personal prior may expose the investigator to a very high maximum risk, which may well be realized if the prior has badly misjudged the situation. It is possible to avoid the worst consequences of both these approaches through a compromise which permits the use of personal judgment and yet provides adequate protection against unacceptably high risks.

Suppose that M is the maximum risk of the minimax estimator. Then one may be willing to consider estimators whose maximum risk exceeds M, if the excess is controlled, say, if

$$R(\theta, \delta) \leqslant M(1 + \varepsilon) \qquad \text{for all} \quad \theta \tag{48}$$

where ε is the proportional increase in risk that one is willing to tolerate. A *restricted Bayes estimator* is then obtained by minimizing, subject to (48), the average risk (1) for the prior Λ of one's choice.

Such restricted Bayes estimators are typically quite difficult to evaluate. There is, however, one class of situations in which the evaluation is trivial: If the maximum risk of the unrestricted Bayes estimator satisfies (48), it of course coincides with the restricted Bayes estimator. This possibility is illustrated by the following example.

Example 2.7. In Example 1.1, suppose we believe p to be near zero (it may, for instance, be the probability of a rarely occurring disease or accident). As a prior distribution for p, we therefore take $B(1, b)$ with a fairly high value of b. The Bayes estimator (1.12) is then $\delta = (X + 1)/(n + b + 1)$ and its risk is

$$E(\delta - p)^2 = \frac{npq + (q - bp)^2}{(n + b + 1)^2}. \tag{49}$$

At $p = 1$, the risk is $[b/(n + b + 1)]^2$, which for fixed n and sufficiently large b can be arbitrarily close to 1, while the constant risk of the minimax estimator is only $1/4\ (1 + \sqrt{n})^2$. On the other hand, for fixed b an easy calculation shows that (Problem 2.24)

$$4(1 + \sqrt{n})^2 \cdot \sup R(p, \delta) \to 1 \qquad \text{as} \quad n \to \infty.$$

For any given b and $\varepsilon > 0$, δ therefore will satisfy (48) for sufficiently large values of n.

A quite different, and perhaps more typical, situation is illustrated by the normal case.

Example 2.8. Normal. If in the situation of Example 1.3, without loss of generality we put $\sigma = 1$ and $\mu = 0$, the Bayes estimator (1.14) reduces to $c\bar{X}$ with $c = nb^2/(1 + nb^2)$. Since its risk function is unbounded for all n, while the minimax risk is $1/n$, no such Bayes estimator can be restricted Bayes.

As a compromise, Efron and Morris propose an estimator of the form

$$\delta = \begin{cases} \bar{X} + (1 - c)\, d & \text{if} \quad \bar{X} < -d \\ c\bar{X} & \text{if} \quad |\bar{X}| \leqslant d \\ \bar{X} - (1 - c)\, d & \text{if} \quad \bar{X} > d. \end{cases} \tag{50}$$

The risk of these estimators is bounded (Problem 2.25) with the maximum risk tending toward $1/n$ as $d \to 0$. On the other hand, for large d values, (50) is close to the Bayes estimator. Although (50) is not the exact optimum solution of the restricted Bayes problem, Efron and Morris (1971) and Marazzi (1980) show it to be close to optimal.

3. MINIMAXITY AND ADMISSIBILITY IN EXPONENTIAL FAMILIES

It was seen in Section 2.5 that a UMVU estimator δ need not be admissible. If a biased estimator δ' has uniformly smaller risk, the choice between δ and δ' is not clear-cut: one must balance the advantage of unbiasedness against the drawback of larger risk. The situation is, however, different for minimax estimators. If δ' dominates a minimax estimator δ, then δ' is also minimax and thus definitely preferred. It is therefore particularly important to ascertain whether a proposed minimax estimator is admissible. In the present section we shall obtain some admissibility results (and in the process some minimax results) for exponential families, and in the next section shall consider the corresponding problem for group families.

To prove inadmissibility of an estimator δ, it is sufficient to produce an estimator δ' which dominates it. An example was given in Lemma 2.5.1. The following is another instance.

Lemma 3.1. *Let the range of the estimand* $g(\theta)$ *be an interval with end points* a *and* b, *and suppose that the loss function* $L(\theta, d)$ *is positive when* $d \neq g(\theta)$ *and zero when* $d = g(\theta)$, *and that for any fixed* θ, $L(\theta, d)$ *is increasing as* d *moves away from* $g(\theta)$ *in either direction. Then any estimator* δ *taking on values outside the closed interval* $[a, b]$ *with positive probability is inadmissible.*

Proof. δ is dominated by the estimator δ', which is a or b when $\delta < a$ or $> b$, and which otherwise is equal to δ.

A principal method for proving admissibility is the following result.

Theorem 3.1. *Any unique* Bayes estimator is admissible.*

Proof. If δ is unique Bayes with respect to the prior distribution Λ and is dominated by δ', then

$$\int R(\theta, \delta') d\Lambda(\theta) \leqslant \int R(\theta, \delta)\, d\Lambda(\theta),$$

which contradicts uniqueness.

An example is provided by the binomial minimax estimator (2.10) of Example 2.1. For the corresponding nonparametric minimax estimator (2.36) of Example 2.5 admissibility was proved by Hjort (1976) who showed that it is the essentially unique minimax estimator with respect to a class of Dirichlet-process priors described by Ferguson (1973).

We shall in the present section illustrate a number of ideas and results concerning admissibility on the estimation of the mean and variance of a normal distribution and then indicate some of their generalizations. Unless stated otherwise, the loss function will be assumed to be squared error.

Example 3.1. **Normal Mean.** Let $X_1, \ldots, X_n$ be iid according to $N(\theta, \sigma^2)$, with σ^2 known. In the preceding Section, $\bar{X}$ was seen to be minimax for estimating θ. Is it admissible? Instead of attacking this question directly, we shall consider the admissibility of an arbitrary linear function $a\bar{X} + b$.

*Uniqueness here means that any two Bayes estimators differ only on a set N with $P_\theta(N) = 0$ for all θ.

From Example 1.3 it follows that the unique Bayes estimator with respect to the normal prior for θ with mean μ and variance τ^2 is

$$\frac{n\tau^2}{\sigma^2 + n\tau^2}\bar{X} + \frac{\mu\sigma^2}{\sigma^2 + n\tau^2} \tag{1}$$

and that the associated Bayes risk is finite (Problem 3.1). It follows that $a\bar{X} + b$ is unique Bayes and hence admissible whenever

$$0 < a < 1. \tag{2}$$

To see what can be said about other values of a, we shall now prove an inadmissibility result for linear estimators, which is quite general and in particular does not require the assumption of normality.

Theorem 3.2. *Let* X *be a random variable with mean* θ *and variance* σ^2. *Then* aX + b *is an inadmissible estimator of* θ *whenever*

(i) $a > 1$, *or*
(ii) $a < 0$, *or*
(iii) $a = 1, b \neq 0$.

Proof. Let the risk of $aX + b$ be

$$\rho(a, b) = E(aX + b - \theta)^2 = a^2\sigma^2 + [(a-1)\theta + b]^2. \tag{3}$$

(i) If $a > 1$, then

$$\rho(a, b) \geqslant a^2\sigma^2 > \sigma^2 = \rho(1, 0)$$

so that $aX + b$ is dominated by X.

(ii) If $a < 0$, then $(a-1)^2 > 1$ and hence

$$\rho(a, b) \geqslant [(a-1)\theta + b]^2 = (a-1)^2\left[\theta + \frac{b}{a-1}\right]^2$$

$$> \left(\theta + \frac{b}{a-1}\right)^2 = \rho\left(0, -\frac{b}{a-1}\right).$$

Thus, $aX + b$ is dominated by the constant estimator $\delta \equiv -b/(a-1)$.

(iii) In this case, $aX + b = X + b$ is dominated by X (see Lemma 2.5.1).

Example 3.1. *(Continued).* Combining the results of Example 3.1 and Theorem 3.2 we see that the estimator $a\bar{X} + b$ is admissible in the strip $0 < a < 1$ in the (a, b)-plane, that it is inadmissible to the left ($a < 0$) and to the right ($a > 1$).

The left boundary $a = 0$ corresponds to the constant estimators $\delta = b$ which are admissible since $\delta = b$ is the only estimator with zero risk at $\theta = b$. Finally, the right boundary $a = 1$ is inadmissible by (iii) of Theorem 3.2, with the possible exception of the point $a = 1$, $b = 0$.

We have thus settled the admissibility of $a\bar{X} + b$ for all cases except $\bar{X}$ itself, which was the estimator of primary interest. In the next example, we shall prove that $\bar{X}$ is indeed admissible.

Example 3.2. Admissibility of $\bar{X}$. The admissibility of $\bar{X}$ for estimating the mean of a normal distribution is not only of great interest in itself but can also be regarded as the starting point of many other admissibility investigations. For this reason we shall now give two proofs of this fact—they represent two principal methods for proving admissibility and are seen particularly clearly in this example because of its great simplicity.

First Proof of Admissibility (the Limiting Bayes Method). Suppose that $\bar{X}$ is not admissible, and without loss of generality assume that $\sigma = 1$. Then there exists δ^* such that

$$R(\theta, \delta^*) \leqslant \frac{1}{n} \qquad \text{for all} \quad \theta$$

$$R(\theta, \delta^*) < \frac{1}{n} \qquad \text{for at least some} \quad \theta.$$

Now $R(\theta, \delta)$ is a continuous function of θ for every δ [see, for example, Ferguson (1967), Th. 2 on p. 139], so that there exists $\varepsilon > 0$ and $\theta_0 < \theta_1$ such that

$$R(\theta, \delta^*) < \frac{1}{n} - \varepsilon \qquad \text{for all} \quad \theta_0 < \theta < \theta_1.$$

Let r_τ^* be the average risk of δ^* with respect to the prior distribution $\Lambda_\tau = N(0, \tau^2)$, and let r_τ be the Bayes risk, that is, the average risk of the Bayes solution with respect to Λ_τ. Then, by (2.30) with $\sigma = 1$ and τ in place of b,

$$\frac{\dfrac{1}{n} - r_\tau^*}{\dfrac{1}{n} - r_\tau} = \frac{\dfrac{1}{\sqrt{2\pi}\,\tau}\displaystyle\int_{-\infty}^{\infty}\left[\frac{1}{n} - R(\theta, \delta^*)\right]e^{-\theta^2/2\tau^2}d\theta}{\dfrac{1}{n} - \dfrac{\tau^2}{1 + n\tau^2}}$$

$$\geqslant \frac{n(1 + n\tau^2)\varepsilon}{\tau\sqrt{2\pi}}\int_{\theta_0}^{\theta_1} e^{-\theta^2/2\tau^2}\, d\theta.$$

The integrand converges monotonely to 1, as $\tau \to \infty$. By the Lebesgue monotone convergence theorem (TSH, p. 35), the integral therefore converges to $\theta_1 - \theta_0$, and hence

$$\frac{1/n - r_\tau^*}{1/n - r_\tau} \to \infty .$$

Thus, there exists τ_0 such that $r_{\tau_0}^* < r_{\tau_0}$, which contradicts the fact that r_{τ_0} is the Bayes risk for Λ_{τ_0}. This completes the proof.

Second Proof of Admissibility (the Information Inequality Method). It follows from the information inequality (2.7.27) and the fact that

$$R(\theta, \delta) = E(\delta - \theta)^2 = \text{var}_\theta(\delta) + b^2(\theta)$$

where $b(\theta)$ is the bias of δ, that

(4) $$R(\theta, \delta) \geqslant b^2(\theta) = \frac{[1 + b'(\theta)]^2}{nI(\theta)}.$$

In the present case, since $\sigma^2 = 1$, $I(\theta) = 1$ by Table 2.6.1.

Suppose, now, that δ is any estimator satisfying

(5) $$R(\theta, \delta) \leqslant \frac{1}{n} \quad \text{for all} \quad \theta$$

and hence

(6) $$b^2(\theta) + \frac{[1 + b'(\theta)]^2}{n} \leqslant \frac{1}{n} \quad \text{for all} \quad \theta.$$

We shall then show that (6) implies

(7) $$b(\theta) \equiv 0,$$

that is, that δ is unbiased.

(i) Since $|b(\theta)| \leqslant 1/\sqrt{n}$, the function b is bounded.
(ii) From the fact that

$$1 + 2b'(\theta) + [b'(\theta)]^2 \leqslant 1$$

it follows that $b'(\theta) \leqslant 0$, so that b is nonincreasing.

(iii) We shall show, next, that there exists a sequence of values θ_i tending to ∞ and such that $b'(\theta_i) \to 0$. For suppose that $b'(\theta)$ were bounded away from 0 as $\theta \to \infty$, say $b'(\theta) \leqslant -\varepsilon$ for all $\theta > \theta_0$. Then $b(\theta)$ cannot be bounded as $\theta \to \infty$, which contradicts (i).

(iv) Analogously it is seen that there exists a sequence of values $\theta_i \to -\infty$ and such that $b'(\theta_i) \to 0$ (Problem 3.2).

Inequality (6) together with (iii) and (iv) shows that $b(\theta) \to 0$ as $\theta \to \pm\infty$, and (7) now follows from (ii).

Since (7) implies that $b(\theta) = b'(\theta) = 0$ for all θ, it implies by (4) that

$$R(\theta, \delta) \geqslant \frac{1}{n} \qquad \text{for all} \quad \theta$$

and hence that

$$R(\theta, \delta) \equiv \frac{1}{n}.$$

This proves that $\bar{X}$ is admissible and minimax. That it is in fact the only minimax estimator is an immediate consequence of Theorem 1.6.5.

Admissibility (and hence minimaxity) of $\bar{X}$ holds not only for squared error loss but for large classes of loss functions $L(\theta, d) = \rho(d - \theta)$. In particular, it holds if $\rho(t)$ is nondecreasing as t moves away from 0 in either direction and satisfies the growth condition

$$\int |t| \rho(2|t|) \phi(t)\, dt < \infty,$$

with the only exceptions being the loss functions

$$\rho(0) = a, \qquad \rho(t) = b \qquad \text{for} \quad |t| \neq 0, \qquad a < b.$$

This result* follows from Brown (1966, Theorem 2.1.1); it is also proved under somewhat stronger conditions in Hájek (1972).

Example 3.3. ***Truncated Normal Mean.*** In Example 3.2, suppose it is known that $\theta > \theta_0$. Then it follows from Lemma 3.1 that $\bar{X}$ is no longer admissible. However, using the method of the second proof of Example 3.2, it is easy to show that $\bar{X}$ continues to be minimax. If it were not, there would exist an estimator δ and an $\varepsilon > 0$ such that

$$R(\theta, \delta) \leqslant \frac{1}{n} - \varepsilon \qquad \text{for all} \quad \theta > \theta_0$$

and hence

$$b^2(\theta) + \frac{[1 + b'(\theta)]^2}{n} \leqslant \frac{1}{n} - \varepsilon \qquad \text{for all} \quad \theta > \theta_0.$$

*Communicated to me by L. Brown.

As a consequence, $b(\theta)$ would be bounded and satisfy $b'(\theta) \leqslant -\varepsilon n/2$ for all $\theta > \theta_0$, and these two statements are contradictory.

This example provides an instance in which the minimax estimator is not unique, and the constant risk estimator $\bar{X}$ is inadmissible. A uniformly better estimator which a fortiori is also minimax is $\max(\theta_0, \bar{X})$, but it, too, is inadmissible [see Sacks (1963), in which a characterization of all admissible estimators is given.] Admissible minimax estimators in this case were found by Katz (1961) and Sacks (1963); see also Gupta and Rohatgi (1980).

If θ is further restricted to satisfy $a \leqslant \theta \leqslant b$, $\bar{X}$ is not only inadmissible but also no longer minimax. If $\bar{X}$ were minimax, the same would be true of its improvement

$$\delta^*(X) = \begin{matrix} a & \text{if} & \bar{X} < a \\ \bar{X} & \text{if} & a \leqslant \bar{X} \leqslant b \\ b & \text{if} & \bar{X} > b \end{matrix}$$

so that

$$\sup_{a \leqslant \theta \leqslant b} R(\theta, \delta^*) = \sup_{a \leqslant \theta \leqslant b} R(\theta, \bar{X}) = \frac{1}{n}.$$

However, $R(\theta, \delta^*) < R(\theta, \bar{X}) = \dfrac{1}{n}$ for all $a \leqslant \theta \leqslant b$. Furthermore, $R(\theta, \delta^*)$ is a continuous function of θ and hence takes on its maximum at some point $a \leqslant \theta_0 \leqslant b$. Thus

$$\sup_{a \leqslant \theta \leqslant b} R(\theta, \delta^*) = R(\theta_0, \delta^*) < \frac{1}{n},$$

which provides a contradiction.

It follows from Wald's general decision theory (see, for example, Theorem 5.9 of Wald, 1950) that in the present situation there exists a probability distribution Λ over $[a, b]$ which satisfies (2.2) and (2.4). We shall now prove that the associated set ω_Λ of (2.3) consists of a finite number of points. Suppose the contrary were true. Then ω_Λ contains an infinite sequence of points with a limit point. Since $R(\theta, \delta_\Lambda)$ is constant over these points and since it is an analytic function of θ, it follows that $R(\theta, \delta_\Lambda)$ is constant, not only in $[a, b]$ but for all θ. Example 3.2 then shows that $\delta_\Lambda = \bar{X}$, which is in contradiction to the fact that $\bar{X}$ is not minimax for the present problem.

Casella and Strawderman (1981) show that, for $b - a \leqslant 2.1/\sqrt{n}$, a least favorable Λ puts probability $1/2$ on each of the points $\theta = a$ and $\theta = b$ and that the decrease in the minimax risk in this range of $b - a$ is substantial. (They also show that for $b - a < 2/\sqrt{n}$, the risk of the minimax estimator is uniformly smaller than that of the MLE for $a \leqslant \theta \leqslant b$). As $b - a$ increases, so does the number of points in ω_Λ, with Λ approximating Lebesgue measure as $b - a \to \infty$. Some interesting results concerning Λ and δ_Λ when $b - a$ is large are given by Bickel (1981).

Example 3.4. Linear Model. Consider the general linear model of Section 3.4 and suppose we wish to estimate some linear function of the ξ's. Without loss of generality, we can assume that the model is expressed in the canonical form (3.4.8) so that $Y_1, \ldots, Y_n$ are independent, normal, with common variance σ^2 and $E(Y_i) = \eta_i$ $(i = 1, \ldots, s)$; $E(Y_{s+1}) = \cdots = E(Y_n) = 0$. The estimand can be taken to be η_1. If $Y_2, \ldots, Y_n$ were not present it would follow from Example 3.2 that Y_1 is admissible for estimating η_1. It is obvious from the Rao–Blackwell theorem (Section 1.6) that the presence of $Y_{s+1}, \ldots, Y_n$ cannot affect this result. The following lemma shows that, as one would expect, the same is true for $Y_2, \ldots, Y_s$.

Lemma 3.2. *Let* X, Y *be independent (possibly vector-valued) with distributions* F_ξ, G_η, *respectively, where* ξ *and* η *vary independently. Then, if* $\delta(X)$ *is admissible for estimating* ξ *when* Y *is not present, it continues to be so in the presence of* Y.

Proof. Suppose, to the contrary, that there exists an estimator $T(X, Y)$ satisfying

$$R(\xi, \eta; T) \leqslant R(\xi; \delta) \qquad \text{for all} \quad \xi, \eta;$$

$$R(\xi_0, \eta_0; T) < R(\xi_0; \delta) \qquad \text{for some} \quad \xi_0, \eta_0.$$

Consider the case in which it is known that $\eta = \eta_0$. Then $\delta(X)$ is admissible on the basis of X and Y by the proof of Theorem 1.5.1. On the other hand,

$$R(\xi, \eta_0; T) \leqslant R(\xi; \delta) \qquad \text{for all} \quad \xi,$$

$$R(\xi_0, \eta_0; T) < R(\xi_0; \delta) \qquad \text{for some} \quad \xi_0,$$

and this is a contradiction.

Thus the UMVU estimators derived in Section 3.4 are admissible.

The examples so far have been concerned with normal means. Let us now turn to the estimation of a normal variance.

Example 3.5. Normal Variance. Under the assumptions of Example 1.4, let us consider the admissibility of linear estimators $aY + b$ of $1/\tau = 2\sigma^2$. The Bayes solutions

$$\frac{\alpha + Y}{r + g - 1}, \tag{8}$$

derived there for the prior distributions $\Gamma(g, 1/\alpha)$, appear to prove admissibility of $aY + b$ with

$$0 < a < \frac{1}{r - 1}, \qquad 0 < b. \tag{9}$$

In particular, this includes the estimators $(1/r)Y + b$ for any $b > 0$. On the other hand, it follows from (1.18) that $E(Y) = r/\tau$, so that $(1/r)Y$ is an unbiased estimator of $1/\tau$, and hence from Lemma 2.5.1 that $(1/r)Y + b$ is inadmissible for any $b > 0$. What went wrong?

Conditions (i) and (ii) of Corollary 1.2 indicate two ways in which the uniqueness (and hence admissibility) of a Bayes estimator may be violated. The second of these clearly does not apply here since the gamma prior assigns positive density to all values $\tau > 0$. This leaves the first possibility as the only visible suspect. Let us, therefore, consider the Bayes risk of the estimator (8).

Given τ, we find [by adding and subtracting the expectation of $Y/(g + r - 1)$], that

$$E\left(\frac{Y+\alpha}{g+r-1} - \frac{1}{\tau}\right)^2 = \frac{1}{(g+r-1)^2}\left[\frac{r}{\tau^2} - \left(\alpha - \frac{g-1}{\tau}\right)^2\right].$$

The Bayes risk will therefore be finite if and only if $E(1/\tau^2) < \infty$, where the expectation is taken with respect to the prior, and hence if and only if $g > 2$. Applying this condition to (8), we see that admissibility has not been proved for the region (9), as seemed the case originally, but only for the smaller region

$$(10) \qquad 0 < a < \frac{1}{r+1}, \qquad 0 < b.$$

In fact, it is not difficult to prove inadmissibility for all $a > 1/(r + 1)$ (Problem 3.6) whereas for $a < 0$, and for $b < 0$, it of course follows from Lemma 3.1.

The left boundary $a = 0$ of the strip (10) is admissible as it was in Example 3.1; the bottom boundary $b = 0$ was seen to be inadmissible for any positive $a \neq 1/(r + 1)$ in Example 2.5.4. This leaves in doubt only the point $a = b = 0$, which is admissible (Problem 3.6), and the right boundary, corresponding to the estimators

$$(11) \qquad \frac{1}{r+1}Y + b, \qquad 0 \leqslant b < \infty.$$

Admissibility of (11) for $b = 0$ was first proved by Karlin (1958), who considered the case of general one-parameter exponential families. His proof was extended to other values of b by Ping (1964) and Gupta (1966). We shall follow Ping's proof, which uses the second method of Example 3.2 while Karlin (1958) and Stone (1967) employed the first method.

Let X have probability density

$$(12) \qquad p_\theta(x) = \beta(\theta)e^{\theta T(x)} \qquad (\theta, T \text{ real valued})$$

with respect to μ and let Ω be the natural parameter space. Then Ω is an interval, with end points, say, $\underline{\theta}$ and $\bar{\theta}$ $(-\infty \leqslant \underline{\theta} \leqslant \bar{\theta} \leqslant \infty)$ (see Section 1.4). For estimating $E_\theta(T)$, the estimator $aT + b$ is inadmissible if $a < 0$ or $a > 1$ and is a constant for $a = 0$. To state Karlin's sufficient condition in the remaining cases, it is convenient to write the estimator as

$$\frac{1}{1+\lambda}T + \frac{\gamma\lambda}{1+\lambda}, \tag{13}$$

with $0 \leqslant \lambda < \infty$ corresponding to $0 < a \leqslant 1$.

Theorem 3.3. *Under the above assumptions, a sufficient condition for the admissibility of the estimator* (13) *for estimating* $g(\theta) = E_\theta(T)$ *with squared error loss is that the integral of* $e^{-\gamma\lambda\theta}[\beta(\theta)]^{-\lambda}$ *diverges at* $\underline{\theta}$ *and* $\bar{\theta}$, *that is, that for some (and hence for all)* $\underline{\theta} < \theta_0 < \bar{\theta}$, *the two integrals*

$$\int_{\theta_0}^{\theta} \frac{e^{-\gamma\lambda\theta}}{[\beta(\theta)]^{\lambda}} d\theta \quad \text{and} \quad \int_{\theta}^{\theta_0} \frac{e^{-\gamma\lambda\theta}}{[\beta(\theta)]^{\lambda}} d\theta \tag{14}$$

tend to infinity as θ *tends to* $\bar{\theta}$ *and* $\underline{\theta}$, *respectively.*

Proof. It is seen from (1.4.11) and (1.4.12) that

$$g(\theta) = E_\theta(T) = \frac{-\beta'(\theta)}{\beta(\theta)} \tag{15}$$

and

$$g'(\theta) = \text{var}_\theta(T) = I(\theta), \tag{16}$$

where $I(\theta)$ is the Fisher information defined in (2.6.10). For any estimator $\delta(X)$ we have by (2.7.27) and (16)

$$E_\theta[\delta(X) - g(\theta)]^2 \geqslant b^2(\theta) + \frac{[I(\theta) + b'(\theta)]^2}{I(\theta)} \tag{17}$$

where $b(\theta) = E_\theta[\delta(X)] - g(\theta)$ is the bias of δ. Suppose that there exists an estimator δ_0 such that

$$E_\theta\left[\frac{T+\gamma\lambda}{1+\lambda} - g(\theta)\right]^2 \geqslant E_\theta[\delta_0(X) - g(\theta)]^2 \quad \text{for all} \quad \theta. \tag{18}$$

By (16), the left side is equal to

$$E_\theta\left[\frac{T+\gamma\lambda}{1+\lambda} - g(\theta)\right]^2 = \frac{I(\theta)}{(1+\lambda)^2} + \frac{\lambda^2[g(\theta)-\gamma]^2}{(1+\lambda)^2}. \tag{19}$$

Application of (17) to δ_0 shows that

$$b_0^2(\theta) + \frac{[I(\theta) + b_0'(\theta)]^2}{I(\theta)} \leqslant \frac{I(\theta)}{(1+\lambda)^2} + \frac{\lambda^2[g(\theta) - \gamma]^2}{(1+\lambda)^2}. \tag{20}$$

Let us now choose δ to be the estimator (13), put

$$h(\theta) = b_0(\theta) - b(\theta), \tag{21}$$

and use the fact that

$$b(\theta) = \frac{\lambda}{1+\lambda}[\gamma - g(\theta)], \qquad b'(\theta) = \frac{-\lambda}{1+\lambda} g'(\theta).$$

Then (20) is seen to reduce to

$$h^2(\theta) - \frac{2\lambda}{1+\lambda} h(\theta)[g(\theta) - \gamma] + \frac{2}{1+\lambda} h'(\theta) + \frac{[h'(\theta)]^2}{I(\theta)} \leqslant 0. \tag{22}$$

which implies

$$h^2(\theta) - \frac{2\lambda}{1+\lambda} h(\theta)[g(\theta) - \gamma] + \frac{2}{1+\lambda} h'(\theta) \leqslant 0. \tag{23}$$

Finally, let

$$\kappa(\theta) = h(\theta)\beta^{\lambda}(\theta)e^{\gamma\lambda\theta}.$$

Differentiation of $\kappa(\theta)$ and use of (15) reduce (23) to (Problem 3.7)

$$\kappa^2(\theta)\beta^{-\lambda}(\theta)e^{-\gamma\lambda\theta} + \frac{2}{1+\lambda}\kappa'(\theta) \leqslant 0. \tag{24}$$

We shall now show that (24) with $\lambda \geqslant 0$ implies that $\kappa(\theta) \geqslant 0$ for all θ. Suppose to the contrary that $\kappa(\theta_0) < 0$ for some θ_0. Then $\kappa(\theta) < 0$ for all $\theta \geqslant \theta_0$ since $\kappa'(\theta) < 0$, and for $\theta > \theta_0$ we can write (24) as

$$\frac{d}{d\theta}\left[\frac{1}{\kappa(\theta)}\right] \geqslant \frac{1+\lambda}{2}\beta^{-\lambda}(\theta)e^{-\gamma\lambda\theta}.$$

Integrating both sides from θ_0 to θ leads to

$$\frac{1}{\kappa(\theta)} - \frac{1}{\kappa(\theta_0)} \geqslant \frac{1+\lambda}{2}\int_{\theta_0}^{\theta} \beta^{-\lambda}(\theta)e^{-\gamma\lambda\theta}\,d\theta.$$

As $\theta \to \bar{\theta}$, the right-hand side tends to infinity, and this provides a contradiction since the left side is $< -1/\kappa(\theta_0)$.

Similarly, $\kappa(\theta) \leqslant 0$ for all θ. It follows that $\kappa(\theta)$ and hence $h(\theta)$ is zero for all θ. This shows that for all θ equality holds in (22), (20), and thus (18). This proves the admissibility of (13).

Under some additional restrictions, it is shown by Diaconis and Ylvisaker (1979) that when the sufficient condition of Theorem 3.3 holds, $aX + b$ is Bayes with respect to a proper prior distribution (a member of the conjugate family) and has finite risk. This, of course, implies that it is admissible.

Karlin (1958) conjectured that the sufficient condition of Theorem 3.3 is also necessary for the admissibility of (13). Despite further work on this problem [Morton and Raghavachari (1966), Stone (1967)], Joshi (1969), this conjecture has not yet been settled.

Let us now see whether Theorem 3.3 settles the admissibility of $Y/(r+1)$, which was left open in Example 3.5.

Example 3.5. (*Continued*). The density of Example 1.4 is of the form (12) with

$$\theta = -r\tau, \qquad \beta(\theta) = \left(\frac{-\theta}{r}\right)^r, \qquad \frac{Y}{r} = T(X), \qquad \underline{\theta} = -\infty, \qquad \bar{\theta} = 0.$$

Here the parametrization is chosen so that

$$E_\theta[T(X)] = \frac{1}{\tau}$$

coincides with the estimand of Example 3.5. An estimator

$$\frac{1}{1+\lambda} \cdot \frac{Y}{r} + \frac{\gamma\lambda}{1+\lambda} \tag{25}$$

is therefore admissible provided the integrals

$$\int_{-\infty}^{-c} e^{-\gamma\lambda\theta}\left(\frac{-\theta}{r}\right)^{-r\lambda} d\theta = C\int_c^\infty e^{\gamma\lambda\theta}\theta^{-r\lambda}\, d\theta$$

and

$$\int_0^c e^{\gamma\lambda\theta}\theta^{-r\lambda}\, d\theta$$

are both infinite.

The conditions for the first integral to be infinite are that either

$$\gamma = 0 \quad \text{and} \quad r\lambda \leqslant 1, \quad \text{or} \quad \gamma\lambda > 0.$$

For the second integral, the factor $e^{\gamma\lambda\theta}$ plays no role and the condition is simply

$$r\lambda \geqslant 1.$$

Combining these conditions, we see that the estimator (25) is admissible if either

$$(a) \qquad \gamma = 0 \quad \text{and} \quad \lambda = \frac{1}{r}$$

or

$$(b) \qquad \lambda \geqslant \frac{1}{r} \quad \text{and} \quad \gamma > 0 \ (\text{since } r > 0).$$

If we put $a = 1/(1+\lambda)r$, $b = \gamma\lambda/(1+\lambda)$, it follows that $aY + b$ is admissible if either

$$(a') \qquad b = 0 \quad \text{and} \quad a = \frac{1}{1+r}$$

or

$$(b') \qquad b > 0 \quad \text{and} \quad 0 < a \leqslant \frac{1}{1+r}.$$

The first of these results settles the one case that was left in doubt in Example 3.5; the second confirms the admissibility of the interior of the strip (10), which had already been established in that example. The admissibility of $Y/(r+1)$ for estimating $1/\tau = 2\sigma^2$ means that

$$(26) \qquad \frac{1}{2r+2}Y = \frac{1}{n+2}Y$$

is admissible for estimating σ^2. The estimator (26) is the MRE estimator for σ^2 found in Section 3.3.

Admissibility of the estimator $\Sigma X_i^2/(n+2)$ when the X's are from $N(0, \sigma^2)$ naturally raises the corresponding question for

$$(27) \qquad \Sigma(X_i - \bar{X})^2/(n+1),$$

the MRE estimator of σ^2 when the X's are from $N(\xi, \sigma^2)$ with ξ unknown. The surprising answer, due to Stein (1964), is that (27) is not admissible. An estimator with uniformly smaller risk is

$$(28) \qquad \delta = \min\left(\frac{\Sigma(X_i - \bar{X})^2}{n+1}, \frac{\Sigma X_i^2}{n+2}\right).$$

To motivate δ, suppose that it is thought a priori likely but by no means certain that $\xi = 0$. One might then wish to test the hypothesis H: $\xi = 0$, by the usual t-test. If

$$\frac{|\sqrt{n}\,\bar{X}|}{\sqrt{\Sigma(X_i - \bar{X})^2/(n-1)}} < c, \tag{29}$$

one would accept H and correspondingly estimate σ^2 by $\Sigma X_i^2/(n+2)$; in the contrary case, H would be rejected and σ^2 estimated by (27). For the value $c = \sqrt{(n-1)/(n+1)}$, it is easily checked (Problem 3.7) that (29) is equivalent to

$$\frac{1}{n+2}\Sigma X_i^2 < \frac{1}{n+1}\Sigma(X_i - \bar{X})^2, \tag{30}$$

and the resulting estimator then reduces to (28).

While (27) is inadmissible, it is clear that no substantial improvement is possible, since $\Sigma(X_i - \bar{X})^2/\sigma^2$ has the same distribution as $\Sigma(X_i - \xi)^2/\sigma^2$ with n replaced by $n-1$ so that ignorance of σ^2 can be compensated for by one additional observation. For this reason and because we shall consider very similar but more important situations in Section 4.5, we shall omit the proof of Stein's result. [A proof can be found in Stein's paper and in Zacks (1971, pp. 396–397). For generalizations, see Brown (1968).]

Let us now return to Theorem 3.3 and as another illustration apply it to the binomial case.

***Example 3.6. Binomial* p.** Let X have the binomial distribution $b(p, n)$, which we shall write as

$$P(X = x) = \binom{n}{x} q^n e^{(x/n)n\log(p/q)}. \tag{31}$$

Putting $\theta = n\log(p/q)$, we have

$$\beta(\theta) = q^n = [1 + e^{\theta/n}]^{-n}$$

and

$$g(\theta) = E_\theta\left(\frac{X}{n}\right) = p = \frac{e^{\theta/n}}{1 + e^{\theta/n}}.$$

Furthermore, as p ranges from 0 to 1, θ ranges from $\underline{\theta} = -\infty$ to $\bar{\theta} = +\infty$.

The integral in question is then

$$\int e^{-\gamma\lambda\theta}(1 + e^{\theta/n})^{\lambda n}\,d\theta \tag{32}$$

and the estimator $X/[n(1+\lambda)] + \gamma\lambda/(1+\lambda)$ is admissible provided this integral diverges at both $-\infty$ and $+\infty$. If $\lambda < 0$, the integrand is $\leqslant e^{-\gamma\lambda\theta}$ and the integral cannot diverge at both limits, while for $\lambda = 0$ the integral does diverge at both limits. Suppose, therefore, that $\lambda > 0$. Near infinity, the dominating term (which is also a lower bound) is

$$\int e^{-\gamma\lambda\theta + \lambda\theta}\, d\theta,$$

which diverges provided $\gamma \leqslant 1$. At the other end, we have

$$\int_{-\infty}^{-c} e^{-\gamma\lambda\theta}(1 + e^{\theta/n})^{\lambda n}\, d\theta = \int_{c}^{\infty} e^{\gamma\lambda\theta}\left(1 + \frac{1}{e^{\theta/n}}\right)^{\lambda n} d\theta.$$

The factor in parentheses does not affect the convergence or divergence of this integral, which therefore diverges if and only if $\gamma\lambda \geqslant 0$. The integral will therefore diverge at both limits provided

(33) $$\lambda > 0 \quad \text{and} \quad 0 \leqslant \gamma \leqslant 1, \quad \text{or} \quad \lambda = 0.$$

With $a = 1/(1 + \lambda)$ and $b = \gamma\lambda/(1 + \lambda)$, this condition is seen to be equivalent (Problem 3.7) to

(34) $$0 < a \leqslant 1, \quad 0 \leqslant b, \quad a + b \leqslant 1.$$

The estimator, of course, is also admissible when $a = 0$ and $0 \leqslant b \leqslant 1$, and it is easy to see that it is inadmissible for the remaining values of a and b (Problem 3.8). The region of admissibility is therefore the closed triangle $\{(a, b)\colon a \geqslant 0, b \geqslant 0, a + b \leqslant 1\}$.

Theorem 3.3 provides a simple condition for the admissibility of T as an estimator of $E_\theta(T)$.

Corollary 3.1. *If the natural parameter space of* (12) *is the whole real line so that* $\underline{\theta} = -\infty$, $\bar{\theta} = \infty$, *then* T *is admissible for estimating* $E_\theta(T)$ *with squared error loss.*

Proof. With $\lambda = 0$, $\gamma = 1$, the two integrals (14) clearly tend toward infinity as $\theta \to \pm\infty$.

The condition of this corollary is satisfied by the normal (variance known), binomial, and Poisson distribution but not in the gamma or negative binomial case (Problem 3.15).

The starting point of this section was the question of admissibility of some minimax estimators. In the opposite direction, it is sometimes possible to use the admissibility of an estimator to prove that it is minimax.

Lemma 3.3. *If an estimator has constant risk and is admissible, it is minimax.*

Proof. If it were not, another estimator would have smaller maximum risk and hence uniformly smaller risk.

This lemma together with Corollary 3.1 yields the following minimax result.

Corollary 3.2. *Under the assumptions of Corollary* 3.1, T *is the unique minimax estimator of* $g(\theta) = E_\theta(T)$ *for the loss function* $[d - g(\theta)]^2/\text{var}_\theta(T)$.

Proof. For this loss function, T is a constant risk estimator which is admissible by Corollary 3.1 and unique by Theorem 1.6.5.

Example 3.7. (i) If X has the binomial distribution $b(p, n)$, then X/n is the unique minimax estimator of p for the loss function $(d - p)^2/pq$. This had already been seen in Example 2.1.

(ii) If $X_1, \dots, X_n$ are iid according to the Poisson distribution $P(\lambda)$, then $\bar{X}$ is unique minimax for estimating λ with loss function $(d - \lambda)^2/\lambda$.

The estimation of a normal variance with unknown mean provides a surprising example of a reasonable estimator which is inadmissible. We shall conclude this section with an example of a totally unreasonable estimator that is admissible.

Example 3.8. ***Two Binomials.*** Let X and Y be independent binomial random variables with distributions $b(p, m)$ and $b(\pi, n)$, respectively. It was shown by Makani (1972, 1977) that a necessary and sufficient condition for

$$a\frac{X}{m} + b\frac{Y}{n} + c \tag{35}$$

to be admissible for estimating p with squared error loss is that either

$$\begin{gathered} 0 \leqslant a < 1, \quad 0 \leqslant c \leqslant 1, \quad 0 \leqslant a + c \leqslant 1, \\ 0 \leqslant b + c \leqslant 1, \quad 0 \leqslant a + b + c \leqslant 1, \end{gathered} \tag{36}$$

or

$$a = 1 \quad \text{and} \quad b = c = 0. \tag{37}$$

We shall now prove the sufficiency part which is the result of interest; for necessity, see Problem 3.12.

Suppose there exists another estimator $\delta(X, Y)$ with risk uniformly at least as small as that of (35) so that

$$E\left(a\frac{X}{m} + b\frac{Y}{n} + c - p\right)^2 \geqslant E[\delta(X, Y) - p]^2 \qquad \text{for all } p.$$

Then

$$\begin{aligned} &\sum_{x=0}^{m}\sum_{k=0}^{n}\left(a\frac{x}{m} + b\frac{k}{n} + c - p\right)^2 P(X = x, Y = k) \\ &\qquad \geqslant \sum_{x=0}^{m}\sum_{k=0}^{n}[\delta(x, k) - p]^2 P(X = x, Y = k). \end{aligned} \tag{37}$$

Letting $\pi \to 0$, this leads to

$$\sum_{x=0}^{m}\left(a\frac{x}{m}+c-p\right)^2 P(X=x) \geqslant \sum_{x=0}^{m}[\delta(x,0)-p]^2 P(X=x) \qquad \text{for all} \quad p.$$

However, $a(X/m)+c$ is admissible by Example 3.6, and hence $\delta(x,0)=a(x/m)+c$ for all $x=0,1,\ldots,m$.

The terms in (38) with $k=0$, therefore, cancel. The remaining terms contain a common factor π which can also be cancelled and one can now proceed as before. Continuing in this way by induction over k, one finds at the $(k+1)^{\text{st}}$ stage that

$$\sum_{x=0}^{m}\left(a\frac{x}{m}+b\frac{k}{n}+c-p\right)^2 P(X=x) \geqslant \sum_{x=0}^{m}[\delta(x,k)-p]^2 P(X=x)$$

$$\text{for all} \quad p.$$

However, $aX/m+bk/n+c$ is admissible by Example 3.6 since

$$a+b\frac{k}{n}+c \leqslant 1,$$

and hence

$$\delta(x,k)=a\frac{x}{m}+b\frac{k}{n}+c \qquad \text{for all} \quad x.$$

This shows that (38) implies

$$\delta(x,y)=a\frac{x}{m}+b\frac{y}{n}+c \qquad \text{for all} \quad x \quad \text{and} \quad y,$$

and hence that (35) is admissible.

Putting $a=0$ in (35), we see that estimates of the form $b(Y/n)+c$ $(0 \leqslant c \leqslant 1,$ $0 \leqslant b+c \leqslant 1)$ are admissible for estimating p despite the fact that only the distribution of X depends on p and that X and Y are independent. This paradoxical result suggests that admissibility is an extremely weak property. While it is somewhat embarrassing for an estimator to be inadmissible, the fact that it is admissible in no way guarantees that it is a good or even halfway reasonable estimator.

The result of Example 3.8 is not isolated. An exactly analogous result holds in the Poisson case (Problem 3.13) and a very similar one due to Brown for normal distributions (for the latter, see Makani, 1977); that an exactly analogous example is not possible in the normal case follows from Cohen (1965).

4. EQUIVARIANCE, ADMISSIBILITY, AND THE MINIMAX PROPERTY

The two preceding sections dealt with minimax estimators and their admissibility in exponential families. Let us now consider the corresponding problems for group families. As was seen in Section 3.2, in these families there typically exists an MRE estimator δ_0 for any invariant loss function, and it is a constant risk estimator. If δ_0 is also a Bayes estimator it is minimax by Corollary 2.1 and admissible if it is unique Bayes.

A group family is a family of distributions which is invariant under a group G of transformations for which $\overline{G}$ is transitive over the parameter space. To see whether δ_0 is a Bayes estimator, it seems natural to look for prior distributions with invariance properties. We shall say that a distribution Λ for θ is *invariant* with respect to $\overline{G}$ if the distribution of $\bar{g}\theta$ is also Λ for all $\bar{g} \in \overline{G}$, that is, if for all $\bar{g} \in \overline{G}$ and all measurable B

$$P_\Lambda(\bar{g}\theta \in B) = P_\Lambda(\theta \in B) \tag{1}$$

or equivalently

$$\Lambda(\bar{g}^{-1}B) = \Lambda(B). \tag{2}$$

Suppose now that such a Λ exists, and that the Bayes solution δ_Λ with respect to it is unique. By (1) any δ then satisfies

$$\int R(\theta, \delta)\, d\Lambda(\theta) = \int R(\bar{g}\theta, \delta)\, d\Lambda(\theta). \tag{3}$$

Now

$$\begin{aligned} R(\bar{g}\theta, \delta) &= E_{\bar{g}\theta}\{L[\bar{g}\theta, \delta(X)]\} = E_\theta\{L[\bar{g}\theta, \delta(gX)]\} \\ &= E_\theta\{L[\theta, g^{*-1}\delta(gX)]\}. \end{aligned} \tag{4}$$

Here, the second equality follows from (3.2.4) and the third from (3.2.14). On substituting this last expression into the right side of (3), we see that if $\delta_\Lambda(x)$ minimizes (3) so does the esimtator $g^{*-1}\delta_\Lambda(gx)$. Hence, if the Bayes estimator is unique, the two must coincide. By (3.2.15), this appears to prove δ_Λ to be equivariant. However, at this point a technical difficulty arises. Uniqueness can be asserted only up to null sets, that is, sets N with $P_\theta(N) = 0$ for all θ. Moreover, the set N may depend on g. An estimator δ satisfying

$$\delta(x) = g^{*-1}\delta(gx) \qquad \text{for all} \quad x \notin N_g \tag{5}$$

where $P_\theta(N_g) = 0$ for all θ is said to be *almost equivariant*. We have therefore proved the following result.

Theorem 4.1. *Suppose that an estimation problem is invariant under a group and that there exists a distribution* Λ *over* Ω *such that* (2) *holds for all (measurable) subsets* B *of* Ω *and all* $g \in G$. *Then, if the Bayes estimator* δ_Λ *is unique, it is almost equivariant.*

It follows that under the assumptions of the theorem there exists an almost equivariant estimator which is admissible. Furthermore, it turns out that under very weak additional assumptions, given any almost equivariant estimator δ, there exists an equivariant estimator δ' which differs from δ only on a null set N. The existence of such a δ' is obvious in the simplest case, that of a finite group. We shall not prove it here for more general groups (a precise statement and proof can be found in TSH, Chap. 6, Th. 4). Since δ and δ' then have the same risk function, this establishes the existence of an equivariant estimator that is admissible.

Theorem 4.1 does not require $\bar{G}$ to be transitive over Ω. If we add the assumption of transitivity, it follows that the MRE estimator is admissible and, since it has constant risk, that it is minimax.

The crucial assumption in this approach is the existence of an invariant prior distribution. The following example illustrates the rather trivial case in which the group is finite.

Example 4.1. Let $X_1, \ldots, X_n$ be iid according to the normal distribution $N(\xi, 1)$. Then the problem of estimating ξ with squared error loss remains invariant under the two-element group G, which consists of the identity transformation e and the transformation

$$g(x_1, \ldots, x_n) = (-x_1, \ldots, -x_n); \qquad \bar{g}\xi = -\xi; \qquad g^*d = -d.$$

In the present case, any distribution Λ for ξ, which is symmetric with respect to the origin, clearly satisfies (2). It follows from Theorem 4.1 and the discussion following it, that for any such Λ there is a version of the Bayes solution which is equivariant, that is, which satisfies $\delta(-x_1, \ldots, -x_n) = -\delta(x_1, \ldots, x_n)$. The group $\bar{G}$ in this case is, of course, not transitive over Ω.

As an example of (2) in which G is not finite, we shall consider the following version of the location problem on the circle.

Example 4.2. Circular Location Family. Let $U_1, \ldots, U_n$ be iid on $(0, 2\pi)$ according to a distribution F with density f. We shall interpret these variables as n points chosen at random on the unit circle according to F. Suppose that each point is translated on the circle by an amount $\theta (0 \leqslant \theta < 2\pi)$ (i.e., the new positions are those obtained by rotating the circle by an amount, θ). When a value $U_i + \theta$ exceeds 2π, it is, of course, replaced by $U_i + \theta - 2\pi$. The resulting values are the observations $X_1, \ldots, X_n$. It is then easily seen (Problem 4.2) that the density of X_i is

$$(6) \qquad \begin{array}{ll} f(x_i - \theta + 2\pi) & \text{when } 0 < x_i < \theta; \\ f(x_i - \theta) & \text{when } \theta < x_i < 2\pi. \end{array}$$

This can also be written as

$$(7) \qquad f(x_i - \theta)I(\theta, x_i) + f(x_i - \theta + 2\pi)I(x_i, \theta)$$

where $I(a, b)$ is 1 when $a < b$, and 0 otherwise.

If we straighten the circle to a straight line segment of length 2π, we can also represent this family of distributions in the following form. Select n points at random on $(0, 2\pi)$ according to F. Cut the line segment at an arbitrary point $\theta\,(0 < \theta < 2\pi)$. Place the upper segment so that its end points are $(0, 2\pi - \theta)$ and the lower segment so that its end points are $(2\pi - \theta, 2\pi)$, and denote the coordinates of the n points in their new positions by $X_1, \ldots, X_n$. Then the density of X_i is given by (6).

As an illustration of how such a family of distributions might arise, suppose that in a study of gestation in rats, n rats are impregnated by artificial insemination at a given time, say at midnight on day zero. The observations are the n times $Y_1, \ldots, Y_n$ to birth, recorded as the number of days plus a fractional day. It is assumed that the Y's are iid according to $G(y - \eta)$ where G is known and η is an unknown location parameter. A scientist who is interested in the time of day at which births occur abstracts from the data the fractional parts $X_i' = Y_i - [Y_i]$. The variables $X_i = 2\pi X_i'$ have a distribution of the form (6) where θ is 2π times the fractional part of η.

Let us now return to (7) and consider the problem of estimating θ. The model as originally formulated remains invariant under rotations of the circle. To represent these transformations formally, consider for any real number a the unique number a^*, $0 \leqslant a^* < 2\pi$, for which $a = 2\kappa\pi + a^*$ (κ an integer). Then the group G of rotations can be represented by

$$x_i' = (x_i + c)^*, \qquad \theta' = (\theta + c)^*, \qquad d' = (d + c)^*.$$

A loss function $L(\theta, d)$ remains invariant under G if and only if it is of the form $L(\theta, d) = \rho[(d - \theta)^*]$ (Problem 4.3.). Typically, one would want it to depend only on $(d - \theta)^{**} = \min\{(d - \theta)^*, (2\pi - (d - \theta))^*\}$, which is the difference between d and θ along the smaller of the two arcs connecting them. Thus, the loss might be $((d - \theta)^{**})^2$ or $|(d - \theta)^{**}|$. It is important to notice that neither of these is convex (Problem 4.4).

The group G is transitive over Ω and an invariant distribution for θ is the uniform distribution over $(0, 2\pi)$. By applying an obvious extension of the construction (25) or (26) below, one obtains an admissible equivariant (and hence constant risk) Bayes estimator, which a fortiori is also minimax. If the loss function is not convex in d, only the extension of (25) is available, and the equivariant Bayes procedure may be randomized.

The existence of a proper invariant prior distribution is rather special. More often, the invariant measure for θ will be improper, and the situation is then considerably more complicated. In particular (i) the integral (3) may not be finite, and the argument leading to Theorem 4.1 is thus no longer valid; (ii) it becomes necessary to distinguish between left and right invariant measures Λ. A good discussion of these complications can be found in Berger (1980b). The following example, in which (ii) does not arise, shows how the present point of view may provide an alternative interpretation of the MRE estimators derived in Chapter 3.

Example 4.3. Location Family on the Line. Let us return to the location problem considered in Section 3.1 and ask whether the MRE estimators derived there are admissible or at least minimax. Suppose that $\underline{X} = (X_1, \ldots, X_n)$ has density

$$f(\underline{x} - \theta) = f(x_1 - \theta, \ldots, x_n - \theta), \tag{8}$$

and let G and $\bar{G}$ be the groups of translations $x_i' = x_i + a$ and $\theta' = \theta + a$. The parameter space Ω is the real line and there is a measure over Ω which is invariant under translations. It is the measure ν which to any interval I assigns its length, that is, Lebesgue measure. Unfortunately, it is not a probability distribution since $\nu(\Omega) = \infty$.

Let us for a moment ignore this difficulty and proceed formally as if ν, which assigns to θ the "uniform distribution on $(-\infty, \infty)$," were a proper prior distribution. Then the posterior density of θ given $\underline{x}$ would be given by

$$\frac{f(\underline{x} - \theta)}{\int f(\underline{x} - \theta)\, d\theta}. \tag{9}$$

This quantity is non-negative and its integral, with respect to θ, is equal to 1. It therefore defines a proper distribution for θ and by Section 4.1 the Bayes solution to the problem of estimating θ with loss function L would be obtained by minimizing the posterior expected loss

$$\int L[\theta, \delta(\underline{x})] f(\underline{x} - \theta)\, d\theta \Big/ \int f(\underline{x} - \theta)\, d\theta. \tag{10}$$

In particular, if L is squared error, the minimizing value of $\delta(\underline{x})$ is the expectation of θ under (9), which is exactly the Pitman estimator (3.1.18). The agreement of the estimator minimizing (10) with that obtained in Section 3.1 of course holds also for all other invariant loss functions.

Up to this point, the development is completely analogous to that of Example 4.2. However, since ν is not a probability distribution, Theorem 4.1 is not applicable and we cannot conclude that the Pitman estimator is admissible or even minimax. The minimax character of the Pitman estimator was established in the normal case in Example 2.4 by the use of a least favorable sequence of prior distributions. We shall now consider more generally the minimax and admissibility properties of MRE estimators in group families, beginning with the case of a general location family.

Theorem 4.2. *Suppose* $\underline{X} = (X_1, \ldots, X_n)$ *is distributed according to the density* (8) *and that the Pitman estimator* δ^* *given by* (3.1.18) *has finite variance. Then* δ^* *is minimax for squared error loss.*

Proof. As in Example 2.4, we shall utilize Theorem 2.2, and for this purpose require a least favorable sequence of prior distributions. In view of the discussion at the beginning of Example 4.3, one would expect a sequence of priors that approximates Lebesgue measure to be suitable. The sequence of normal distributions with variance tending toward infinity used in Example 2.4 was of this kind. Here it will be more convenient to use instead a sequence of uniform densities

$$\pi_T(u) = \begin{cases} 1/2T & \text{if } |u| < T \\ 0 & \text{otherwise,} \end{cases} \tag{11}$$

with T tending to infinity. If δ_T is the Bayes estimator with respect to (11) and r_T its Bayes risk, the minimax character of δ^* will follow if it can be shown that r_T tends to the constant risk $r^* = E_0\delta^{*2}(\underline{X})$ of δ^* as $T \to \infty$. Since $r_T \leqslant r^*$ for all T, it is enough to show

$$\liminf r_T \geqslant r^*. \tag{12}$$

We begin by establishing for r_T the lower bound

$$r_T \geqslant (1-\varepsilon) \inf_{\substack{a \leqslant -\varepsilon T \\ b \geqslant \varepsilon T}} E_0\delta_{a,b}^2(\underline{X}), \tag{13}$$

where ε is any number between 0 and 1, and $\delta_{a,b}$ is the Bayes estimator with respect to the uniform prior on (a, b) so that in particular $\delta_T = \delta_{-T,T}$. Then for any c (Problem 4.7),

$$\delta_{a,b}(\underline{X}+c) = \delta_{a-c,b-c}(\underline{X}) + c \tag{14}$$

and hence

$$E_\theta[\delta_{-T,T}(\underline{X}) - \theta]^2 = E_0[\delta_{-T-\theta,T-\theta}(\underline{X})]^2.$$

It follows that for any $0 < \varepsilon < 1$,

$$\begin{aligned} r_T &= \frac{1}{2T}\int_{-T}^{T} E_0[\delta_{-T-\theta,T-\theta}(\underline{X})]^2\,d\theta \\ &\geqslant (1-\varepsilon) \inf_{|\theta| \leqslant (1-\varepsilon)T} E_0[\delta_{-T-\theta,T-\theta}(\underline{X})]^2. \end{aligned}$$

Since $-T-\theta \leqslant -\varepsilon T$ and $T-\theta \geqslant \varepsilon T$ when $|\theta| \leqslant (1-\varepsilon)T$, this implies (13).

Next we show that

$$\lim_{T\to\infty} \inf r_T \geqslant E_0\left[\liminf_{\substack{a\to-\infty\\ b\to\infty}} \delta_{a,b}^2(\underline{X})\right] \tag{15}$$

where the lim inf on the right side is defined as the smallest limit point of all sequences $\delta_{a_n,b_n}^2(\underline{X})$ with $a_n \to \infty$, $b_n \to \infty$. To see this, note that for any function h of two real arguments, one has (Problem 4.8)

$$\liminf_{T\to\infty}\left[\inf_{\substack{a\leqslant -T\\ b\geqslant T}} h(a,b)\right] = \liminf_{\substack{a\to-\infty\\ b\to\infty}} h(a,b). \tag{16}$$

Taking the lim inf of both sides of (13), and using (16) and Fatou's lemma (1.2.16), proves (15).

We shall finally show that as $a \to -\infty$, $b \to \infty$,

$$\delta_{a,b}(\underline{X}) \to \delta^*(\underline{X}) \qquad \text{with probability 1.} \tag{17}$$

From this it follows that the right side of (15) is r^*, which completes the proof.

The limit (17) is seen from the fact that

$$\delta_{a,b}(\underline{x}) = \int_a^b uf(\underline{x}-u)\,du \Big/ \int_a^b f(\underline{x}-u)\,du$$

and that by Problems 3.1.19 and 3.1.20 the set of points $\underline{x}$ for which

$$0 < \int_{-\infty}^{\infty} f(\underline{x}-u)\,du < \infty \quad \text{and} \quad \int_{-\infty}^{\infty} |u| f(\underline{x}-u)\,du < \infty$$

has probability 1.

Theorem 4.2 is due to Girshick and Savage (1951), who proved it somewhat more generally without assuming a probability density and under the sole assumption that there exists an estimator (not necessarily equivariant) with finite risk. The streamlined proof given here is due to Peter Bickel.

Of course, one would like to know whether the constant risk minimax estimator δ^* is admissible. This question was essentially settled by Stein (1959). We state without proof the following special case of his result.

Theorem 4.3. *If* $X_1, \ldots, X_n$ *are independently distributed with common probability density* $f(x-\theta)$, *and if there exists an equivariant estimator* δ_0 *of* θ *for which* $E_0|\delta_0(\underline{X})|^3 < \infty$, *then the Pitman estimator* δ^* *is admissible under squared error loss.*

It was shown by Perng (1970) that this admissibility result need not hold when the third-moment condition is dropped.

In Example 4.3, we have so far restricted attention to squared error loss. Admissibility of the MRE estimator has been proved for large classes of loss functions by Farrell (1964), Brown (1966), and Brown and Fox (1974b). A key assumption is the uniqueness of the MRE estimator. An early counterexample when that assumption does not hold was given by Blackwell (1951). A general inadmissibility result in the case of nonuniqueness is due to Farrell (1964).

Examples 4.2 and 4.3 involved a single parameter θ. That an MRE estimator of θ may be inadmissible in the presence of nuisance parameters, when the corresponding estimator of θ with known values of the nuisance parameters is admissible, is illustrated by the estimator (3.27). Other examples of this type have been studied by Brown (1968), Zidek (1973), and Berger (1976b, c) among others. An important illustration of the inadmissibility of the MRE estimator of a vector-valued parameter constitutes the principal subject of the next two sections.

Even when the best equivariant estimator is not admissible, it may still be—and frequently is—minimax. Conditions for an MRE estimator to be minimax are given by Kiefer (1957). The general treatment of admissibility and minimaxity of MRE estimators is beyond the scope of this book. However, roughly speaking, MRE estimators will typically not be admissible except in the simplest situations but have a much better chance of being minimax.

The difference can be seen by comparing Example 2.4 and the proof of Theorem 4.2 with the first admissibility proof of Example 3.2. If there exists an invariant measure over the parameter space of the group family (or equivalently over the group, see Section 3.2) which can be suitably approximated by a sequence of probability distributions, one may hope that the corresponding Bayes estimators will tend to the MRE estimator and Theorem 4.2 will become applicable. In comparison, the corresponding proof in Example 3.2 is much more delicate because it depends on the rate of convergence of the risks (this is well illustrated by the attempted admissibility proof at the beginning of the next section).

As a contrast to Theorem 4.2, we shall now give some examples in which the MRE estimator is not minimax.

Example 4.4. Consider once more the estimation of Δ in Example 3.2.3 with loss 1 when $|d - \Delta|/\Delta > 1/2$ and $= 0$ otherwise. The problem remains invariant under the group G of transformations

$$X_1' = a_1X_1 + a_2X_2, \qquad Y_1' = c(a_1Y_1 + a_2Y_2)$$
$$X_2' = b_1X_1 + b_2X_2, \qquad Y_2' = c(b_1Y_1 + b_2Y_2)$$

with $a_1b_2 \neq a_2b_1$ and $c > 0$. The only equivariant estimator is $\delta(x, y) \equiv 0$ and its risk is 1 for all values of Δ. On the other hand, the risk of the estimator $k^*Y_2^2/X_2^2$ obtained in Example 3.2.3 is clearly less than 1.

Example 4.5. A Random Walk.* Consider a walk in the plane. The walker at each step goes one unit either right, left, up, or down and these possibilities will be denoted by a, a^-, b, b^-, respectively. Such a walk can be represented by a finite "path" such as

$$bba^-b^-a^-a^-a^-a^-.$$

In reporting a path, we shall, however, cancel any pair of successive steps which reverse each other, such as a^-a or bb^-. The resulting set of all finite paths constitutes the parameter space Ω. A typical element of Ω will be denoted by

$$\theta = \pi_1 \cdots \pi_m,$$

its length by $l(\theta) = m$. Being a parameter, θ (as well as m) is assumed to be unknown. What is observed is the path X obtained from θ by adding one more step, which is taken in one of the four possible directions at random, that is, with probability 1/4 each. If this last step is π_{m+1}, we have

$$X = \theta\pi_{m+1} \qquad \text{if} \quad \pi_m \text{ and } \pi_{m+1} \text{ do not cancel each other,}$$

$$X = \pi_1 \cdots \pi_{m-1} \qquad \text{otherwise.}$$

A special case occurs if θ or X, after cancellation, reduce to a path of length 0; this happens, for example, if $\theta = a^-$ and the random step leading to X is a. The resulting path will then be denoted by e.

The problem is to estimate θ, having observed $X = x$; the loss will be 1 if the estimated path $\delta(x)$ is $\neq \theta$, and 0 if $\delta(x) = \theta$.

If we observe X to be

$$x = \pi_1 \cdots \pi_k,$$

the natural estimate is

$$\delta_0(x) = \pi_1 \cdots \pi_{k-1}.$$

An exception occurs when $x = e$. In that case, which can arise only when $l(\theta) = 1$, let us arbitrarily put $\delta_0(e) = a$. The estimator defined in this way clearly satisfies

$$R(\theta, \delta_0) \leqslant \tfrac{1}{4} \qquad \text{for all} \quad \theta.$$

Now consider the transformations that modify the paths θ, x, and $\delta(x)$ by having each preceded by an initial segment $\pi_{-r} \cdots \pi_{-1}$ on the left so that, for

*A more formal description of this example which is due to Peisakoff (1950), is given in TSH, Chap. 1, Problem 11 (ii).

example, $\theta = \pi_1 \cdots \pi_m$ is transformed into

$$\bar{g}\theta = \pi_{-r} \cdots \pi_{-1}\pi_1 \cdots \pi_m$$

where, of course, some cancellations may occur. The group G is obtained by considering the addition in this manner of all possible initial path segments. Equivariance of an estimator δ under this group is expressed by the condition

(18)

$$\delta(\pi_{-r} \cdots \pi_{-1}x) = \pi_{-r} \cdots \pi_{-1}\delta(x) \qquad \begin{array}{l}\text{for all } x,\\ \text{and all } \pi_{-1},\ldots,\pi_{-r}, r = 1, 2 \cdots.\end{array}$$

This implies in particular that

(19)
$$\delta(\pi_{-r} \cdots \pi_{-1}) = \pi_{-r} \cdots \pi_{-1}\delta(e),$$

and this condition is sufficient as well as necessary for δ to be equivariant because (19) implies that

$$\pi_{-r} \cdots \pi_{-1}\delta(x) = \pi_{-r} \cdots \pi_{-1}x\delta(e) = \delta(\pi_{-r} \cdots \pi_{-1}x).$$

Since $\bar{G}$ is clearly transitive over Ω, the risk function of any equivariant estimator is constant. Let us now determine the MRE estimator. Suppose that $\delta(e) = \pi_{10} \cdots \pi_{k0}$, so that by (19)

$$\delta(x) = x\pi_{10} \cdots \pi_{k0}.$$

The only possibility of $\delta(x)$ being equal to θ occurs when π_{10} cancels the last element of x. The best choice for k is clearly $k = 1$, and the choice of π_{10} (fixed or random) is then immaterial; in any case the probability of cancellation with the last element of X is $1/4$ so that the risk of the MRE estimator (which is not unique) is $3/4$. Comparison with δ_0 shows that a best equivariant estimator in this case is not only not admissible but not even minimax.

The following example, in which the MRE estimator is again not minimax but where G is simply the group of translations on the real line is due to Blackwell and Girshick (1954).

Example 4.6. Discrete Location Family. Let $X = U + \theta$ where U takes on the values $1, 2, \ldots$ with probabilities

$$P(U = k) = p_k.$$

We observe x and wish to estimate θ with loss function

(20)
$$\begin{aligned} L(\theta, d) &= d - \theta && \text{if } d > \theta \\ &= 0 && \text{if } d \leqslant \theta. \end{aligned}$$

The problem remains invariant under arbitrary translation of X, θ, and d by the same amount. It follows from Section 3.1 that the only equivariant estimators are those of the form $X - c$. The risk of such an estimator, which is constant, is given by

$$\sum_{k>c} (k - c) p_k. \tag{21}$$

If the p_k tend to 0 sufficiently slowly, any equivariant estimator will have infinite risk. This is the case, for example, when

$$p_k = \frac{1}{k(k+1)} \tag{22}$$

(Problem 4.12). The reason is that there is a relatively large probability of substantially overestimating θ for which there is a heavy penalty. This suggests a deliberate policy of grossly underestimating θ, for which by (20) there is no penalty. One possible such estimator (which, of course, is not equivariant) is

$$\delta(X) = X - M|X|, \qquad M > 1, \tag{23}$$

and it is not hard to show that its maximum risk is finite (Problem 4.13).

The ideas of the present section have relevance beyond the transitive case for which they were discussed so far. If $\bar{G}$ is not transitive, we can no longer ask whether the UMRE estimator is minimax since a UMRE estimator will then typically not exist. Instead, we can ask whether there exists a minimax estimator which is equivariant. Similarly, the question of the admissibility of the UMRE estimator can be rephrased by asking whether an estimator which is admissible among equivariant estimators is also admissible within the class of all estimators.

The conditions for affirmative answers to these two questions are essentially the same as in the transitive case. In particular, the answer to both questions is affirmative when G is finite. A proof along the lines of Theorem 4.1 is possible, but not very convenient because it would require a characterization of all admissible (within the class of equivariant estimators) equivariant estimators as Bayes solutions with respect to invariant prior distributions. Instead, we shall utilize the fact that for every estimator δ there exists an equivariant estimator whose average risk (to be defined below) is no worse than that of δ.

Let the elements of the finite group G be $g_1, \ldots, g_N$ and consider the estimators

$$\delta_i(X) = g_i^{*-1}\delta(g_i X). \tag{24}$$

When δ is equivariant, of course, $\delta_i(x) = \delta(x)$ for all i. Consider the randomized estimator δ^* for which

$$\delta^*(X) = \delta_i(X) \quad \text{with probability } 1/N \text{ for each } i = 1, \ldots, N, \tag{25}$$

and, assuming the set $\mathcal{D}$ of possible decisions to be convex, the estimator

$$\delta^{**}(X) = \frac{1}{N}\sum_{i=1}^{N} \delta_i(X) \tag{26}$$

which for given x is the expected value of $\delta^*(x)$. Then $\delta^{**}(x)$ is equivariant, and so is $\delta^*(x)$ in the sense that $g^{*-1}\delta^*(gx)$ again is equal to $\delta_i(x)$ with probability $1/N$ for each i (Problem 4.14). For these two estimators it is easy to prove that (Problem 4.15),

(i) for any loss function L

$$R(\theta, \delta^*) = \frac{1}{N}\Sigma R(\bar{g}_i\theta, \delta) \tag{27}$$

and (ii) for any loss function $L(\theta, d)$ which is convex in d,

$$R(\theta, \delta^{**}) \leqslant \frac{1}{N}\Sigma R(\bar{g}_i\theta, \delta). \tag{28}$$

From (27) and (28), it follows immediately that

$$\sup R(\theta, \delta^*) \leqslant \sup R(\theta, \delta) \quad \text{and} \quad \sup R(\theta, \delta^{**}) \leqslant \sup R(\theta, \delta),$$

which proves the existence of an equivariant minimax estimator provided a minimax estimator exists.

Suppose, next, that δ_0 is admissible among all equivariant estimators. If δ_0 is not admissible within the class of all estimators, it is dominated by some δ. Let δ^*, δ^{**} be as above. Then (27) and (28) imply that δ^* and δ^{**} dominate δ_0, which is a contradiction.

Of the two constructions, δ^{**} has the advantage of not requiring randomization while δ^* has the advantage of greater generality since it does not require L to be convex. Both constructions easily generalize to groups that satisfy a condition analogous to (2), namely,

$$\Lambda(Bg) = \Lambda(B) \tag{29}$$

where, however, the set B is now a subset of G rather than of Ω, so that Λ is a distribution over G. (In the transitive case, this distinction essentially disappears since for any given "origin" θ_0 the relation $\theta = \bar{g}\theta_0$ establishes a $1:1$ relationship between the elements θ of Ω and $\bar{g}$ of $\bar{G}$.) A general treatment of the relationship of equivariance to admissibility and the minimax property can be found in Kiefer (1957). See also Berger (1980b) and Bondar and Milnes (1982).

5. SIMULTANEOUS ESTIMATION

So far, we have been concerned with the estimation of a single real-valued parameter $g(\theta)$. However, one may wish to estimate several parameters simultaneously, for example, several physiological constants of a patient, several quality characteristics of an industrial or agricultural product, or several dimensions of musical ability. One is then dealing with a vector-valued estimand

$$g(\theta) = [g_1(\theta), \dots, g_r(\theta)]$$

and a vector-valued estimator

$$\delta = (\delta_1, \dots, \delta_r).$$

A natural generalization of squared error as a measure of accuracy is

$$\Sigma[d_i - g_i(\theta)]^2, \tag{1}$$

which we shall continue to call squared error loss. More generally, we shall consider loss functions $L(\theta, d)$ where $d = (d_1, \dots, d_r)$, and then denote the risk of an estimator δ by

$$R(\theta, \delta) = E_\theta L[\theta, \delta(X)]. \tag{2}$$

Another generalization of expected squared error loss is the matrix $M(\theta, \delta)$ whose (i, j)th element is

$$E\{[\delta_i(X) - g_i(\theta)][\delta_j(X) - g_j(\theta)]\}. \tag{3}$$

We shall say that δ is more *concentrated* about $g(\theta)$ than δ' if

$$M(\theta, \delta') - M(\theta, \delta) \tag{4}$$

is positive semidefinite (but not identically zero). This definition differs from that based on (2) by providing only a partial ordering of estimators, since (4) may be neither positive nor negative semidefinite.

Lemma 5.1. (i) *δ is more concentrated about* $g(\theta)$ *than* δ' *if and only if*

$$E\{\Sigma k_i[\delta_i(X) - g_i(\theta)]\}^2 \leqslant E\{\Sigma k_i[\delta_i'(X) - g_i(\theta)]\}^2 \tag{5}$$

for all constants $k_1, \ldots, k_r$.

(ii) *In particular, if δ is more concentrated about* $g(\theta)$ *than* δ', *then*

$$E[\delta_i(X) - g_i(\theta)]^2 \leqslant E[\delta_i'(X) - g_i(\theta)]^2 \quad \textit{for all} \quad i. \tag{6}$$

(iii) *If* $R(\theta, \delta) \leqslant R(\theta, \delta')$ *for all convex loss functions, then δ is more concentrated about* $g(\theta)$ *than* δ'.

Proof. (i) If $E\{\Sigma k_i[\delta_i(X) - g_i(\theta)]\}^2$ is expressed as a quadratic form in the k_i, its matrix is $M(\theta, \delta)$.

(iii) This follows from the fact that $\{\Sigma k_i[d_i - g_i(\theta)]\}^2$ is a convex function of $d = (d_1, \ldots, d_r)$.

Let us now consider the extension of some of the earlier theory to the case of simultaneous estimation.

1) *The Rao–Blackwell theorem.* The proof of this theorem shows that its results remain valid when δ and g are vector-valued. In particular, for any convex loss function, the risk of any estimator is reduced by taking its expectation given a sufficient statistic. It follows that for such loss functions one can dispense with randomized estimators. Also, Lemma 5.1 shows that an estimator δ is always less concentrated about $g(\theta)$ than the expectation of $\delta(X)$ given a sufficient statistic.

2) *Unbiased estimation.* In the vector-valued case, an estimator δ of $g(\theta)$ is said to be unbiased if

$$E_\theta[\delta_i(X)] = g_i(\theta) \quad \text{for all} \quad i \quad \text{and} \quad \theta. \tag{7}$$

For unbiased estimators, the concentration matrix M defined by (3) is just the covariance matrix of δ.

From the Rao–Blackwell theorem, it follows as in Theorem 2.1.2 for the case $r = 1$, that if L is convex and if a complete sufficient statistic T exists, then any U-estimable g has a unique unbiased estimator depending only on T, and this estimator uniformly minimizes the risk among all unbiased estimators and thus is also more concentrated about $g(\theta)$ than any other unbiased estimator.

3) *Equivariant estimation.* The definitions and concepts of Section 3.2 apply without changes. They are illustrated by the following example, which will be considered in more detail later in the section.

Example 5.1. Several Normal Means. Let $X = (X_1, \ldots, X_r)$, with the X_i independently distributed as $N(\theta_i, 1)$, and consider the problem of estimating the vector mean $\theta = (\theta_1, \ldots, \theta_r)$ with squared error loss. This problem remains invariant under the group G_1 of translations

$$gX = (X_1 + a_1, \ldots, X_r + a_r), \tag{8}$$

$$\bar{g}\theta = (\theta_1 + a_1, \ldots, \theta_r + a_r),$$

$$g^*d = (d_1 + a_1, \ldots, d_r + a_r).$$

The only equivariant estimators are those of the form

$$\delta(X) = (X_1 + c_1, \ldots, X_r + c_r) \tag{9}$$

and an easy generalization of Example 3.1.3 shows that X is the MRE estimator of θ.

The problem also remains invariant under the group G_2 of orthogonal transformations

$$gX = X\Gamma, \qquad \bar{g}\theta = \theta\Gamma, \qquad g^*d = d\Gamma \tag{10}$$

where Γ is an orthogonal $r \times r$ matrix. An estimator δ is equivariant if and only if it is of the form (Problem 5.1)

$$\delta(X) = u(X) \cdot X, \tag{11}$$

where $u(X)$ is any scalar satisfying

$$u(X\Gamma) = u(X) \quad \text{for all orthogonal} \quad \Gamma \quad \text{and all} \quad X \tag{12}$$

and hence is an arbitrary function of ΣX_i^2 (Problem 5.2). The group $\bar{G}$ defined by (10) is not transitive over the parameter space, and a UMRE estimator of θ therefore cannot be expected.

4) *Bayes estimators.* The following result frequently makes it possible to reduce Bayes estimation of a vector-valued estimand to that of its components.

Lemma 5.2. *Suppose that $\delta_i^*(X)$ is the Bayes estimator of $g_i(\theta)$ when θ has the prior distribution Λ and the loss is squared error. Then $\delta^* = (\delta_1^*, \ldots,$*

δ_r^*) *is more concentrated about* $g(\theta)$ *in the Bayes sense that it minimizes*

$$(13) \qquad E[\Sigma k_i(\delta_i(X) - g_i(\Theta))]^2 = E[\Sigma k_i \delta_i(X) - \Sigma k_i g_i(\Theta)]^2$$

for all k_i, *where the expectation is taken over both* Θ *and* X.

Proof. The result follows from the fact that the estimator $\Sigma k_i \delta_i(X)$ minimizing (13) is

$$E[\Sigma k_i g_i(\Theta)|X] = \Sigma k_i E[g(\Theta_i)|X] = \Sigma k_i \delta_i^*(X).$$

Example 5.2. Multinomial. Let $X = (X_0, \ldots, X_s)$ have the multinomial distribution $M(n; p_0, \ldots, p_s)$, and consider the Bayes estimation of the vector $p = (p_0, \ldots, p_s)$ when the prior distribution of p is the Dirichlet distribution Λ with density

$$(14) \qquad \frac{\Gamma(a_0, \ldots, a_s)}{\Gamma(a_0) \ldots \Gamma(a_s)} p_0^{a_0 - 1} \cdots p_s^{a_s - 1} (0 \leqslant p_i, \quad \Sigma p_j = 1).$$

The Bayes estimator of p_i for squared error loss is (Problem 5.3)

$$(15) \qquad \delta_i(X) = \frac{a_i + X_i}{\Sigma a_j + n}$$

and by Lemma 5.2, the estimator $[\delta_0(X), \ldots, \delta_s(X)]$ is then most concentrated in the Bayes sense. As a check, note that $\Sigma \delta_i(X) = 1$ as, of course, it must since Λ assigns probability 1 to $\Sigma p_i = 1$.

5) *Minimax estimators.* In generalization of the binomial minimax problem treated in Example 2.1, let us now determine the minimax estimator of $(p_0, \ldots, p_s)$ for the multinomial model of Example 5.2.

Example 5.2. (*Continued*). Suppose the loss function is squared error. In light of Example 2.1, one might guess as least favorable distribution the Dirichlet distribution (14) with $a_0 = \cdots = a_s = a$. The Bayes estimator (15) reduces to

$$(16) \qquad \delta_i(X) = \frac{a + X_i}{(s+1)a + n}.$$

The estimator $\delta(X)$ with components (16) has constant risk over the support of (14) provided $a = \sqrt{n}/(s+1)$, and for this value of a, $\delta(X)$ is therefore minimax by Corollary 2.1. [Various versions of this problem are discussed by Steinhaus (1957), Trybula (1958), Rutkowska (1977), and Olkin and Sobel (1979).]

Example 5.3. Independent Experiments. Suppose the components X_i of $X = (X_1, \ldots, X_r)$ are independently distributed according to distributions P_{θ_i}, where the

θ_i vary independently over Ω_i, so that the parameter space for $\theta = (\theta_1, \ldots, \theta_r)$ is $\Omega = \Omega_1 \times \cdots \times \Omega_r$. Suppose, further, that for the ith component problem of estimating θ_i with squared error loss, Λ_i is least favorable for θ_i and the minimax estimator δ_i is the Bayes solution with respect to Λ_i, satisfying condition (2.3) with $\omega_i = \omega_{\Lambda_i}$. Then $\delta = (\delta_1, \ldots, \delta_r)$ is minimax for estimating θ with squared error loss. This follows from the facts that (i) δ is a Bayes estimator with respect to the prior distribution Λ for θ, according to which the components θ_i are independently distributed with distribution Λ_i; (ii) $\Lambda(\omega) = 1$ where $\omega = \omega_1 \times \cdots \times \omega_r$; and (iii) the set of points θ at which $R(\theta, \delta)$ attains its maximum is exactly ω.

The analogous result holds if the component minimax estimators δ_i are not Bayes solutions with respect to least favorable priors but have been obtained through a least favorable sequence by Theorem 2.2. As an example, suppose that X_i $(i = 1, \ldots, r)$ are independently distributed as $N(\theta_i, 1)$. Then it follows that $(X_1, \ldots, X_r)$ is minimax for estimating $(\theta_1, \ldots, \theta_r)$ with squared error loss.

The extensions so far brought no great surprises. The results for general r were fairly straightforward generalizations of those for $r = 1$. This will no longer always be the case for the last topic to be considered.

6) *Admissibility.* The multinomial minimax estimator (16) was seen to be a unique Bayes estimator and hence is admissible. To investigate the admissibility of the minimax estimator X for the case of r normal means considered at the end of Example 5.3, one might try the argument suggested following Theorem 4.1. It was seen in Example 5.1 that the problem under consideration remains invariant under the group G_1 of translations and the group G_2 of orthogonal transformations, given by (8) and (10), respectively. Of these, G_1 is transitive; if there existed an invariant probability distribution over G_1, the remark following Theorem 4.1 would lead to an admissible estimator, hopefully, X. However, the measures $c\nu$, where ν is Lebesgue measure, are the only invariant measures (Problem 4.19) and they are not finite. Let us instead consider G_2. An invariant probability distribution over G_2 does exist (TSH, Example 6 of Chap. 8). However, the approach now fails because $\bar{G}_2$ is not transitive. Equivariant estimators do not necessarily have constant risk and, in fact, in the present case a UMRE estimator does not exist [Strawderman (1971)].

Since neither of these two attempts works, let us try instead the limiting Bayes method (Example 3.2, first proof), which was successful in the case $r = 1$. For the sake of convenience, we shall take the loss to be the average squared error,

$$L(\theta, d) = \frac{1}{r} \Sigma (d_i - \theta_i)^2. \tag{17}$$

If X is not admissible, there exists an estimator δ^*, a number $\varepsilon > 0$ and intervals $(\theta_{i0}, \theta_{i1})$ such that

$$R(\theta, \delta^*)\begin{cases} \leqslant 1 & \text{for all } \theta \\ < 1 - \varepsilon & \text{for } \theta \quad \text{satisfying} \quad \theta_{i0} < \theta_i < \theta_{i1} \quad \text{for all } i. \end{cases}$$

A computation analogous to that of Example 3.2 now shows that

$$\text{(18)} \qquad \frac{1 - r_\tau^*}{1 - r_\tau} \geqslant \frac{\varepsilon(1 + \tau^2)}{(\sqrt{2\pi}\,\tau)^r} \int_{\theta_{10}}^{\theta_{11}} \cdots \int_{\theta_{r0}}^{\theta_{r1}} \exp(-\Sigma\theta_i^2/2\tau^2)\, d\theta_1 \cdots d\theta_r.$$

Unfortunately, the factor preceding the integral no longer tends to infinity when $r > 1$, and so this proof too breaks down.

It was shown by Stein (1956) that X is in fact no longer admissible when $r \geqslant 3$ although admissibility continues to hold for $r = 2$. For $r \geqslant 3$, there are many different estimators whose risk is uniformly less than that of X. A particularly simple class, due to James and Stein (1961), is given by

$$\text{(19)} \qquad \delta_i = \mu_i + \left(1 - \frac{r-2}{S^2}\right)(X_i - \mu_i)$$

where $\mu_1, \ldots, \mu_r$ are any given numbers and

$$\text{(20)} \qquad S^2 = \Sigma(X_i - \mu_i)^2.$$

A motivation for the general structure of the estimator (19), although not its specific functional form, can be obtained by the following consideration. Suppose it were thought a priori likely, though not certain, that $\theta_i = \mu_i$ $(i = 1, \ldots, r)$. Then it might be reasonable first to test

$$H: \theta_1 = \mu_1, \ldots, \theta_r = \mu_r$$

and to estimate θ by μ when H is accepted and otherwise by X. The best acceptance region has the form $\Sigma(X_i - \mu_i)^2 \leqslant C$ so that the estimator becomes

$$\text{(21)} \qquad \delta = \begin{cases} \mu & \text{if } S^2 \leqslant C \\ X & \text{if } S^2 > C. \end{cases}$$

[For the case of unknown σ, the corresponding estimator has been investigated by Sclove, Morris, and Radhakrishnan (1972). They show that it does not provide a uniform improvement over X and that its risk is uniformly greater than that of the corresponding James–Stein estimator.]

A smoother approach is provided by an estimator with components of the form

$$\delta_i = \psi(S)X_i + [1 - \psi(S)]\mu_i \tag{22}$$

where ψ, instead of being two-valued as in (21), is a function increasing continuously from 0 at $S = 0$ to 1 as $S \to \infty$. The estimator (19) is of the form (22) (although with $\psi(0) = -\infty$), but the argument given above provides no explanation of the particular choice for ψ. Such a motivation will be given at the beginning of the next section.

The risk function of the estimator (19) is

$$R(\theta, \delta) = 1 - \frac{r-2}{r} E_\theta\left(\frac{r-2}{S^2}\right). \tag{23}$$

We shall postpone the proof of this formula to the next section and shall here consider only some of the basic properties of δ.

Unlike X, the estimator δ is, of course, biased. An aspect that in some circumstances is disconcerting is the fact that the estimator of θ_i depends not only on X_i but also on the other (independent) X's. Do we save enough in risk to make up for these drawbacks?

Since S^2 has a noncentral χ^2-distribution with noncentrality parameter

$$\lambda = \Sigma(\theta_i - \mu_i)^2,$$

the risk function (23) is an increasing function of λ. (This follows from Chap. 7, Problem 4, and Chap. 3, Lemma 2 of TSH.) It tends to 1 as $\lambda \to \infty$, and takes on its minimum value at $\lambda = 0$. For this value, S^2 has a χ^2-distribution with r degrees of freedom, and it follows from (2.2.5) that (Problem 5.5)

$$E\left(\frac{1}{S^2}\right) = \frac{1}{r-2}$$

and hence $R(\mu, \delta) = 2/r$. Particularly for large values of r, the savings can therefore be substantial.

We thus have the surprising result that X is not only inadmissible when $r \geqslant 3$ but that even substantial risk savings are possible. This is the case not only for squared error loss but for a wide variety of loss functions which in a suitable way combine the losses resulting from the r component problems. In particular, Brown (1966, Theorem 3.1.1) proves that X is inadmissible for $r \geqslant 3$ when $L(\theta, d) = \rho(d - \theta)$ where ρ is a convex function satisfying, in addition to some mild conditions, the requirement that the $r \times r$ matrix M with (i, j)th element

$$E_0\left[X_i \frac{\partial}{\partial X_j}\rho(X)\right] \tag{24}$$

is nonsingular. Here the derivative in (24) is replaced by zero whenever it does not exist.

Example 5.4. Consider the following loss functions $\rho_1, \ldots, \rho_4$.

$$\rho_1(t) = \Sigma \nu_i t_i^2 \qquad (\text{all } \nu_i > 0);$$

$$\rho_2(t) = \max_i t_i^2;$$

$$\rho_3(t) = t_1^2;$$

$$\rho_4(t) = \left(\frac{1}{r}\Sigma t_i\right)^2.$$

All four are convex, and M is nonsingular for ρ_1, ρ_2 but singular for ρ_3, ρ_4 (Problem 5.6). For $r \geqslant 3$ it follows from Brown's theorem that X is inadmissible for ρ_1 and ρ_2. On the other hand, it is admissible for ρ_3 and ρ_4 (Problem 5.8).

Other ways in which the admissibility of X depends on the loss function are indicated by the following example (Brown, 1980b) in which $L(\theta, d)$ is not of the form $\rho(d - \theta)$.

Example 5.5. Let X_i $(i = 1, \ldots, r)$ be independently distributed as $N(\theta_i, 1)$ and consider the estimation of θ with loss function

$$L(\theta, d) = \sum_{i=1}^{r} \frac{v(\theta_i)}{\Sigma v(\theta_j)}(\theta_i - d_i)^2. \tag{25}$$

Then the following results hold.

(i) When $v(t) = e^{kt}$ $(k \neq 0)$, X is inadmissible if and only if $r \geqslant 2$.
(ii) When $v(t) = (1 + t^2)^{k/2}$, then
(*a*) X is admissible for $k < 1, 1 \leqslant r < (2 - k)/(1 - k)$ and for $k \geqslant 1$, all r;
(*b*) X is inadmissible for $k < 1, r > (2 - k)/(1 - k)$.
Parts (i) and [ii(b)] will not be proved here. For the proof of [ii(a)], see Problem 5.9.

In the formulations considered so far, the loss function in some way combines the losses resulting from the different component problems. Suppose, however, that the problems of estimating $\theta_1, \ldots, \theta_r$ are quite unrelated and that it is important to control the error on each of them. It might then be of interest to minimize

$$\max_i\left[\sup E(\delta_i - \theta_i)^2\right]. \tag{26}$$

It is easy to see that X is the unique estimator minimizing (26) and hence is admissible from this point of view. This follows from the fact that X_i is the unique estimator for which

$$\sup E(\delta_i - \theta_i)^2 \leqslant 1.$$

[On the other hand, it follows from Example 5.4 that X is inadmissible for $r \geqslant 3$ when $L(\theta, d) = \max_i(d_i - \theta_i)^2$.]

The performance measure (26) is not a risk function in the sense defined in Chapter 1 because it is not the expected value of some loss but the maximum of a number of such expectations. An interesting way of looking at such a criterion was proposed by Brown (1975). He considers a family $\mathcal{L}$ of loss functions L, with the thought that it is not clear which of these loss functions will be most appropriate. (It may not be clear to which use the data will be put, or they may be destined for multiple uses. In this connection see also Rao, 1977.) If

$$R_L(\theta, \delta) = E_\theta L[\theta, \delta(X)], \tag{27}$$

Brown defines δ to be admissible with respect to the class $\mathcal{L}$ if there exists no δ' such that

$$R_L(\theta, \delta') \leqslant R_L(\theta, \delta) \qquad \text{for all} \quad L \in \mathcal{L} \quad \text{and all} \quad \theta$$

with strict inequality holding for at least one $L = L_0$ and $\theta = \theta_0$.

The argument following (26) shows that X is admissible when $\mathcal{L}$ contains the r loss functions $L_i(\theta, d) = (d_i - \theta_i)^2$, $i = 1, \ldots, r$, and hence in particular when $\mathcal{L}$ is the class of all loss functions

$$\sum_{i=1}^{r} c_i(\delta_i - \theta_i)^2, \qquad 0 \leqslant c_i < \infty, \tag{28}$$

On the other hand, Brown shows that if the ratios of the weights c_i to each other are bounded,

$$c_i/c_j < M, \qquad i, j = 1, \ldots, r, \tag{29}$$

then no matter how large M, the estimator X is inadmissible with respect to the class $\mathfrak{L}$ of loss functions (28) satisfying (29).

The above considerations make it clear that the choice between X and competitors such as (19) must depend on the circumstances. (In this connection see also Robinson 1979 a, b) A more detailed discussion of some of the issues will be given in the next section.

6. SHRINKAGE ESTIMATORS

The simultaneous consideration of a number of similar estimation problems involving independent variables and parameters (X_i, θ_i) often occurs in repetitive situations in which it may be reasonable to view the θ's themselves as random variables. Suppose that the X_i are independent normal with mean θ_i and variance σ^2, but suppose in addition that the θ's are also normal, say with mean ξ and variance A. This is essentially the model II version of the one-way layout considered in Section 3.5. However, there interest centered on the variances σ^2 and A, while we now wish to estimate the θ's. To simplify the problem we shall begin by assuming that σ and ξ are known, say $\sigma = 1$ (as in the preceding section) and $\xi = 0$, so that only A and the θ's are unknown.

Under these assumptions, it follows from (1.14) and (1.15) that the Bayes estimator of θ_i and its Bayes risk are

$$\delta_i = E(\theta_i|X) = (1 - B)X_i \quad \text{and} \quad \text{var}(\theta_i|X) = 1 - B \tag{1}$$

where $B = 1/(A + 1)$. Unfortunately, since B is unknown, δ_i is not an estimator. It can, however, be approximated if a suitable estimator of B is available. To see how to estimate B, let us determine the marginal distribution of X_i. Writing $X_i = \theta_i + U_i$ where U_i is $N(0, 1)$ and independent of θ_i, we see that the X_i are iid as $N(0, 1/B)$. The natural estimator of B is therefore of the form $\hat{B} = c/S^2$ where $S^2 = \Sigma X_i^2$, and this suggests an estimator of θ_i:

$$\hat{\delta}_i = (1 - \hat{B})X_i. \tag{2}$$

For $c = r - 2$ (which, below, will be seen to be the value of c minimizing the risk of $\hat{\delta}$), this is just the James–Stein estimator (5.19) with $\mu_i = 0$. We have here obtained it as an *empirical Bayes estimator*, that is, as a Bayes estimator with respect to a prior the parameters of which have been estimated from the data. For an introduction to the literature on empirical Bayes estimation see Susarla (1982)

Having motivated the estimator, we shall now prove that its risk is given by (5.23) as claimed in the preceding section. Slightly more generally, we shall prove the following result.

Theorem 6.1. *Let* $X_i, i = 1, \ldots, r, (r > 2)$ *be independent, with distributions* $N(\theta_i, 1)$ *and let the estimator* δ_c *of* θ *be given by*

$$\delta_{ic} = \left(1 - c\frac{r-2}{S^2}\right)X_i, \qquad S^2 = \Sigma X_j^2. \tag{3}$$

Then the risk function of δ_c, *with loss function* (5.17) *is*

$$R(\theta, \delta_c) = 1 - \frac{(r-2)^2}{r} E_\theta\left(\frac{2c - c^2}{S^2}\right). \tag{4}$$

Proof. Let

$$f(\theta) = \frac{1}{r}\sum_{i=1}^{r} E_\theta[\delta_{ic}(X) - \theta_i]^2 \tag{5}$$

denote the left side of (4) and let $g(\theta)$ denote the right side. The proof that $f(\theta) = g(\theta)$ for all θ will be given in the following three principal steps.

1. Both $f(\theta)$ and $g(\theta)$ depend on θ only through $\Sigma\theta_i^2$.

2. The average of $f(\theta)$ and $g(\theta)$, respectively, with respect to the weight function

$$w(\theta) = \frac{1}{(\sqrt{2\pi A})^r}\exp\left(-\frac{1}{2A}\Sigma\theta_i^2\right), \qquad 0 < A \tag{6}$$

is given for all $B = 1/(A + 1)$, by

$$\int f(\theta)w(\theta)\,d\theta = 1 - \frac{kB}{r} = \int g(\theta)w(\theta)\,d\theta \tag{7}$$

with $k = (r - 2)(2c - c^2)$. These two averages can, of course, be interpreted as the expected values of $f(\theta)$ and $g(\theta)$ when $\theta = (\theta_1, \ldots, \theta_r)$ are random variables with joint density (6).

3. According to Step 1, we can write

$$f(\theta) = f^*(\Sigma\theta_i^2), \qquad g(\theta) = g^*(\Sigma\theta_i^2).$$

For the family of distributions (6), $\Sigma\theta_i^2$ is a complete sufficient statistic, and it follows from (7) that $f^*(\Sigma\theta_i^2) = g^*(\Sigma\theta_i^2)$ a.e. and hence, by the continuity of these functions ensured by Theorem 1.4.1 that $f(\theta) = g(\theta)$ for all θ as was to be proved.

It remains to prove Steps 1 and 2, and we begin with the latter, using the interpretation of $w(\theta)$ as a probability density for θ. Then

$$E[g(\theta)] = 1 - \frac{r-2}{r} E\left[E\left(\frac{k}{S^2}\middle|\theta\right)\right] \tag{8}$$

where the inner expectation is with respect to the conditional distribution $N(\theta_i, 1)$ of the X_i given θ, and the outer expectation is with respect to the density (6) of θ. Then, by equation (1.5.1)

$$E[g(\theta)] = 1 - \frac{r-2}{r} E\left(\frac{k}{S^2}\right), \tag{9}$$

where the expectation now is with respect to the marginal distribution $N(0, 1/B)$ of the X_i. This distribution implies that BS^2 is distributed as χ_r^2 and hence by Problem 5.5 that $E(1/S^2) = B/(r-2)$, which proves the second equation in (7).

To evaluate the expectation of $f(\theta)$, we proceed in the opposite order, obtaining first the conditional expectation

$$\frac{1}{r}\Sigma E\{[(1-\hat{B})X_i - \theta_i]^2 | x\} \tag{10}$$

where

$$\hat{B} = c(r-2)/S^2. \tag{11}$$

Now the expectation, which is being averaged in (10), is equal to

$$\text{var}[\theta_i | X] + [E(\theta_i | X) - (1-\hat{B})X_i]^2 = (1-B) + (\hat{B}-B)^2 X_i^2,$$

so that (10) reduces to

$$1 - B + \frac{1}{r}(\hat{B}-B)^2 S^2. \tag{12}$$

Expansion of $(\hat{B}-B)^2$ with $\hat{B}$ given by (11) and use of the relations

$$E(S^2) = r/B \quad \text{and} \quad E(1/S^2) = B/(r-2)$$

then proves the first equation in (7).

It remains to prove Step 1. To show that $f(\theta)$ is a function of $\Sigma\theta_i^2$ only, note that $f(\theta)$ is the expectation of a function of θ and X which depends only on $S^2 = \Sigma X_i^2$, $\Sigma\theta_i X_i$ and $\Sigma\theta_i^2$. Make an orthogonal transformation from $(X_1, \ldots, X_r)$ to $(Y_1, \ldots, Y_r)$ such that $Y_1 = \Sigma\theta_i X_i / \sqrt{\Sigma\theta_i^2}$. Then $Y_1, \ldots, Y_r$ are again independent with unit variance but with expectations $E(Y_1) = \sqrt{\Sigma\theta_i^2}$ and $E(Y_j) = 0$ for $j > 1$. In terms of the Y's, $f(\theta)$ is the expectation of a function of $(Y_1, \Sigma_{j=2}^r Y_j^2, \Sigma\theta_j^2)$, and the result follows. It also follows for $g(\theta)$, which is the expectation of a function of S^2.

Corollary 6.1. *The estimator* δ_c *defined by* (3) *dominates* X (δ_c *when* $c = 0$) *provided* $0 < c < 2$ *and* $r \geqslant 3$.

Proof. For these values, $2c - c^2 > 0$ and hence $R(\theta, \delta_c) < 1$ for all θ.

Corollary 6.2. *The James–Stein estimator* δ, *which equals* δ_c *with* $c = 1$, *dominates all estimators* δ_c *with* $c \neq 1$.

Proof. The factor $2c - c^2$ takes on its maximum value 1 if and only if $c = 1$.

For $c = 1$, of course, formula (4) verifies the risk formula (5.23).

Since the James–Stein estimator dominates all estimators δ_c with $c \neq 1$, one might hope that it is admissible. However, unfortunately this is not the case, as is shown by the following theorem.

Theorem 6.2. *Let* $\hat{\delta}$ *be any estimator of the form* (2) *with* $\hat{B}$ *any strictly decreasing function of the* X*'s and suppose that*

$$P_\theta(\hat{B} > 1) > 0. \tag{13}$$

Then

$$R(\theta, \hat{\hat{\delta}}) < R(\theta, \hat{\delta})$$

where

$$\hat{\hat{\delta}}_i = max[(1 - \hat{B}), 0]X_i. \tag{14}$$

Proof. By (5.17)

$$R(\theta, \hat{\delta}) - R(\theta, \hat{\hat{\delta}}) = \frac{1}{r}\Sigma\left[E\left(\hat{\delta}_i^2 - \hat{\hat{\delta}}_i^2\right) - 2\theta_i E\left(\hat{\delta}_i - \hat{\hat{\delta}}_i\right)\right].$$

To show that the expression in brackets is always > 0, calculate the expectations by first conditioning on $\hat{B}$. For any value $\hat{B} \leqslant 1$, we have

$\hat{\delta}_i = \hat{\hat{\delta}}_i$ so that it is enough to show that the right side is positive when conditioned on any value $\hat{B} = b > 1$. Since in that case $\hat{\hat{\delta}}_i = 0$, it is finally enough to show that for any $b > 1$

$$\theta_i E[\hat{\delta}_i | \hat{B} = b] = \theta_i (1 - b) E(X_i | \hat{B} = b) \leqslant 0$$

and hence that $\theta_i E(X_i | \hat{B} = b) \geqslant 0$. Now $\hat{B} = b$ is equivalent to $S^2 = c$ for some c and hence to $X_1^2 = c - (X_2^2 + \cdots + X_n^2)$. Conditioning further on $X_2, \ldots, X_n$ we find that

$$E(\theta_1 X_1 | S^2 = c, x_2, \ldots, x_n) = \frac{\theta_1 y (e^{\theta_1 y} - e^{-\theta_1 y})}{e^{\theta_1 y} + e^{-\theta_1 y}}$$

where $y = \sqrt{c - (x_2^2 + \cdots + x_n^2)}$. The right side is an increasing function of $|\theta_1 y|$, which is zero when $\theta_1 y = 0$, and this completes the proof.

Theorem 6.2 shows in particular that the James–Stein estimator is dominated by

$$\delta_i^+ = \left(1 - \frac{r-2}{S^2}\right)_+ X_i \tag{15}$$

where $(\cdot)_+$ indicates that the quantity in parentheses is replaced by 0 whenever it is negative. We shall call

$$(\cdot)_+ = \max[(\cdot), 0]$$

the positive part of $(\cdot)$. Unfortunately, it can be shown that even δ^+ is inadmissible because it is not smooth enough to be either Bayes or limiting Bayes. In order to exhibit at least one admissible minimax estimator, we now first state without proof the following generalization of Corollary 6.1, due to Baranchik (1970) [see also Strawderman (1971) and Efron and Morris (1976).]

Theorem 6.3. *An estimator of the form*

$$\delta_i = \left[1 - c(S)\frac{r-2}{S^2}\right] X_i \tag{16}$$

is minimax for $r \geqslant 3$ *provided*

(i) $0 \leqslant c(S) \leqslant 2$

and

(ii) *the function* c *is nondecreasing.*

It is interesting to note how very different the situation for $r \geqslant 3$ is from the typical one-dimensional problem discussed in Sections 4.2 and 4.3. There, minimax estimators were unique (although recall Example 31.3); here, they constitute an enormously rich collection. It follows from Theorem 6.2 that the estimators (16) are inadmissible whenever $c(S)/S^2 > 1/(r-2)$ with positive probability. On the other hand, the family (16) does contain some admissible members, as is shown by the following example, due to Strawderman (1971).

Example 6.1. Let X_i be independent normal with mean θ_i and unit variance, and suppose that the θ's are themselves random variables with the following two-stage prior distribution. For a fixed value of λ, let the θ_i be iid according to $N[0, \lambda/(1-\lambda)]$. In addition, suppose that λ itself is a random variable with uniform distribution $U(0, 1)$. Then a straightforward calculation (Problem 6.2) shows that the Bayes estimator δ is given by (16) with

$$c(S) = \frac{1}{r-2}\left[r + 2 - \frac{2\exp(-\frac{1}{2}S^2)}{\int_0^1 \lambda^{r/2}\exp(-\lambda S^2/2)\,d\lambda}\right]. \tag{17}$$

It follows from Problem 6.2 that $E[(1-\lambda)|X] = (r-2)c(S)/S^2$ and hence that $c(S) \geqslant 0$ since $\lambda < 1$. On the other hand, $c(S) \leqslant (r+2)/(r-2)$ and hence $c(S) \leqslant 2$ provided $r \geqslant 6$. It remains to show that $c(S)$ is nondecreasing or, equivalently, that

$$\int_0^1 \lambda^{r/2}\exp\left[\tfrac{1}{2}S^2(1-\lambda)\right]d\lambda$$

is nondecreasing in S. This is obvious since $0 < \lambda < 1$.

Thus the estimator (16) with $c(S)$ given by (17) is an admissible minimax estimator for $r \geqslant 6$.

Although neither the James–Stein estimator nor the estimator (15) are admissible, it appears that no substantial improvements over the latter are possible (see Efron and Morris, 1973a), and we shall now turn to some modifications of these estimators in situations slightly different from that assumed at the beginning of this section. In the first of these, we shall consider a class of estimators that shrink the observations toward $\bar{X}$ rather than toward a given point μ.

The estimator (5.19) was seen to provide a substantial improvement over X when $\theta = (\theta_1, \ldots, \theta_r)$ is close to the point $\mu = (\mu_1, \ldots, \mu_r)$.* When the r

*A minimax estimator which achieves the principal improvement in any specified elliptical region is discussed in Berger (1982a) and a choice with desirable Bayesian properties in Berger (1982b).

components of the multiparameter problem are similar in nature, one may wish to choose the μ_i to be equal but may not be able to assume that their common value is known. A development completely analogous to that leading to Theorem 6.1 is then possible. To derive an empirical Bayes estimator, assume the θ's to be iid as $N(\xi, A)$, which results in the Bayes estimator

$$\delta_i = \xi + (1 - B)(X_i - \xi).$$

Taking both ξ and $B = 1/(A + 1)$ to be unknown, we can estimate them from the data. The fact that the marginal distribution of the X's is $N(\xi, 1/B)$ suggests $\bar{X}$ and c/S'^2 as estimators of ξ and B, where

$$S'^2 = \Sigma(X_i - \bar{X})^2,$$

and this leads to the estimator

$$\delta'_{ic} = \bar{X} + \left(1 - c\frac{r-3}{S'^2}\right)(X_i - \bar{X}). \tag{18}$$

The argument used to prove (4) applies with only the obvious modifications also to (18) and then shows (Problem 6.3) that the risk function of δ'_c is given by

$$R(\theta, \delta'_c) = 1 - \frac{(r-3)^2}{r} E_\theta\left(\frac{2c - c^2}{S'^2}\right). \tag{19}$$

The estimator δ'_c therefore dominates X (and hence is minimax) provided $0 < c < 2$ and $r \geqslant 4$. The uniformly best of these estimators as in Corollary 6.2 corresponds to $c = 1$, so that its risk is given by

$$R(\theta, \delta'_1) = 1 - \frac{r-3}{r} E_\theta\left(\frac{r-3}{S'^2}\right). \tag{20}$$

Since S'^2 has a noncentral χ^2 distribution with $r - 1$ degrees of freedom and noncentrality parameter $\lambda' = \Sigma(\theta_i - \bar{\theta})^2$, the properties of (5.23) established in Section 4.5 immediately extend to

$$\delta'_{i,1} = \bar{X} + \left(1 - \frac{r-3}{S'^2}\right)(X_i - \bar{X}) \tag{21}$$

and in particular show that the minimum risk of this estimator, which now occurs at all points θ with $\theta_1 = \cdots = \theta_r$, is $3/r$ instead of the value $2/r$ achieved by (5.23).

It naturally follows from Theorem 6.2 that (21) is dominated by the estimator in which $1 - (r-3)/S'^2$ is replaced by its positive part.

So far, the variance of the X's has been assumed to be equal to 1. If instead $\text{var}(X_i) = \sigma^2$, where σ^2 is known, the problem can be reduced to that with unit variance by considering the variables X_i/σ and estimating their means θ_i/σ and then multiplying the estimator of θ_i/σ by σ to obtain an estimator of θ_i. This argument leads to replacing (5.19) by

$$\delta_i = \mu_i + \left(1 - \frac{r-2}{S^2/\sigma^2}\right)(X_i - \mu_i) \tag{22}$$

with risk function (Problem 6.4)

$$R(\theta, \delta) = \sigma^2\left[1 - \frac{r-2}{r}E\left(\frac{r-2}{S^2/\sigma^2}\right)\right]. \tag{23}$$

Since the distribution of S^2/σ^2 is independent of σ^2, this means that the risk (5.23) is multiplied by σ^2 as one would expect.

Suppose now that σ^2 is unknown. We shall then suppose that there exists a random variable S_0^2 independent of X and such that S_0^2/σ^2 is distributed as χ_f^2 and shall in (22) replace σ^2 by $\hat{\sigma}^2 = kS_0^2$. The estimator is then modified to

$$\begin{aligned}\delta_i &= \mu_i + \left(1 - \frac{r-2}{S^2/\hat{\sigma}^2}\right)(X_i - \mu_i) \\ &= \mu_i + \left(1 - \frac{r-2}{S^2/\sigma^2}\cdot\frac{\hat{\sigma}^2}{\sigma^2}\right)(X_i - \mu_i).\end{aligned} \tag{24}$$

The conditional risk of δ given $\hat{\sigma}$ is given by (4) with S^2 replaced by S^2/σ^2 and $c = \hat{\sigma}^2/\sigma^2$ and because of the independence of S_0^2 and S^2 we thus have

(25)

$$R(\theta, \delta) = 1 - \frac{(r-2)^2}{r}E_\theta\left(\frac{1}{S^2/\sigma^2}\right)E_\theta\left[2k(S_0^2/\sigma^2) - k^2(S_0^2/\sigma^2)^2\right].$$

Now $E(S_0^2/\sigma^2) = f$ and $E[(S_0^2/\sigma^2)^2] = f(f+2)$, so that the second expectation is equal to

$$2kf - k^2f(f+2),$$

which is minimized for $k = 1/(f+2)$ and for this value reduces to $f/(f+2)$.

The best choice of k in $\hat{\sigma}^2$ thus leads to

$$\hat{\sigma}^2 = S_0^2/(f+2) \tag{26}$$

and the risk of the resulting estimator (24) is

$$R(\theta, \delta) = 1 - \frac{f}{f+2}\frac{(r-2)^2}{r}E_\theta(\sigma^2/S^2). \tag{27}$$

The improvement over the risk of X is thus reduced from that of (5.23) by a factor of $f/(f+2)$.

An analogous modification for unknown σ is possible for the estimator (21), which is then replaced by

$$\bar{X} + \left(1 - \frac{r-3}{S'^2/\hat{\sigma}^2}\right)(X_i - \bar{X}) \tag{28}$$

with risk function (Problem 6.5)

$$1 - \frac{f}{f+2}\frac{r-3}{r}E_\theta\left(\frac{r-3}{S'^2}\right). \tag{29}$$

The principal application of (24) [and (28)] is to the case of sn variables Y_{ij}, $i = 1, \dots, s$; $j = 1, \dots, n$ independently distributed as $N(\theta_i, \tau^2)$. If we then put $X_i = Y_{i\cdot}$, $S_0^2 = \Sigma\Sigma(Y_{ij} - Y_{i\cdot})^2$, $\sigma^2 = \tau^2/n$ and $f = (n-1)s$, we have exactly the situation assumed for (24). The case of unequal sample sizes n_i is not covered by what we have done so far since the variances of the X_i are then no longer equal. This problem is considered in Efron and Morris (1973b).

The estimator can, of course, be improved again by replacing the factor of $(X_i - \mu_i)$ in (24) by its positive part. This is seen by applying Theorem 6.2 to the conditional risk given S_0^2.

Let us now return to the problem in which the X_i are $N(\theta_i, 1)$ and the risk function is

$$\bar{R}(\theta, \delta) = \frac{1}{r}\Sigma R(\theta_i, \delta_i) \tag{30}$$

with $R(\theta_i, \delta_i) = E(\delta_i - \theta_i)^2$. It was seen in Section 4.3 that it is not possible to find a δ_i for which $R(\theta_i, \delta_i)$ is uniformly better than $R(\theta_i, X_i) = 1$. Thus

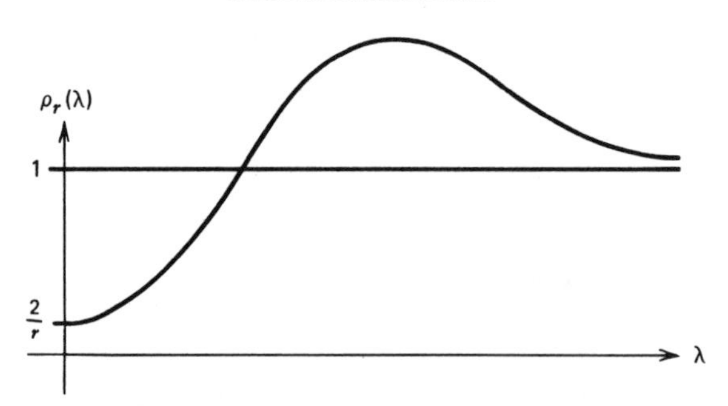

Figure 6.1. Maximum component risk $\rho_r(\lambda)$.

the improvement in the average risk can be achieved only through increasing some of the component risks, and it becomes of interest to consider the maximum possible component risk

$$\max_i \sup_{\theta_i} R(\theta_i, \delta_i). \tag{31}$$

For given $\lambda = \Sigma\theta_j^2$, it can be shown [Baranchik (1964)] that (31) attains its maximum when all but one of the θ's are zero, say $\theta_2 = \cdots = \theta_r = 0$, $\theta_1 = \sqrt{\lambda}$, and that this maximum risk $\rho_r(\lambda)$ as a function of λ increases from a minimum of $2/r$ at $\lambda = 0$ to a maximum and then decreases and tends to 1 as $\lambda \to \infty$.

The following values of max $\rho_r(\lambda)$ and the value λ_r at which the maximum is attained are given by Efron and Morris (1972).

Table 6.1 suggests that shrinkage estimators will typically not be appropriate when the component problems concern different clients. No one wants his or her blood test or Pap smear subjected to the possibility of large errors in order to improve a laboratory's average performance.

To get a feeling for the behavior of the James–Stein estimator (5.19) with $\mu = 0$, in a situation in which most of the θ's are at or near zero (representing the standard or normal situation of no effect) but a few relatively large θ's are present, consider 20 observations $U_1, \ldots, U_{20}$ from a table of random

Table 6.1. Maximum Component Risk

r	3	5	10	20	30	∞
λ_r	2.49	2.85	3.62	4.80	5.75	
$\rho_r(\lambda_r)$	1.24	1.71	2.93	5.40	7.89	$r/4$

normal deviates taken from Wold (1948, p. 15) with the following results:

$$-.69, 1.40, 1.79, -.83, .34, .19, 1.46, .21, .73, -1.00, .81, .01,$$
$$-1.13, -1.53, -.07, -.22, -.50, -1.59, .82, .71.$$

From these, three sets of observations $X_i = \theta_i + U_i$ were created. In the first: $\theta_1 = \cdots = \theta_{19} = 0, \theta_{20} = 2, 3, 4$; in the second: $\theta_1 = \cdots = \theta_{18} = 0, \theta_{19} = i, \theta_{20} = j$ with $2 \leqslant i \leqslant j \leqslant 4$; in the third: $\theta_1 = \cdots = \theta_{17} = 0, \theta_{18} = i, \theta_{19} = j, \theta_{20} = k$ with $2 \leqslant i \leqslant j \leqslant k \leqslant 4$. The resulting factor $[1 - (r - 2)/S^2]$, by which the observation is multiplied to obtain the estimators δ_i of θ_i, is shown in the following table.

To see the effect of the shrinkage explicitly, consider the observation $X_{20} = 2.71$ representing an observation on $\theta_{20} = 2$. The modified estimate ranges from $2.71 \times .08 = .22$ (when $\theta_1 = \cdots = \theta_{19} = 0$) to $2.71 \times .69 = 1.87$ when $\theta_1 = \cdots = \theta_{17} = 0, \theta_{18} = \theta_{19} = 4$).

What is seen in this example can be generalized roughly as follows.

(i) If all the θ's are at or fairly close to zero, then the James–Stein estimator will reduce the X's very substantially in absolute value and thereby typically will greatly improve the accuracy of the estimated values.

(ii) If there are some very large θ's or a substantial number of moderate ones, the factor by which the X's are multiplied will not be very far from 1, and the modification will not have a great effect.

Neither of these situations causes much of a problem: In (ii) the modification presents an unnecessary but not particularly harmful complication; in (i) it is clearly very beneficial. The danger arises in the following intermediate situation.

(iii) Most of the θ's are close to zero, but there are a few moderately large θ's (of the order of 2 to 4 standard deviations, say). These represent the cases in which something is going on, which we will usually want to know about. However, in these cases the estimated values are heavily shrunk toward the norm with the resulting risk of their being found "innocent by association."

If one is interested in minimizing the average risk (30) but is concerned about the possibility of large component risks, a compromise is possible along the lines of restricted Bayes estimation mentioned in Section 4.2. One

Table 6.2.

θ's $\neq 0$	2	3	4	22	23	33	24	34	44	222	223	224	233	234	244	333	334	344	444
Factor	.08	.22	.37	.12	.24	.34	.38	.45	.53	.42	.54	.63	.57	.65	.69	.61	.68	.71	.73

can impose an upper bound on the maximum component risk, say 10 or 25 percent above the minimax risk of 1 (when $\sigma = 1$). Subject to this restriction, one can then try to minimize the average risk, for example, in the sense of obtaining a Bayes or empirical Bayes solution. An approximation to such an approach has been developed by Efron and Morris (1971/1972).

The results discussed in this and the preceding section for the simultaneous estimation of normal means have been extended, particularly to exponential families and to general location parameter families, with and without nuisance parameters. This work forms a rich literature, to which access can be found through the references in Section 4.8. The titles of most of these papers are self-explanatory, but special mention should be made of the extension to regression problems. Here the use of biased estimators in an effort to improve the classic least-squares estimators links up with the theory of ridge regression. Good entries into this literature, which is not covered in the references, are Draper and Van Nostrand (1979), Brown and Zidek (1980), and Casella (1980).

7. PROBLEMS

Section 1

1.1 Give examples of pairs of values (a, b) for which the beta density $B(a, b)$ is (i) decreasing, (ii) increasing, (iii) increasing for $p < p_0$ and decreasing for $p > p_0$, and (iv) decreasing for $p < p_0$ and increasing for $p > p_0$.

1.2 In Example 1.1, if p has the improper prior density $1/pq$, show that the posterior density of p given x is proper provided $0 < x < n$.

1.3 Determine the Jeffreys prior for p in Example 1.1, and the associated Bayes estimator δ_Λ.

1.4 For the estimator δ_Λ of Problem 1.3,

(i) calculate the bias and maximum bias;

(ii) calculate the expected squared error and compare it with that of the UMVU estimator.

1.5 In Example 1.1, find the Bayes estimator δ of pq when p has the prior $B(a, b)$.

1.6 (i) Compare the estimator δ of Problem 1.5 with the UMVU estimator δ' of pq given in Example 2.3.1.

(ii) Compare the expected squared error of the estimator of pq for the Jeffreys prior in Example 1.1 with that of δ' given in Problem 2.3.1.

1.7 Show that the estimator (1.16) tends in probability (i) to θ as $n \to \infty$,

(ii) to μ as $b \to 0$, (iii) to θ as $b \to \infty$.

1.8 In analogy with Problem 1.1, determine the possible shapes of the gamma density $\Gamma(g, 1/\alpha)$, $\alpha, g > 0$.

1.9 Let $X_1, \ldots, X_n$ be iid according to the Poisson distribution $P(\lambda)$ and let λ have a gamma distribution $\Gamma(g, \alpha)$.

(i) For squared error loss, determine the Bayes estimator $\delta_{\alpha, g}$ of λ, and show that it has a representation analogous to (1.13).

(ii) What happens to $\delta_{\alpha, g}$ as (a) $n \to \infty$, (b) $\alpha \to \infty, g \to 0$, or both?

1.10 For the situation of the preceding problem, solve the two parts corresponding to Problem 1.4(i) and (ii).

1.11 In Problem 1.9, if λ has the improper prior density $d\lambda/\lambda$ (corresponding to $\alpha = g = 0$), under what circumstances is the posterior distribution proper?

1.12 Solve the problems analogous to 1.9 and 1.10 when the observations consist of a single random variable X having a negative binomial distribution $Nb(p, m)$, p has the beta prior $B(a, b)$, and the estimand is (i) p, (ii) $1/p$.

1.13 Verify the estimator (1.21).

1.14 In Example 1.5 verify that the posterior distribution of τ is $\Gamma(r + g - \frac{1}{2}, 1/(\alpha + z))$.

1.15 In Example 1.5 with $\alpha = g = 0$, show that the posterior distribution given the X's of $\sqrt{n}(\theta - \bar{X})/\sqrt{Z/(n-1)}$ is Student's t-distribution with $n - 1$ degrees of freedom.

1.16 In Example 1.5, show that the posterior distribution of θ is symmetric about $\bar{x}$ when the joint prior of θ and σ is of the form $h(\sigma)\, d\sigma\, d\theta$, where h is an arbitrary probability density on $(0, \infty)$.

1.17 Let X, Y be independently distributed according to distributions P_ξ and Q_η, respectively. Suppose that ξ and η are real-valued and independent according to some prior distributions Λ and Λ'. If with squared error loss δ_Λ is the Bayes estimator of ξ on the basis of X, and $\delta'_{\Lambda'}$ that of η on the basis of Y,

(i) show that $\delta'_{\Lambda'} - \delta_\Lambda$ is the Bayes estimator of $\eta - \xi$ on the basis of (X, Y);

(ii) if $\eta > 0$ and $\delta^*_{\Lambda'}$ is the Bayes estimator of $1/\eta$ on the basis of Y, show that $\delta_\Lambda \cdot \delta^*_{\Lambda'}$ is the Bayes estimator of ξ/η on the basis of (X, Y).

1.18 For the density (1.24) and improper prior $(d\sigma/\sigma) \cdot (d\sigma_A/\sigma_A)$, show that the posterior distribution of (σ, σ_A) continues to be improper.

1.19 (i) In Example 1.6, obtain the Jeffreys prior distribution of (σ, τ).

(ii) Show that for the prior of part (i), the posterior distribution of (σ, τ) is proper.

1.20 Verify the Bayes estimator (1.25).

Section 2

2.1 In Example 2.1 let $\delta^*(X) = X/n$ with probability $1 - \varepsilon$ and $= \frac{1}{2}$ with probability ε. Determine the risk function of δ^* and show that for $\varepsilon = 1/(n + 1)$ its risk is constant and less than $\sup R(p, X/n)$.

2.2 Find the bias of the minimax estimator (2.10) and discuss its direction.

2.3 In Example 2.1,

(i) determine c_n and show that $c_n \to 0$ as $n \to \infty$;
(ii) show that $R_n(\frac{1}{2})/r_n \to 1$ as $n \to \infty$.

2.4 In Example 2.1, graph the risk functions of X/n and the minimax estimator (2.10) for $n = 1, 4, 9, 16$, and indicate the relative positions of the two graphs for large values of n.

2.5 (i) Find two points $0 < p_0 < p_1 < 1$ such that the estimator (2.10) for $n = 1$ is Bayes with respect to a distribution Λ for which $P_\Lambda(p = p_0) + P_\Lambda(p = p_1) = 1$.
(ii) For $n = 1$, show that (2.10) is a minimax estimator of p even if it is known that $p_0 \leqslant p \leqslant p_1$.
(iii) In (ii), find the values p_0 and p_1 for which $p_1 - p_0$ is as small as possible.

2.6 Evaluate (2.15) and show that its maximum is $1 - \alpha$.

2.7 Let $X = 1$ or 0 with probabilities p and q, respectively, and consider the estimation of p with loss $= 1$ when $|d - p| \geqslant \frac{1}{4}$ and $= 0$ otherwise. The most general randomized estimator is $\delta = U$ when $X = 0$ and $\delta = V$ when $X = 1$ where U, and V are two random variables with known distributions.

(i) Evaluate the risk function and the maximum risk of δ when U and V are uniform on $(0, \frac{1}{2})$ and $(\frac{1}{2}, 1)$, respectively.
(ii) Show that the estimator δ of (i) is minimax by considering the three values $p = 0, \frac{1}{2}, 1$.

[*Hint:* (ii) The risk at $p = 0, \frac{1}{2}, 1$ is, respectively, $P(U > \frac{1}{4})$, $\frac{1}{2}[P(U < \frac{1}{4}) + P(V > \frac{3}{4})]$, and $P(V < \frac{3}{4})$].

2.8 The problem of Example 2.2 remains invariant under the transformations

$$X' = n - X, \qquad p' = 1 - p, \qquad d' = 1 - d.$$

This illustrates that randomized equivariant estimators may have to be considered when $\bar{G}$ is not transitive.

2.9 Let r_Λ be given by (2.1). If $r_\Lambda = \infty$ for some Λ, then any estimator δ has unbounded risk.

2.10 In Example 2.3, show that no linear estimator has constant risk.

2.11 The risk function of (2.21) depends on p_1 and p_2 only through $p_1 + p_2$ and takes on its maximum when $p_1 + p_2 = 1$.

2.12 (i) In Example 2.3 determine the region in the (p_1, p_2) unit square in which (2.21) is better than the UMVU estimator of $p_2 - p_1$ for $m = n = 2, 8, 18$, and 32.

(ii) Extended Problems 2.3 and 2.4 to Example 2.3.

2.13 In Example 2.4, show that $\bar{X}$ is minimax for the loss function $(d - \theta)^2/\sigma^2$ without any restrictions on σ.

2.14 (i) Verify (2.37).

(ii) Show that equality holds in (2.39) if and only if

$$P(X_i = 0) + P(X_i = 1) = 1.$$

2.15 In Example 2.5(ii), for any $k > 0$, the estimator

$$\delta = \frac{\sqrt{n}}{1 + \sqrt{n}} \frac{1}{n} \sum_{i=1}^{n} X_i^k + \frac{1}{2(1 + \sqrt{n})}$$

is a Bayes estimator for the prior distribution Λ over $\mathcal{F}_0$ for which (2.36) was shown to be Bayes.

2.16 Let X_i $(i = 1, \ldots, n)$ and Y_j $(j = 1, \ldots, n)$ be independent with distributions F and G, respectively. If $F(1) - F(0) = G(1) - G(0) = 1$ but F and G are otherwise unknown, find a minimax estimator for $E(Y_j) - E(X_i)$ with squared error loss.

2.17 Let X_i $(i = 1, \ldots, n)$ be iid with unknown distribution F. Show that

$$\delta = \frac{\text{No. of } X_i \leqslant 0}{\sqrt{n}} \cdot \frac{1}{1 + \sqrt{n}} + \frac{1}{2(1 + \sqrt{n})}$$

is minimax for estimating $F(0) = P(X_i \leqslant 0)$ with squared error loss. [*Hint:* Consider the risk function of δ.]

2.18 Let $X_1, \ldots, X_m$ and $Y_1, \ldots, Y_n$ be independently distributed as $N(\xi, \sigma^2)$ and $N(\eta, \tau^2)$, respectively, and consider the problem of estimating $\Delta = \eta - \xi$ with squared error loss.

(i) If σ and τ are known, $\bar{Y} - \bar{X}$ is minimax.

(ii) If σ and τ are restricted by $\sigma^2 \leqslant A, \tau^2 \leqslant B$ $(A, B$ known and finite), $\bar{Y} - \bar{X}$ continues to be minimax.

2.19 In the linear model (3.4.4), $\Sigma a_i \hat{\xi}_i$ (in the notation of Th. 3.4.2) is minimax for estimating $\theta = \Sigma a_i \xi_i$ with squared error loss, under the restriction $\sigma^2 \leqslant M$. [*Hint:* Treat the problem in its canonical form.]

2.20 For the random variable X whose distribution is (2.42), show that x must satisfy the inequalities stated below (2.42).

2.21 Show that the estimator defined by (2.43)

(i) has constant risk;
(ii) is Bayes with respect to the prior distribution specified by (2.44) and (2.45).

2.22 Show that for fixed X and n, (2.43) $\to$ (2.10) as $N \to \infty$.

2.23 Show that var($\bar{Y}$) given by (3.6.6) takes on its maximum value subject to (2.41) when all the a's are 0 or 1.

2.24 (i) If $R(p, \delta)$ is given by (2.49), show that $\sup R(p, \delta) \cdot 4(1 + \sqrt{n})^2 \to 1$ as $n \to \infty$.
(ii) Determine the smallest value of n for which the Bayes estimator of Example 2.7 satisfies (2.48) for $r = 1$ and $b = 5$, 10, and 20.

2.25 Show that the risk of the estimator (2.50) is bounded.

Section 3

3.1 Determine the Bayes risk of the estimator (3.1) when θ has the prior distribution $N(\mu, \tau^2)$.

3.2 Prove part (iv) in the second proof of Example 3.2.

3.3 An estimator $aX + b$ $(0 \leqslant a \leqslant 1)$ of $E_\theta(X)$ is inadmissible (with squared error loss) under each of the following conditions:

(i) if $E_\theta(X) \geqslant 0$ for all θ, and $b < 0$;
(ii) if $E_\theta(X) \leqslant k$ for all θ, and $ak + b > k$.

[*Hint:* In (ii) replace X by $X' = k - X$ and $aX + b$ by $k - (aX + b) = aX' + k - b - ak$, respectively and use (i).]

3.4 An estimator $[1/(1 + \lambda) + \varepsilon]X$ of $E_\theta(X)$ is inadmissible (with squared error loss) under each of the following conditions:

(i) if $\text{var}_\theta(X)/E_\theta^2(X) > \lambda > 0$ and $\varepsilon > 0$;
(ii) if $\text{var}_\theta(X)/E_\theta^2(X) < \lambda$ and $\varepsilon < 0$.

[*Hint:* (i) Differentiate the risk function of the estimator with respect to ε to show that it decreases as ε decreases.] (Karlin, 1958).

3.5 If $\text{var}_\theta(X)/E_\theta^2(X) > \lambda > 0$, an estimator $[1/(1 + \lambda) + \varepsilon]X + b$ is inadmissible (with squared error loss) under each of the following conditions:

(i) if $E_\theta(X) > 0$ for all θ, $b > 0$ and $\varepsilon > 0$;
(ii) if $E_\theta(X) < 0$ for all θ, $b < 0$ and $\varepsilon > 0$ (Gupta, 1966).

3.6 For the estimator $aY + b$ in Example 3.5, prove

(i) inadmissibility when $a > 1/(r + 1)$;
(ii) admissiblity when $a = b = 0$.

[*Hint:* (i) Problems 3.3–3.5; (ii) Note that $\delta \geqslant 0$, and > 0 with positive probability, implies $\delta/\sigma^2 \to \infty$ in probability, and hence $E(\delta/\sigma^2) \to \infty$ as

$\sigma \to 0$. It follows that $\lim_{\sigma\to 0} E[(\delta - 2\sigma^2)^2/\sigma^4] < \infty$ if and only if $\delta \geqslant 0$ implies $\delta = 0$ with probability 1.]

3.7 Show the equivalence of the following relationships:

(i) (3.23) and (3.24);
(ii) (3.29) and (3.30) when $c = \sqrt{(n-1)/(n+1)}$;
(iii) (3.33) and (3.34).

3.8 In Example 3.6 show that the estimator $aX/n + b$ is inadmissible for all (a, b) outside the triangle (3.34). [*Hint:* Problems 3.3–3.5.]

3.9 Prove admissibility of the estimators corresponding to the interior of the triangle (3.34) by applying Theorem 3.1 and using the results of Example 1.1.

3.10 Use Theorem 3.3 to provide an alternative proof for the admissibility of the estimator $a\bar{X} + b$ satisfying (3.3) in Example 3.1.

3.11 Determine which estimators $aX + b$ are admissible for estimating $E(X)$ in the following situations, for squared error loss:

(i) X has a Poisson distribution.
(ii) X has a negative binomial distribution (Gupta, 1966).

3.12 Show that the conditions (3.36) and (3.37) of Example 3.8 are not only sufficient but also necessary for admissibility of (3.35).

3.13 Let X, Y be independently distributed according to Poisson distributions with $E(X) = \xi$ and $E(Y) = \eta$, respectively. Show that $aX + bY + c$ is admissible for estimating ξ with squared error loss if and only if either

$$0 \leqslant a < 1, \qquad b \geqslant 0, \qquad c \geqslant 0$$

or

$$a = 1, \qquad b = c = 0$$

(Makani, 1972).

3.14 Let X be distributed with density $\frac{1}{2}\beta(\theta)e^{\theta x}e^{-|x|}$, $|\theta| < 1$.

(i) Show that $\beta(\theta) = 1 - \theta^2$.
(ii) Show that $aX + b$ is admissible for estimating $E_\theta(X)$ with squared error loss if and only if $0 \leqslant a \leqslant \frac{1}{2}$.

[*Hint:* (ii) To see necessity, let $\delta = (\frac{1}{2} + \varepsilon)X + b$ $(0 < \varepsilon \leqslant \frac{1}{2})$ and show that δ is dominated by $\delta' = (1 - \frac{1}{2}\alpha + \alpha\varepsilon)X + (b/\alpha)$ for some α with $0 < \alpha < 1/(\frac{1}{2} - \varepsilon)$.]

3.15 Show that the natural parameter space of the family (3.12) is $(-\infty, \infty)$ for the normal (variance known), binomial and Poisson distribution but not in the gamma or negative binomial case.

Section 4

4.1 Theorem 3.2.1 remains valid for almost equivariant estimators.

4.2 Verify the density (4.6).

4.3 In Example 4.2, a loss function remains invariant under G if and only if it is a function of $(d - \theta)^*$.

4.4 In Example 4.2, neither of the loss functions $[(d - \theta)^{**}]^2$ or $|(d - \theta)^{**}|$ is convex.

4.5 Let Y be distributed as $G(y - \eta)$. If $T = [Y]$ and $X = Y - T$, find the distribution of X and show that it depends on η only through $\eta - [\eta]$.

4.6 In model (3.3.1), let the measure ν assign $\int_a^b (1/\tau)\, d\tau$ to the interval $a < \tau < b$.

(i) Show that ν is invariant in the sense of (4.2).

(ii) Determine the posterior distribution of τ given $\underline{x}$ as if ν were a proper distribution.

(iii) Show that the MRE estimator (3.3.22) with $r = 1$ is the "Bayes" estimator with respect to ν for the loss function (3.3.17) with $r = 1$.

4.7 Prove formula (4.14).

4.8 Prove (4.16).
[*Hint:* In the term on the left side, lim inf can be replaced by lim. Let the left side of (4.16) be A and the right side B, and let $A_N = \inf h(a, b)$, where the inf is taken over $a \leqslant -N, b \geqslant N, N = 1, 2, \ldots,$ so that $A_N \to A$. There exist (a_N, b_N) such that $|h(a_N, b_N) - A_N| \leqslant 1/N$. Then $h(a_N, b_N) \to A$ and $A \geqslant B$.]

4.9 For the model (3.3.24) find a measure ν in the (ξ, τ)-plane which remains invariant under the transformations (3.3.25).

4.10 In Example 4.5, let $h(\theta)$ be the length of the path θ after cancellation. Show that h does not satisfy conditions (3.2.11).

4.11 Discuss Example 4.5 for the case that the random walk instead of being in the plane is (i) on the line, (ii) in three-space.

4.12 (i) Show that the probabilities (4.22) add up to 1.

(ii) With p_k given by (4.22), show that the risk (4.21) is infinite.

[*Hint*: (i) $1/k(k + 1) = (1/k) - 1/(k + 1)$.]

4.13 Show that the risk $R(\theta, \delta)$ of (4.23) is finite.
[*Hint:* $R(\theta, \delta) < \Sigma_{k > M|k+\theta|} 1/(k + 1) \leqslant \Sigma_{c<k<d} 1/(k + 1) < \int_c^{d+1} dx/x$, where $c = M|\theta|/(M + 1)$, $d = M|\theta|/(M - 1)$. The reason for the second inequality is that values of k outside (c, d) make no contribution to the sum.]

4.14 Show that the two estimators δ^* and δ^{**}, defined by (4.25) and (4.26), respectively, are equivariant.

4.15 Prove the relations (4.27) and (4.28).

4.16 Let the distribution of X depend on parameters θ and ϑ, let the risk function of an estimator $\delta = \delta(x)$ of θ be $R(\theta, \vartheta; \delta)$, and let $r(\theta, \delta) = \int R(\theta, \vartheta; \delta)\, dP(\vartheta)$ for some distribution P. If δ_0 minimizes $\sup_\theta r(\theta, \delta)$ and satisfies

$$\sup_\theta r(\theta, \delta_0) = \sup_{\theta, \vartheta} R(\theta, \vartheta; \delta_0),$$

then δ_0 minimizes $\sup_{\theta, \vartheta} R(\theta, \vartheta; \delta)$.

4.17 A measure Λ over a group G is said to be right invariant if it satisfies (4.29) and left invariant if it satisfies

$$\Lambda(gB) = \Lambda(B).$$

Note that if G is commutative, the two definitions agree.

(i) If the elements $g \in G$ are real numbers $(-\infty < g < \infty)$ and group composition is $g_2 \cdot g_1 = g_1 + g_2$, the measure ν defined by

$$\nu(B) = \int_B dx$$

(i.e., Lebesgue measure) is both left and right invariant.

(ii) If the elements $g \in G$ are the positive real numbers, and composition of g_2 and g_1 is multiplication of the two numbers, the measure ν defined by

$$\nu(B) = \int_B \frac{1}{y} dy$$

is both left and right invariant.

4.18 If the elements $g \in G$ are pairs of real numbers (a, b), $b > 0$ corresponding to the transformations $gx = a + bx$, group composition by (1.3.8) is

$$(a_2, b_2) \cdot (a_1, b_1) = (a_2 + a_1 b_2, b_1 b_2).$$

Of the measures defined by

$$\nu(B) = \iint_B \frac{1}{y} dx\, dy \quad \text{and} \quad \nu(B) = \iint_B \frac{1}{y^2} dx\, dy$$

the first is right but not left invariant, and the second is left but not right invariant.

4.19 The four densities defining the measures ν of Problems 4.17 and 4.18 $(dx, (1/y)\, dy, (1/y)\, dx\, dy, (1/y^2)\, dx\, dy)$ are the only densities (up to multiplicative constants) for which ν has the stated invariance properties in the situations of these problems.

[*Hint:* In each case consider the equation

$$\int_B \pi(\theta)\,d\theta = \int_{gB} \pi(\theta)\,d\theta.$$

In the right integral, make the transformation to new variable or variables $\theta' = g^{-1}\theta$. If J is the Jacobian of this transformation, it follows that

$$\int_B [\pi(\theta) - J\pi(g\theta)]\,d\theta = 0 \qquad \text{for all} \quad B$$

and hence that $\pi(\theta) = J\pi(g\theta)$ for all θ except in a null set N_g. The proof of Theorem 4 in Chapter 6 of TSH shows that N_g can be chosen independent of g. This proves in Problem 4.17(i) that for all $\theta \notin N$, $\pi(\theta) = \pi(\theta + c)$, and hence that $\pi(c) =$ constant a.e. The other three cases can be treated analogously.]

4.20 Let X be distributed as $N(\theta, 1)$ and let θ have the improper prior density $\pi(\theta) = e^{\theta}(-\infty < \theta < \infty)$. For squared error loss, the formal Bayes estimator of θ is $X + 1$, which is neither minimax nor admissible (due to Farrell; see Kiefer, 1966).
(Conditions under which the formal Bayes estimator corresponding to an improper prior distribution for θ in Example 4.3 is admissible are given by Zidek, 1970).

Section 5

5.1 In Example 5.1, an estimator δ is equivariant if and only if it satisfies (5.11) and (5.12).

5.2 A function u satisfies (5.12) if and only if it depends only on ΣX_i^2.

5.3 Verify the Bayes estimator (5.15).

5.4 Let X_i be independent with binomial distribution $b(p_i, n_i)$, $i = 1, \ldots, r$. For estimating $p = (p_1, \ldots, p_r)$ with average squared error loss (5.17) find the minimax estimator of p, and determine whether it is admissible.

5.5 If S^2 is distributed as χ_r^2, use (2.2.5) to show that $E(S^{-2}) = 1/(r-2)$.

5.6 In Example 5.4, show that M is nonsingular for ρ_1, ρ_2 and singular for ρ_3, ρ_4.

5.7 Show that the function ρ_2 of Example 5.4 is convex.

5.8 In Example 5.4, show that X is admissible for (i) ρ_3, (ii) ρ_4.
[*Hint*: (i) It is enough to show that X_1 is admissible for estimating θ_1 with loss $(d_1 - \theta_1)^2$. This can be shown by letting $\theta_2, \ldots, \theta_r$ be known. (ii) Note that X is admissible minimax for $\theta = (\theta_1, \ldots, \theta_r)$ when $\theta_1 = \cdots = \theta_r$.]

5.9 In Example 5.5, show that X is admissible under the assumptions (ii(a)). [*Hint:*

1. If $v(t) > 0$ is such that

$$\int \frac{1}{v(t)} e^{-t^2/2\tau^2}\, dt < \infty,$$

show that there exists a constant $k(\tau)$ for which

$$\lambda_\tau(\theta) = k(\tau)\left[\Sigma v(\theta_j)\right] \exp\left(-\frac{1}{2\tau^2}\Sigma\theta_j^2\right) \Big/ \Pi v(\theta_j)$$

is a probability density for $\theta = (\theta_1, \ldots, \theta_r)$.

2. If the X_i are independent $N(\theta_i, 1)$ and θ has the prior $\lambda_\tau(\theta)$, the Bayes estimator of θ with loss function (5.25) is $\tau^2 X/(1 + \tau^2)$.

3. To prove X admissible, use (5.18) with $\lambda_\tau(\theta)$ instead of a normal prior.]

5.10 Let $\mathfrak{L}$ be a family of loss functions and suppose there exists $L_0 \in \mathfrak{L}$ and a minimax estimator δ_0 with respect to L_0 such that in the notation of (5.27)

$$\sup_{L,\theta} R_L(\theta, \delta_0) = \sup_\theta R_{L_0}(\theta, \delta_0).$$

Then δ_0 is minimax with respect to $\mathfrak{L}$, that is, it minimizes $\sup_{L,\theta} R_L(\theta, \delta)$.

5.11 Assuming (5.23), show that $E = 1 - [(r-2)^2/rS^2]$ is the unique unbiased estimator of the risk (5.23), and that E is inadmissible.

Section 6

6.1 Show that the estimator δ_c defined by (6.3) with $0 < c = 1 - \Delta < 1$ is dominated by any δ_d with $|d - 1| < \Delta$.

6.2 In Example 6.1 suppose that λ has probability density $\lambda^{-a}(1-a)$ on $(0, 1)$ where $0 \leqslant a < 1$. Determine the Bayes estimator for θ and show that it reduces to (6.16) with $c(S)$ given by (6.17) when $a = 0$.

6.3 Prove formula (6.19).

6.4 Verify formula (6.23).

6.5 Prove formula (6.29).

6.6 Let X_i, Y_j be independent $N(\xi_i, 1)$ and $N(\eta_j, 1)$, respectively ($i = 1, \ldots, r$; $j = 1, \ldots, s$).

(i) Find an estimator of $(\xi_1, \ldots, \xi_r; \eta_1, \ldots, \eta_s)$ that would be good near $\xi_1 = \cdots = \xi_r = \xi, \eta_1 = \cdots = \eta_s = \eta$, with ξ and η unknown, if the variability of the ξ's and η's is about the same.

(ii) When the loss function is (5.17), determine the risk function of your estimator. [*Hint:* Consider the Bayes situation in which ξ_i: $N(\xi, A)$, η_j: $N(\eta, A)$.]

8. REFERENCES

Following the basic paper by Bayes (published posthumously in 1763), Laplace initiated a widespread use of Bayes procedures, particularly with noninformative priors (for example, in his paper of 1774 and the fundamental book of 1820). However, Laplace also employed non-Bayesian methods, without always making a clear distinction. A systematic theory of statistical inference based on noninformative (locally invariant) priors, generalizing and refining Laplace's approach, was developed by Jeffreys in his book on probability theory (1st ed., 1939). A corresponding subjective theory owes its modern impetus principally to the work of de Finetti (for example, 1937, 1970) and that of L. J. Savage, particularly his book on the *Foundations of Statistics* (1954). The idea of selecting an appropriate prior from the conjugate family was put forward and implemented by Raiffa and Schlaifer (1961). Interest in Bayes procedures (although not from a Bayesian point of view) also received support from Wald's result (for example, 1950) that all admissible procedures are either Bayes or limiting Bayes.

Deliberate efforts to develop statistical inference and decision making not based on "inverse probability" (i.e., without assuming prior distributions) were mounted by R. A. Fisher [for example, (1922, 1930, and 1935); see also Lane (1980)], by Neyman and Pearson (for example, 1933*a* and *b*) and by Wald (1950). The latter's general decision theory introduced as central notions the minimax principle and least favorable distributions in close parallel to the corresponding concepts of the theory of games. Many of the examples of Section 4.2 were first worked out by Hodges and Lehmann (1950). Admissibility is another basic concept of Wald's decision theory. The admissibility proofs in Example 4.3.1 are due to Blyth (1951) and Hodges and Lehmann (1951). A general necessary and sufficient condition for admissibility was obtained by Stein (1955). (For a discussion of this and related conditions see Berger, 1980b, Section 8.8.) Theorem 3.3 is due to Karlin (1958), and the surprising inadmissibility results of Section 4.5 had their origin in Stein's seminal paper (1956). The relationship between equivariance and the minimax property was foreshadowed in Wald (1939) and was developed for point estimation by Peisakoff (1950), Girshick and Savage (1951), Blackwell and Girshick (1954), Kudo (1955), and Kiefer (1957).

Akaike, H.
(1980). "Ignorance prior distribution of a hyperparameter and Stein's estimator." *Ann. Inst. Statist. Math.* **32**, 171–178.

Alam, K.
(1973). "A family of admissible minimax estimators of the mean of a multivariate normal distribution." *Ann. Statist.* **1**, 517–525.

Alam, K.
(1975). "Minimax and admissible minimax estimators of the mean of a multivariate normal distribution for unknown convariance matrix." *J. Multiv. Anal.* **5**, 83–95.

Alam, K.
(1979). "Estimation of multinomial probabilities." *Ann. Statist.* **7**, 282–283.

Arnold, B. C.
(1970). "Inadmissibility of the usual scale estimate for a shifted exponential distribution." *J. Amer. Statist. Assoc.* **65**, 1260–1264.

Baranchik, A. J.
(1964). "Multiple regression and estimation of the mean of a multivariate normal distribution." Stanford University Technical Report No. 51.

Baranchik, A. J.
(1970). "A family of minimax estimators of the mean of a multivariate normal distribution." *Ann. Math. Statist.* **41**, 642–645.

Baranchik, A. J.
(1973). "Inadmissibility of maximum likelihood estimators in some multiple regression problems with three or more independent variables." *Ann. Statist.* **1**, 312–321.

Bayes, T.
(1763). "An essay toward solving a problem in the doctrine of chances." *Phil. Trans. Roy. Soc.*, **153**, 370–418. Reprinted in (1958) *Biometrika* **45**, 293–315.

Berger, J.
(1975). "Minimax estimation of location vectors for a wide class of densities." *Ann. Statist.* **3**, 1318–1328.

Berger, J.
(1976a). "Tail minimaxity in location vector problems and its applications." *Ann. Statist.* **4**, 33–50.

Berger, J.
(1976b). "Admissible minimax estimation of a multivariate normal mean with arbitrary quadratic loss." *Ann. Statist.* **4**, 223–226.

Berger, J.
(1976c). "Inadmissibility results for generalized Bayes estimators of coordinates of a location vector." *Ann. Statist.* **4**, 302–333.

Berger, J.
(1976d). "Admissibility results for generalized Bayes estimators of coordinates of a location vector." *Ann. Statist.* **4**, 334–356.

Berger, J.
(1976e). "Inadmissibility results for the best invariant estimator of two coordinates of a location vector." *Ann. Statist.* **4**, 1065–1076.

Berger, J.
(1976f). "Minimax estimation of a multivariate normal mean with arbitrary quadratic loss." *J. Multiv. Anal.* **6**, 256–264.

Berger, J.
(1978). "Minimax estimation of a multivariate normal mean under polynomial loss." *J. Multiv. Anal.* **8**, 173–180.

Berger, J.
(1980a). "Improving on inadmissible estimators in continuous exponential families with applications to simultaneous estimation of gamma scale parameters." *Ann. Statist.* **8**, 545–571.

Berger, J.
(1980b). *Statistical Decision Theory*. Springer, New York.

Berger, J.
(1982a). "Selecting a minimax estimator of a multivariate normal mean." *Ann. Statist.* **10**, 81–92.

Berger, J.
(1982b). "Bayesian robustness and the Stein effect." *J. Amer. Statist. Assoc.* **77**, 358–368.

Berger, J. and Bock, M. E.
(1976). "Combining independent normal mean estimation problems with unknown variances." *Ann. Statist.* **4**, 642–648.

Berger, J., Bock, M. E., Brown, L. D., Casella, G., and Gleser, L. (1977). "Minimax estimation of a normal mean vector for arbitrary quadratic loss and unknown covariance matrix. *Ann. Statist.* **5**, 763–771.

Berger, J. and Srinivasan, C.
(1978). "Generalized Bayes estimators in multivariate problems." *Ann. Statist.* **6**, 783–801.

Bhattacharya, P. K.
(1966). "Estimating the mean of a multivariate normal population with general quadratic loss function." *Ann. Math. Statist.* **37**, 1819–1824.

Bickel, P. J.
(1981). "Minimax estimation of the mean of a normal distribution when the parameter space is restricted." *Ann. Statist.* **9**, 1301–1309.

Bickel, P. J. and Blackwell, D.
(1967). "A note on Bayes estimates." *Ann. Math. Statist.* **38**, 1907–1911.

Bickel, P. J. and Lehmann, E. L.
(1981). "A minimax property of the sample mean." *Ann. Statist.* **9**, 1119–1122.

Bishop, Y. M., Fienberg, S. E., and Holland, P. W.
(1975). *Discrete Multivariate Analysis*. M.I.T. Press, Cambridge.

Blackwell, D.
(1951). "On the translation parameter problem for discrete variables." *Ann. Math. Statist.* **22**, 393–399.

Blackwell, D. and Girshick, M. A.
(1954). *Theory of Games and Statistical Decisions*. Wiley, New York.

Bloch, D. A. and Watson, G. S.
(1967). "A Bayesian study of the multinomial distribution." *Ann. Math. Statist.* **38**, 1423–1435.

Blyth, C. R.
(1951). "On minimax statistical decision procedures and their admissibility." *Ann. Math. Statist.* **22**, 22–42.

Blyth, C. R. and Roberts, D. M.
(1972). "On inequalities of Cramér-Rao type and admissibility proofs." *Proc. Sixth Berkeley Symp. Math. Statist. Prob.*, University of California Press, **1**, 17–30.

Bock, M. E.
(1974). "Minimax estimators of the mean of a multivariate normal distribution." *Ann. Statist.* **3**, 209–218.

Bondar, J. V. and Milnes, P.
(1981). "Amenability: A survey for statistical applications of Hunt–Stein and related conditions on groups." *Zeitschr. Wahrsch. Verw. Geb.* **57**, 103–128.

Box, G. E. and Tiao, G. C.
(1973). *Bayesian Inference in Statistical Analysis.* Addison-Wesley, Reading, Mass.

Brandwein, A. C.
(1979). "Minimax estimation of the mean of spherically symmetric distributions under general quadratic loss." *J. Multiv. Anal.* **9**, 579–588.

Brandwein, A. C. and Strawderman, W. E.
(1978). "Minimax estimation of location parameters for spherically symmetric unimodal distributions." *Ann. Statist.* **6**, 377–416.

Brandwein, A. C. and Strawderman, W. E.
(1980). "Minimax estimation of location parameters for spherically symmetric distributions with concave loss." *Ann. Statist.* **8**, 279–284.

Brewster, J. and Zidek, J.
(1974). "Improving on equivariant estimators." *Ann. Statist.* **2**, 21–38.

Brown, L. D.
(1966). "On the admissibility of invariant estimators of one or more location parameters." *Ann. Math. Statist.* **37**, 1087–1136.

Brown, L. D.
(1968). "Inadmissibility of the usual estimators of scale parameters in problems with unknown scale and location parameters." *Ann. Math. Statist.* **39**, 29–48.

Brown, L. D.
(1971). "Admissible estimators, recurrent diffusions, and insoluble boundary value problems." *Ann. Math. Statist.* **42**, 855–903.

Brown, L. D.
(1975). "Estimation with incompletely specified loss functions (the case of several location parameters)." *J. Amer. Statist. Assoc.* **70**, 417–427.

Brown, L. D.
(1979). "A heuristic method for determining admissibility of estimators—with applications." *Ann. Statist.* **7**, 960–994.

Brown, L. D.
(1980a). "A necessary condition for admissibility." *Ann. Statist.* **8**, 540–544.

Brown, L. D.
(1980b). "Examples of Berger's phenomenon in the estimation of independent normal means." *Ann. Statist.* **8**, 572–585.

Brown, L. D. and Fox, M.
(1974a). "Admissibility of procedures in two-dimensional location parameter problems." *Ann. Statist.* **2**, 248–266.

Brown, L. D. and Fox, M.
(1974b). "Admissibility in statistical problems involving a location or scale parameter." *Ann. Statist.* **2**, 807–814.

Brown, P. J. and Zidek, J. V.
(1980). "Adaptive multivariate ridge regression." *Ann. Statist.* **8**, 64–74.

Casella, G.
(1980). "Minimax ridge regression estimation." *Ann. Statist.* **8**, 1036–1056.

Casella, G. and Strawderman, W. E.
(1981). "Estimating a bounded normal mean." *Ann. Statist.* **9**, 870–878.

Clevenson, M. and Zidek, J.
(1975). "Simultaneous estimation of the mean of independent Poisson laws." *J. Amer. Statist. Assoc.* **70**, 698–705.

Cohen, A.
(1965). "Estimates of linear combinations of the parameters in the mean vector of a multivariate distribution." *Ann. Math. Statist.* **36**, 78–87.

de Finetti, B.
(1937). "La prévision: Ses lois logiques, ses source subjectives." *Ann. de l'Inst. Henri Poincaré* **7**, 1–68. (Translated in *Studies in Subjective Probability*. Kyburg and Smokler, Eds. Wiley, New York.)

de Finetti, B.
(1970). *Teoria Delle Probabilità*. English Translation (1974) *Theory of Probability*. Wiley, New York.

De Groot, M., H. and Rao, M. M.
(1963). "Bayes estimation with convex loss." *Ann. Math. Statist.* **34**, 839–846.

Diaconis, P. and Ylvisaker, D.
(1979). "Conjugate priors for exponential families." *Ann. Statist.* **7**, 269–281.

Draper, N. R. and Van Nostrand, R. C.
(1979). "Ridge regression and James–Stein estimation: Review and comments." *Technometrics* **21**, 451–466.

Efron, B. and Morris, C.
(1971/1972). "Limiting the risk of Bayes and empirical Bayes estimators." *J. Amer. Statist. Assoc.* **66**, 807–815, **67**, 130–139.

Efron, B. and Morris, C.
(1972). "Empirical Bayes on vector observations—An extension of Stein's method." *Biometrika* **59**, 335–347.

Efron, B. and Morris, C.
(1973a). "Stein's estimation rule and its competitors—An empirical Bayes approach." *J. Amer. Statist. Assoc.* **68**, 117–130.

Efron, B. and Morris, C.
(1973b). "Combining possibly related estimation problems (with discussion)." *J. Roy. Statist. Soc. B* **35**, 379–421.

Efron, B. and Morris, C.
(1975). "Data analysis using Stein's estimator and its generalizations." *J. Amer. Statist. Assoc.* **70**, 311–319.

Efron, B. and Morris, C.
(1976). "Families of minimax estimators of the mean of a multivariate normal distribution." *Ann. Statist.* **4**, 11–21.

Ericson, W. A.
(1969). "Subjective Bayesian models in sampling finite populations (with discussion)." *J. Roy. Statist. Soc. B* **31**, 195–233.

Farrell, R.
(1964). "Estimators of a location parameter in the absolutely continuous case." *Ann. Math. Statist.* **35**, 949–998.

Ferguson, T. S.
(1967). *Mathematical Statistics*. Academic Press, New York.

Ferguson, T. S.
(1973). "A Bayesian analysis of some nonparametric problems." *Ann. Statist.* **1**, 209–230.

Fisher, R. A.
(1922). "On the mathematical foundations of theoretical statistics." *Phil. Trans. Roy. Soc. A* **222**, 309–368.

Fisher, R. A.
(1930). "Inverse probability." *Proc. Camb. Phil. Soc.* **26**, 528–535.

Fisher, R. A.
(1935). "The fiducial argument in statistical inference." *Ann. Eugenics* **6**, 391–398.

Fox, M.
(1981). "An inadmissible best invariant estimator: The i.i.d. case." *Ann. Statist.* **9**, 1127–1129.

Ghosh, M. N.
(1964). "Uniform approximation of minimax point estimates." *Ann. Math. Statist.* **35**, 1031–1047.

Ghosh, M. and Meeden, G.
(1977). "Admissibility of linear estimators in the one parameter exponential family." *Ann. Statist.* **5**, 772–778.

Ghosh, M. and Parsian, A.
(1980). "Admissible and minimax multiparameter estimation in exponential families." *J. Multiv. Anal.* **10**, 551–564.

Girshick, M. A. and Savage, L. J.
(1951). "Bayes and minimax estimates for quadratic loss functions." *Proc. 2nd Berkeley Symp.* 53–73.

Godambe, V. P.
(1982). "Estimation in survey sampling: Robustness and optimality." *J. Amer. Statist. Assoc.* **77**, 393–406.

Good, I. J.
(1965). *The Estimation of Probabilities: An Essay on Modern Bayesian Methods.* M.I.T. Press, Cambridge.

Gupta, A. K. and Rohatgi, V. K.
(1980). "On the estimation of restricted mean." *J. Statist. Planning and Inference* **4**, 369–379.

Gupta, M. K.
(1966). "On the admissibility of linear estimates for estimating the mean of distributions of the one parameter exponential family." *Calc. Statist. Assoc. Bull.* **15**, 14–19.

Guttman, S.
(1982). "Stein's paradox is impossible with a finite sample space." *Ann. Statist.* **10**, 1017–1020.

Hájek, J.
(1972). Local asymptotic minimax and admissibility in estimation. *Proc. Sixth Berkeley Symp. Math. Statist. Prob.*, **1**, 175–194.

Hill, B.
(1965). "Inference about variance components in a one-way model." *J. Amer. Statist. Assoc.* **60**, 806–825.

Hjort, N. L.
(1976). "Applications of the Dirichlet process to some nonparametric problems." (Norwegian). Univ. of Tromsφ, Inst. for math. and phys. sciences.

Hodges, J. L., Jr., and Lehmann, E. L.
(1950). "Some problems in minimax point estimation." *Ann. Math. Statist.* **21**, 182–197.

Hodges, J. L., Jr., and Lehmann, E. L.
(1951). "Some applications of the Cramér–Rao inequality." *Proc. 2nd Berkeley Symp.* 13–22.

Hodges, J. L., Jr., and Lehmann, E. L.
(1952). "The use of previous experience in reaching statistical decisions." *Ann. Math. Statist.* **23**, 396–407.

Hodges, J. L., Jr., and Lehmann, E. L.
(1981). "Minimax estimation in simple random sampling." In *Essays in Statistics* (In honor of C. R. Rao), Krishnaiah, Ed. North Holland, Amsterdam, pp. 323–327.

Hudson, H. M.
(1978). A natural identity for exponential families with applications in multiparameter estimation." *Ann. Statist.* **6**, 473–484.

Hudson, H. M. and Tsui, K.-W.
(1981). "Simultaneous Poisson estimators for a priori hypotheses about means." *J. Amer. Statist. Assoc.* **76**, 182–187.

Hwang, J. T.
(1982). "Improving upon standard estimators in discrete exponential families with applications to Poisson and negative binomial cases." *Ann. Statist.* **10**, 857–867.

James, W. and Stein, C.
(1961). "Estimation with quadratic loss." *Proc. Fourth Berkeley Symp. Math. Statist. Prob.* **1**, 311–319.

Jaynes, E. T.
(1979). "Where do we stand on maximum entropy?" In *The Maximum Entropy Formalism*, R. D. Levine and M. Tribus, Eds., M.I.T. Press, Cambridge, pp. 15–118.

Jeffreys, H.
(1939, 1948, 1961). *The Theory of Probability*. Oxford University Press, Oxford.

Johnson, B. McK.
(1971). "On the admissible estimators for certain fixed sample binomial problems." *Ann. Math. Statist.* **42**, 1579–1587.

Joshi, V. M.
(1969). "On a theorem of Karlin regarding admissible estimates for exponential populations." *Ann. Math. Statist.* **40**, 216–223.

Joshi, V. M.
(1979). "Joint admissibility of the sample means as estimators of the means of finite populations." *Ann. Statist.* **7**, 995–1002.

Karlin, S.
(1958). "Admissibility for estimation with quadratic loss." *Ann. Math. Statist.* **29**, 406–436.

Katz, M. W.
(1961). "Admissible and minimax estimates of parameters in truncated spaces." *Ann. Math. Statist.* **32**, 136–142.

Kiefer, J.
(1957). "Invariance, minimax sequential estimation, and continuous time processes." *Ann. Math. Statist.* **28**, 573–601.

Kiefer, J.
(1966). "Multivariate optimality results." In Multivariate Analysis. Krishnaiah, Ed., Academic Press, New York, pp. 255–274.

Klotz, J., Milton, R. C., and Zacks, S.
(1969). "Mean square efficiency of estimators of variance components." *J. Amer. Statist. Assoc.* **64**, 1383–1402.

Kudo, H.
(1955). "On minimax invariant estimates of the translation parameter." *Natural Sci. Report Ochanomizu Univ.* **6**, 31–73.

Lane, D. A.
(1980). "Fisher, Jeffreys, and the nature of probability." In *R. A. Fisher: An Appreciation.* Fienberg and Hinkley, Eds. Lecture Notes in Statistics, 1. Springer, New York.

Laplace, P. S.
(1774). "Mémoire sur la probabilité des causes par évènements." *Mém. Acad. Sci. Sav. Etranger* **6**, 621–656.

Laplace, P. S.
(1820). *Theorie Analytique des Probabilités.* 3rd ed. Paris.

Lindley, D. V.
(1964). "The Bayesian analysis of contingency tables." *Ann. Math. Statist.* **35**, 1622–1643.

Lindley, D. V.
(1965). *Introduction to Probability and Statistics.* Cambridge University Press, Cambridge.

Lindley, D. V. and Smith, A. F. M.
(1972). "Bayes estimates for the linear model (with discussion)." *J. Roy. Statist. Soc. B* **34**, 1–41.

Makani, S. M.
(1972). "Admissibility of linear functions for estimating sums and differences of exponential parameters." Ph.D. thesis. University of California, Berkeley.

Makani, S. M.
(1977). "A paradox in admissibility." *Ann. Statist.* **5**, 544–546.

Marazzi, A.
(1980). "Robust Bayesian estimation for the linear model." Res. Report No. 27, Fachgruppe f. Stat., Eidg. Tech. Hochsch., Zurich.

Morton, R. and Raghavachari, M.
(1966). "On a theorem of Karlin regarding admissibility of linear estimates in exponential populations." *Ann. Math. Statist.* **37**, 1809–1813.

Neyman, J. and Pearson, E. S.
(1933a). "The testing of statistical hypotheses in relation to probabilities a priori." *Proc. Camb. Phil. Soc.* **24**, 492–510.

Neyman, J. and Pearson, E. S.
(1933b). "On the problem of the most efficient tests of statistical hypotheses." *Phil. Trans. Roy. Soc. A.* **231**, 289–337.

Novick, M. R. and Jackson, P. H.
(1974). *Statistical Methods for Educational and Psychological Research.* McGraw-Hill, New York.

Olkin, I. and Sobel, M.
(1979). "Admissible and minimax estimation for the multinomial distribution and for k independent binomial distributions." *Ann. Statist.* **7**, 284–290.

Peele, L. and Ryan, T. P.
(1982). "Minimax linear regression estimators with application to ridge regression. *Techn.* **24**, 157–159.

Peisakoff, M. P.
(1950). "Transformation parameters." Unpublished Ph.D. thesis. Princeton University, Princeton, N.J.

Perng, S. K.
(1970). "Inadmissibility of various 'good' statistical procedures which are translation invariant." *Ann. Math. Statist.* **41**, 1311–1321.

Ping, C.
(1964). "Minimax estimates of parameters of distributions belonging to the exponential family." *Chinese Math.* **5**, 277–299.

Portnoy, S.
(1971). "Formal Bayes estimation with application to a random effects model." *Ann. Math. Statist.* **42**, 1379–1402.

Portnoy, S.
(1975). "Admissibility of the best invariant estimator of one coordinate of a location vector." *Ann. Statist.* **3**, 448–450.

Raiffa, H. and Schlaifer, R.
(1961). *Applied Statistical Decision Theory*. Harvard University Press, Cambridge.

Rao, C. R.
(1977). "Simultaneous estimation of parameters—A compound decision problem." In *Decision Theory and Related Topics*. S. S. Gupta and D. S. Moore, Eds. Academic Press, New York, pp. 327–350.

Rao, C. R. and Shinozaki, N.
(1978). "Precision of individual estimators in simultaneous estimation of parameters." *Biometrika* **65**, 23–30.

Robinson, G. K.
(1979a). "Conditional properties of statistical procedures." *Ann. Statist.* **7**, 742–755.

Robinson, G. K.
(1979b). "Conditional properties of statistical procedures for location and scale parameters." *Ann. Statist.* **7**, 756–771.

Rothenberg, T. J.
(1977). "The Bayesian approach and alternatives in econometrics." In *Studies in Bayesian Econometrics and Statistics*. Vol. 1. Fienberg and Zellner, Eds. North Holland, Amsterdam, pp. 55–75.

Rutkowska, M.
(1977). "Minimax estimation of the parameters of the multivariate hypergeometric and multinomial distributions." *Zastos. Mat.* **16**, 9–21.

Sacks, J.
(1963). "Generalized Bayes solutions in estimation problems." *Ann. Math. Statist.* **34**, 751–768.

Savage, L. J.
(1954). *The Foundations of Statistics*. Wiley, New York.

Sclove, S. L.
(1968). "Improved estimators for coefficients in linear regression." *J. Amer. Statist. Assoc.* **63**, 596–606.

Sclove, S. L., Morris, C., and Radhakrishnan, R.
(1972). "Nonoptimality of preliminary-test estimators for the mean of a multivariate normal distribution." *Ann. Math. Statist.* **43**, 1481–1490.

Shinozaki, N.
(1979). "Estimation of a multivariate normal mean with a class of quadratic loss functions." *J. Amer. Statist. Assoc.* **75**, 973–976.

Stein, C.
(1955). "A necessary and sufficient condition for admissibility." *Ann. Math. Statist.* **26**, 518–522.

Stein, C.
(1956). "Inadmissibility of the usual estimator for the mean of a multivariate distribution." *Proc. Third Berkeley Symp. Math. Statist. Prob.*, University of California Press, **1**, 197–206.

Stein, C.
(1959). "The admissibility of Pitman's estimator for a single location parameter." *Ann. Math. Statist.* **30**, 970–979.

Stein, C.
(1964). "Inadmissibility of the usual estimator for the variance of a normal distribution with unknown mean." *Ann. Inst. Statist. Math.* **16**, 155–160.

Stein, C.
(1965). "Approximation of improper prior measures by prior probability measures." In *Bernoulli, Bayes, Laplace Anniversary Volume*. Springer, New York.

Stein, C.
(1973). "Estimation of the mean of a multivariate distribution." *Proc. Prague Symp. on Asymptotic Statistics*, 345–381.

Stein, C.
(1981). "Estimation of the mean of a multivariate normal distribution." *Ann. Statist.* **9**, 1135–1151.

Steinhaus, H.
(1957). "The problem of estimation." *Ann. Math. Statist.* **28**, 633–648.

Stone, M.
(1967). "Generalized Bayes decision functions, admissibility and the exponential family." *Ann. Math. Statist.* **38**, 818–822.

Stone, M. and Springer, B. G. F.
(1965). "A paradox involving quasi prior distributions." *Biometrika* **59**, 623–627.

Strawderman, W. E.
(1971). "Proper Bayes minimax estimators of the multivariate normal mean." *Ann. Statist.* **42**, 385–388.

Strawderman, W. E.
(1973). "Proper Bayes minimax estimators of the multivariate normal mean vector for the case of common unknown variances." *Ann. Statist.* **1**, 1189–1194.

Strawderman, W. E. and Cohen, A.
(1971). "Admissibility of estimators of the mean vector of a multivariate normal distribution with quadratic loss." *Ann. Math. Statist.* **42**, 270–296.

Susarla, V.
(1982). "Empirical Bayes theory." Article in vol. 2 of Encyclopedia of Statistical Sciences. (Kotz, Johnson, and Read, Eds), Wiley. New York.

Takada, Y.
(1979). "A family of minimax estimators in some multiple regression problems." *Ann. Statist.* **7**, 1144–1147.

Tiao, G. C. and Tan, W. Y.
(1965). "Bayesian analysis of random effects models in analysis of variance, I. Posterior distribution of variance components." *Biometrika* **52**, 37–53.

Trybula, S.
(1958). "Some problems of simultaneous minimax estimation." *Ann. Math. Statist.* **29**, 245–253.

Tsui, K.-W.
(1981). "Simultaneous estimation of several Poisson parameters under squared error loss." *Ann. Inst. Statist. Math.* **33**, 215–224.

Wald, A.
(1939). "Contributions to the theory of statistical estimation and hypothesis testing." *Ann. Math. Statist.* **10**, 299–326.

Wald, A.
(1950). *Statistical Decision Functions*. Wiley, New York.

Wold, H.
(1948). *Random Normal Deviates* (Tracts for Computers XXV). Cambridge University Press, London.

Zacks, S.
(1971). *The Theory of Statistical Inference*. Wiley, New York.

Zidek, J. V.
(1970). "Sufficient conditions for the admissibility under squared error loss of formal Bayes estimators." *Ann. Math. Statist.* **41**, 446–456.

Zidek, J. V.
(1971). "Inadmissibility of a class of estimators of a normal quantile." *Ann. Math. Statist.* **42**, 1444–1447.

Zidek, J. V.
(1973). "Estimating the scale parameter of the exponential distribution with unknown location." *Ann. Statist.* **1**, 264–278.

CHAPTER 5

Large-Sample Theory

1. CONVERGENCE IN PROBABILITY AND IN LAW

The mathematical precision of the optimality results obtained in Chapters 2 through 4 tends to obscure the fact that statistically they are only rough approximations in view of the approximate nature of both the assumed models and the loss functions. In this and the next chapter a further approximation is added based on the assumption that the sample size is "large." Although this injects an additional element of uncertainty into the conclusions, it mitigates the effects of the earlier approximations because the results become less dependent on the loss function and, in large samples, it is much easier to study the dependence on the model.

Large-sample theory considers a sample $\underline{X} = (X_1, \ldots, X_n)$ not for fixed n, but as a member of a sequence corresponding to $n = 1, 2, \ldots$ (or more generally $n = n_0, n_0 + 1, \ldots$), and assessing the performance of estimator sequences as $n \to \infty$. Mathematically, the results are thus limit theorems. In applications, the limiting result is used as an approximation to the situation obtaining for the actual, finite n. It is a weakness of this approach that, typically, no good estimates are available for the accuracy of the approximation. To obtain at least some idea of the accuracy, we shall sometimes give numerical checks for selected values of n.

Suppose for a moment that $X_1, \ldots, X_n$ are iid according to a distribution P_θ, $\theta \in \Omega$ and that the estimand is $g(\theta)$. As n increases, more and more information about θ becomes available, and one would expect that for sufficiently large values of n, it would typically be possible to estimate $g(\theta)$ very closely. If $\delta_n = \delta_n(X_1, \ldots, X_n)$ is a reasonable estimator, of course, it can not be expected to be close to $g(\theta)$ for every sample point $(x_1, \ldots, x_n)$ since the values of a particular sample may always be atypical (e.g., a fair coin may fall heads in 1000 successive spins). What one can hope for is that δ_n will be close to $g(\theta)$ with high probability.

This idea is captured in the following definitions, which no longer assume the random variables to be iid.

A sequence of random variables Y_n defined over sample spaces $(\mathcal{Y}_n, \mathcal{B}_n)$ tends *in probability* to a constant c ($Y_n \xrightarrow{P} c$) if for every $a > 0$

$$P[|Y_n - c| \geqslant a] \to 0 \quad \text{as} \quad n \to \infty. \tag{1}$$

A sequence of estimators δ_n of $g(\theta)$ is *consistent* if for every $\theta \in \Omega$

$$\delta_n \xrightarrow{P_\theta} g(\theta). \tag{2}$$

The following condition frequently provides a convenient method for proving consistency.

Lemma 1.1. *A sufficient condition for* Y_n *to converge in probability to* c *is that*

$$E(Y_n - c)^2 \to 0. \tag{3}$$

Proof. This is immediate from Chebyshev's inequality (Problem 1.1), which states that for any $a > 0$ and c,

$$E(Y_n - c)^2 \geqslant a^2 P(|Y_n - c| \geqslant a). \tag{4}$$

The lemma can be restated in statistical terms as follows.

Theorem 1.1. (i) *Let* $\{\delta_n\}$ *be a sequence of estimators of* $g(\theta)$ *with risk* $R(\theta, \delta_n) = E[\delta_n - g(\theta)]^2$. *Then*

$$R(\theta, \delta_n) \to 0 \quad \textit{for all} \quad \theta \tag{5}$$

implies that δ_n *is consistent for estimating* $g(\theta)$.

(ii) *Equivalently to* (5), δ_n *is consistent if*

$$b_n(\theta) \to 0 \quad \textit{and} \quad var_\theta(\delta_n) \to 0 \quad \textit{for all} \quad \theta, \tag{6}$$

where b_n *is the bias of* δ_n.

(iii) *In particular,* δ_n *is consistent if it is unbiased for each* n *and if*

$$var_\theta(\delta_n) \to 0 \quad \textit{for all} \quad \theta. \tag{7}$$

Example 1.1. Consistency of the Mean. Let $X_1, \ldots, X_n$ be iid with expectation $E(X_i) = \xi$ and variance $\sigma^2 < \infty$. Then $\bar{X}$ is an unbiased estimator of ξ with variance σ^2/n, and hence is consistent by Theorem 1.1(iii). Actually, it was proved by

Khinchin [see, for example, Feller (1970, Chap. X, Sect. 1, 2)] that consistency of $\bar{X}$ already follows from the existence of the expectation, so that the assumption of finite variance is not needed.

Note. The statement that $\bar{X}$ is consistent is shorthand for the fuller assertion that the sequence of estimators $\bar{X}_n = (X_1 + \cdots + X_n)/n$ is consistent. This type of shorthand is used very commonly and will be used here. However, the full meaning should be kept in mind.

Example 1.2. Consistency of S^2. Let $X_1, \ldots, X_n$ be iid with finite variance σ^2. Then the unbiased estimator

$$S_n^2 = \Sigma(X_i - \bar{X})^2/(n-1)$$

is a consistent estimator of σ^2. To see this, assume without loss of generality that $E(X_i) = 0$, and note that

$$S_n^2 = \frac{n}{n-1}\left[\frac{1}{n}\Sigma X_i^2 - \bar{X}^2\right].$$

By Example 1.1, $\Sigma X_i^2/n \xrightarrow{P} \sigma^2$ and $\bar{X}^2 \xrightarrow{P} 0$. Since $n/(n-1) \to 1$, it follows from Problem 1.4 that $S_n^2 \xrightarrow{P} \sigma^2$.

Example 1.3. Markov Chains. As an illustration of a situation involving dependent random variables, consider a two-state Markov chain. The variables $X_1, X_2, \ldots$ each take on the values 0 and 1, with the joint distribution determined by the initial probability $P(X_1 = 1) = p_1$, and the transition probabilities

$$P(X_{i+1} = 1|X_i = 0) = \pi_0, \qquad P(X_{i+1} = 1|X_i = 1) = \pi_1,$$

of which we shall assume $0 < \pi_0, \pi_1 < 1$. For such a chain, the probability

$$p_k = P(X_k = 1)$$

typically depends on k and the initial probability p_1 (but see Problem 1.9). However, as $k \to \infty$, p_k tends to a limit p, which is independent of p_1. It is easy to see what the value of p must be. For consider the recurrence relation

$$p_{k+1} = p_k\pi_1 + (1 - p_k)\pi_0 = p_k(\pi_1 - \pi_0) + \pi_0. \tag{8}$$

If

$$p_k \to p, \tag{9}$$

this implies

$$p = \frac{\pi_0}{1 - \pi_1 + \pi_0}. \tag{10}$$

To prove (9), it is only necessary to iterate (8) starting with $k = 1$ to find (Problem 1.5)

$$p_k = (p_1 - p)(\pi_1 - \pi_0)^{k-1} + p. \tag{11}$$

Since $|\pi_1 - \pi_0| < 1$, the result follows.

For estimating p, after n trials, the natural estimator is $\bar{X}_n$, the frequency of ones in these trials. Since

$$E(\bar{X}_n) = (p_1 + \cdots + p_n)/n,$$

it follows from (9) that $E(\bar{X}_n) \to p$ (Problem 1.6), so that the bias of $\bar{X}_n$ tends to zero. Consistency of $\bar{X}_n$ will therefore follow if we can show that $\text{var}(\bar{X}_n) \to 0$. Now

$$\text{var}(\bar{X}_n) = \sum_{i=1}^{n} \sum_{j=1}^{n} \text{cov}(X_i, X_j)/n^2.$$

This average of n^2 terms will go to zero only if $\text{cov}(X_i, X_j) \to 0$ sufficiently fast as $|j - i| \to \infty$. The covariance of X_i and X_j can be obtained by a calculation similar to that leading to (11) and satisfies

$$|\text{cov}(X_i, X_j)| \leqslant M|\pi_1 - \pi_0|^{j-i}. \tag{12}$$

From (12) one finds that $\text{var}(\bar{X}_n)$ is of order $1/n$ and hence that $\bar{X}_n$ is consistent (Problem 1.11).

Instead of p, one may be interested in estimating π_0 and π_1 themselves. Again it turns out that the natural estimator $[N_{01}/(N_{00} + N_{01})$ for π_0, where N_{0j} is the number of pairs (X_i, X_{i+1}) with $X_i = 0$, $X_{i+1} = j$, $j = 0, 1]$ is consistent. For further discussion and references, see, for example, Anderson and Goodman (1957, Sect. 5) and Billingsley (1961).

Consider, on the other hand, the estimation of p_1. It does not appear that observations beyond the first provide any information about p_1, and one would therefore not expect to be able to estimate p_1 consistently. To obtain a formal proof, suppose for a moment that the π's are known, so that p_1 is the only unknown parameter. If a consistent estimator δ_n exists for the original problem, then δ_n will continue to be consistent under this additional assumption. However, when the π's are known, X_1 is a sufficient statistic for p_1 and the problem reduces to that of estimating a success probability from a single trial. That a consistent estimator of p_1 cannot exist under these circumstances follows from the definition of consistency.

When $X_1, \ldots, X_n$ are iid according to a distribution P_θ, $\theta \in \Omega$, consistent estimators of real-valued functions of θ will exist in most of the situations we shall encounter (see, for example, Problem 1.7). There is, however, an important exception. Suppose the X's are distributed according to $F(x_i - \theta)$ where F is $N(\xi, \sigma^2)$, with θ, ξ, and σ^2 unknown. Then no consistent estimator of θ exists. To see this, note that the X's are distributed as $N(\xi + \theta, \sigma^2)$. Thus $\bar{X}$ is consistent for estimating $\xi + \theta$, but ξ and θ cannot be estimated separately because they are not uniquely defined. This idea is formalized in the following definition. Let X be distributed according to $P_{\theta, \vartheta}$. If there exist pairs (θ_1, ϑ_1) and (θ_2, ϑ_2) with $\theta_1 \neq \theta_2$ for which $P_{\theta_1, \vartheta_1} = P_{\theta_2, \vartheta_2}$, the parameter θ is said to be *unidentifiable*. Such a parameter cannot be estimated consistently since $\delta(X_1, \ldots, X_n)$ cannot simultaneously be close to both θ_1 and θ_2.

Consistency is too weak a property to be of much interest in itself. It tells us that for large n, the error $\delta_n - g(\theta)$ is likely to be small but not whether the order of the error is $1/n$, $1/\sqrt{n}$, $1/\log n$, and so on. To obtain an idea of the rate of convergence of a consistent estimator δ_n, consider the probability

$$P_n(a) = P\left\{ |\delta_n - g(\theta)| \leqslant \frac{a}{k_n} \right\}. \tag{13}$$

If k_n is bounded, then $P_n(a) \to 1$. On the other hand, if $k_n \to \infty$ sufficiently fast, $P_n(a) \to 0$. This suggests that for a given $a > 0$ there might exist an intermediate sequence $k_n \to \infty$ for which $P_n(a)$ tends to a limit strictly between 0 and 1. This will be the case for most of the estimators with which we are concerned. Commonly, there will exist a sequence $k_n \to \infty$ and a limit function H which is a continuous distribution function such that for all a

$$P\{k_n[\delta_n - g(\theta)] \leqslant a\} \to H(a) \quad \text{as} \quad n \to \infty. \tag{14}$$

We shall then say that the error $|\delta_n - g(\theta)|$ tends to zero at rate $1/k_n$. The rate, of course, is not uniquely determined by this definition. If $1/k_n$ is a possible rate, so is $1/k_n'$ for any sequence k_n' for which k_n'/k_n tends to a finite nonzero limit. On the other hand, if k_n' tends to ∞ more slowly (or faster) than k_n, that is, if $k_n'/k_n \to 0$ (or ∞), then $k_n'[\delta_n - g(\theta)]$ tends in probability to zero (or ∞) (Problem 1.12).

One can think of the normalizing constants k_n in (14) in another way. If δ_n is consistent, the errors $\delta_n - g(\theta)$ tend to zero as $n \to \infty$. Multiplication by constants k_n tending to infinity magnifies these minute errors—it acts as a microscope. If (14) holds, then k_n is just the right degree of magnification to give a well-focused picture of the behavior of the errors.

In generalization of (14), suppose that $\{Y_n\}$ is a sequence of random variables with cdf

$$H_n(a) = P(Y_n \leqslant a)$$

and that there exists a cdf H such that

$$H_n(a) \to H(a) \tag{15}$$

at all points a at which H is continuous. Then we shall say that the distribution functions H_n *converge weakly* to H, and that the Y_n have the *limit distribution* H, or *converge in law* to any random variable Y with distribution H. This will be denoted by $Y_n \xrightarrow{\mathcal{L}} Y$ or by $\mathcal{L}(Y_n) \to H$. We may also say that Y_n tends in law to H and write $Y_n \xrightarrow{\mathcal{L}} H$. The crucial assumption in (15) is that $H(-\infty) = 0$ and $H(+\infty) = 1$, that is, that no probability mass escapes to $\pm\infty$ (see Problem 1.17).

The following example illustrates the reason for requiring (15) to hold only for the continuity points of H.

Example 1.4. (i) Let Y_n be normally distributed with mean zero and variance σ_n^2 where $\sigma_n \to 0$ as $n \to \infty$.

(ii) Let Y_n be a random variable taking on the value $1/n$ with probability 1.

In both cases, it seems natural to say that Y_n tends in law to a random variable Y which takes on the value 0 with probability 1. The cdf $H(a)$ of Y is zero for $a < 0$ and 1 for $a \geqslant 0$. The cdf $H_n(a)$ of Y_n in both (i) and (ii) tends to $H(a)$ for all $a \neq 0$, but not for $a = 0$ (Problem 1.14).

An important property of weak convergence is given by the following theorem.

Theorem 1.2. *The sequence* Y_n *converges in law to* Y *if and only if* $E[f(Y_n)] \to E[f(Y)]$ *for every bounded continuous real-valued function* f.

For a proof, see, for example, Billingsley (1979, p. 288).

A basic tool for obtaining the limit distribution of many estimators of interest is the central limit theorem (CLT), of which the following is the simplest case.

Theorem 1.3. *Let* X_i $(i = 1, \ldots, n)$ *be iid with* $E(X_i) = \xi$ *and* $var(X_i) = \sigma^2 < \infty$. *Then* $\sqrt{n}(\bar{X} - \xi)$ *tends in law to* $N(0, \sigma^2)$ *and hence* $\sqrt{n}(\bar{X} - \xi)/\sigma$ *to the standard normal distribution* $N(0, 1)$.

A proof can be found in most texts on probability theory.

The usefulness of this result is greatly extended by Theorems 1.4 and 1.5 below.

Theorem 1.4. *If* $Y_n \xrightarrow{\mathcal{L}} Y$, *and* A_n, B_n *tend in probability to* a *and* b, *respectively, then* $A_n + B_n Y_n \xrightarrow{\mathcal{L}} a + bY$.

A proof can be found, for example, in Bickel and Doksum (1977) and in Cramér (1946).

When Y_n converges to a distribution H, it is often required to evaluate probabilities of the form $P(Y_n \leqslant y_n)$ where $y_n \to y$, and one may hope that these probabilities will tend to $H(y)$.

Corollary 1.1. *If* $Y_n \xrightarrow{\mathcal{L}} H$, *and* y_n *converges to a continuity point* y *of* H, *then* $P(Y_n \leqslant y_n) \to H(y)$.

Proof. $P(Y_n \leqslant y_n) = P[Y_n + (y - y_n) \leqslant y]$ and the result follows from Theorem 1.4 with $B_n = 1$, $A_n = y - y_n$.

Theorem 1.5. *If*

$$\sqrt{n}\,[T_n - \theta] \xrightarrow{\mathcal{L}} N(0, \tau^2), \tag{16}$$

then

$$\sqrt{n}\,[f(T_n) - f(\theta)] \xrightarrow{\mathcal{L}} N\big(0, \tau^2 [f'(\theta)]^2\big) \tag{17}$$

provided $f'(\theta)$ *exists and is not zero.*

Proof. Consider the Taylor expansion of $f(T_n)$ around $f(\theta)$:

$$f(T_n) = f(\theta) + (T_n - \theta)[f'(\theta) + R_n], \tag{18}$$

where $R_n \to 0$ as $T_n \to \theta$. It follows from (16) that $T_n \to \theta$ in probability and hence that $R_n \to 0$ in probability. The result now follows by applying Theorem 1.4 to $\sqrt{n}\,[f(T_n) - f(\theta)]$.

Example 1.5. Let $X_1, \ldots, X_n$ be iid $N(\theta, \sigma^2)$, and let the estimand be θ^2. Recall the estimators (Problem 2.2.1 and 2.2.2).

$$\delta_{1n} = \bar{X}^2 - \frac{\sigma^2}{n} \qquad \text{(UMVU when } \sigma \text{ is known)}$$

$$\delta_{2n} = \bar{X}^2 - \frac{S^2}{n(n-1)} \qquad \text{(UMVU when } \sigma \text{ is unknown) where } S^2 = \Sigma(X_i - \bar{X})^2$$

$$\delta_{3n} = \bar{X}^2 \qquad \text{(MLE in either case).}$$

For each of these three sequences of estimators $\delta(\underset{\sim}{X})$, let us find the limiting distribution of $[\delta(\underset{\sim}{X}) - \theta^2]$ suitably normalized. Now, $\sqrt{n}(\bar{X} - \theta) \to N(0, \sigma^2)$ in law by the central limit theorem. Using $f(u) = u^2$ in Theorem 1.5, it follows that

$$\sqrt{n}(\bar{X}^2 - \theta^2) \to N(0, 4\sigma^2\theta^2), \tag{19}$$

provided $f'(\theta) = 2\theta \neq 0$. This establishes the limiting behavior of δ_{3n}.

Next consider δ_{1n}. Since

$$\sqrt{n}\left(\bar{X}^2 - \frac{\sigma^2}{n} - \theta^2\right) = \sqrt{n}(\bar{X}^2 - \theta^2) - \frac{\sigma^2}{\sqrt{n}},$$

it follows from Theorem 1.4 that

$$\sqrt{n}(\delta_{1n} - \theta^2) \to N(0, 4\sigma^2\theta^2). \tag{20}$$

Finally, consider

$$\sqrt{n}\left(\bar{X}^2 - \frac{S^2}{n(n-1)} - \theta^2\right) = \sqrt{n}(\bar{X}^2 - \theta^2) - \frac{1}{\sqrt{n}}\left(\frac{S^2}{n-1}\right).$$

Now, $S^2/(n-1)$ tends to σ^2 in probability, so $S^2/\sqrt{n}(n-1)$ tends to 0 in probability. Thus,

$$\sqrt{n}(\delta_{2n} - \theta^2) \to N(0, 4\sigma^2\theta^2).$$

Hence when $\theta \neq 0$, all three estimators have the same limit distribution.

There remains the case $\theta = 0$. It is seen from (18) that if (16) holds and $f'(\theta) = 0$, then

$$\sqrt{n}[f(T_n) - f(\theta)] \to 0 \qquad \text{in probability.}$$

Thus, in particular in the present situation, when $\theta = 0$, $\sqrt{n}[\delta(\underset{\sim}{X}) - \theta^2] \to 0$ in probability for all three estimators. When $f'(\theta) = 0$, $\sqrt{n}$ is no longer the appropriate normalizing factor; it tends to infinity too slowly.

When the dominant term in the Taylor expansion (18) vanishes, it is natural to carry the expansion one step further to obtain

$$f(T_n) = f(\theta) + (T_n - \theta)f'(\theta) + \tfrac{1}{2}(T_n - \theta)^2[f''(\theta) + R_n],$$

where $R_n \to 0$ in probability as $T_n \to \theta$, or, since $f'(\theta) = 0$,

$$f(T_n) - f(\theta) = \tfrac{1}{2}(T_n - \theta)^2[f''(\theta) + R_n]. \tag{21}$$

In view of (16), the distribution of $[\sqrt{n}(T_n - \theta)]^2$ tends to a nondegenerate limit distribution, namely (after division by τ^2), to a χ^2-distribution with 1 degree of freedom,

$$n(T_n - \theta)^2 \to \tau^2 \cdot \chi_1^2. \tag{22}$$

The same argument as that leading to (17) shows that when $f'(\theta) = 0$ but $f''(\theta) \neq 0$,

$$n[f(T_n) - f(\theta)] \to \tfrac{1}{2}\tau^2 f''(\theta)\chi_1^2. \tag{23}$$

Example 1.5. (*Continued*). Let us apply (23) to the three estimators of Example 1.5 when $\theta = 0$. Since $f''(0) = 2$, it follows that for the MLE $\bar{X}^2$,

$$n(\bar{X}^2 - \theta^2) = n(\bar{X}^2 - 0^2) \to \tfrac{1}{2}\sigma^2 \cdot 2 \cdot \chi_1^2 = \sigma^2\chi_1^2.$$

Actually, since the distribution of $\sqrt{n}\,\bar{X}$ is $N(0, \sigma^2)$ for each n, the statistic $n\bar{X}^2$ is distributed exactly as $\sigma^2\chi_1^2$, so that no asymptotic argument would have been necessary.

For δ_{1n} we find

$$n\left(\bar{X}^2 - \frac{\sigma^2}{n} - \theta^2\right) = n\bar{X}^2 - \sigma^2,$$

and the right-hand side tends in law to $\sigma^2(\chi_1^2 - 1)$. In fact, here too this is the exact rather than just a limit distribution. Finally, consider δ_{2n}. Here

$$n\left(\bar{X}^2 - \frac{S^2}{n(n-1)} - \theta^2\right) = n\bar{X}^2 - \frac{S^2}{n-1},$$

and since $S^2/(n-1)$ tends in probability to σ^2, the limiting distribution is again $\sigma^2(\chi_1^2 - 1)$.

Although for $\theta \neq 0$, the three sequences of estimators had the same limit distribution, this is no longer true when $\theta = 0$. In this case the limit distribution of $n(\delta - \theta^2)^2$ is $\sigma^2(\chi_1^2 - 1)$ for δ_{1n} and δ_{2n} but $\sigma^2\chi_1^2$ for the MLE δ_{3n}. These two distributions differ only in their location. The distribution of $\sigma^2(\chi_1^2 - 1)$ is centered so that its expectation is zero, while that of $\sigma^2\chi_1^2$ has expectation σ^2. When the limit distribution of the error $[\delta_n - g(\theta)]$ of a sequence of estimators (normalized so that it is not degenerate) has an expectation different from 0, the sequence $\{\delta_n\}$ is said to be asymptotically biased. Thus for $\theta = 0$, the MLE is asymptotically biased (but not for $\theta \neq 0$).

The typical behavior of estimator sequences as sample sizes tend to infinity is that suggested by Theorem 1.5; that is, if δ_n is the estimator of $g(\theta)$ based on n observations, one may expect that $\sqrt{n}\,[\delta_n - g(\theta)]$ will tend to a normal distribution with mean zero and variance, say $\tau^2(\theta)$. It is in this sense that the large-sample behavior of such estimators can be studied without reference to a specific loss function. The asymptotic behavior of δ_n is governed solely by $\tau^2(\theta)$ since knowledge of $\tau^2(\theta)$ determines the probability of the error $\delta_n - g(\theta)$ lying in any given interval. In particular, $\tau^2(\theta)$ provides a basis for the large-sample comparison of different estimators to be taken up in the next section.

Instead of working with the variance τ^2 or $\tau^2[f'(\theta)]^2$ of the limit distribution (16) or (17), the *asymptotic variance*, one may consider directly the limiting behavior of the expected squared error $E[\delta_n - g(\theta)]^2$, which in the case of unbiased estimators is the variance. This was the principal approach used in Section 2.5. To illustrate it in the present context, let us return to Example 1.5.

Example (1.5). *(completed).* The estimators δ_{1n} and δ_{3n} for θ^2 are special cases of the sequence

$$\delta_n^{(c)} = \bar{X}^2 - \frac{c\sigma^2}{n} \tag{24}$$

with $c = 1$ or 0. It is not difficult to calculate the expected squared error of (24) to find that

$$E\left(\bar{X}^2 - \frac{c\sigma^2}{n} - \theta^2\right)^2 = \frac{4\sigma^2\theta^2}{n} + \frac{(c^2 - 2c + 3)\sigma^4}{n^2} \tag{25}$$

Formula (25) has some interesting consequences. Note, first, that

$$nE\left(\bar{X}^2 - \frac{c\sigma^2}{n} - \theta^2\right)^2 \to 4\sigma^2\theta^2. \tag{26}$$

Comparing this with (20), we see that the asymptotic variance (that is, the variance of the normal limit distribution) is equal to the limit of the risk (and the variance) when normalized by the same factor.

By (25) the expected squared error of the MLE and of the UMVU estimator of θ^2 agree up to terms of order $1/n$ but differ in the $1/n^2$ terms. Up to terms of order $1/n^2$, the UMVU estimator minimizes the risk among the estimators (24). This follows from the fact that the quadratic $c^2 - 2c + 3$ has its minimum at $c = 1$.

We note finally that, as $\theta \to 0$, the right side of (26) tends to zero as would be expected since the asymptotic distribution for $\theta = 0$ requires the normalizing factor n rather than $\sqrt{n}$.

The equality of the asymptotic variance with the limit of the variance observed in this example does not always hold (Problem 1.22). What can be stated quite generally is that the appropriately normalized limit of the variance is greater than or equal to the asymptotic variance. To see this, let us first state the following lemma.

Lemma 1.2. *Let* $Y_n, n = 1, 2, \ldots$ *be a sequence of random variables tending in law to a random variable* Y *with cdf* H *and with* $E(Y^2) = v^2$. *Let* Y_{nA} *be the random variable* Y_n *truncated at* $\pm A$, *so that* $Y_{nA} = Y_n$ *if* $|Y_n| \leqslant A$, *and* $Y_{nA} = +A$ *or* $-A$ *if* $Y_n > A$ *or* $< -A$.

(i) *Then*

$$\lim_{A\to\infty}\lim_{n\to\infty} E\left(Y_{nA}^2\right) = \lim_{A\to\infty}\lim_{n\to\infty} E\left[\min\left(Y_n^2, A^2\right)\right] \tag{27}$$

exists and is equal to v^2.

(ii) *If, in addition,*

$$E\left(Y_n^2\right) \to w^2 \quad \text{as} \quad n \to \infty, \tag{28}$$

then $v^2 \leqslant w^2$.

Proof. (i) By Theorem 1.2,

$$\lim_{n\to\infty} E\left(Y_{nA}^2\right) = \int_{-A}^{A} y^2\, dH(y) + A^2 P(|Y| > A),$$

and as $A \to \infty$, the right side tends to v^2.

(ii) It follows from Problem 1.26 that

$$\lim_{A\to\infty}\lim_{n\to\infty} E\left(Y_{nA}^2\right) \leqslant \lim_{n\to\infty}\lim_{A\to\infty} E\left(Y_{nA}^2\right) \tag{29}$$

provided the indicated limits exist. Now

$$\lim_{A\to\infty} E\left(Y_{nA}^2\right) = E\left(Y_n^2\right)$$

so that the right side of (29) is w^2, while the left side is v^2 by (i).

Suppose now that T_n is a sequence of statistics for which $Y_n = k_n[T_n - E(T_n)]$ tends in law to a random variable Y with zero expectation. Then the asymptotic variance $v^2 = \text{var}(Y)$ and the limit of the variances $w^2 = \lim E(Y_n^2)$, if it exists, satisfy $v^2 \leqslant w^2$ as was claimed. Conditions for v^2 and w^2 to coincide are given by Chernoff (1956). For the special case that T_n is a function of a sample mean of iid variables, the two coincide under the assumptions of Theorems 2.5.1 and 2.5.2.

Limit theorems such as Theorem 1.5 refer to sequences of situations as $n \to \infty$. However, in a given problem one is dealing with a specific large value of n. Any particular situation can be embedded in many different sequences, which lead to different approximations.

Suppose, for example, that it is desired to find an approximate value for

$$P(|T_n - g(\theta)| \geqslant a) \tag{30}$$

when $n = 100$ and $a = 0.2$. If $\sqrt{n}[T_n - g(\theta)]$ is asymptotically normally distributed as $N(0, 1)$, one might want to put $a = c/\sqrt{n}$ (so that $c = 2$) and consider (30) as a member of the sequence

$$P\left(|T_n - g(\theta)| \geqslant \frac{2}{\sqrt{n}}\right) \approx 2[1 - \Phi(2)]. \tag{31}$$

Alternatively, one could keep $a = 0.2$ fixed and consider (30) as a member of the sequence

$$P(|T_n - g(\theta)| \geqslant 0.2). \tag{32}$$

Since $T_n - g(\theta) \to 0$, this sequence of probabilities tends to zero, and in fact does so at a very fast rate. In this approach the normal approximation is no longer useful (it only tells us that (32) $\to 0$ as $n \to \infty$). The study of the limiting behavior of sequences such as (32) is called *large deviation theory* since, for large n, a deviation of T_n from $g(\theta)$ greater than some fixed value is very unlikely to occur. For an exposition of large deviation theory, see Bahadur (1971).

We would, of course, like to choose the approximation that comes closer to the true value. It seems plausible that for values of (30) not extremely close to 0 and for moderate sample sizes, (31) would tend to do better than that obtained from the sequence (32). Some numerical comparisons in the context of hypothesis testing can be found in Groeneboom and Oosterhoff (1982).

We conclude this section by extending some of the basic probability results for random variables to vectors of random variables. The definitions of convergence in probability and in law generalize very naturally as follows.

A sequence of random vectors $\underline{Y}_n = (Y_{1n}, \ldots, Y_{rn})$, $n = 1, 2, \ldots$ tends *in probability* toward a constant vector $\underline{c} = (c_1, \ldots, c_r)$ if $Y_{in} \xrightarrow{P} c_i$ for each $i = 1, \ldots, r$, and it converges *in law* (or *weakly*) to a random vector $\underline{Y}$ with cdf H if

$$H_n(\underline{a}) \to H(\underline{a}) \tag{33}$$

at all continuity points $\underline{a}$ of H, where

$$H_n(\underline{a}) = P[Y_{1n} \leqslant a_1, \ldots, Y_{rn} \leqslant a_r] \tag{34}$$

is the cdf of $\underline{Y}_n$.

Theorem 1.2 extends to the present case.

Theorem 1.6. *The sequence* $\{\underline{Y}_n\}$ *converges in law to* $\underline{Y}$ *if and only if* $E[f(\underline{Y}_n)] \to E[f(\underline{Y})]$ *for every bounded continuous real-valued* f.

[For a proof of this and Theorem 1.7, see Billingsley (1979, Th. 29.1).]

Weak convergence of $\underline{Y}_n$ to $\underline{Y}$ does not imply

$$P(\underline{Y}_n \in S) \to P(\underline{Y} \in S) \tag{35}$$

for all sets S for which these probabilities are defined since this is not even true for the set of S defined by

$$T_1 \leqslant a_1, \ldots, T_r \leqslant a_r$$

unless H is continuous at $\underline{a}$.

Theorem 1.7. *The sequence* $\{Y_n\}$ *converges in law to* $\underline{Y}$ *if and only if* (35) *holds for all sets* S *for which the probabilities in question are defined and for which the boundary of* S *has probability zero under the distribution of* $\underline{Y}$.

As in the one-dimensional case, the central limit theorem provides a basic tool for multivariate asymptotic theory.

Theorem 1.8 (Multivariate CLT). *Let* $\underline{X}_\nu = (X_{1\nu}, \ldots, X_{r\nu})$ *be iid with mean vector* $\underline{\xi} = (\xi_1, \ldots, \xi_r)$ *and covariance matrix* $\underline{\Sigma} = \|\sigma_{ij}\|$, *and let* $\overline{X}_{in} = (X_{i1} + \cdots + X_{in})/n$. *Then*

$$\left[\sqrt{n}(\overline{X}_{1n} - \xi_1), \ldots, \sqrt{n}(\overline{X}_{rn} - \xi_r)\right]$$

tends in law to the multivariate normal distribution with mean vector $\underline{0}$ *and covariance matrix* $\underline{\Sigma}$.

As a last result, we mention a generalization of Theorem 1.5 and Theorem 2.5.3.

Theorem 1.9. *Suppose that*

$$\left[\sqrt{n}\,(Y_{1n} - \theta_1), \ldots, \sqrt{n}\,(Y_{rn} - \theta_r)\right]$$

tends in law to the multivariate normal distribution with mean vector $\underline{0}$ *and covariance matrix* Σ, *and suppose that* $f_1, \ldots, f_r$ *are* r *real-valued functions of* $\underline{\theta} = (\theta_1, \ldots, \theta_r)$, *defined and continuously differentiable in a neighborhood* ω *of the parameter point* $\underline{\theta}$ *and such that the matrix* $B = \|\partial f_i / \partial \theta_j\|$ *of partial derivatives is nonsingular in* ω. *Then* (*Problem* 1.28)

$$\left[\sqrt{n}\,[f_1(\underline{Y}_n) - f_1(\underline{\theta})], \ldots, \sqrt{n}\,[f_r(\underline{Y}_n) - f_r(\underline{\theta})]\right]$$

tends in law to the multivariate normal distribution with mean vector $\underline{0}$ *and with covariance matrix* $B\Sigma B'$.

2. LARGE-SAMPLE COMPARISONS OF ESTIMATORS

In the present section, we shall show how the large-sample behavior of estimators discussed in Section 5.1 leads to a simple method for comparing different estimators. This approach will then be used in the following sections to compare $\bar{X}$ with some of its more robust competitors.

Let $\{\delta_n\}$ be a sequence of estimators of $g(\theta)$ based on n observations $X_1, \ldots, X_n$, such that*

$$\sqrt{n}\,[\delta_n - g(\theta)] \xrightarrow{\mathcal{L}} N(0, \tau^2) \tag{1}$$

and let δ_n' be an alternative sequence of estimators of $g(\theta)$. Suppose that when δ' is based on $n' = n'(n)$ observations, then also

$$\sqrt{n}\,[\delta_{n'}' - g(\theta)] \xrightarrow{\mathcal{L}} N(0, \tau^2). \tag{2}$$

We shall say that the *asymptotic relative efficiency* (ARE) of $\{\delta_n\}$ with respect to $\{\delta_n'\}$ is*

$$e_{\delta, \delta'} = \lim_{n \to \infty} \frac{n'(n)}{n} \tag{3}$$

*We suppress the dependence of τ^2 and e on θ.

provided this limit exists, and is independent of the particular sequence $n'(n)$ chosen to achieve (2).

Suppose, for example that $e = 1/2$. Then for large values of n, n' is approximately equal to $\frac{1}{2}n$. To obtain the same limit distribution, half as many observations are therefore required with δ' as with δ. It is then reasonable to say that δ' is twice as efficient as δ or that δ is half as efficient as δ'.

Theorem 2.1. *If*

$$\sqrt{n}\,[\delta_{in} - g(\theta)] \xrightarrow{\mathcal{L}} N(0, \tau_i^2), \qquad i = 1, 2, \tag{4}$$

then the ARE of $\{\delta_{2n}\}$ *with respect to* $\{\delta_{1n}\}$ *exists and is*

$$e_{2,1} = \tau_1^2/\tau_2^2. \tag{5}$$

Proof. Since

$$\sqrt{n}\,[\delta_{2n'} - g(\theta)] = \sqrt{\frac{n}{n'}}\sqrt{n'}\,[\delta_{2n'} - g(\theta)],$$

it follows from Theorem 1.4 that the left side has the same limit distribution $N(0, \tau_1^2)$ as $\sqrt{n}\,[\delta_{1n} - g(\theta)]$ if and only if $\lim[n/n'(n)]$ exists and

$$\tau_2^2 \lim \frac{n}{n'(n)} = \tau_1^2,$$

as was to be proved.

Example 2.1. If $X_1, \dots, X_n$ are iid according to $N(\theta, 1)$, it was found in Example 2.2.2 that the UMVU estimator of

$$p = P(X_1 \leq a) \tag{6}$$

is

$$\delta_{1n} = \Phi\left[\sqrt{\frac{n}{n-1}}\,(a - \bar{X})\right]. \tag{7}$$

Suppose now that we do not trust the assumption of normality; then we might instead of (7) prefer to use the nonparametric UMVU estimator derived in Section

2.4, namely,

$$\delta_{2n} = \frac{1}{n}(\text{No. of } X_i \leqq a). \tag{8}$$

What do we lose by using (8) instead of (7) if the X's are $N(\theta, 1)$ after all? Note that p is then given by

$$p = \Phi(a - \theta) \tag{9}$$

and that

$$\sqrt{n}(\delta_{1n} - p) \to N[0, \phi^2(a - \theta)]. \tag{10}$$

On the other hand, $n\delta_{2n}$ is the number of successes in n binomial trials with success probability p, so that

$$\sqrt{n}(\delta_{2n} - p) \to N(0, pq). \tag{11}$$

It thus follows from (5) that

$$e_{2,1} = \frac{\phi^2(a - \theta)}{\Phi(a - \theta)[1 - \Phi(a - \theta)]}. \tag{12}$$

At $a = \theta$ (when $p = 1/2$), this efficiency takes on the value $(1/2\pi)$: $(1/4) = 2/\pi \sim 0.637$. As $a - \theta \to \infty$, the efficiency tends to zero (Problem 2.2). It can be shown, in fact, that (12) is a decreasing function of $|a - \theta|$ (for a proof, see Sampford, 1953). The efficiency loss resulting from the use of δ_{2n} instead of δ_{1n} is therefore quite severe.

Example 2.2. If X has the binomial distribution $b(p, n)$, it was seen in Example 4.1.1 that the Bayes estimator of p corresponding to the beta prior $B(a, b)$ is

$$\delta_n(X) = (a + X)/(a + b + n).$$

Thus

$$\sqrt{n}[\delta_n(X) - p] = \sqrt{n}\left(\frac{X}{n} - p\right) + \frac{\sqrt{n}}{a + b + n}\left[a - (a + b)\frac{X}{n}\right]$$

and it follows from Theorem 1.4 that $\sqrt{n}[\delta_n(X) - p]$ has the same limit distribution as $\sqrt{n}(X/n - p)$, namely, the normal distribution $N(0, pq)$. The ARE of the Bayes estimator relative to X/n is therefore 1, irrespective of the choice of a and b. This is a special case of a very general result concerning the large-sample behavior of Bayes estimators to be proved in Section 6.7.

Consider next the minimax estimator δ_n of p given by (4.2.10) which corresponds to the sequence of beta priors with $a = b = \sqrt{n}/2$. Then

$$\sqrt{n}(\delta_n - p) = \sqrt{n}\left(\frac{X}{n} - p\right) + \frac{\sqrt{n}}{1 + \sqrt{n}}\left(\frac{1}{2} - \frac{X}{n}\right),$$

and the limit distribution of $\sqrt{n}(\delta_n - p)$ is therefore $N(\frac{1}{2} - p, pq)$. It is seen that δ_n has the same asymptotic variance as X/n but that for $p \neq \frac{1}{2}$ it is asymptotically biased. The ARE of δ_n relative to X/n does not exist except in the case $p = \frac{1}{2}$ when it is 1.

It is desirable to generalize the definition of the ARE and Theorem 2.1 also to cover cases in which the normalizing factor is not $\sqrt{n}$ and the common limit distribution not normal. To avoid the possibility of non-uniqueness of the ARE (Problem 2.6), we impose the following two restrictions.

1. The normalizing constant $\sqrt{n}$ in (1) and (2) is replaced by n^α for some $\alpha > 0$.
2. The common limit distribution $N(0, \tau^2)$ in (1) and (2) is replaced by a distribution H which has as its support an interval $-\infty \leqslant A < B \leqslant \infty$ and a cdf which is continuous and strictly increasing on (A, B).

If restrictions 1 and 2 hold, and if the limit (3) exists and is independent of the particular sequence $n'(n)$ chosen, we shall continue to call it the ARE of $\{\delta_n\}$ with respect to $\{\delta_n'\}$.

Theorem 2.2. *Let* $\{\delta_{in}\}$ *be two sequences of estimators of* $g(\theta)$ *such that*

$$n^\alpha[\delta_{in} - g(\theta)] \xrightarrow{\mathcal{L}} \tau_i T, \qquad \alpha > 0, \tau_i > 0, \qquad i = 1, 2; \tag{13}$$

where the distribution H *of* T *satisfies condition* 2. *Then the ARE of* $\{\delta_{2n}\}$ *with respect to* $\{\delta_{1n}\}$ *exists and is*

$$e_{2,1} = (\tau_1/\tau_2)^{1/\alpha}. \tag{14}$$

The proof is exactly analogous to that of Theorem 2.1.

Example 2.3. In Example 1.5, the errors of the three estimators δ_{in} $(i = 1, 2, 3)$ of θ^2 when normalized by $\sqrt{n}$ were seen to have the common limit distribution $N(0, 4\sigma^2\theta^2)$ when $\theta \neq 0$. The associated AREs $e_{i,j}$ $(i \neq j)$ are therefore all equal to 1. It is interesting to note that to the accuracy provided by the ARE, no loss of efficiency results when the UMVU estimator for unknown σ^2 is compared with that for known σ^2.

When $\theta = 0$, Theorem 2.2 with $\alpha = 1$ and H the shifted χ^2-distribution with 1 degree of freedom, shows that the ARE of $\{\delta_{1n}\}$ with respect to $\{\delta_{2n}\}$ continues to be 1. On the other hand, the ARE of $\{\delta_{2n}\}$ to $\{\delta_{3n}\}$ does not exist since their limit distributions cannot be matched by adjusting the sample sizes.

Theorems 2.1 and 2.2 still leave open the possibility that different values of the ARE might be obtained by choosing normalizing constants different from $\sqrt{n}$ or n^α with the possible consequence of this resulting in a different limit distribution from $N(0, \tau^2)$ or H. That this cannot occur under the assumptions of these theorems follows from the following lemma.

Lemma 2.1. *Let* $k_n[\delta_n - g(\theta)]$ *and* $k'_n[\delta_n - g(\theta)]$ *tend in law, respectively to* H *and* H′, *both of which satisfy condition* 2. *Then there exists a constant* c *such that*

$$H'(x) = H(x/c) \quad \textit{for all } x, \quad \textit{and} \quad k'_n/k_n \to c. \tag{15}$$

Proof. Without loss of generality, suppose that k'_n/k_n tends to some limit c. (If not, apply the following argument to a convergent subsequence.) If $0 < c < \infty$, it follows from Problem 1.12 that $H'(x) = H(x/c)$ for all x. The possibilities $c = 0$ and $c = \infty$ are ruled out by the fact that then

$$k'_n[\delta_n - g(\theta)] = \frac{k'_n}{k_n} \cdot k_n[\delta_n - g(\theta)] \xrightarrow{P} 0 \quad \text{or } \infty,$$

respectively.

Under the assumptions of Theorem 2.2, different sequences of normalizing constants thus have to satisfy $k_{in}/n_i^\alpha \to c_i$ and the limit distribution of $k_{in}[\delta_{in} - g(\theta)]$ is that of $c_i \tau_i T$. The argument of Theorem 2.1 then shows that the ARE is still given by (14).

The comparisons considered so far were based on the asymptotic distributions of the estimators. An alternative approach, closely related to that of Section 2.5, is based on the limit of the normalized risk. In analogy with (1), (2), and the definition of ARE preceding Theorem 2.2, suppose that $\{\delta_n\}$ and $\{\delta'_n\}$ are two sequences of estimators of $g(\theta)$ whose risk functions are such that

$$n^r R(\theta, \delta_n) \quad \text{and} \quad n^r R\big(\theta, \delta'_{n'(n)}\big)$$

tend to a common limit τ^2. We shall then say that the *limiting risk efficiency* (LRE) of $\{\delta_n\}$ with respect to $\{\delta'_n\}$ is $\lim[n'(n)/n]$ provided this limit exists as $n \to \infty$ and is independent of the particular sequence $n'(n)$ chosen.

Theorem 2.3. *If*

$$\lim_{n \to \infty} n^r R(\theta, \delta_{in}) = \tau_i^2 > 0 \qquad (i = 1, 2), \tag{16}$$

the LRE of $\{\delta_{2n}\}$ *with respect to* $\{\delta_{1n}\}$ *is* $(\tau_1^2/\tau_2^2)^{1/r}$.

The proof is analogous to that of Theorem 2.1.

The LRE will, of course, depend on the loss function. For the rest of this section, we shall take the loss to be squared error.

When the ARE exists, typically the LRE will also and the two will agree. (For a discussion and suggestions for counterexamples, see Section 2.5.)

Example 2.2. (*Continued*). Consider once more the comparison of $\delta_n'(X) = X/n$ with the minimax estimator δ_n of binomial p, which was discussed at the end of Example 2.2. By (4.2.11)

$$nE(\delta_n - p)^2 = \frac{n}{4(1+\sqrt{n})^2} \to \frac{1}{4}$$

while

$$nE(\delta_n' - p)^2 = pq$$

so that the LRE of $\{\delta_n\}$ with respect to $\{\delta_n'\}$ exists and is $4pq$. This is $\leqslant 1$ for all p, and $= 1$ when $p = \frac{1}{2}$. In contrast, the ARE was seen not to exist when $p \neq \frac{1}{2}$. (For $p = \frac{1}{2}$ it exists and agrees with the LRE.)

Example 2.3. (*Continued*). The AREs of the three estimators $\{\delta_{in}\}$ in Example 2.3, with respect to each other, were seen to equal 1 when $\theta \neq 0$, and the LREs (for squared error loss) agree with these values (Problem 2.10). When $\theta = 0$, the ARE of $\{\delta_{1n}\}$ with respect to $\{\delta_{2n}\}$ is also 1 and the LRE again agrees. On the other hand, the ARE of $\{\delta_{3n}\}$ with respect to $\{\delta_{1n}\}$ does not exist. To determine the LRE in this case, recall (1.25). For $\theta = 0$, this yields

$$n^2E(\delta_{3n} - \theta^2)^2 = n^2E(\bar{X}^4) \to 3\sigma^4 \qquad (c = 0)$$

and

$$n^2E(\delta_{1n} - \theta^2)^2 = n^2E\left(\bar{X}^2 - \frac{\sigma^2}{n}\right)^2 \to 2\sigma^4 \qquad (c = 1).$$

Hence when $\theta = 0$, the LRE of the MLE $(c = 0)$ to the UMVU estimator $(c = 1)$ of θ^2 is

$$\left(\frac{2\sigma^4}{3\sigma^4}\right)^{1/2} = \sqrt{\frac{2}{3}} \sim 0.816. \tag{17}$$

When $\theta = 0$, the UMVU estimator is therefore more efficient than the MLE in this sense. For $\theta \neq 0$, however, both the ARE and the LRE are 1 and thus provide no indication as to which of the two estimators is to be preferred.

Consider now quite generally two sequences $\{\delta_{in}\}$ with risk functions $R(\theta, \delta_{in}) = R_{in}$ $(i = 1, 2)$ and suppose that their LRE is 1. Recall that the definition of the LRE concerns the behavior of the ratio $n'(n)/n$ when n^rR_{1n} and $n^rR_{2n'}$ tend to the same limit so that R_{1n} and $R_{2n'}$ agree up to terms of order $1/n^r$. To obtain a finer distinction between the estimators in that case, let us carry the approximation one step further and consider the difference $d(n) = n'(n) - n$ when n' is chosen so that R_{1n} and $R_{2n'}$ agree up to terms of order $1/n^{r+1}$. If $d = \lim d(n)$ exists and is independent of the particular sequence $n'(n)$ chosen, it will be called the *limiting risk deficiency* (LRD) of $\{\delta_{2n}\}$ with respect to $\{\delta_{1n}\}$. Suppose, for example, that $d = 3$. Then for large values of n, n' is approximately equal to $n + 3$. To obtain the same risk (to terms of order $1/n^{r+1}$) requires three more observations with δ_2 than with δ_1.

To formulate a result analogous to Theorem 2.3, note that (16) is equivalent to

$$R_{in} = \frac{\tau_i^2}{n^r} + o\left(\frac{1}{n^r}\right), \qquad \tau_i^2 > 0. \tag{18}$$

If the LRE of $\{\delta_{2n}\}$ with respect to $\{\delta_{1n}\}$ is 1, it follows from Theorem 2.3 that $\tau_1^2 = \tau_2^2 = a > 0$. Expanding R_{in} one step further, we shall assume that

$$R_{in} = \frac{a}{n^r} + \frac{b_i}{n^{r+1}} + o\left(\frac{1}{n^{r+1}}\right), \qquad a > 0, \qquad i = 1, 2. \tag{19}$$

For the following, it is useful to note that $n'(n)/n \to 1$ implies that for any k

$$o\left(\frac{1}{n'^k}\right) = o\left(\frac{1}{n^k}\right). \tag{20}$$

Theorem 2.4. *If* (19) *holds, then the LRD of* $\{\delta_{2n}\}$ *with respect to* $\{\delta_{1n}\}$ *exists and is*

$$d = \frac{b_2 - b_1}{ar}. \tag{21}$$

Proof. Equating R_{1n} and $R_{2n'}$ up to terms of order $1/n^{r+1}$ gives

$$\frac{1}{n^r}\left[a + \frac{b_1 + o(1)}{n}\right] = \frac{1}{n'^r}\left[a + \frac{b_2 + o(1)}{n'}\right]$$

and hence

$$\frac{n'}{n} = \left[1 + \frac{b_2 + o(1)}{an'}\right]^{1/r}\left[1 + \frac{b_1 + o(1)}{an}\right]^{-1/r}.$$

Substituting $n' = n + d(n)$ and expanding the factors on the right side by Taylor's theorem leads to

$$1 + \frac{d(n)}{n} = 1 + \frac{b_2}{arn'} - \frac{b_1}{arn} + o\left(\frac{1}{n}\right)$$

and hence to

$$d(n) = \frac{b_2}{ar} \cdot \frac{n}{n'} - \frac{b_1}{ar} + o(1) \to \frac{b_2 - b_1}{ar}$$

as was to be proved.

Example 2.3. (*Completed*). Let us now return to the normal one-sample problem which motivated this discussion. The risk of the estimator $\bar{X}^2 - c\sigma^2/n$ of θ^2 is given by (17). For $\theta \neq 0$, this satisfies the conditions of the theorem with $r = 1$ and $a = 4\theta^2\sigma^2$. For the UMVU estimator, $c = 1$ and $b_1 = 2\sigma^4$; for the MLE, $c = 0$ and $b_2 = 3\sigma^4$. Hence the limiting risk deficiency of $\bar{X}^2$ relative to $\bar{X}^2 - \sigma^2/n$ is

$$\frac{b_2 - b_1}{ra} = \frac{\sigma^2}{4\theta^2}.$$

This gives an idea of the number of additional observations required by the MLE to obtain the same expected squared error (up to terms of order $1/n^2$) as the UMVU estimator. As $\theta \to 0$, the deficiency approaches ∞. This is consistent with the fact that when $\theta = 0$, the limiting risk efficiency is less than 1.

That in this example the expected squared error of the MLE has a larger $1/n^2$ term than the UMVU estimator appears to contradict the frequently asserted opinion that "MLEs are second-order efficient," that is, that among all suitably regular consistent estimator sequences they minimize not only the $1/n$ term but also the $1/n^2$ term in the variance (and other measures of accuracy). However, the disagreement turns out to be one of terminology. As in the above example, the MLE frequently has a bias of order $1/n$; the MLE in the statement in quotes refers not to the estimator that maximizes the likelihood but to this estimator after it has been corrected for bias.

The limiting deficiency (21) need not be an integer but the difference of two samples sizes must be integral. If, for example, the limiting deficiency is 2.3, then two additional observations are not enough to equalize the limiting risks and three are too many. The difficulty can be avoided by continualizing the sample size through randomization between successive sample size. Thus an addition of two observations with probability .7 and three observations with probability .3 leads to an expected number of additional observations exactly equal to 2.3. (For details see Hodges and Lehmann, 1970).

3. THE MEDIAN AS AN ESTIMATOR OF LOCATION

Let us return once more to the simple case of n measurements $X_1, \dots, X_n$ of some unknown quantity θ. We shall assume that $X_1, \dots, X_n$ are iid according to some distribution

$$P(X_i \leq x) = F(x - \theta). \tag{1}$$

If F is normal, it was shown in Section 2.2 that $\bar{X}$ is UMVU. The same property was seen in Section 2.4 to hold for $\bar{X}$ when F is an unknown distribution which is only assumed to have a probability density and expectation zero, but no longer when F in addition is assumed to be symmetric (Problem 2.4.4). The reason is that the class of unbiased estimators of θ is vastly larger under the symmetry restriction than without it (Problem 3.1).

Consider now the problem of estimating the center of symmetry θ on the basis of a sample $X_1, \dots, X_n$ from (1). That $\bar{X}$ may be a very poor estimator in this situation is seen from the Cauchy case in which the density f of F is given by

$$f(x) = \frac{1}{\pi} \frac{1}{1 + x^2}. \tag{2}$$

The distribution of $\bar{X}$ is then the same as that of X_1 (Problem 1.1.8); that is, the average of n observations is as variable as a single observation. Even more extreme examples exist (see Brown and Tukey, 1946), in which a single observation is less variable than an average of n. [See also Stigler (1980).]

Why is it that $\bar{X}$ works so well with the normal distribution and so poorly with the Cauchy, although both have smooth bell-shaped densities? An important difference between the two distributions is their tail behavior. The density (2) tends to zero at the rate $1/x^2$ as $x \to \infty$; the normal density tends at the much faster rate $e^{-x^2/2}$. One result of the heavy tails of the Cauchy distribution is that $E(X_i^2)$ and even $E|X_i|$ are infinite. The great variability of $\bar{X}$ thus seems due to the higher probability of relatively large observations. This suggests the possibility of obtaining a more stable estimator by discarding the more extreme observations. An estimator which is particularly insensitive to the tail behavior of F is the median of the X's, which for odd n is the central observation and for even n the average of the two central observations. The remainder of this section is concerned with properties of the median as an estimator of θ.

For simplicity, let us suppose at first that n is odd, say

$$n = 2m - 1 \tag{3}$$

so that the median $\tilde{X}_n$ is equal to $X_{(m)}$, which is an unbiased estimator of θ (Problem 3.1). In order to study its variability, let us begin by finding its distribution.

Theorem 3.1. *Let* $X_1, \ldots, X_n$ *be iid with distribution* F *and density* f, *and let the ordered* X's *be denoted by* $X_{(1)} < \cdots < X_{(n)}$. *Then the density of* $X_{(k)}$ *is given by*

$$n\binom{n-1}{k-1}F^{k-1}(x)[1 - F(x)]^{n-k}f(x). \tag{4}$$

Intuitively, this formula is quite easy to see. The first factor is the number of ways of choosing one of the n variables to be the kth smallest and dividing the remaining $n - 1$ into the $k - 1$ that are to be smaller and $n - k$ that are to be larger than $X_{(k)}$. The factor $F^{k-1}(x)$ is the probability that the $k - 1$ smallest variables are $\leqq x$, and the factor $[1 - F(x)]^{n-k}$ the probability that the $n - k$ largest are larger than x. Finally, $f(x)\,\Delta$ is the probability that $X_{(k)}$ will fall into a small interval of length Δ around x.

Proof. The cdf of $X_{(k)}$ is given by

$$P[X_{(k)} \leqslant x] = \sum_{j=k}^{n}\binom{n}{j}[F(x)]^j[1 - F(x)]^{n-j}$$

and the density then follows easily from the fact that

$$\frac{d}{dp}\sum_{j=k}^{n}\binom{n}{j}p^j(1-p)^{n-j} = n\binom{n-1}{k-1}p^{k-1}(1-p)^{n-k}.$$

Let us now ask whether the median $\tilde{X}_n$, unlike the mean, has finite second moment in the Cauchy case. When $n = 1$, of course, this cannot be the case. From (4) it follows that $E(\tilde{X}_n^2) < \infty$ when $n \geqq 5$ (Problem 3.2), while $E(\tilde{X}_n^2) = \infty$ for $n < 5$.

The Cauchy distribution, though frequently used as a test case, is often not representative of tail behavior likely to occur in practice. Here we are interested primarily in distributions with tails heavier than the normal but not as extreme as the Cauchy.

To obtain a more detailed comparison between median and mean, we shall next determine the asymptotic distribution of $\tilde{X}_n$. As in Theorem 3.1, we do not require F to be symmetric.

Theorem 3.2. *Let* $X_1, \ldots, X_n$ *be iid with distribution* $F(x - \theta)$. *Suppose that* $F(0) = 1/2$ *and that at zero* F *has a density* $f(0) > 0$. *Then*

$$\sqrt{n}\,(\tilde{X}_n - \theta) \xrightarrow{\mathcal{L}} N\left(0, \frac{1}{4f^2(0)}\right). \tag{5}$$

Note that the assumptions imply that θ is the unique median of the distribution of the X's.

Proof. We begin with the case (3) that n is odd and note further that the distribution of $\tilde{X}_n - \theta$ does not depend on θ, so that

$$P_\theta\left[\sqrt{n}\,(\tilde{X}_n - \theta) \leqq a\right] = P_0\left[\sqrt{n}\,\tilde{X}_n \leqq a\right] = P_0\left[X_{(m)} \leqq \frac{a}{\sqrt{n}}\right]. \tag{6}$$

Let S_n be the number of X's that exceed $a/\sqrt{n}$. Then

$$X_{(m)} \leqq \frac{a}{\sqrt{n}} \quad \text{if and only if} \quad S_n \leqq m - 1 = \frac{n-1}{2}.$$

Now S_n has the binomial distribution $b(p_n, n)$ with $p_n = 1 - F(a/\sqrt{n})$. The probability (6) is therefore equal to

$$P_0\left[S_n \leqq \frac{n-1}{2}\right] = P_0\left[\frac{S_n - np_n}{\sqrt{np_nq_n}} \leqq \frac{\frac{1}{2}(n-1) - np_n}{\sqrt{np_nq_n}}\right]. \tag{7}$$

If p_n were independent of n, the asymptotic normality of the binomial distribution could now be used to pass to the limit. Because of the dependence of p_n on n, this step requires the following bound on the error in the central limit theorem.*

The Berry–Esseen Theorem. *If* $X_1, \ldots, X_n$ *are iid with distribution* F *and if* $S_n = X_1 + \cdots + X_n$, *then there exists a constant* c (*independent of* F) *such that for all* x

$$\left| P\left[\frac{S_n - E(S_n)}{\sqrt{\operatorname{var} S_n}} \leqq x\right] - \Phi(x)\right| \leqq \frac{c}{\sqrt{n}} \frac{E|X_1 - E(X_1)|^3}{(\operatorname{var} X_1)^{3/2}}, \tag{8}$$

for all F *with finite third moment.*

A proof of this result can be found in standard books on probability theory.

*An alternative easy proof is obtained by applying the Lindeberg central limit theorem.

The important aspect of the theorem in the present context is that the constant c is independent of the distribution F of the X's. If we let X_i be 1 or 0 with probability p_n and $1 - p_n$, respectively, the distribution of S_n is just the binomial distribution under consideration. Clearly $E|X_1 - E(X_1)|^3 \leqq 1$ while $\text{var}(X_1) = p_n(1 - p_n) \to 1/4$ since $p_n = 1 - F(a/\sqrt{n}) \to 1 - F(0) = 1/2$. Thus the right-hand side of (8) applied to the probability (7) tends to zero, so that the difference

$$P_0\left(S_n \leqq \frac{n-1}{2}\right) - \Phi\left[\frac{\frac{1}{2}(n-1) - np_n}{\sqrt{np_nq_n}}\right] \to 0 \quad \text{as} \quad n \to \infty.$$

The argument of Φ can be written as

$$x_n = \frac{\sqrt{n}\left(\frac{1}{2} - p_n\right) - 1/(2\sqrt{n})}{\sqrt{p_nq_n}}$$

so that

$$\lim x_n = 2\lim\left[\sqrt{n}\left(\tfrac{1}{2} - p_n\right)\right]$$

$$= 2a\lim\frac{F(a/\sqrt{n}) - F(0)}{a/\sqrt{n}} = 2af(0).$$

We have thus shown that

$$P\left[\sqrt{n}\left(X_{(m)} - \theta\right) \leqq a\right] \to \Phi[2f(0)a], \tag{9}$$

and this is equivalent to (5) (Problem 3.4).

It remains to consider the case that n is even. If $n = 2m$, the median $\tilde{X}_n$ is defined to be

$$\tilde{X}_n = \tfrac{1}{2}\left[X_{(m)} + X_{(m+1)}\right]. \tag{10}$$

The limit result (9) was proved for the case that $n = 2m - 1$ so that $m = \frac{1}{2}(n + 1)$. However, the proof applies equally well for any sequence (n, m_n) for which $n \to \infty$ and $m_n/n = \frac{1}{2} + R_n$ where $\sqrt{n}\,R_n \to 0$ (Problem 3.5). It follows that in the present case both

$$P\left[\sqrt{n}\left(X_{(m)} - \theta\right) \leqq a\right] \quad \text{and} \quad P\left[\sqrt{n}\left(X_{(m+1)} - \theta\right) \leqq a\right]$$

tend to the right-hand side of (9). Since $P(\sqrt{n}(\tilde{X}_n - \theta) \leq a)$ lies between these two probabilities (5) also holds for even values of n.

To get an idea of the advantages and disadvantages of the median relative to the mean, we shall now determine the asymptotic relative efficiency of $\tilde{X}_n$ to $\bar{X}_n$ for a variety of distributions. From Theorem 3.2 and the fact that

$$(11) \qquad \sqrt{n}(\bar{X}_n - \theta) \to N(0, \sigma^2)$$

where $\sigma^2 = \text{var}(X_1)$, it follows that

$$(12) \qquad e_{\tilde{X}, \bar{X}}(F) = 4f^2(0)\sigma^2.$$

It is interesting to note that this efficiency is not affected by scale changes in F (Problem 3.8).

The distribution of greatest interest is, of course, the normal. Putting $f(x) = \phi(x)$, we have $\sigma^2 = 1$ and $f(0) = 1/\sqrt{2\pi}$, so that $e_{\tilde{X}, \bar{X}}(\text{normal}) = 4/2\pi = 2/\pi \sim 0.637$. A distribution whose density looks very much like the normal density is the logistic distribution $L(0, 1)$ with density and cdf

$$(13) \qquad f(x) = \frac{e^{-x}}{(1 + e^{-x})^2} \quad \text{and} \quad F(x) = \frac{1}{1 + e^{-x}}.$$

It is easily checked (Problem 3.9) that for this distribution

$$e_{\tilde{X}, \bar{X}}(\text{logistic}) = \frac{\pi^2}{12} \sim 0.82.$$

The somewhat higher relative efficiency of the median in this case reflects in part the fact that the tail of the logistic distribution is somewhat heavier than that of the normal distribution.

A family of symmetric distributions with greatly varying heaviness in the tails is the family of t-distributions with ν degrees of freedom, whose density is

$$(14) \qquad f_\nu(x) = \frac{\Gamma[(\nu + 1)/2]}{\sqrt{\nu\pi}\,\Gamma(\nu/2)} \frac{1}{(1 + x^2/\nu)^{(\nu+1)/2}}.$$

For $\nu = 1$, this reduces to the Cauchy distribution for which $E|X| = \infty$; for $\nu = 2$ the expectation exists but the variance is still infinite. For $\nu \geq 3$, the variance is finite and is given by (Problem 3.7)

$$(15) \qquad \sigma_\nu^2 = \frac{\nu}{\nu - 2}.$$

Table 3.1. Efficiency of $\tilde{X}$ to $\bar{X}$ for t-Distribution with ν $d.f.$

ν	3	4	5	8	∞
e	1.62	1.12	0.96	0.80	0.64

As $\nu \to \infty$, the distribution tends to the normal. From (14) and (15), it is easy to calculate the efficiency (12) (Problem 3.9); Table 3.1 gives a few values ($e = \infty$ for $\nu = 1, 2$).

Consider as another family of distributions, the contaminated normal family suggested by Tukey (1960) as a model for observations, which usually follow a normal distribution but where occasionally something goes wrong with the experiment or its recording so that the resulting observation is a gross error. Under the *Tukey model* $T(\varepsilon, \tau)$, F takes the form

$$F(x) = (1 - \varepsilon)\Phi(x) + \varepsilon\Phi\left(\frac{x}{\tau}\right), \tag{16}$$

that is, in the gross error cases, the observations are assumed to be normally distributed with the same mean θ but a different (larger) variance τ^2.* For such mixtures, F

$$f(0) = \frac{1}{\sqrt{2\pi}}\left(1 - \varepsilon + \frac{\varepsilon}{\tau}\right)$$

and

$$\sigma^2 = (1 - \varepsilon) + \varepsilon\tau^2.$$

The values of the efficiency $e_{\tilde{X},\bar{X}}$ for a number of different combinations of ε and τ are shown in Table 3.2. Again, it is seen that $e_{\tilde{X},\bar{X}}$ increases as the

Table 3.2. Efficiency of $\tilde{X}$ to $\bar{X}$ for Tukey Model $T(\varepsilon, \tau)$.

τ \ ε	.01	.05	.1
2	.65	.70	.75
3	.68	.83	1.00
4	.72	1.03	1.36

*As has been pointed out by Stigler (1973b) such models for heavy-tailed distributions had already been proposed much earlier in forgotten work by Newcomb (1882, 1886).

tail of the distribution gets heavier.

In this discussion, the term *heavy tails* should not be interpreted too literally. In particular, it does not necessarily reflect the rate at which $f(x) \to 0$ as $x \to \infty$. To see this, consider a Cauchy distribution truncated at $\pm A$, so that the density is $f(x)/P(|X| \leq A)$ when $|x| \leq A$ and 0 otherwise, with f given by (2). Then by (12), $e_{\tilde{X}, \bar{X}} \to \infty$ as $A \to \infty$, although for any finite A, no matter how large, f has no tail at all. In general, it is easy to see how changes in f affect the efficiency (12) since this depends only on the value of the density at zero and on the variance. Thus, for example, if the density is fixed in the center so that in particular $f(0)$ is fixed, it will be increased by any change in the tails that increases the variance; such a change, of course, need not occur in the extreme tails of a distribution. In addition, the efficiency depends on $f(0)$.

The comparisons of this section have all been asymptotic, and this raises the question of their relevance for moderate sample sizes. Some indications are obtained by comparing the actual variance of $\sqrt{n}(\tilde{X} - \theta)$ with the asymptotic variance $1/4f^2(0)$. Table 3.3 provides such a comparison for a number of distributions when $n = 20$.

It is seen that the asymptotic variance gives a reasonable approximation to the actual variance except in the Cauchy case when ($n = 40$ the variance in that case is 2.40, so that by then the approximation is much better) and the double exponential distribution with density

$$f(x) = \tfrac{1}{2}e^{-|x|}. \tag{17}$$

The slow convergence in the latter case presumably is the result of the lack of smoothness of f at zero.

A universally preferred choice between $\tilde{X}$ and $\bar{X}$ is not possible since the efficiency $e_{\tilde{X}, \bar{X}}(F)$ given by (12) can be as low as zero [when $f(0) = 0$] and as high as infinity (when $\sigma^2 = \infty$). The latter possibility arises for distributions with sufficiently heavy tails, and high values of e (favoring $\tilde{X}$) can therefore not be ruled out in practice. On the other hand, distributions with $f(0) = 0$ (corresponding to a dip of the density in the center) are less

Table 3.3. Variance and Asymptotic Variance of $\sqrt{n}(\tilde{X} - \theta)$.*

	$N(0,1)$	$T(.01,3)$	$T(.05,3)$	$(.1,3)$	$L(0,1)$	t_3	$C(0,1)$	$DE(0,1)$
$n = 20$	1.47	1.49	1.58	1.70	3.82	1.82	2.90	1.33
$n = \infty$	1.57	1.59	1.68	1.80	4.00	1.85	2.47	1.00

*Variance of t_3 from Andrews et al. (1972). The remaining figures are from Gastwirth and Cohen (1970).

common. This raises the question of how low the efficiency $e_{\tilde{X},\bar{X}}(F)$ can get when f is restricted to be unimodal.

Theorem 3.3. *For unimodal symmetric densities or more generally densities symmetric about zero and satisfying*

$$f(x) \leqq f(0) \quad \text{for all} \quad x, \tag{18}$$

the efficiency of $\tilde{X}$ to $\bar{X}$ satisfies

$$e_{\tilde{X},\bar{X}}(F) \geqq 1/3. \tag{19}$$

The lower bound 1/3 is attained for the uniform distributions and no other.

Proof. Since the efficiency is independent of scale, we shall assume without loss of generality that $f(0)$ is fixed, say $f(0) = 1$. The problem then becomes that of minimizing

$$\sigma^2 = E(X^2) = \int x^2 f(x)\, dx \tag{20}$$

subject to

$$0 \leqq f(x) \leqq 1, \qquad f(0) = 1$$

and

$$\int f(x)\, dx = 1. \tag{21}$$

We use the method of undetermined multipliers. To eliminate the side condition (21), we minimize instead of (20)

$$\int (x^2 - a^2) f(x)\, dx \tag{22}$$

and then determine a so as to satisfy (21). Now (22) is minimized by making $f(x)$ as large as possible when $x^2 < a^2$ and as small as possible when $x^2 > a^2$, that is, by putting

$$f(x) = \begin{cases} 1 & \text{when } |x| < a \\ 0 & \text{when } |x| > a. \end{cases}$$

The side condition (21) determines a to be $1/2$. The minimizing distribution is therefore the uniform distribution, for which $f(0) = 1$, $\sigma^2 = 1/12$ and hence $e_{\tilde{X},\bar{X}}(F) = 1/3$ as was to be proved.

We have so far restricted attention to symmetric distributions F. If the assumption of symmetry is dropped, $\bar{X}$ and $\tilde{X}$ no longer estimate the same quantity. They then no longer seem comparable, and we shall therefore not consider this possibility here any further. However, for another point of view, see the next section.

On the basis of the comparisons made in this section, a choice between mean and median is difficult. As we have seen, $\bar{X}$ is highly sensitive to heavy tails. [For the distribution $T(.1, 3)$, for example, the variance of $\sqrt{n}(\bar{X} - \theta)$ is 1.8, nearly twice what it is for the uncontaminated normal distribution.] By discarding all but the central observation or observations, the median provides much better protection for distributions like the Cauchy or t_2. However, for most practical situations in which one would like a safeguard against moderately heavy tails but expects something close to normal, the efficiency loss represented by $e_{\tilde{X}, \bar{X}} = 0.637$ is just too large, since it means that in the normal case the median would require about $n = 157$ observations to achieve the accuracy that the mean achieves with $n = 100$.

Fortunately, the choice is not limited to the mean and the median. In the next section, we will consider some alternatives that provide reasonable safeguards against gross errors and still give good efficiency even in the normal case.

4. TRIMMED MEANS

In the preceding section the mean and the median were compared as estimators of the center θ of a symmmetric distribution. The median was introduced because of its obvious insensitivity to outlying observations, but it turned out that it went too far in discarding observations. These considerations suggest as a natural compromise between mean and median, to discard only a certain number of the observations, say the k largest and k smallest, and to use as estimator the mean of the remaining observations, a so-called *trimmed mean*. If $k/n = \alpha < \frac{1}{2}$, we shall denote the resulting estimator by $\bar{X}_\alpha$, which is therefore given by

$$\bar{X}_\alpha = \frac{1}{n - 2k}\left[X_{(k+1)} + \cdots + X_{(n-k)}\right]. \tag{1}$$

More generally, $\bar{X}_\alpha$ is defined for any α by (1) with $k = [n\alpha]$ the largest integer $\leq n\alpha$. The mean and the median are the two extreme cases of symmetric trimming in which either no observation is discarded or all except the one or two observations in the center of the sample are. To determine the efficiency of a trimmed mean (for fixed α) relative to the mean and median, we require its asymptotic distribution, which is given by the following theorem.

Theorem 4.1. *Let* F *be symmetric about zero and suppose there exists a constant* $0 < c \leqq \infty$ *such that* $F(-c) = 0$, $F(c) = 1$ *and that* F *possesses a density* f *which is continuous and positive for all* $-c < x < c$. *If* $X_1, \ldots, X_n$ *are iid according to* $F(x - \theta)$ *then for any* $0 < \alpha < 1/2$,

$$\sqrt{n}\,(\bar{X}_\alpha - \theta) \to N(0, \sigma_\alpha^2)$$

where

$$\sigma_\alpha^2 = \frac{2}{(1-2\alpha)^2}\left[\int_0^{\xi(1-\alpha)} t^2 f(t)\,dt + \alpha\xi^2(1-\alpha)\right]. \tag{2}$$

Here $\xi(\alpha)$ *is the (unique) value for which*

$$F[\xi(\alpha)] = \alpha. \tag{3}$$

We shall not prove this theorem here. A sketch of a proof and references to other methods of proof are given by Bickel (1965) (see also Theorem 5.1 below). The asymptotic behavior of $\bar{X}_\alpha$ under more general assumptions is discussed by Stigler (1973c).

On the basis of Theorem 4.1 it is now possible to calculate the asymptotic relative efficiency of $\bar{X}_\alpha$ to $\bar{X}$ given by

$$e_{\bar{X}_\alpha, \bar{X}}(F) = \sigma^2/\sigma_\alpha^2. \tag{4}$$

Table 4.1 shows this efficiency for the case that F is the t-distribution with ν degrees of freedom ($\nu = \infty$ corresponds to the normal distribution) and for several values of α; for $\alpha = 0.5$ the estimator $\bar{X}_\alpha$ is the median which was discussed in the preceding section but is included here to facilitate the comparison.

This table provides impressive support for the idea that a moderate amount of trimming can provide much better protection than $\bar{X}$ against

Table 4.1. ARE of $\bar{X}_\alpha$ to $\bar{X}$ for t-Distributions with ν $d.f.$*

ν \ α	.05	.125	.25	.375	.5
3	1.70	1.91	1.97	1.85	1.62
5	1.20	1.24	1.21	1.10	.96
∞	.99	.94	.84	.74	.64

*Adapted from Siddiqui and Raghunandanan (1967).

fairly heavy tails, as represented for example by t_3, while at the same time giving up little in the normal case. The table suggests that the efficiency is a fairly flat function of α in the neighborhood of the optimal value and that, for example, $\alpha = 0.125$ would do quite well for both t_3 and t_5 at little efficiency loss at the normal.

As a second example, consider $e_{\bar{X}_\alpha, \bar{X}}$ for the Tukey model $T(\varepsilon, 3)$ with $\tau = 3$ and varying ε. The results, shown in Table 4.2, are quite similar to those for the t-family and continue to support a trimmed mean with $\alpha \sim 0.125$.

Despite the great range of tail behavior of the distributions considered in Tables 4.1 and 4.2, it is seen that the efficiencies $e_{\bar{X}_\alpha, \bar{X}}$ for $\alpha = 0.1$ or $\alpha = 0.125$ are uniformly high, and this raises the question of how low this efficiency can fall under unrestricted variation of F. Subject to the conditions of Theorem 4.1, the answer is given by the following theorem.

Theorem 4.2. *The efficiency* $e_{\bar{X}_\alpha, \bar{X}}(F)$ *satisfies*

(5) $$e_{\bar{X}_\alpha, \bar{X}}(F) \geqq (1 - 2\alpha)^2$$ *for all* F *satisfying the assumptions of Theorem* 4.1

and

(6) $$e_{\bar{X}_\alpha, \bar{X}}(F) \geqq 1/(1 + 4\alpha)$$ *for all unimodal* F *satisfying the assumptions of Theorem* 4.1.

In both cases the lower bound is sharp.

Proof of (5). This inequality follows from the fact that

$$\tfrac{1}{2}\sigma^2 = \int_0^\infty t^2 f(t)\,dt = \int_0^{\xi(1-\alpha)} t^2 f(t)\,dt + \int_{\xi(1-\alpha)}^\infty t^2 f(t)\,dt$$

$$\geqq \int_0^{\xi(1-\alpha)} t^2 f(t)\,dt + \alpha\xi^2(1-\alpha) = \tfrac{1}{2}\sigma_\alpha^2(1-2\alpha)^2.$$

Table 4.2. ARE of $\bar{X}_\alpha$ to $\bar{X}$ for Tukey Model $T(\varepsilon, 3)$.*

ε \ α	.05	.1	.125	.25	.375	.5
.25	1.40	1.62	1.66	1.67	1.53	1.33
.05	1.20	1.21	1.19	1.09	.97	.83
.01	1.04	1.03	.98	.89	.79	.68
0	.99	.97	.94	.84	.74	.64

**Adapted from Siddiqui and Raghunandanan* (1967) *and Gastwirth and Cohen* (1970).

That the lower bound is sharp can be seen by considering a sequence of distributions that can be arbitrary inside the interval $[\xi(\alpha), \xi(1-\alpha)]$ but concentrates the mass outside of this interval more and more closely to the end points (Problem 4.4).

We shall not prove (6) here; a proof is given in Bickel (1965) where these two bounds were first derived. The values of the bounds for various α's are shown in Table 4.3.

The relatively high values of these bounds, particularly for $\alpha = 0.05$, are impressive, although taken by themselves they might simply reflect the fact that $\bar{X}_\alpha$ for small values of α does not differ all that much from $\bar{X}$, so that the ARE of one to the other should always be fairly close to 1. This is not the case, however, since $e_{\bar{X}_\alpha, \bar{X}}(F)$ can be infinite (for example, when F is Cauchy or t_2). Thus the situation is quite one-sided: $\bar{X}_{.05}$ can never be much worse than $\bar{X}$ but it can be infinitely better.

So far, we have considered the efficiency of $\bar{X}_\alpha$ *relative* to $\bar{X}$; another interesting aspect of performance is the *absolute* efficiency, that is, the asymptotic efficiency relative to a best estimator. Since we are dealing with a location family and the trimmed means are location equivariant, let us define absolute efficiency as the efficiency of the given estimator relative to the Pitman estimator δ_n. It was shown by Stone (1974) that

$$\sqrt{n}\,(\delta_n - \theta) \xrightarrow{\mathcal{L}} N(0, 1/I_f)$$

where I_f is the Fisher information in X_1 defined under suitable regularity conditions by (2.6.21) and in complete generality by Port and Stone (1974). The asymptotic absolute efficiency of $\bar{X}_\alpha$ is therefore

$$e_{\bar{X}_\alpha} = 1/\sigma_\alpha^2 I_f. \tag{7}$$

Table 4.4 shows $e_{\bar{X}_\alpha}$ for a number of different distributions and $\alpha = 0$, 0.125, 0.25, 0.375, and 0.5, so that the first column corresponds to the mean and the last column to the median.

The efficiencies for $\alpha = 0.125$, for example, are surprisingly high (particularly if we exclude the Cauchy distribution) over a wide range of distribu-

Table 4.3. Lower Bounds for $e_{\bar{X}_\alpha, \bar{X}}(F)$.

α	.05	.1	.125	.25	.375	.5
$(1-2\alpha)^2$	.81	.64	.56	.25	.06	0
$1/(1+4\alpha)$	.83	.71	.67	.50	.40	.33

Table 4.4. Absolute (Asymptotic) Efficiency of $\bar{X}_\alpha$.*

α	0	.125	.25	.375	.5
Normal	1.00	.94	.84	.74	.64
$T(.01, 3)$	.95	.94	.85	.75	.63
$T(.05, 3)$	.81	.96	.88	.78	.60
Logistic	.91	.99	.95	.86	.75
t_5	.80	.99	.96	.88	.77
t_3	.50	.96	.98	.92	.81
Cauchy	.00	.50	.79	.88	.81
DE	.50	.70	.82	.91	1.00

*From Crow and Siddiqui (1967), Siddiqui and Rhagunandanan (1967), and Gastwirth and Cohen (1970).

tions. However, for sufficiently extreme distributions, $e_{\bar{X}_\alpha}$ can become arbitrarily small (Problem 4.5). In addition, the absolute efficiency of $\bar{X}_\alpha$ is zero for distributions such as the uniform where the density is discontinuous at the end points and as a result $I_f = \infty$.

In view of the asymptotic nature of the results of this section, the question again arises of the applicability to finite sample sizes. Some comparisons by Gastwirth and Cohen (1970) show that for $n = 20$ the agreement is even closer for $\bar{X}_\alpha$ with $\alpha < 1/2$ than it was for $\tilde{X}$. As an example, Table 4.5 shows the actual variance for $n = 20$ and the asymptotic variance of $\sqrt{n}(\bar{X}_\alpha - \theta)$ for $\alpha = 0.1$ and a number of distributions. The third line shows their ratio.

The discussion of this section suggests that a suitably trimmed mean may have considerable advantages over the traditional estimator $\bar{X}$ of θ. There is a further argument in favor of the trimmed mean.

Suppose that $n = 20$ measurements of a quantity θ are taken and that 19 of them turn out to give values near 1, say all lying between 0.85 and 1.15, and that the remaining value is 4,000,000. Even if it is planned to use $\bar{X}$, not many investigators would average all 20 observations to produce an esti-

Table 4.5. Variance of $\sqrt{n}(\bar{X}_\alpha - \theta)$ for $n = 20$ and $n = \infty$; $\alpha = 0.1$.*

	Normal	T(0.01, 3)	T(0.05, 3)	T(0.1, 3)	DE
$n = 20$	1.055	1.078	1.178	1.332	1.556
$n = \infty$	1.026	1.050	1.155	1.317	1.494
$n = 20/n = \infty$	1.03	1.03	1.02	1.02	1.04

*From Table 4 of Gastwirth and Cohen (1970).

mated value of θ in the neighborhood of 200,000. It would be taken for granted that something had gone wrong with this observation (perhaps with its recording or transcribing), and if an original "true" observation could not be recovered, it is likely that the investigator would discard the outlier and form the average of the remaining observations. The example is extreme, but presumably the same procedure would have been followed if the value of the outlier had been 400,000 or 40,000 or 4000 or 400 or 40—perhaps also if it had been 20, 10, or maybe even 4. However, the line would be drawn somewhere, just where depending on the judgment of the investigator.

The resulting estimator is not $\bar{X}$ but a trimmed mean, however, with a trimming rule that is not clearly defined. Such a subjectively trimmed mean suffers from two disadvantages: it opens the door to the exercise of investigator biases and it is not capable of having its performance evaluated. Thus, in practice, the trimmed means discussed in this section, might replace not so much the mean but an ill-defined subjectively trimmed mean.

Throughout this discussion, it has been assumed that the primary concern is with heavy-tailed distributions, whose centers are best estimated by the central observations. The situation is just the reverse for distributions with very thin tails. An extreme example is the uniform distribution on $(\theta - \frac{1}{2}, \theta + \frac{1}{2})$ where the best estimator of θ is the midrange $\frac{1}{2}[X_{(1)} + X_{(n)}]$. For distributions of this kind, trimmed means tend to be very inefficient (Problem 4.9). It is sometimes argued that such distributions are realistic since measurements often have natural bounds. However, in such cases it is usually unreasonable to assume the location model (3.1); in situations in which such a model is appropriate, heavy-tailed distributions seem to correspond to the more common experience.

Perhaps a more serious concern is the assumption of symmetry. Without this assumption, a conceptual difficulty arises. What do we wish to estimate? The expectation of the distribution of the X's? Its median? Or a trimmed expectation? One possible approach [taken by Bickel and Lehmann (1975)], is to say that it does not really matter, because each of these quantities provides an equally valid measure of location; what does matter is how efficiently the quantity in question can be estimated. If the mean is taken as the natural estimator of the expectation, the sample median as the natural estimator of the population median, and the trimmed sample mean as that of the corresponding trimmed expectation, each of these estimators is asymptotically normally distributed about its estimand, and the precision of the estimator is again measured by the asymptotic variance. With this reinterpretation, the asymptotic relative efficiency (4) retains its relevance although the two estimators are now estimating two different quantities. It turns out that the inequality (5) for the efficiency of $\bar{X}_\alpha$ relative to $\bar{X}$ found above for symmetric F continues to hold when the assumption of symmetry is dropped (Problem 4.7). Thus $\bar{X}_\alpha$ retains its advantage over $\bar{X}$ in this case. When F is no longer required to be symmetric, it may sometimes be preferable to trim asymmetrically if the tail is expected to be heavier in one direction than the other. If the trimming proportions are α_1 on the left and α_2 on the right, and if the distribution F is centered so that

$$\int_{\xi(\alpha_1)}^{\xi(1-\alpha_2)} tf(t)\,dt = 0, \tag{8}$$

the asymptotic variance of $\sqrt{n}$ times the α_1, α_2-trimmed mean $\bar{X}_{\alpha_1,\alpha_2}$ is

$$\frac{1}{(1-\alpha_1-\alpha_2)^2}\left\{\int_{\xi(\alpha_1)}^{\xi(1-\alpha_2)} t^2 f(t)\,dt + \alpha_1\xi^2(\alpha_1) + \alpha_2\xi^2(1-\alpha_2) - \left[\alpha_1\xi(\alpha_1) + \alpha_2\xi(1-\alpha_2)\right]^2\right\}. \tag{9}$$

In generalization of (5) the ARE of $\bar{X}_{\alpha_1,\alpha_2}$ to the untrimmed mean $\bar{X}$ satisfies (Problem 4.7)

$$e_{\bar{X}_{\alpha_1,\alpha_2},\bar{X}}(F) \geqq (1-\alpha_1-\alpha_2)^2 \tag{10}$$

not only for symmetric distributions F but also for asymmetric distributions for which the assumptions of Theorem 4.1 hold excepting the assumption of symmetry. Thus, $\bar{X}_\alpha$ and $\bar{X}_{\alpha_1,\alpha_2}$ retain the advantages of $\bar{X}_\alpha$ over $\bar{X}$ even in the asymmetric case.

To use a trimmed mean to estimate θ, one must decide on a value of α. The numerical results presented in the preceding tables suggest that a value of about 0.1 would be a good general choice. If one is particularly concerned with protection against gross errors and from past experience has an idea of the frequency of occurrence of such errors (5 to 10% is typical for many types of data) one would choose a value α somewhat above the expected proportion of gross errors.

Of course, if one knew F (but still wanted to use a trimmed mean), there would be little difficulty. There would exist an α minimizing $\sigma_\alpha^2(F)$, which we assume to be unique and which would be the α to use. It turns out that without knowing F (and if n is not too small) one can do just about as well by estimating the best value of α from the data. Such an approach is called *adaptive* estimation (the estimator adapts itself to the true F) and is a case of eating your cake and having it too. Suppose that

$$(11) \qquad 0 < \alpha_0 \leqq \alpha \leqq \alpha_1 < 1/2.$$

Then for each of the finite number of possible values of $\alpha = k/n$ ($\alpha_0 \leqq \alpha \leqq \alpha_1$), one estimates $\sigma^2_{k/n}$ by

$$(12) \quad S^2_{k/n} = \frac{1}{(1-2k/n)^2}\left\{\frac{1}{n}\sum_{i=k+1}^{n-k}\left[X_{(i)} - \bar{X}_{k/n}\right]^2 + \frac{k}{n}\left[X_{(k+1)} - \bar{X}_{k/n}\right]^2 + \frac{k}{n}\left[X_{(n-k)} - \bar{X}_{k/n}\right]^2\right\}$$

and determines the value $\hat{k}$ for which $S^2_{k/n}$ is minimum. The estimator is the corresponding $\bar{X}_{\hat{k}/n}$. It was shown by Jaeckel (1971b) for symmetric F, under only mild additional restrictions, that the asymptotic distribution is the same as that of the $\bar{X}_\alpha$ which is best for the true F, regardless of what this F is. An idea of how well this procedure works for moderate n is obtained from Table 4.6, which lists the variances of $\sqrt{n}(\delta - \theta)$ for $n = 20$ and for the cases that δ is, respectively, the mean, $\bar{X}_\alpha$ for $\alpha = 0.05, 0.1, 0.15, 0.25$, and the adaptive estimator J of Jaeckel. Several modifications of the Jaeckel estimator are discussed by Andrews et al. (1972, Sect. 2B3).

Two comments may help clarify the significance of condition (11). It is, first of all, a restriction on F which is not satisfied by the normal distribution for which $\alpha = 0$. Actually, it is suggested by Jaeckel (1969) that under some additional conditions on F, his result seems capable of extension to the case $\alpha_0 = 0$. Second, the Jaeckel estimator is affected by the choice of α_0 and α_1. In particular, the estimator J in Table 4.6 corresponds to the case $\alpha_0 = 0$, $\alpha_1 = 0.25$.

In the next section, we shall briefly discuss three large classes of estimators of θ, one of which contains the trimmed means as special cases. A

Table 4.6. Variances of $\sqrt{n}(\delta - \theta)$ for $n = 20$.

δ	Normal	T(0.01, 3)	T(0.05, 3)	T(0.1, 3)	t_3	Cauchy	DE
Mean	1.00	1.08	1.43	1.84	3.14	∞	2.10
$\bar{X}_{.05}$	1.02	1.05	1.19	1.41	1.83	24.0	1.77
$\bar{X}_{.1}$	1.06	1.08	1.18	1.33	1.68	7.3	1.60
$\bar{X}_{.15}$	1.09	1.12	1.20	1.32	1.60	4.6	1.48
$\bar{X}_{.25}$	1.20	1.22	1.29	1.39	1.59	3.1	1.33
J	1.10	1.13	1.22	1.35	1.61	3.5	1.48

*From Exhibits 5-7D and 5-7E of Andrews et al. (1972).

number of studies comparing many of these estimators either in theoretical terms (Bickel and Lehmann, 1975) or on real data (Stigler, 1977; Spjøtvoll and Aastreit, 1980, Hill and Dixon, 1982, Rocke, Downs, and Rocke, 1982) suggest that a trimmed mean with α of the order of 10 percent is an estimator which, though simple, combines high efficiency near the normal distribution with good protection against outlying observations.

Finding a satisfactory competitor to $\bar{X}$ in the one-sample problem is the first step in a search for a corresponding alternative to least squares estimators in linear models (Section 3.4). Trimmed means have been extended to such models by Ruppert and Carroll (1980).

5. LINEAR COMBINATIONS OF ORDER STATISTICS (*L*-ESTIMATORS)*

Although the trimmed means $\bar{X}_\alpha$ appear to provide a fairly satisfactory approach to the estimation of the center θ of a symmetric distribution that is not significantly shorter tailed than the normal, several other approaches are also available, and we shall sketch some of them in this and the next section. Alternatives are of interest—even if they should prove not to be much better for the problem of estimating θ—if one views the problem of estimating θ not in isolation but as the simplest special case of the large class of problems (linear models) that are traditionally approached through least squares but for which more robust methods seem desirable. It is not yet clear which of the three alternative approaches mentioned here will turn out to be most satisfactory in this more general context.

The trimmed means (4.1) are linear combinations of the order statistics giving zero weight to the k extreme observations at either end, and equal weight $1/(n-2k)$ to the $n-2k$ central observations. Instead, one may prefer a smoother weighting in which (typically) the observations in the center are given the greatest weight and the weights decrease symmetrically in either direction.

To study the large sample behavior of *L-estimators*, that is, estimators of the form

$$(1) \qquad \Sigma w_i X_{(i)} \quad (\Sigma w_i = 1),$$

consider a sequence of such estimators, defined through a sequence of weights w_{in} $(i = 1,\ldots, n;\ n = 1, 2,\ldots)$. For the purposes of large-sample theory, one would want the sequences $(w_{1n},\ldots, w_{nn})$ to have a smooth

*Sections 5 and 6 continue the search for a satisfactory estimator of the center of a symmetric distribution. This material will not be used in the sequel.

limiting behavior. It is convenient to specify this by means of a probability distribution Λ defined on (0, 1) with density

$$\lambda(t) = \Lambda'(t) \tag{2}$$

and to let

$$w_{in} \sim \lambda\left(\frac{i}{n+1}\right) \tag{3}$$

where the proportionality factor is determined so that the w's add up to 1.

The trimmed mean $\bar{X}_\alpha$ corresponds to the uniform density on $(\alpha, 1-\alpha)$

$$\lambda(t) = \frac{1}{1-2\alpha} \quad \text{if} \quad \alpha < t < 1 - \alpha. \tag{4}$$

As another example, suppose that $\lambda(t)$ is a symmetric beta density, so that it is proportional to $t^{a-1}(1-t)^{a-1}$. For the case $a = 2$, for example,

$$w_{in} \sim \frac{i}{n+1}\left(1 - \frac{i}{n+1}\right). \tag{5}$$

The estimators (1) will be relatively insensitive to outlying observations if Λ assigns most of its mass to the central part of the unit interval. A technical definition of robustness requires the existence of $0 < \alpha < 1/2$ such that Λ assigns probability 1 to $(\alpha, 1-\alpha)$ (see, for example, Bickel and Lehmann, 1975).

Theorem 5.1. *Let* $X_1, \ldots, X_n$ *be iid over an interval* (a, b), $-\infty \leqslant a < b \leqslant \infty$ *according to a distribution* F *for which*

$$E(X_i^2) < \infty \tag{6}$$

and which possesses a density f *with*

$$0 < f(x) \quad \text{for all} \quad a < x < b. \tag{7}$$

(i) *Let*

$$L_n = \frac{1}{n}\sum_{i=1}^{n} \lambda\left(\frac{i}{n+1}\right) X_{(i)} \tag{8}$$

where λ *is a bounded function defined over* (0, 1) *which is continuous a.e.* (*with respect to Lebesgue measure) and satisfies* $\int_0^1 \lambda(t)\,dt = 1$. *Let*

(9) $$\mu(F, \lambda) = \int_0^1 \lambda(u) F^{-1}(u)\,du$$

and

(10) $$\sigma^2(F, \lambda) = \int_0^1 A^2(t)\,dt - \left(\int_0^1 A(t)\,dt\right)^2$$

where A is any function with derivative

(11) $$A'(t) = \frac{\lambda(t)}{f[F^{-1}(t)]}.$$

Then the distribution of

(12) $$\frac{\sqrt{n}\,[L_n - \mu(F, \lambda)]}{\sigma(F, \lambda)}$$

tends to the standard normal distribution $N(0, 1)$ as $n \to \infty$, provided $\sigma^2(F, \lambda) > 0$.

(ii) The result of part (i) remains valid if in (8) the density λ is replaced by λ_n, provided the sequence λ_n is uniformly bounded and

(13) $$\lambda_n(t) \to \lambda(t)$$

where λ is continuous a.e. and the convergence is uniform in a neighborhood of any continuity point of λ.

The conclusion of this theorem, which we shall not prove here, has been established also under various other sets of assumptions. [See Serfling (1980), Chap. 8, for a number of different approaches.] The version given here can be found in Stigler (1974) [see also Bickel (1965)], in which references to alternative assumptions and further discussion are also provided. In particular, Stigler does not require the existence of a density f, assumed here for the sake of simplicity; he also provides weaker alternatives to (6). Conditions for the asymptotic normality of L_n when λ is unbounded are given by Shorack (1972).

The statistic L_n does not quite agree with that defined by (1) since the weights $(1/n)\lambda[i/(n + 1)]$ need not add up to 1. The extension of Theorem 5.1 to the estimators (1) is provided by the following corollary.

Corollary 5.1. *Under the assumptions of Theorem* 5.1(*i*) *let*

(14) $$\delta_n = \Sigma w_{in} X_{(i)}$$

where the w_{in} *are proportional to* $\lambda(i/(n+1))$ *and add up to* 1. *Then*

$$\sqrt{n}\,(\delta_n - \mu) \xrightarrow{\mathcal{L}} N(0, \sigma^2) \tag{15}$$

where μ *and* σ^2 *are defined by* (9) *and* (10), *respectively.*

Proof. Let

$$w_{in} = \frac{1 + c_n}{n}\lambda\left(\frac{i}{n+1}\right) \quad \text{so that} \quad \delta_n = \frac{1}{n}\Sigma\lambda_n\left(\frac{i}{n+1}\right)X_{(i)}$$

with $\lambda_n(t) = (1 + c_n)\lambda(t)$. Then the conditions of Theorem 5.1(ii) are satisfied provided $c_n \to 0$. Now

$$1 = \Sigma w_{in} = (1 + c_n)\Sigma\frac{1}{n}\lambda\left(\frac{i}{n+1}\right)$$

and

$$\Sigma\frac{1}{n}\lambda\left(\frac{i}{n+1}\right) \to \int\lambda(t)\,dt = 1,$$

and hence $c_n \to 0$ as was to be proved.

Theorem 5.1 and Corollary 5.1 were stated without any symmetry conditions. However, we are here primarily concerned with the case that both F and λ are symmetric.

Corollary 5.2. *Under the assumptions of Corollary* 5.1, *if* F *is symmetric about* θ *and* λ *about* $1/2$, *then* $\sqrt{n}\,(\delta_n - \theta) \to N(0, \sigma^2)$, *with* σ^2 *given by* (10).

Proof. It is easy to check that under the stated symmetry conditions $\mu(F, \lambda) = \theta$ (Problem 5.5).

Let us return for a moment to the situation of Theorem 5.1 without the symmetry assumptions. For fixed λ, let $\xi(F) = \mu(F, \lambda)$ define a functional over the space of all distributions F for which the integral

$$\xi(F) = \int_0^1 \lambda(t)F^{-1}(t)\,dt = \int_{-\infty}^{\infty} x\lambda[F(x)]\,dF(x) \tag{16}$$

exists. Then a natural estimator of $\xi(F)$ is $\delta_n' = \xi(F_n)$ where F_n is the sample cdf of $X_1, \ldots, X_n$. This estimator is given by (Problem 5.10)

$$\delta_n' = \frac{1}{n}\Sigma\lambda\left(\frac{i}{n}\right)X_{(i)},$$

which differs only trivially from L_n and which has the same limit distribution. This representation of the estimator as a (differentiable) functional of F evaluated at the sample cdf provides another approach to asymptotic normality, which is discussed by Boos (1979); see also Serfling (1980) and Staudte (1980).

For the remainder of this section we shall be concerned with the estimation of the center of symmetry θ of a symmetric distribution F, and shall suppose that λ is symmetric about $1/2$. Corollary 5.2 then provides the AREs of pairs of L-estimators of θ corresponding to different λ's. If the λ's satisfy

$$0 \leqslant \lambda_2(t)/\lambda_1(t) \leqslant c \qquad (1 < c) \tag{17}$$

one can obtain a lower bound for the efficiency $e_{2,1}(F)$ of δ_{2n} relative to δ_{1n}.

Corollary 5.3. *If* δ_{in}, λ_i $(i = 1,2)$ *and* F *satisfy the assumptions of Corollary* 5.2, *and if in addition the* λ_i *are* $\geqslant 0$ *and satisfy* (17), *then*

$$e_{2,1}(F) \geqslant 1/c^2 \tag{18}$$

and the inequality is sharp when c *is the sharp upper bound of* λ_2/λ_1.

Proof of (18). From (10) it follows that

$$\sigma^2(F, \lambda) = \text{var}[A(T)] \tag{19}$$

where T is uniformly distributed on (0, 1) and hence, if T_1 and T_2 are independent copies of T, that

$$\begin{aligned} 2\sigma^2(F, \lambda) &= \text{var}[A(T_2) - A(T_1)] \\ &= E[A(T_2) - A(T_1)]^2 = E\left[\int_{T_1}^{T_2} A'(s)\,ds\right]^2. \end{aligned} \tag{20}$$

Since $0 \leqslant A_2'(s) \leqslant cA_1'(s)$, where the A_i' are defined by (11), we have that

$$\sigma_2^2(F) \leqslant c^2\sigma_1^2(F),$$

and hence (18).

For a proof of the fact that there exist distributions F for which the left side of (18) is arbitrarily close to the right side, see Bickel and Lehmann (1975).

As an illustration, let

$$\lambda(t) = 6t(1 - t) \tag{21}$$

so that w_{in} is given by (5). Then $\sup \lambda(t) = 3/2$ and $e_{\delta_n, \bar{X}}(F) \geqslant 4/9$ for all symmetric F satisfying (6) and (7).

For a given F, the most efficient L-estimator of θ is obtained by minimizing the asymptotic variance (10). This minimization is achieved in the following theorem.

Theorem 5.2. *Suppose that the density* f *has two derivatives,* f′, f″ *a.e. and that* f(x) → 0 *as* x → ± ∞, *and let*

$$\gamma(x) = -\frac{f'(x)}{f(x)}. \tag{22}$$

Then the variance $\sigma^2(F, \lambda)$ *given by* (10) *and* (11) *is minimized by*

$$\lambda_0(t) = \gamma'[F^{-1}(t)] \Big/ \int \gamma^2(x) f(x)\, dx \tag{23}$$

and for this λ_0, *the variance is equal to* $1/I_f$, *where* I_f *is Fisher's information defined by* (2.6.21).

Proof. Since $\sigma^2(F, \lambda)$ is unchanged by adding a constant to $A(t)$ (Problem 5.3), we can assume without loss of generality that

$$\int_0^1 A(t)\, dt = 0. \tag{24}$$

The problem then reduces to determining the function λ minimizing

$$\int_0^1 A^2(t)\, dt \tag{25}$$

subject to (24) and to

$$\int_0^\infty \lambda(t)\, dt = 1. \tag{26}$$

By (11)

$$\lambda(t) = A'(t) f[F^{-1}(t)] \tag{27}$$

so that, on integrating by parts,

$$\int_0^1 \lambda(t)\, dt = A(t) f[F^{-1}(t)]\Big|_0^1 - \int_0^1 A(t) \frac{f'[F^{-1}(t)]}{f[F^{-1}(t)]}\, dt.$$

The first term vanishes since f is zero at $\pm\infty$, and condition (26) thus becomes

$$\int_0^1 A(t)\gamma[F^{-1}(t)]\, dt = 1. \tag{28}$$

Application of the Schwarz inequality yields

$$\int A^2(t)\, dt \cdot \int \gamma^2[F^{-1}(t)]\, dt \geqslant \left[\int A(t)\gamma[F^{-1}(t)]\, dt\right]^2 = 1,$$

and hence

$$\int A^2(t)\, dt \geqq \frac{1}{\int \gamma^2(x) f(x)\, dx} = \frac{1}{I_f}. \tag{29}$$

To complete the proof, we need only show that for $\lambda = \lambda_0$ the left side of (29) attains the lower bound given by the right side and that λ_0 satisfies conditions (24) and (26). If A_0 is given by (11) with $\lambda = \lambda_0$,

$$A_0'(t) = \frac{\gamma'[F^{-1}(t)]}{f[F^{-1}(t)]} \cdot \frac{1}{\int \gamma^2(x) f(x)\, dx}$$

and hence

$$A_0(t) = \frac{\gamma[F^{-1}(t)]}{\int \gamma^2(x) f(x)\, dx} + k. \tag{30}$$

Since

$$\int_0^1 \gamma[F^{-1}(t)]\, dt = \int_{-\infty}^{\infty} \gamma(x) f(x)\, dx = -\int_{-\infty}^{\infty} f'(x)\, dx = 0,$$

condition (24) will hold provided $k = 0$, in which case (28) is also satisfied. Analogously

$$\int_0^1 \gamma^2[F^{-1}(t)]\, dt = \int_{-\infty}^{\infty} \gamma^2(x) f(x)\, dx$$

so that A_0 given by (30) with $k = 0$ satisfies

$$\int A_0^2(t)\, dt = 1 / \int \gamma^2(x) f(x)\, dx \tag{31}$$

as was to be proved.

Instead of verifying that equality obtains in (29) for $A = A_0$, we could have derived the minimizing λ_0 from the condition for equality in Schwarz's inequality (Problem 5.4).

Example 5.1. ***Normal.*** Suppose that F is the standard normal distribution, so that

$$\frac{f'(x)}{f(x)} = \frac{d}{dx}[\log f(x)] = -x,$$

and hence

$$\gamma(x) = x \quad \text{and} \quad \gamma'(x) = 1.$$

It follows that $\lambda_0(t) = 1$ for $0 < t < 1$ and that the most efficient L-estimator is $\bar{X}$.

Example 5.2. ***Logistic.*** Let F be the logistic distribution $L(0,1)$. An easy calculation shows that (Problem 5.7)

$$\gamma(x) = 2F(x) - 1, \qquad \gamma'(x) = 2f(x)$$

and

$$\gamma'[F^{-1}(t)] = 2t(1-t). \tag{32}$$

The weights which minimize the asymptotic variance are therefore proportional to (5).

The optimal λ_0 given by (23) may turn out to be negative for some values of t, and some of the resulting weights w_{in} will also be negative. Although this may seem surprising, the theory of L-estimators requires no changes to accommodate this possibility.

Example 5.3. ***The Student t-Distribution.*** If F is the Student t-distribution with ν degrees of freedom, an examination of (23) shows (Problem 5.8) that $\lambda_0(t) < 0$ for $|t - \frac{1}{2}| > k_\nu$ where k_ν increases with ν and tends to $\frac{1}{2}$ as $\nu \to \infty$. This is related to the heavy-tails property of t_ν and the fact that this decreases with ν.

A necessary and sufficient condition for $\lambda_0(t)$ to be non-negative for all t, is that $\gamma'(x) \geq 0$ for all x. By (22) this is equivalent to $-\log f(x)$ being convex. Distributions with this property are called *strongly unimodal.*

That the asymptotic variance of the best L-estimator for a given F is $1/I_f$ shows that its absolute efficiency, that is, its ARE with respect to the Pitman estimator for that F, is 1. This is unlike the situation for trimmed means in which some values of the absolute efficiency were shown in Table 4.4. The difference, of course, is that the class of L-estimators is much larger and has the flexibility through proper choice of λ to adapt itself to a given F.

If F is unknown but sufficiently smooth, it is possible, as in the preceding section, to use an adaptive estimator which estimates F from the data and asymptotically does as well as the L-estimator which is best when F is known. Since the latter does asymptotically as well as the Pitman estimator, this adaptive estimator has the remarkable property that for all sufficiently smooth F its asymptotic performance equals that of the Pitman estimator. (This program is carried out by Sacks, 1975; see also Randles, 1982).

6. *M*- AND *R*-ESTIMATORS*

The trimmed means $\bar{X}_\alpha$ considered in Section 5.4 constitute a compromise between the mean $\bar{X}$ and the median $\tilde{X}$. Another such compromise, proposed by Huber (1964), is suggested by the fact that mean and median are the quantities that minimize, respectively, $\Sigma(X_i - a)^2$ and $\Sigma|X_i - a|$. Huber suggested minimizing instead

$$\sum_{i=1}^{n} \rho(X_i - a) \tag{1}$$

where ρ is given by

$$\rho(x) = \begin{cases} \frac{1}{2}x^2 & \text{if } |x| \leqq k \\ k|x| - \frac{1}{2}k^2 & \text{if } |x| \geqq k. \end{cases} \tag{2}$$

This function is proportional to x^2 for $|x| \leqq k$ but outside this interval replaces the parabolic arcs by straight lines. The pieces fit together so that ρ and its derivative ρ' are continuous (Problem 6.1). As k gets larger, ρ will agree with $\frac{1}{2}x^2$ over most of its range, so that the estimator comes close to $\bar{X}$; as k gets smaller, the estimator will become closer to the median. As a moderate compromise, the value $k = 1.5$ is sometimes suggested.

Just as the trimmed means are a one-parameter subset of the class of linear functions of order statistics, the *Huber estimators* minimizing (1) with ρ given by (2) are a subset of the class of *M-estimators* obtained by minimizing (1) for arbitrary ρ. If ρ is convex and even, as is the case for (2), it follows from Theorem 1.6.8 that the minimizing values of (1) constitute a closed interval; if ρ is strictly convex, the minimizing value is unique. If ρ has a derivative $\rho' = \psi$, the M-estimators M_n may be defined as the

*The material of this section will not be used in the sequel.

solutions of the equation

$$\Sigma\psi(x_i - a) = 0. \tag{3}$$

If $X_1, \ldots, X_n$ are iid according to $F(x - \theta)$ where F is symmetric about zero and has density f, it turns out under weak assumptions on ψ and F that*

$$\sqrt{n}(M_n - \theta) \to N[0, \sigma^2(F, \psi)] \tag{4}$$

where

$$\sigma^2(F, \psi) = \frac{\int\psi^2(x)f(x)\,dx}{[\int\psi'(x)f(x)\,dx]^2}, \tag{5}$$

provided both numerator and denominator on the right side are finite, and the denominator is positive.

Proofs of (4) can be found in Huber (1981), in which a detailed account of the theory of M-estimators is given not only for location parameters but also in more general settings, and in Chapter 7 of Serfling (1980).

For

$$\rho(x) = -\log f(x), \tag{6}$$

minimizing (1) is equivalent to maximizing $\Pi f(x_i - a)$ and the M-estimator then coincides with the maximum likelihood estimator (MLE). Maximum likelihood estimation constitutes the principal topic of the next chapter. In particular, it will be proved there (under suitable regularity conditions) that for known F, the M-estimator of θ corresponding to (6) satisfies (4) with $\sigma^2 = 1/I_f$. (For an alternative proof see Problem 6.2.) Thus, as was the case for L-estimators, for a given F the ARE of the most efficient M-estimator with respect to the Pitman estimator for that F is 1.

Rather than assuming F to be known, it may be more realistic to assume it to be known only approximately. Such approximate models have been formulated by Huber, who assumes that F lies within a stated distance of a known distribution G, for example, that it is of the form

$$F(x) = (1 - \varepsilon)G(x) + \varepsilon H(x) \tag{7}$$

where ε and G are given, and H is an arbitrary (unknown) distribution, both

*For more accurate approximations, see Field and Hampel (1982).

G and H being symmetric about zero. If the family satisfying (7) is denoted by $\mathcal{F}$, then any distribution of the form $(1-\varepsilon')G(x)+\varepsilon' H(x)$ with $0 \leqslant \varepsilon' \leqslant \varepsilon$ is also in $\mathcal{F}$ (Problem 6.3).

Corresponding to the minimax estimators of Chapter 4, one may now wish to determine the M-estimator (i.e., the function ψ) which minimizes

$$\sup_{F\in\mathcal{F}} \sigma^2(F, \psi). \tag{8}$$

For the particular case that $G = \Phi$, the standard normal distribution with density ϕ, we shall now prove that the solution to this mimimax problem is one of the Huber estimators corresponding to (2), so that the minimax ψ is given by

$$\psi_0(x) = \begin{cases} -k & \text{for } x \leqslant -k \\ x & \text{for } |x| < k \\ k & \text{for } x \geqslant k, \end{cases} \tag{9}$$

where k is determined by ε through (14) below.

To prove this, we shall show (i) that there exists a distribution $F_0 \in \mathcal{F}$ such that ψ_0 is the function ψ that minimizes $\sigma^2(F_0, \psi)$. This means that ψ_0 is the Bayes solution with respect to the prior distribution Λ over $\mathcal{F}$ which assigns probability 1 to the single "point" F_0. The minimax character of ψ_0 will then follow from Theorem 4.2.1 if we can also show (ii) that $\sigma^2(F, \psi_0)$ takes on its maximum value over $\mathcal{F}$ at F_0.

(i) For given F_0, it was stated above that $\sigma^2(F_0, \psi) \geqslant I_{F_0}^{-1} = \sigma^2(F_0, \psi_0)$ with $\psi_0 = \rho_0'$ given by (6) and hence proportional to $f_0'(x)/f_0(x)$, where f_0 is the density of F_0, which we shall assume to exist and be differentiable. Thus, the given ψ_0 will minimize $\sigma^2(F_0, \psi)$ provided

$$\frac{f_0'(x)}{f_0(x)} = -A\psi_0(x) \tag{10}$$

for some constant A. Using the fact that $\psi_0 = \rho_0'$ and $f_0'(x)/f_0(x) = (d/dx)[\log f_0(x)]$, we see that ψ_0 minimizes $\sigma^2(F_0, \psi)$ if

$$f_0(x) = Be^{-A\rho_0(x)}. \tag{11}$$

For any $A > 0$, we can determine B so that f_0 is a probability density provided it is integrable, as is the case when $\rho = \rho_0$ is given by (2).

(ii) Let us now see what conditions on $F_0 \in \mathcal{F}$ are required so that $\sigma^2(F, \psi_0)$ will take on its maximum value over $\mathcal{F}$ at F_0. The numerator of (5) is

$$\int \psi_0^2\, dF = (1-\varepsilon)\int \psi_0^2\, d\Phi + \varepsilon \int \psi_0^2\, dH.$$

Since $\psi_0^2(x) < k^2$ for $|x| < k$ and $= k^2$ for $|x| > k$, the numerator will take on its maximum value

$$(1-\varepsilon)\int \psi_0^2\, d\Phi + \varepsilon k^2$$

for any distribution H which assigns probability 1 to the set $|x| > k$. Consider next the denominator of (5) which is the square of

$$\int \psi_0'\, dF = (1-\varepsilon)\int \psi_0'\, d\Phi + \varepsilon \int \psi_0'\, dH.$$

Since

$$\psi_0'(x) = \begin{cases} 0 & \text{for } |x| > k \\ 1 & \text{for } |x| < k, \end{cases}$$

the denominator will take on its minimum value for any distribution H which assigns probability 1 to the set $|x| > k$. Therefore, a distribution F_0 given by (11) will maximize $\sigma^2(F, \psi_0)$ over $\mathcal{F}$ and thereby prove ψ_0 to be a minimax solution of our problem provided it satisfies (7) with $G = \Phi$ for some distribution H assigning probability 1 to the set $|x| > k$. Equations (7) and (11) give

$$(1-\varepsilon)\phi(x) + \varepsilon h(x) = \begin{cases} Be^{-Ax^2/2} & \text{if } |x| < k \\ Be^{-A(k|x| - k^2/2)} & \text{if } |x| > k. \end{cases}$$

Thus, for $|x| < k$,

$$\varepsilon h(x) = Be^{-Ax^2/2} - \frac{1-\varepsilon}{\sqrt{2\pi}} e^{-x^2/2}$$

which is zero provided

$$A = 1 \quad \text{and} \quad B = (1-\varepsilon)/\sqrt{2\pi}\,. \tag{12}$$

For these values it also follows that $h(x) > 0$ when $|x| > k$. That $\int h(x)\, dx = 1$, is a consequence of (7) and the fact that f_0 is a probability density.

Putting $A = 1$ in (11) and integrating we find that $\int f_0(x)\, dx = 1$ requires

$$\frac{1}{B} = \int_{-k}^{k} e^{-x^2/2}\, dx + \frac{2}{k} e^{-k^2/2}.$$

Thus, finally,

$$f_0(x) = \frac{1 - \varepsilon}{\sqrt{2\pi}} e^{-\rho_0(x)} \tag{13}$$

is a probability density in $\mathcal{F}$ with H assigning probability 1 to the set $|x| > k$ provided ε and k are connected by the equation

$$\frac{1}{1 - \varepsilon} = \int_{-k}^{k} \phi(x)\, dt + \frac{2}{k}\phi(k). \tag{14}$$

The Huber estimator corresponding to ρ given by (2) with k determined by (14) thus minimizes (8) among all M-estimators. In fact, if the asymptotic variance of any equivariant estimator δ is denoted by $\sigma_\delta^2(F)$, this Huber estimator minimizes $\sup_{\mathcal{F}} \sigma_\delta^2(F)$ among all equivariant estimators δ.*

To see this, let us examine parts (i) and (ii) of the minimax proof for M-estimators. Part (i) stated that ψ_0 minimizes $\sigma^2(F_0, \psi)$ among all ψ. We now need instead that it minimizes $\sigma_\delta^2(F_0)$ among all equivariant estimators, that is, that the M-estimator corresponding to ψ_0 has ARE 1 with respect to the Pitman estimator for F_0. This is so since it was stated earlier that $\sigma^2(F_0, \psi_0) = 1/I_{f_0}$. Part (ii) stated that F_0 maximizes $\sigma^2(F, \psi_0)$ over $\mathcal{F}$ and requires no change. This completes the proof.

We have so far considered the cases that F is either known or approximately known. If it is completely unknown but sufficiently smooth, it is possible—as in the case of L-estimators—to estimate F from the data, and thus to obtain a fully efficient adaptive estimator. This is shown by Stone (1975). Some numerical values in his paper suggest that for $n = 40$ the method works fairly well. The resulting estimators, however, are considerably more complex.

There is an interesting relationship between L- and M-estimators. Given any symmetric (about 0) distribution F, which for simplicity will be assumed to be strictly increasing, with density f, and any odd function ψ for which (5) is defined and finite, let

$$\lambda_\psi(t) \sim \psi'[F^{-1}(t)]. \tag{15}$$

Then the variances of the L- and M-estimators, corresponding respectively to λ_ψ and ψ, satisfy

$$\sigma_L^2(F, \lambda_\psi) = \sigma_M^2(F, \psi). \tag{16}$$

*For some related results, see Sacks and Ylvisaker (1982).

To see this, it is convenient to normalize ψ, which can be multiplied by an arbitrary constant, so that

$$\int \psi'(x) f(x)\, dx = 1. \tag{17}$$

It then follows (Problem 6.6) that $\lambda_\psi(t) = \psi'[F^{-1}(t)]$ and that the function A defined by (5.11) is

$$A(t) = \int_{1/2}^{t} A'(u)\, du = \psi(x), \tag{18}$$

where $x = F^{-1}(t)$, $dx/dt = 1/f[F^{-1}(t)]$. This shows that

$$\int A^2(t)\, dt = \int \psi^2(x) f(x)\, dx$$

while

$$\int A(t)\, dt = \int \psi(x) f(x)\, dx = 0$$

and hence proves (16).

Under additional regularity assumptions, it can be shown (Jaeckel, 1971a) that not only do the estimators L and M related through (15) satisfy (16) but $\sqrt{n}(M - L) \to 0$ in probability.

As an example of the correspondence (15), consider the Huber estimators for which $\psi'(x) = 1$ if $|x| < k$ and $= 0$ otherwise. Then $\lambda_\psi(t)$ is proportional to

$$\psi'[F^{-1}(t)] = 1 \qquad \text{if} \quad |F^{-1}(t)| < k$$

and zero otherwise. Since $|F^{-1}(t)| < k$ if and only if $F(-k) < t < F(k)$, $\lambda_\psi(t)$ is constant over $\alpha < t < 1 - \alpha$ and zero outside that interval, and the associated L-estimator is therefore the α-trimmed mean where

$$\alpha = F(-k). \tag{19}$$

The correspondence between the two one-parameter families—the Huber estimators and the trimmed means—is thus much closer than might have been expected from (15). For every F, (15) establishes a 1 : 1 correspondence between the two families through (19).

This relationship raises the question whether the L-estimator minimizing

$$\sup_{F \in \mathfrak{F}} \sigma_L^2(F, \lambda)$$

for the normal neighborhood $\mathcal{F}$ defined by (7) with $G = \Phi$, might be the trimmed mean corresponding to the Huber estimator with k given by (14). That this is so was proved by Jaeckel (1971b).

Let us next consider more briefly a third class of estimators of the center θ of a symmetric distribution. These so-called *R-estimators* were originally derived from rank-tests of the hypothesis specifying the value of θ (Hodges and Lehmann, 1963). We shall here adopt a different (equivalent) definition by Jaeckel (1969) which lacks the motivation of the earlier derivation but is more convenient in the present context.

Let $X_{(1)} < \cdots < X_{(n)}$ denote the ordered observations as before, and let $d_1, \ldots, d_n$ be n non-negative constants. Consider the

$$n + \binom{n}{2} = \frac{n(n+1)}{2}$$

averages $[X_{(j)} + X_{(k)}]/2$ for $j \leqq k$. (Note that for $j = k$, these are just the order statistics themselves.) To each such average assign the weight

$$w_{jk} = d_{n-(k-j)} \Big/ \sum_{i=1}^{n} i d_i. \tag{20}$$

It is easily checked that these weights add up to 1 (Problem 6.7). Consider the discrete distribution which assigns probability w_{jk} to the value $\frac{1}{2}[X_{(j)} + X_{(k)}]$. The estimator R_n we wish to consider is the median of this distribution.

Before giving some examples it may be worth noting that this estimator is similar to the L-estimator (5.1), which is based on the discrete distribution assigning probabilities w_i to the values $X_{(i)}$. However, while (5.1) is the expectation of that earlier distribution, we are now instead concerned with the median.

Example 6.1. Median. Suppose that $d_1 = \cdots = d_{n-1} = 0, d_n = 1$. Then $\Sigma i d_i = n$ and

$$w_{jk} = \begin{cases} 1/n & \text{if } j = k \\ 0 & \text{otherwise.} \end{cases}$$

The discrete distribution on which the estimator is based assigns probability $1/n$ to each of $X_{(1)}, \ldots, X_{(n)}$ and R_n is the median of the X's.

At the other extreme, suppose that $d_1 = 1, d_2 = \cdots = d_n = 0$. Then w_{jk} is positive only when $k - j = n - 1$, that is, when $j = 1$ and $k = n$. The discrete distribution thus assigns probability 1 to the value $\frac{1}{2}[X_{(1)} + X_{(n)}]$ so that R_n is the midrange of the observations.

Example 6.2. Hodges–Lehmann Estimator. If $d_1 = \cdots = d_n = 1$, the discrete distribution assigns equal probability to each of the values $[X_{(j)} + X_{(k)}]/2$ and R_n

is therefore the median of these $n(n+1)/2$ values. Because of its relationship to the Wilcoxon test (TSH, p. 241), we shall denote this estimator (the Hodges–Lehmann estimator) by W_n. If instead, $d_1 = \cdots = d_{n-1} = 1$, $d_n = 0$, the estimator is the median of the averages $[X_{(j)} + X_{(k)}]/2$ with $j < k$, which is asymptotically equivalent to W_n.

Consider now a sequence of such estimators corresponding to constants $d_{1n}, d_{2n}, \ldots, d_{nn}$; $n = 2, 3, \ldots$. To obtain an asymptotic distribution for R_n we shall assume a special form for the d's analogous to (5.3), which has its origin in the relationship of the estimators to rank tests but which seems difficult to motivate here. Specifically, we shall suppose that the d's are given in terms of a function K defined over (0, 1), which is nondecreasing and satisfies

$$K(1-t) = -K(t). \tag{21}$$

The d's are then defined by

$$d_{in} = K\left(\frac{i+1}{2n+1}\right) - K\left(\frac{i}{2n+1}\right), \tag{22}$$

and under sufficient regularity conditions on K and F (assumed to be symmetric)

$$\sqrt{n}\,(R_n - \theta) \to N\left[0, \sigma^2(F, K)\right] \tag{23}$$

where

$$\sigma^2(F, K) = \frac{\int_0^1 K^2(t)\,dt}{\left[\int_{-\infty}^{\infty} K'[F(x)]\,f^2(x)\,dx\right]^2}. \tag{24}$$

As was the case for L- and M-estimators, if F is known there exists an asymptotically efficient R-estimator. This corresponds to

$$K(t) = \gamma\left[F^{-1}(t)\right] \tag{25}$$

with γ given by (5.22).

Example 6.3. Logistic. If F is the logistic distribution $L(0, 1)$, an asymptotically efficient L-estimator was obtained in Example 5.2. It was found there that $\gamma(x) = 2F(x) - 1$, so that

$$K(t) = 2t - 1 \tag{26}$$

Table 6.1. ARE of W to $\bar{X}$.

	Normal	T(0.01, 3)	T(0.05, 3)	Logistic	t_5	t_3	DE	Cauchy
$e_{W,\bar{X}}$	$\frac{3}{\pi} = .955$	1.01	1.19	1.10	1.24	1.90	1.50	∞

and hence $d_{in} = 2/(2n+1)$ for all $i = 1,\dots,n$. Since an R-estimator by (20) is unchanged if all the d's are multiplied by the same positive constant, the resulting R-estimator is the estimator W_n of Example 6.2, which is therefore asymptotically efficient when F is logistic.

It follows from (24) and (26) that the asymptotic variance of $\sqrt{n}(W_n - \theta)$ is (Problem 6.9)

$$\sigma_W^2(F) = \frac{1}{12\left(\int f^2(x)\,dx\right)^2}. \tag{27}$$

This formula makes it possible to compute asymptotic efficiencies. Corresponding to Tables 4.1 and 4.2 for $e_{\bar{X}_a,\bar{X}}$, the asymptotic efficiency $e_{W,\bar{X}}$ of W_n to the mean $\bar{X}$ is shown in Table 6.1 for a number of distributions.

The surprisingly high values of the efficiencies in Table 6.1 raise the question as to how low $e_{W,\bar{X}}(F)$ can fall under unrestricted variation of F. The answer is given by the following result, which corresponds to Theorem 4.2 for trimmed means.

Theorem 6.1. *The asymptotic efficiency* $e_{W,\bar{X}}(F)$ *satisfies*

$$e_{W,\bar{X}}(F) \geqq .864 \quad \textit{for all symmetric } F. \tag{28}$$

Proof. For symmetric F the efficiency $e_{W,\bar{X}}(F)$ is given by*

$$e_{W,\bar{X}}(F) = 12\sigma^2\left(\int f^2(x)\,dx\right)^2 \tag{29}$$

where σ^2 is the variance of the X's and hence also the asymptotic variance of $\sqrt{n}\,\bar{X}$. Since (29) is independent of location and scale (Problem 6.10) suppose without loss of generality that $E(X_i) = 0$, $\sigma^2 = 1$. The minimization of $e_{W,\bar{X}}(F)$ then reduces to the problem of minimizing $\int f^2(x)\,dx$ subject to the condition

$$f(x) \geqq 0 \tag{30}$$

and

$$\int f(x)\,dx = 1, \qquad \int x f(x)\,dx = 0, \qquad \int x^2 f(x)\,dx = 1. \tag{31}$$

**Note:* Formula (29) does not apply to asymmetric F.

Using the method of undetermined multipliers we may instead minimize

$$(32) \qquad \int [f^2(x) + 2ax^2 f(x) + 2bf(x)]\, dx$$

subject to (30) and then determine a and b so as to satisfy (31). [Note that (32) should really include also a term $2cxf(x)$ corresponding to the middle condition of (31). However, it turns out that the minimizing density will automatically satisfy this condition, so that c can be taken to be zero.]

The integral (32) is minimized by minimizing the integrand for each x. Writing f for $f(x)$, the problem therefore reduces to minimizing for each x

$$(33) \qquad y = f^2 + 2ax^2 f + 2bf$$

subject to $f \geqq 0$. It is easily seen that the minimizing value of f is given by (Problem 6.12)

$$(34) \qquad f(x) = \begin{cases} -(ax^2 + b) & \text{if } ax^2 + b < 0 \\ 0 & \text{if } ax^2 + b > 0, \end{cases}$$

and that the side conditions (31) are satisfied if $a = 3/20\sqrt{5}$ and $b = -3\sqrt{5}/20$. For these values, (29) is equal to 0.864, as was to be proved (Problem 6.12).

The density f which minimizes (29) is unimodal, so that restriction to unimodal distributions does not lead to an improvement in the lower bound.

The lower bound 0.864 is substantially higher than the corresponding bounds for the trimmed means $\bar{X}_\alpha$ for $\alpha = 0.1$, or 0.125, as shown in Table 4.3. A direct comparison of W with $\bar{X}_\alpha$ shows that for $\alpha = 0.1$, $e_{W, \bar{X}_\alpha}(F) \geqq 0.865$ for all distributions F (see Bickel, 1965) but that $e_{W, \bar{X}_\alpha}$ can take on arbitrarily large values (Problem 6.13).

Although this comparison favors W over $\bar{X}_\alpha$ in the present context, the situation is reversed when F is not restricted to be symmetric. It was mentioned in Section 5.4 that the lower bound (4.5) for $e_{\bar{X}_\alpha, \bar{X}}(F)$ continues to hold for asymmetric F. The corresponding result is not true for $e_{W, \bar{X}}(F)$, which can take on values arbitrarily close to zero when F is not restricted to be symmetric (see Hoyland, 1965). General discussions of the robustness of R-estimators are provided by Bickel and Lehmann (1975) and by Hampel (1982).

Let us now return to the case of symmetric F. It is again of interest to compare W_n not only with $\bar{X}$ but with the best possible estimator for a given F. This absolute efficiency of W_n, which is given by

$$(35) \qquad e_W = I_f^{-1} 12\left(\int f^2(x)\, dx\right)^2,$$

Table 6.2. Absolute Efficiency of W_n.

Normal	T(0.01, 3)	T(0.05, 3)	T(0.1, 3)	logistic	t_5	t_3	Cauchy	DE
.955	.962	.966	.958	1.00	.993	.950	.608	.75

is shown for a number of distributions in Table 6.2. A comparison with Table 4.4 shows that despite the very different nature of the two estimators, the efficiency of W_n is quite comparable to that of $\bar{X}_{.125}$ although in most cases slightly higher.

In view of the asymptotic nature of these comparisons, again it seems important to get an idea of the closeness of the asymptotic variance of $\sqrt{n}(W_n - \theta)$ as an approximation to the actual variance. Table 6.3 gives such a comparison for $n = 20$ (and two values for $n = 40$). It suggests that for typical distributions and sample sizes above 20, the asymptotic results can be trusted to give reasonable approximations.

The excellent performance of W raises the question whether there might exist R-estimators that do even better, for example, that improve the lower bound (28). There is, in fact, an R-estimator N satisfying

$$e_{N,\bar{X}}(F) \geqq 1 \qquad \text{for all symmetric} \quad F \tag{36}$$

with strict inequality except when F is normal. Since this R-estimator is fully efficient when F is normal ($\bar{X}$ then being efficient), its score function K is given by (25) with F the standard normal distribution Φ, that is, by

$$K(t) = \Phi^{-1}(t) \tag{37}$$

since $\gamma(x) \equiv x$ in the normal case. Inequality (36) follows from the corresponding result for rank tests due to Chernoff and Savage (1958); see also Gastwirth and Wolff (1968).

In the light of earlier results it is not surprising that it is possible to obtain an adaptive R-estimator that is efficient simultaneously for all

Table 6.3. Variance of $\sqrt{n}(W_n - \theta)$ for $n = 20$ and $n = \infty$.*

	Normal	T(0.01, 3)	T(0.05, 3)	Cauchy	DE
$n = 20$	1.063	1.09	1.20	4.2	1.49
$n = 40$	1.055			3.3	
$n = \infty$	1.047	1.07	1.17	3.0	1.33
Ratio $20/\infty$	1.02	1.02	1.03	1.4	1.1

*From Andrews et al. (1972)

sufficiently smooth F. For an estimator of this kind (and references to earlier work) see Beran (1978).

For the sake of completeness, we briefly mention the alternative definition of R-estimators. Let K be a function satisfying (21). For any given a, consider the $2n$ numbers $\pm(X_1 - a), \dots, \pm(X_n - a)$ and let $\varepsilon_i = 1$ if the ith smallest is of the form $+(X_i - a)$ and $\varepsilon_i = 0$ if it is $-(X_i - a)$. If

$$(38) \qquad W(a) = \sum_{i=1}^{2n} \varepsilon_i K\left(\frac{i}{2n+1}\right),$$

then R is a solution of the equation $W(a) = 0$. When K is increasing, R is essentially unique. The equivalence of this definition with that given earlier is proved in Jaeckel (1969).

A relationship analogous to that given by (15) for L- and M-estimators holds for R- and M-estimators. For any symmetric (about 0) F and any odd function ψ, let

$$(39) \qquad K_\psi(t) = \psi[F^{-1}(t)].$$

Then (Problem 6.14)

$$(40) \qquad \sigma_R^2(F, K_\psi) = \sigma_M^2(F, \psi) = \sigma_L^2(F, \lambda_\psi)$$

where the second equation is just the restatement of (16).

In view of (40), one might expect lower bounds such as (28) and (36) to carry over to the corresponding L- and M-estimators. This is, however, not the case since the functions λ_ψ and K_ψ defined by (15) and (39) depend not only on ψ but also on F. For example, to the R-estimator W_n, which is efficient when F is logistic, corresponds the L-estimator δ_n with λ given by (5.21) (Example 6.3). The efficiency of δ_n relative to $\bar{X}$ has the sharp lower bound $4/9 \sim 0.444$ rather than the lower bound 0.864 for $e_{W_n, \bar{X}}(F)$. (For a more detailed comparison of R-estimators with the corresponding L-estimators see Scholz, (1974).)

The principal motivation for defining the classes of L-, M-, and R-estimators was the need for estimators that are robust against non-normal, particularly heavy-tailed, error distributions. However, here we have not discussed the theory of robust estimation (a book-length treatment of which is provided by Huber, 1981), or even given a formal definition of robustness, or discussed the central concepts of influence curve and breakdown point introduced by Hampel (1968, 1971, and 1974). There has also not been any discussion of robustness against other deviations from assumptions, for example, independence (in this connection, see Gastwirth and Rubin, 1975) and Portnoy (1977).*

*Robust estimation for independent but not identically distributed observations is treated in Beran (1982) and Carroll and Ruppert (1982); for dependent observations in Portnoy (1977).

A crucial assumption for most of this robustness work was the assumption of symmetry, without which even the definition of the estimand is in doubt. (Alternative approaches that avoid this assumption are proposed by Jaeckel 1971b); Bickel and Lehmann, 1975; and Collins, 1976; see also Huber, 1981, Sect. 4.9.) For the measurement problem, which motivated the one-sample problem under discussion, there often seems little justification for assuming symmetry. The assumption, however, is frequently more realistic in another context. Suppose that it is desired to estimate the difference in the effects of two treatments A and B. For this purpose a *paired-comparisons* experiment is carried out, in which the effects of the two treatments are observed on a sample of n pairs of subjects, with A and B assigned to the two members of the pair at random. If $Z_i = Y_i - X_i$, where X_i, Y_i denote responses of the members of the ith pair receiving treatments A and B, respectively, it may be reasonable to assume that the Z's are iid according to a density $f(Z_i - \Delta)$, where f is symmetric about zero and $\Delta = E(Y_i) - E(X_i)$ is the treatment difference being estimated. (For a more detailed discussion of this model, see for example, Lehmann, 1975, Chap. 4.) Except for a change of notation, this is exactly the problem we have been discussing.

For the methods of the last three sections to become widely applicable, it is necessary to extend them to general linear models. A formal extension is easiest for the M-estimators, where it is only necessary to replace the minimization of (1) with that of $\Sigma\rho(X_i - \xi_i)$ as $(\xi_1, \dots, \xi_n)$ varies over the subspace specified by the model. [For details and references, see Huber (1981, Chap. 7). Extensions of trimmed means were mentioned at the end of Sect. 5.4; the more general case of L-estimators is treated by Bickel (1973). R-estimators were proposed for the two-sample problem by Hodges and Lehmann (1963), for analysis of variance by Lehmann (1963) and Spjøtvoll (1968), and for simple linear regression by Theil (1950), Adichie (1967), and Sen (1968). Extensions to more general regression models were proposed by Koul (1969), Jureckova (1971), and Jaeckel (1972). Adaptive estimators for the linear model are discussed by Dionne (1981) and Bickel (1982).]

7. PROBLEMS

Section 1

1.1 *The Chebyshev inequality*. Inequality (1.4) holds for any random variable $Y_n = Y$ and any constants $a > 0$ and c.

1.2 To see that the converse of Theorem 1.1 does not hold, let $X_1, \dots, X_n$ be iid with $E(X_i) = \theta$, $\text{var}(X_i) = \sigma^2 < \infty$, and let $\delta_n = \bar{X}$ with probability $1 - \varepsilon_n$

and $\delta_n = A_n$ with probability ε_n. If ε_n and A_n are constants satisfying

$$\varepsilon_n \to 0 \quad \text{and} \quad \varepsilon_n A_n \to \infty,$$

then δ_n is consistent for estimating θ but $E(\delta_n - \theta)^2$ does not tend to zero.

1.3 Suppose $\rho(x)$ is an even function, nondecreasing and non-negative for $x \geqslant 0$ and positive for $x > 0$. Then $E\{\rho[\delta_n - g(\theta)]\} \to 0$ for all θ implies that δ_n is consistent for estimating $g(\theta)$.

1.4 (i) If A_n, B_n and Y_n tend in probability to a, b, and y, respectively, then $A_n + B_n Y_n$ tends in probability to $a + by$.

(ii) If A_n takes on the constant value a_n with probability 1 and $a_n \to a$, then $A_n \to a$ in probability.

1.5 Verify equation (1.11).

1.6 If $\{a_n\}$ is a sequence of real numbers tending to a, and if $b_n = (a_1 + \cdots + a_n)/n$, then $b_n \to a$.

1.7 (i) If δ_n is consistent for θ, and g is continuous, then $g(\delta_n)$ is consistent for $g(\theta)$.

(ii) Let $X_1, \ldots, X_n$ be iid as $N(\theta, 1)$, and let $g(\theta) = 0$ if $\theta \neq 0$ and $g(0) = 1$. Find a consistent estimator of $g(\theta)$.

1.8 (i) In Example 1.3, find $\text{cov}(X_i, X_j)$ for any $i \neq j$.

(ii) Verify (1.12).

1.9 (i) In Example 1.3 find the value of p_1 for which p_k becomes independent of k.

(ii) If p_1 has the value given in (i), then for any integers $i_1 < \cdots < i_r$ and k, the joint distribution of $X_{i_1}, \ldots, X_{i_r}$ is the same as that of $X_{i_1+k}, \ldots, X_{i_r+k}$.

[*Hint:* Don't calculate but use the definition of the chain.]

1.10 Let $X_1, \ldots, X_n$ be iid with density $f(x_i - \theta)$, and consider the estimation of θ with squared error loss. The Pitman estimator δ_n is consistent if for some integer r there exists an equivariant estimator based on r observations, which has finite variance.
[*Hint:* See Problem 1.29]

1.11 Suppose $X_1, \ldots, X_n$ have a common mean ξ and variance σ^2, and that $\text{cov}(X_i, X_j) = \rho_{j-i}$. For estimating ξ

(i) $\bar{X}$ is not consistent if $\rho_{j-i} = \rho \neq 0$ for all $i \neq j$;

(ii) $\bar{X}$ is consistent if $|\rho_{j-i}| \leqslant M\gamma^{j-i}$ with $|\gamma| < 1$.

[*Hint:* (i) Note that $\text{var}(\bar{X}) > 0$ for all sufficiently large n requires $\rho \geqslant 0$, and determine the distribution of $\bar{X}$ in the multivariate normal case.]

1.12 Suppose that $k_n[\delta_n - g(\theta)]$ tends in law to a continuous limit distribution H.

(i) If $k'_n/k_n \to d \neq 0$ or ∞, then $k'_n[\delta_n - g(\theta)]$ also tends to a continuous limit distribution.

(ii) If $k'_n/k_n \to 0$ or ∞, then $k'_n[\delta_n - g(\theta)]$ tends in probability to zero or infinity, respectively.

(iii) If $k_n \to \infty$, then $\delta_n \to g(\theta)$ in probability.

1.13 If $Y_n \to c$ in probability, then it tends in law to a random variable Y which is equal to c with probability 1.

1.14 (i) In Example 1.4(i) and (ii), $Y_n \to 0$ in probability.

(ii) If H_n denotes the cdf of Y_n in Example 1.4(i) and (ii), then $H_n(a) \to 0$ for all $a < 0$ and $H_n(a) \to 1$ for all $a > 0$.

(iii) Determine $\lim H_n(0)$ for Example 1.4(i) and (ii).

1.15 In Example 1.5 with $\theta = 0$, show that δ_{2n} is not exactly distributed as $\sigma^2(\chi_1^2 - 1)/n$.

1.16 In Example 1.5, let $\delta_{4n} = \max(0, \bar{X}^2 - \sigma^2/n)$, which is an improvement over δ_{1n}.

(i) $\sqrt{n}(\delta_{4n} - \theta^2)$ has the same limit distribution as $\sqrt{n}(\delta_{1n} - \theta^2)$ when $\theta \neq 0$.

(ii) Describe the limit distribution of $n\delta_{4n}$ when $\theta = 0$.

[*Hint:* Write $\delta_{4n} = \delta_{1n} + R_n$ and study the behavior of R_n.]

1.17 Let Y_n be distributed as $N(0, 1)$ with probability π_n and as $N(0, \tau_n^2)$ with probability $1 - \pi_n$. If $\tau_n \to \infty$ and $\pi_n \to \pi$, determine for what values of π the sequence $\{Y_n\}$ does and does not have a limit distribution.

1.18 Let X have the binomial distribution $b(p, n)$, and let $g(p) = pq$. The UMVU estimator of $g(p)$ is $\delta = X(n - X)/n(n - 1)$. Determine the limit distribution of $\sqrt{n}(\delta - pq)$ and $n(\delta - pq)$ when $g'(p) \neq 0$ and $g'(p) = 0$, respectively. Compare your results with those of Example 2.5.2.
[*Hint:* Consider first the limit behavior of $\delta' = X(n - X)/n^2$.]

1.19 Let $X_1, \ldots, X_n$ be iid as $N(\xi, 1)$. Determine the limit behavior of the distribution of the UMVU estimator of $p = P[|X_i| \leqslant u]$.

1.20 Let $X_1, \ldots, X_n$ be iid as $N(\xi, \sigma^2)$. Find the limiting distribution of $\sqrt{n}(S^2 - \sigma^2)$ where $S^2 = \Sigma(X_i - \bar{X})^2/(n - 1)$.

1.21 Determine the limit behavior of the estimator (2.5.27) as $n \to \infty$.
[*Hint:* Consider first the distribution of $\log \delta(T)$.]

1.22 (i) In Problem 1.17 determine to what values $\text{var}(Y_n)$ can tend as $n \to \infty$ if $\pi_n \to 1$ and $\tau_n \to \infty$ but otherwise both are arbitrary.

(ii) Use (i) to show that the limit of the variance need not agree with the variance of the limit distribution.

1.23 Prove Theorem 2.5.3.
[*Hint:* Under the assumptions of the theorem we have the Taylor expansion

$$h(x_1, \ldots, x_s) = h(\xi_1, \ldots, \xi_s) + \Sigma(x_i - \xi_i)\left[\frac{\partial h}{\partial \xi_i} + R_i\right]$$

where $R_i \to 0$ as $x_i \to \xi_i$.]

1.24 A sequence of random variables Y_n is *bounded* in probability if given any $\varepsilon > 0$ there exist M and n_0 such that $P(|Y_n| > M) < \varepsilon$ for all $n > n_0$. Show that if Y_n converges in law, then Y_n is bounded in probability.

1.25 In generalization of the notation o and O, let us say that $Y_n = o_p(1/k_n)$ if $k_n Y_n \to 0$ in probability and that $Y_n = O_p(1/k_n)$ if $k_n Y_n$ is bounded in probability.

Show that the results of Problems 5.33–5.35 of Chapter 2 continue to hold if o and O are replaced by o_p and O_p.

1.26 Let $b_{m,n}$, $m, n = 1, 2, \ldots$ be a double sequence of real numbers, which for each fixed m is nondecreasing in n. Then

(1)
$$\lim_{n\to\infty} \lim_{m\to\infty} b_{m,n} = \lim_{m,n\to\infty} \inf b_{m,n}$$

and

(2)
$$\lim_{m\to\infty} \lim_{n\to\infty} b_{m,n} = \lim_{m,n\to\infty} \sup b_{m,n}$$

provided the indicated limits exist (they may be infinite) and where lim inf $b_{m,n}$ and lim sup $b_{m,n}$ denote, respectively, the smallest and the largest limit points attainable by a sequence b_{m_k,n_k}, $k = 1, 2, \ldots$ with $m_k \to \infty$ and $n_k \to \infty$.

1.27 Let (X_n, Y_n) have a bivariate normal distribution with means $E(X_n) = E(Y_n) = 0$, variances $E(X_n^2) = E(Y_n^2) = 1$ and with correlation coefficient ρ_n tending to 1 as $n \to \infty$.

(i) Show that $(X_n, Y_n) \xrightarrow{\mathcal{L}} (X, Y)$ where X is $N(0, 1)$ and $P(X = Y) = 1$.

(ii) If $S = \{(x, y)\colon x = y\}$, show that (1.35) does not hold.

1.28 Prove Theorem 1.9.
[*Hint:* Make a Taylor expansion as in the proof of Theorem 1.5 and use Problem 1.3.15.]

1.29 Let $X_1, \ldots, X_n$ be iid with distribution P_θ, and suppose δ_n is UMVU for estimating $g(\theta)$ on the basis of $X_1, \ldots, X_n$. If there exists n_0 and an unbiased estimator $\delta_0(X_1, \ldots, X_{n_0})$ which has finite variance for all θ, then δ_n is consistent for $g(\theta)$.
[*Hint:* For $n = kn_0$ (with k an integer), compare δ_n with the estimator

$$\tfrac{1}{k}\left\{\delta_0\left(X_1, \ldots, X_{n_0}\right) + \delta_0\left(X_{n_0+1}, \ldots, X_{2n_0}\right) + \cdots\right\}.]$$

Section 2

2.1 If $k_n[\delta_n - g(\theta)]$ and $k_n'[\delta_n - g(\theta)]$ tend in law to a common distribution H which satisfies condition 2 of Theorem 2.2, then $k_n'/k_n \to 1$.

2.2 Show that the efficiency (2.12) tends to 0 as $|a - \theta| \to \infty$.

2.3 Let $X_1, \ldots, X_n$ be iid as $N(0, \sigma^2)$.

(i) Show that $\delta_n = k\Sigma|X_i|/n$ is a consistent estimator of σ if and only if $k = \sqrt{\pi/2}$.

(ii) Determine the ARE of δ_n with $k = \sqrt{\pi/2}$ with respect to $\sqrt{\Sigma X_i^2/n}$.

2.4 Let $X_1,\ldots, X_n$ be iid according to the Poisson distribution $P(\lambda)$. Find the ARE of δ_{2n} = [No. of $X_i = 0]/n$ to $\delta_{1n} = e^{-\bar{X}_n}$ as estimators of $e^{-\lambda}$.

2.5 Let $X_1,\ldots, X_n$ be iid with $E(X_i) = \theta$, $\text{var}(X_i) = 1$, $E(X_i - \theta)^4 = \mu_4$, and consider the unbiased estimators $\delta_{1n} = (1/n)\Sigma X_i^2 - 1$ and $\delta_{2n} = \bar{X}_n^2 - 1/n$ of θ^2.

(i) Determine the ARE $e_{2,1}$ of δ_{2n} with respect to δ_{1n}.
(ii) Show that $e_{2,1} \geqslant 1$ if the X_i are symmetric about θ.
(iii) Find a distribution for the X_i for which $e_{2,1} < 1$.

2.6 If $(\log n)[T_n - g(\theta)] \xrightarrow{\mathcal{L}} H$, where H satisfies condition 2 of Theorem 2.2, then for any k, $(\log n)[T_{kn} - g(\theta)]$ also tends to H in law. In this case, the ratio of sample sizes cannot be used to measure relative efficiency.

2.7 Let $X_1,\ldots, X_n$ be iid as $N(\theta, 1)$. Construct a sequence of estimators T_n of θ satisfying the assumptions of Problem 2.6 with $g(\theta) = \theta$ by letting $T_n = \bar{X}_{f(n)}$ for suitable f.

2.8 Let X be distributed as $b(p, n)$ and let $g(p) = pq$.

(i) Find the LRE of $X(n - x)/n^2$ with respect to $X(n - X)/n(n - 1)$ for all values of p.
(ii) Find the LRD when the LRE = 1.

2.9 Let $X_1,\ldots, X_n$ be independently distributed as $N(\xi, \sigma^2)$, both parameters unknown, and consider the three estimators $c_n\Sigma(X_i - \bar{X})^2$ of σ^2 corresponding to $c_n = 1/(n - 1)$ (UMVU); $c_n = 1/n$ (MLE); and $c_n = 1/(n + 1)$ (value of c_n minimizing the expected squared error). Show that the ARE and LRE is 1 for each pair, and in each case determine the LRD.
[*Hint:* Example 2.5.4(iii).]

2.10 In Example 2.3, show that the LRE of each pair of estimators δ_{in}, δ_{jn} $(1 \leqslant i < j \leqslant 3)$ with respect to each other is 1.

2.11 If δ_{in}, $i = 1, 2, 3$ are three sequences of estimators satisfying the assumptions of Theorem 2.2 (or 2.3), the following relationships hold for the ARE (or LRE): (i) $e_{3,1} = e_{3,2} \cdot e_{2,1}$; (ii) $e_{2,1} = 1/e_{1,2}$.

2.12 If the LREs in the preceding problem are 1, and the $\{\delta_{in}\}$ satisfy the assumptions of Theorem 2.4, the LRDs satisfy (i) $d_{3,1} = d_{3,2} + d_{2,1}$ and (ii) $d_{2,1} = -d_{1,2}$.

2.13 In Example 2.2, determine the LRD of the Bayes estimator $(a + X)/(a + b + n)$ with respect to X/n.

2.14 Under the assumptions of Example 4.1.3, show that the ARE and LRE of the Bayes estimator (4.1.16) with respect to $\bar{X}$ is 1, and determine the LRD.

2.15 Show that the ARE and LRE of the Bayes estimator of Problem 4.1.9 with respect to $\bar{X}$ is 1, and determine the LRD (loss function = squared error).

2.16 Let $X_1,\ldots, X_n$ be iid according to the uniform distribution $U(0, \theta)$. By Example 2.1.5, the UMVU estimator of θ is $\delta_n = (n + 1)X_{(n)}/n$ while the

MLE is $X_{(n)}$. Determine the limit distribution of (i) $n[\theta - X_{(n)}]$ and (ii) $n[\theta - \delta_n]$. What can you say about the asymptotic bias of the two estimators? [*Hint:* $P[X_{(n)} \leqslant y] = y^n/\theta^n$ for any $0 < y < \theta$.]

2.17 In the preceding problem, find the LRE (squared error loss) or LRD, or both of $X_{(n)}$ with respect to δ_n.

2.18 Let $X_1, \ldots, X_m$ and $Y_1, \ldots, Y_n$ be iid with distributions $F(x - \xi)$ and $F(y - \eta)$, respectively. Suppose that δ_{im}, $i = 1, 2$ are two sequences such that

$$\sqrt{m}\left[\delta_{im}(X_1, \ldots, X_m) - \xi\right] \xrightarrow{\mathcal{L}} N(0, \sigma_i^2).$$

Let $m_N, n_N \to \infty$, $m_N/n_N \to \lambda$ $(0 < \lambda < \infty)$ as $N \to \infty$, and consider the two sequences of estimators

$$D_{iN} = \delta_{in_N}(Y_1, \ldots, Y_{n_N}) - \delta_{im_N}(X_1, \ldots, X_{m_N})$$

of $\Delta = \eta - \xi$. Then the ARE of D_2 with respect to D_1 is σ_1^2/σ_2^2.

Section 3

3.1 Let $X_1, \ldots, X_n$ be iid according to $F(x - \theta)$ where F is symmetric about zero.

(i) Show that $X_{(k)} - \theta$ and $\theta - X_{(n-k+1)}$ have the same distribution.
(ii) Show that $\Sigma w_i X_{(i)}$ is symmetrically distributed about θ when $\Sigma w_i = 1$ and $w_k = w_{n-k+1}$ for all values of k.
(iii) Show that $\Sigma w_i X_{(i)}$ is location equivariant when $\Sigma w_i = 1$.
(iv) Under the assumptions of (ii), show that $\Sigma w_i X_{(i)}$ is not necessarily unbiased for θ when F is not symmetric.

[*Hint:* In parts (i) and (ii) assume without loss of generality that $\theta = 0$.

(i) Use the fact that when $\theta = 0$ the n-tuples $(X_1, \ldots, X_n)$ and $(-X_1, \ldots, -X_n)$ have the same joint distribution.
(ii) That $\Sigma w_i X_{(i)}$ has the same distribution as $-\Sigma w_i X_{(i)}$ when $\theta = 0$ follows from part (i).]

3.2 Let $X_1, \ldots, X_n$ be iid as $C(0, 1)$.

(i) Show that $E[X_{(k)}^2] < \infty$ if and only if $3 \leqq k \leqq n - 2$.
(ii) Use (i) to show that $E(\tilde{X}_n^2) < \infty$ for $n \geqq 5$.

[*Hint:* (i) As $x \to \infty$, $x^2 f(x) F^{k-1}(x) \to 1$ so that convergence or divergence at ∞ of the integral defining $E[X_{(k)}^2]$ is the same as that of $\int [1 - F(x)]^{n-k}\, dx$. The result follows from the fact that $[1 - F(x)]$: $(1/x) \to 1/\pi$ as $x \to \infty$.
(ii) Distinguish between n odd and n even and, in the latter case, use the fact that $[(x + y)/2)]^2 \leqq \frac{1}{2}(x^2 + y^2)$.]

3.3 If $X_1, \ldots, X_n$ are iid according to the t-distribution with 2 degrees of freedom, show that

(i) $E|X| < \infty$ and $E(X^2) = \infty$;
(ii) $\tilde{X}_n$ has finite variance for $n \geqq 3$.

3.4 If $P(Y \leqq a) = \Phi(ka)$, show that Y is normally distributed with mean zero and variance $1/k^2$.

3.5 Let k_n be a sequence of integers such that $k_n/n = \gamma + R_n$ $(0 < \gamma < 1)$ with $\sqrt{n}\, R_n \to 0$ and let $X_1, \ldots, X_n$ be iid with a distribution F for which $F(\xi) = \gamma$ and with density f which is positive at ξ. Show that

$$\sqrt{n}\left[X_{(k_n)} - \xi\right] \to N\left(0, \frac{\gamma(1-\gamma)}{f^2(\xi)}\right).$$

(Note the difference between this and the result of Problem 2.16.)

3.6 Under the assumptions of Problem 3.5, let m_n be a sequence of integers such that $m_n/n = \lambda + R'_n$ where $0 < \gamma < \lambda < 1$, $\sqrt{n}\, R'_n \to 0$, $F(\eta) = \lambda$ and $f(\eta) > 0$. Show that the joint limit distribution of $\{\sqrt{n}\,[X_{(k_n)} - \xi], \sqrt{n}\,[X_{(m_n)} - \eta]\}$ is bivariate normal with zero means and with covariance matrix (σ_{ij}) where

$$(3) \qquad \sigma_{11} = \frac{\gamma(1-\gamma)}{f^2(\xi)}, \qquad \sigma_{22} = \frac{\lambda(1-\lambda)}{f^2(\eta)}, \qquad \sigma_{12} = \frac{\gamma(1-\lambda)}{f(\xi)f(\eta)}.$$

3.7 (i) Show that

$$\int_{-\infty}^{\infty} \frac{x^{2a}}{(1+x^2)^b}\, dx = \int_0^1 y^{b-a-3/2}(1-y)^{a-1/2}\, dy$$

$$= \frac{\Gamma(a+1/2)\Gamma(b-a-1/2)}{\Gamma(b)}$$

by making the substitution $y = 1/(1+x^2)$.

(ii) Verify formula (3.15).

3.8 Prove that the efficiency (3.12) is unchanged if $F(x)$ is replaced by $F(cx)$.

3.9 Evaluate $e_{\tilde{X}, \bar{X}}(F)$ when F is

(i) the logistic distribution $L(0, 1)$;

(ii) the t-distribution with ν degrees of freedom, $(\nu > 2)$;

(iii) the double exponential distribution (3.17).

3.10 Let $X_1, \ldots, X_m$ and $Y_1, \ldots, Y_n$ be independently distributed according to distributions $F(x - \xi)$ and $F(y - \eta)$, respectively, and let $m, n \to \infty$.

(i) Show that both $\bar{Y} - \bar{X}$ and $\tilde{Y} - \tilde{X}$ are unbiased estimators of $\Delta = \eta - \xi$ if $m = n$. What can you say when $m \neq n$?

(ii) If $F(0) = 1/2$, show that the efficiency of $\tilde{Y} - \tilde{X}$ relative to $\bar{Y} - \bar{X}$ is given by (3.12).

3.11 Under the assumptions of Theorem 3.2, show that the sequence $\tilde{X}_n$ is consistent for estimating the median of the distribution of the X's even when F is not restricted to be symmetric.

Section 4

4.1 For $n = 7$, show that $\bar{X}_\alpha$ has finite variance

(i) if F is Cauchy and $\alpha \geqq 2/7$;
(ii) if F is t_2 and $\alpha \geqq 1/7$.

4.2 In the Tukey model $T(\varepsilon, \tau)$ determine for every fixed ε and α whether or not $\xi(\alpha)$, defined by (4.3), remains bounded as $\tau \to \infty$.

4.3 Determine explicit expressions for $\xi(\alpha)$ and σ_α^2 when F is double exponential.

4.4 Prove that the lower bound in (4.5) is sharp.

4.5 Show that there exists a sequence of Tukey models for which the absolute efficiency $e_{\bar{X}_\alpha} \to 0$.

4.6 In preparation for Problem 4.7, prove the following fact. If X is a random variable satisfying $P(X \leqq a) = p$, $P(X \geqq b) = q$ with $p + q = 1$, and $a < b$, and if $P(Y = a) = p$, $P(Y = b) = q$, then $\text{var}(Y) \leqq \text{var}(X)$.
[*Hint:* Let $a' = E(X|X \leqq a)$ and $b' = E(X|X \geqq b)$ and let $P(X' = a') = p$, $P(X' = b') = q$. Show that $\text{var}(X) \geqq \text{var}(X') \geqq \text{var}(Y)$.]

4.7 Prove formula (4.9).
[*Hint:* It is required to prove that the expression in { } of (4.10) does not exceed

$$\sigma^2 = \int_{-\infty}^{\infty} t^2 f(t)\, dt - \mu^2$$

where by (4.8)

$$\mu = \int_{-\infty}^{\xi(\alpha_1)} tf(t)\, dt + \int_{\xi(1-\alpha_2)}^{\infty} tf(t)\, dt.$$

Subtract $\int_{\xi(\alpha_1)}^{\xi(1-\alpha_2)} f(t) t^2\, dt$ from both sides of this inequality and apply the result stated in Problem 4.6.]

4.8 Let $X_1, \ldots, X_n$ be iid according to the uniform distribution $U(\xi, \eta)$. Then $n[X_{(1)} - \xi]$ and $n[\eta - X_{(n)}]$ are asymptotically independently distributed as $E(0, \eta - \xi)$.
[*Hint:* Show that

$$P\left[n\left(X_{(1)} - \xi\right) > a \quad \text{and} \quad n\left(\eta - X_{(n)}\right) > b\right] \to e^{-(a+b)(\eta-\xi)}.]$$

4.9 Under the assumptions of Problem 4.8, if $\xi = \theta - \frac{1}{2}$, $\eta = \theta + \frac{1}{2}$, then, asymptotically, $n[\frac{1}{2}(X_{(1)} + X_{(n)}) - \theta]$ has the distribution of the average of two independent random variables with distribution $E(0, 1)$.
[*Hint:* Theorem 1.7.]

4.10 The trimming proportion of $\bar{X}_\alpha$ is $[\alpha n]/n$. Let

$$\bar{X}'_\alpha = \frac{pX_{(k+1)} + X_{(k+2)} + \cdots + X_{(n-k-1)} + pX_{(n-k)}}{n - 2k} \tag{4}$$

with $k = [n\alpha]$ and $p = k + 1 - \alpha n$.

(i) Show that $\bar{X}'_\alpha$ is of the form $\Sigma w_i X_{(i)}$ with the w_i satisfying the conditions of Problem 3.1(ii).

(ii) Explain in what sense the trimming proportion of $\bar{X}'_\alpha$ is α.

4.11 Show that the absolute asymptotic efficiency of the median is

(i) $8/\pi^2 \sim 0.81$ in the Cauchy case;

(ii) 1 when F is double exponential.

[*Hint:* Table 2.6.2.]

Section 5

5.1 Find the constant of proportionality for the weights (5.5).

5.2 (i) Show that Theorem 5.1 implies Theorem 4.1.

(ii) Use Theorem 5.1 to obtain the asymptotic distribution of an asymmetrically trimmed mean without assuming F to be symmetric.

5.3 Show that $\sigma^2(F, \lambda)$ given by (5.10) is unchanged if a constant is added to A.

5.4 Obtain the expression (5.23) for λ_0 from the condition for equality in (5.29).

5.5 If F is symmetric about θ and λ about 1/2, then $\mu(F, \lambda)$ given by (5.9) reduces to θ.

5.6 Show that (5.21) defines a probability density on (0, 1).

5.7 In Example 5.2, show that $\gamma(x) = 2F(x) - 1$, $\gamma'(x) = 2f(x)$, and $\int \gamma^2(x)f(x)\,dx = 1/3$; and that λ_0 is given by (5.21).

5.8 In Example 5.3, (i) determine k_ν; (ii) show that λ_0 given by (5.23) satisfies $\lambda_0(t) < 0$ for $|t - \frac{1}{2}| > k_\nu$.

5.9 Let λ be the beta density $B(1/2, 1/2)$.

(i) Determine the weights (5.3) up to a constant of proportionality.

(ii) For the L-estimator δ_n with these weights, find the sharp lower bound to the efficiency $e_{\delta_n, \bar{X}}(F)$.

[*Hint:* $\Gamma(1/2) = \sqrt{\pi}$.]

5.10 Evaluate $\xi(F)$ given by (5.16) when F is the sample cdf F_n.

Section 6

6.1 If ρ is defined by (6.2), then ρ and ρ' are everywhere continuous.

6.2 Let F have a differentiable density f and let $\int \psi^2 f < \infty$.

(i) Using integration by parts show that the denominator of (6.5) is equal to

$$[\int \psi(x)f'(x)\,dx]^2. \tag{5}$$

(ii) Applying the Schwarz inequality to (5), show that $\sigma^2(F, \psi) \geq [\int (f'/f)^2 f]^{-1} = I_f^{-1}$.

6.3 If $\mathcal{F}$ is the family of all distributions (6.7) with ε, G given, and G symmetric about zero, show that $(1 - \varepsilon')G + \varepsilon' H \in \mathcal{F}$ if $0 \leqslant \varepsilon' \leqslant \varepsilon$.

6.4 Show that the least favorable density f_0 given by (6.11) is

$$f_0(x) = \begin{cases} (1-\varepsilon)\phi(-k)e^{k(x+k)}, & x < -k \\ (1-\varepsilon)\phi(x), & |x| < k \\ (1-\varepsilon)\phi(k)e^{-k(x-k)}, & x > k. \end{cases} \tag{6}$$

6.5 Generalize the minimax solution (6.9) and the least favorable distribution given in Problem 6.4 by minimizing (6.8) when F is given by (6.7) for any G with differentiable density g, symmetric about zero and strongly unimodal. [*Hint:* In (6) replace ϕ by g and in (6.9), ψ_0 by $-f_0'/f_0$.]

6.6 If ψ is normalized so that (6.17) holds, show that $\int \psi'[F^{-1}(t)]\,dt = 1$ and that (6.18) is valid.

6.7 Show that the weights (6.20) add up to 1.

6.8 Determine the R-estimator when $n = 4$ and (i) $d_1 = 1, d_4 = 2, d_2 = d_3 = 0$; (ii) $d_1 = d_4 = 2, d_2 = d_3 = 0$.
[*Hint:* Distinguish the cases in which $X_{(i)} < \frac{1}{2}[X_{(1)} + X_{(4)}] < X_{(i+1)}$, for $i = 1, 2, 3$.]

6.9 Show that $\sigma^2(F, K)$ given by (6.24) reduces to (6.27) for $K(t) = 2t - 1$.

6.10 Show that the efficiency (6.29) is unchanged when the X_i are replaced by $aX_i + b$.

6.11 Calculate the efficiency of W to $\bar{X}$ given by (6.29) for the following distributions:

(i) the logistic distribution (*Answer*: $\pi^2/9$);
(ii) the t-distribution with ν degrees of freedom;
(iii) the double exponential distribution.

[*Hint:* (ii) Problem 3.7.]

6.12 Show that

(i) (6.33) is minimized subject to $f \geqq 0$ by (6.34);
(ii) (6.31) is satisfied by the density f defined by

$$f(x) = \begin{cases} A(B^2 - x^2) & \text{if } |x| < B \\ 0 & \text{otherwise,} \end{cases}$$

where $A = 3/20\sqrt{5}$, $B = \sqrt{5}$;
(iii) the value of (6.29) for the density f of (ii) is $108/125 = 0.864$.

6.13 Show that there exist distributions F for which $e_{W,\bar{X}_\alpha}$ is arbitrarily large.

6.14 Prove formula (6.40)

8. REFERENCES

As is the case with so many statistical ideas, the concept of asymptotic relative efficiency can be found in early examples treated by Laplace and Gauss (see, for example, Stigler, 1973a). Absolute efficiency seems to have been considered first by Fisher (1922) in his work on maximum likelihood estimation. The second-order concept of deficiency was introduced by Hodges and Lehmann (1970); references to other second-order approaches when the first-order efficiency (ARE) is 1, are given in Section 6.7. Lemma 1.2 is due to Chernoff (1956) and Hodges and Lehmann (1956).

The median, trimmed means, and suitable L-, M-, and R-estimators as robust alternatives to the mean for estimating location have a long history going back to Laplace some of which, particularly the remarkable contributions of Daniell (1920), are sketched in Stigler (1973b). The modern development of robust estimation received an important impetus from the writing of Tukey, for example, in his paper on the effects of contamination (1960). A systematic approach through approximate models (and the associated minimax theorem for M-estimators) is due to Huber (1964). An alternative global approach was suggested by Bickel and Lehmann (1975–1979). The central concepts of influence curve and breakdown point were introduced by Hampel (1968), and by Hodges (1967) and Hampel (1968), respectively. R-estimators were defined by Hodges and Lehmann (1963) and, in the form given here, by Jaeckel (1969) and Jureckova (1971); an early forerunner is the regression estimator of Theil (1950). An appraisal of the progress in this area during the 1960s was obtained during the "Princeton robustness-year," which resulted in the book by Andrews et al. (1972). A systematic account of robust estimation is given by Huber (1981).

Adichie, J.
(1967). "Estimates of regression parameters based on rank tests." *Ann. Math. Statist.* **38**, 894–904.

Anderson, T. W. and Goodman, L. A.
(1957). "Statistical inferences about Markov chains." *Ann. Math. Statist.* **28**, 89–110.

Andrews, D. F., Bickel, P. J., Hampel, F. R., Huber, P. J., Rogers, W. H., and Tukey, J. W.
(1972). *Robust Estimates of Location: Survey and Advances.* Princeton University Press, Princeton, N.J.

Bahadur, R. R.
(1971). *Some Limit Theorems in Statistics.* SIAM, Philadelphia.

Beran, R.
(1978). "An efficient and robust adaptive estimator of location." *Ann. Statist.* **6**, 292–313.

Beran, R.
(1982). "Robust estimation in models for independent nonidentically distributed data." *Ann. Statist.* **10**, 415–428.

Bickel, P. J.
(1965). "On some robust estimates of location." *Ann. Math. Statist.* **36**, 847–858.

Bickel, P. J.
(1973). "On some analogues to linear combinations of order statistics in the linear model." *Ann. Statist.* **1**, 597–616.

Bickel, P. J.
(1982). "On adaptive estimation." *Ann. Statist.* **10**, 647–671.

Bickel, P. J. and Doksum, K. A.
(1977). *Mathematical Statistics*. Holden-Day, San Francisco.

Bickel, P. J. and Lehmann, E. L.
(1975–1979). "Descriptive statistics for nonparametric models. I–III." *Ann. Statist.* **3**, 1038–1045, 1045–1069; **4**, 1139–1158; IV *Contributions to Statistics: Hájek Memorial Volume*, J. Jureckova, Ed. Academia, Prague, pp. 33–40.

Billingsley, P.
(1961). *Statistical Inference for Markov Processes*. University of Chicago Press, Chicago.

Billingsley, P.
(1979). *Probability and Measure*. Wiley, New York.

Boos, D. D.
(1979). "A differential for *L*-statistics." *Ann. Statist.* **7**, 955–959.

Brown, G. W. and Tukey, J. W.
(1946). "Some distributions of sample means." *Ann. Math. Statist.* **17**, 1–12.

Carroll, R. J. and Ruppert, D.
(1982). "Robust estimation in heteroscedastic linear models." *Ann. Statist.* **10**, 429–441.

Chernoff, H.
(1956). "Large-sample theory: Parametric case." *Ann. Math. Statist.* **27**, 1–22.

Chernoff, H. and Savage, I. R.
(1958). Asymptotic normality and efficiency of certain nonparametric test statistics." *Ann. Math. Statist.* **29**, 972–994.

Collins, J. R.
(1976). "Robust estimation of a location parameter in the presence of asymmetry." *Ann. Statist.* **4**, 68–85.

Cramér, H.
(1946). *Mathematical Methods of Statistics*. Princeton University Press, Princeton, N.J.

Crow, E. and Siddiqui M.
(1967). "Robust estimation of location." *J. Amer. Statist. Assoc.* **62**, 353–389.

Daniell, P. J.
(1920). "Observations weighted according to order." *Am. J. Math.* **42**, 222–236.

Dionne, L.
(1981). "Efficient nonparametric estimators of parameters in the general linear hypothesis." *Ann. Statist.* **9**, 457–460.

Feller, W.
(1970). *An Introduction to Probability Theory and Its Applications*. Vol.1, 3d Ed., Rev. Printing. Wiley, New York.

Field, C. A. and Hampel, F. R.
(1982). "Small-sample asymptotic distributions of *M*-estimators of location." *Biometrika* **69**, 29–46.

Fisher, R. A.
(1922). "On the mathematical foundations of theoretical statistics." *Phil. Trans. Roy. Soc.* (*A*) **222**, 309–368.

Freedman, D. A. and Diaconis, P.
(1982). "On inconsistent *M*-estimators." *Ann. Statist.* **10**, 454–461.

Gastwirth, J. and Cohen, M.
(1970). "Small sample behavior of some robust linear estimates of location." *J. Amer. Statist. Assoc.* **65**, 946–973.

Gastwirth, J. and Rubin, H.
(1975). "The behavior of robust estimators on dependent data." *Ann. Statist.* **3**, 1070–1100.

Gastwirth, J. and Wolff, S.
(1968). "An elementary method of obtaining lower bounds on the asymptotic power of rank tests." *Ann. Math. Statist.* **39**, 2128–2130.

Groeneboom, P. and Oosterhoof, J.
(1981). "Bahadur efficiency and small sample efficiency." *Rev. Intern. Statist.* **49**, 127–141.

Hampel, F. R.
(1968). "Contributions to the theory of robust estimation." Ph.D. thesis, University of California, Berkeley.

Hampel, F. R.
(1971). "A general qualitative definition of robustness." *Ann. Math. Statist.* **42**, 1887–1896.

Hampel, F. R.
(1974). "The influence curve and its role in robust estimation." *J. Amer. Statist. Assoc.* **62**, 1179–1186.

Hampel, F. R.
(1982). "The robustness of some nonparametric procedures." In Festschrift for Erich Lehmann, Ed. Bickel, Doksum, Hodges, Wadsworth, Belmont.

Hodges, J. L., Jr.
(1967). "Efficiency in normal samples and tolerance of extreme values for some estimates of location." *Proc. Fifth Berkeley Symp.* **1**, 163–186.

Hodges, J. L., Jr., and Lehmann, E. L.
(1956). "Two approximations to the Robbins–Monro process." *Proc. Third Berkeley Symp.* **1**, 95–104.

Hodges, J. L., Jr., and Lehmann, E. L.
(1963). "Estimates of location based on rank tests." *Ann. Math. Statist.* **34**, 598–611.

Hodges, J. L., Jr., and Lehmann, E. L.
(1970). "Deficiency." *Ann. Math. Statist.* **41**, 783–801.

Hoyland, A.
(1965). "Robustness of the Hodges–Lehmann estimates for shift." *Ann. Math. Statist.* **36**, 174–197.

Huber, P. J.
(1964). "Robust estimation of a location parameter." *Ann. Math. Statist.* **35**, 73–101.

Huber, P. J.
(1973). "Robust regression: Asymptotics, conjectures and Monte Carlo." *Ann. Statist.* **1**, 799–821.

Huber, P. J.
(1981). *Robust Statistics*. Wiley, New York.

Jaeckel, L. A.
(1969). "Robust estimates of location." Ph.D. thesis, University of California, Berkeley.

Jaeckel, L. A.
(1971a). "Robust estimates of location: Symmetry and asymmetric contamination." *Ann. Math. Statist.* **42**, 1020–1034.

Jaeckel, L. A.
(1971b). "Some flexible estimates of location." *Ann. Math. Statist.* **42**, 1540–1552.

Jaeckel, L. A.
(1972). "Estimating regression coefficients by minimizing the dispersion of the residuals. *Ann. Math. Statist.* **43**, 1449–1458.

Jureckova, J.
(1971). "Nonparametric estimates of regression coefficients." *Ann. Math. Statist.* **42**, 1328–1338.

Koul, H. L.
(1969). "Asymptotic behavior of Wilcoxon type confidence regions in multiple linear regression." *Ann. Statist.* **40**, 1950–1979.

Lehmann, E. L.
(1963). "Robust estimation in analysis of variance." *Ann. Math. Statist.* **34**, 957–966.

Lehmann, E. L.
(1975). *Nonparametrics: Statistical Methods Based on Ranks*. Holden-Day, San Francisco.

Newcomb, S.
(1882). "Discussion and results of observations on transits of Mercury from 1677 to 1881." *Astronomical Papers*, Vol. 1. U.S. Nautical Almanac Office, pp. 363–487.

Newcomb, S.
(1886). "A generalized theory of the combination of observations so as to obtain the best result." *Amer. J. Math.* **8**, 343–366.

Port, S. C. and Stone, C. J.
(1974). "Fisher information and the Pitman estimator of a location parameter." *Ann. Statist.* **2**, 225–247.

Portnoy, S. L.
(1977). "Robust estimation in dependent situations." *Ann. Statist.* **5**, 22–43.

Randles, R. H.
(1982). "On the asymptotic normality of statistics with estimated parameters." *Ann. Statist.* **10**, 462–474.

Rocke, D. M., Downs, G. W., and Rocke, A. J.
(1982). "Are robust estimators really necessary?" *Technometrics* **24**, 95–101.

Ruppert, D. and Carroll, R. J.
(1980). "Trimmed least squares estimation in the linear model." *J. Amer. Statist. Assoc.* **75**, 828–838.

Sacks, J.
(1975). "An asymptotically efficient sequence of estimators of a location parameter." *Ann. Statist.* **3**, 285–298.

Sacks, J. and Ylvisaker, D.
(1982). "*L*- and *R*-estimation and the minimax property." *Ann. Statist.* **10**, 643–645.

Sampford, M. R.
(1953). "Some inequalities on Mill's ratio and related functions." *Ann. Math. Statist.* **24**, 130–132.

Scholz, F. W.
(1974). "A comparison of efficient location estimators." *Ann. Statist.* **2**, 1323–1326.

Sen, P. K.
(1968). "Estimates of the regression coefficient based on Kendall's tau." *J. Amer. Statist. Assoc.* **63**, 1379–1389.

Serfling, R. J.
(1980). *Approximation Theorems of Mathematical Statistics*. Wiley, New York.

Shorack, G. R.
(1972). "Functions of order statistics." *Ann. Math. Statist.* **43**, 412–427.

Siddiqui, M. M. and Raghunandanan, K.
(1967). "Asymptotically robust estimators of location." *J. Amer. Statist. Assoc.* **62**, 950–953.

Spjøtvoll, E.
(1968). "A note on robust estimation in analysis of variance." *Ann. Math. Statist.* **39**, 1486–1492.

Spjøtvoll, E. and Aastreit, A. H.
(1980). "Comparison of robust estimators on data from field experiments." *Scand. J. Statist.* **7**, 1–13.

Staudte, R. G., Jr.
(1980). "Robust estimation." *Queen's Papers in Pure and Applied Mathematics.* No. 53. Queen's University, Kingston, Ontario.

Stigler, S. M.
(1973a). "Laplace, Fisher, and the discovery of the concept of sufficiency." *Biometrika* **60**, 439–445.

Stigler, S. M.
(1973b). "Simon Newcomb, Percy Daniell, and the history of robust estimation." *J. Amer. Statist. Assoc.* **68**, 872–879.

Stigler, S. M.
(1973c). "The asymptotic distribution of the trimmed mean." *Ann. Statist.* **1**, 472–477.

Stigler, S. M.
(1974). "Linear functions of order statistics with smooth weight functions." *Ann. Statist.* **2**, 676–693. (Correction, *Ann. Statist.* **7**, 466).

Stigler, S. M.
(1977). "Do robust estimators work with real data?" *Ann. Statist.* **5**, 1055–1077.

Stigler, S. M.
(1980). "An Edgeworth curiosum." *Ann. Statist.* **8**, 931–934.

Stone, C. J.
(1974). "Asymptotic properties of estimators of a location parameter." *Ann. Statist.* **2**, 1127–1137.

Stone, C. J.
(1975). "Adaptive maximum likelihood estimators of a location parameter." *Ann. Statist.* **3**, 267–284.

Theil, H.
(1950). "A rank-invariant method of linear and polynomial regression analysis." *Koninkl. Ned. Weteusch. Proc.* **53**, 386–392, 521–525, 1397–1412.

Tukey, J. W.
(1960). "A survey of sampling from contaminated distributions." In *Contributions to Probability and Statistics*. Olkin, Ed. Stanford University Press, Stanford, Calif.

CHAPTER 6

Asymptotic Optimality

1. ASYMPTOTIC EFFICIENCY

The large sample approximations of the preceding chapter not only provide a convenient method for assessing the performance of an estimator and for comparing different estimators, they also permit a new approach to optimality that is much less restrictive than the theories of unbiased and equivariant estimation developed in Chapters 2 and 3.

It was seen in Sections 5.1 and 5.2 that estimators* of interest typically are consistent as the sample sizes tend to infinity and, suitably normalized, are asymptotically normally distributed about the estimand with a variance $v(\theta)$ (the asymptotic variance), which provides a reasonable measure of the accuracy of the estimator sequence. (In this connection, see Problem 1.1.) Within this class of consistent asymptotically normal estimators, it turns out that under mild additional restrictions there exist estimators that uniformly minimize $v(\theta)$. The present chapter is mainly concerned with the development of fairly explicit methods of obtaining such *asymptotically efficient* estimators.

Before embarking on this program, it may be helpful to note an important difference between the present large-sample approach and the small-sample results of Chapters 2 through 4. Both UMVU and MRE estimators tend to be unique (Theorem 1.6.5) and so are at least some of the minimax estimators derived in Chapter 4. On the other hand, it is in the nature of asymptotically optimal solutions not to be unique since asymptotic results refer to the limiting behavior of sequences, and the same limit is shared by many different sequences. More specifically, if

$$\sqrt{n}\,[\delta_n - g(\theta)] \xrightarrow{\mathcal{L}} N(0, v)$$

*As in the preceding chapter, we shall frequently use *estimator* instead of the more accurate but cumbersome term *estimator sequence*.

and $\{\delta_n\}$ is asymptotically optimal in the sense of minimizing v, then $\delta_n + R_n$ is also optimal provided

$$\sqrt{n}\, R_n \to 0 \qquad \text{in probability.}$$

As we shall see later, asymptotically equivalent optimal estimators can be obtained from quite different starting points.

A central role in the theory of asymptotic efficiency is played by an analog of the information inequality (2.6.28). If $X_1, \dots, X_n$ are iid according to a density $f_\theta(x)$ (with respect to μ) satisfying suitable regularity conditions, this inequality states that the variance of any unbiased estimator δ of $g(\theta)$ satisfies

$$\text{var}_\theta(\delta) \geqslant \frac{[g'(\theta)]^2}{nI(\theta)}, \tag{1}$$

where $I(\theta)$ is the amount of information in a single observation defined by (2.6.10).

Suppose now that $\delta_n = \delta_n(X_1, \dots, X_n)$ is asymptotically normal, say that

$$\sqrt{n}\,[\delta_n - g(\theta)] \xrightarrow{\mathcal{L}} N[0, v(\theta)], \quad v(\theta) > 0. \tag{2}$$

Then it turns out that under some additional restrictions one also has

$$v(\theta) \geqslant \frac{[g'(\theta)]^2}{I(\theta)}. \tag{3}$$

However, although the lower bound (1) is attained only in exceptional circumstances (Sect. 2.6), there exist sequences $\{\delta_n\}$ that satisfy (2) with $v(\theta)$ equal to the lower bound (3) subject only to quite general regularity conditions. A sequence $\{\delta_n\}$ satisfying (2) with

$$v(\theta) = \frac{[g'(\theta)]^2}{I(\theta)} \tag{4}$$

is said to be *asymptotically efficient*.

At first glance, (3) might be thought to be a consequence of (1). Two differences between the inequalities (1) and (3) should be noted, however. (i) The estimator δ in (1) is assumed to be unbiased, while (2) implies consistency of $\{\delta_n\}$ but not that δ_n is unbiased or even that its bias tends to zero (Problem 1.5). (ii) The quantity $v(\theta)$ in (3) is an asymptotic variance and (1) refers to the actual variance of δ. It follows from Lemma 5.1.2 that

$$v(\theta) \leqslant \liminf[n \operatorname{var}_\theta \delta_n] \tag{5}$$

but equality need not hold. Thus, (3) is a consequence of (1) provided

$$\operatorname{var}\{\sqrt{n}\,[\delta_n - g(\theta)]\} \to v(\theta) \tag{6}$$

and if δ_n is unbiased, but not necessarily if these requirements do not hold.

For a long time, (3) was nevertheless believed to be valid subject only to regularity conditions on the densities f_θ. This belief was exploded by the example (due to Hodges; see Le Cam, 1953) given below. Before stating the example, note that in discussing the inequality (3) under assumption (2), if θ is real-valued and $g(\theta)$ is differentiable, it is enough to consider the case $g(\theta) = \theta$, for which (3) reduces to

$$v(\theta) \geqq \frac{1}{I(\theta)}. \tag{7}$$

For if

$$\sqrt{n}\,(\delta_n - \theta) \xrightarrow{\mathcal{L}} N[0, v(\theta)] \tag{8}$$

and if g has derivative g', it was seen in Theorem 5.1.5 that

$$\sqrt{n}\,[g(\delta_n) - g(\theta)] \xrightarrow{\mathcal{L}} N[0, v(\theta) g'^2(\theta)]. \tag{9}$$

After the obvious change of notation, this implies (3).

Example 1.1. Let $X_1, \ldots, X_n$ be iid according to the normal distribution $N(\theta, 1)$ and let the estimand be θ. It was seen in Table 2.6.1 that in this case $I(\theta) = 1$ so that (7) reduces to $v(\theta) \geqq 1$. On the other hand, consider the sequence of estimators,

$$\delta_n = \begin{cases} \bar{X} & \text{if } |\bar{X}| \geqq 1/n^{1/4} \\ a\bar{X} & \text{if } |\bar{X}| < 1/n^{1/4}. \end{cases}$$

Then (Problem 1.2)

$$\sqrt{n}\,(\delta_n - \theta) \xrightarrow{\mathcal{L}} N[0, v(\theta)]$$

There with $v(\theta) = 1$ when $\theta \neq 0$ and $v(\theta) = a^2$ when $\theta = 0$. If $a < 1$, inequality (3) is therefore violated at $\theta = 0$.

This phenomenon is quite general (Problems 1.3–1.4). There will typically exist estimators satisfying (8) but with $v(\theta)$ violating (7) for at least some values of θ, called points of *superefficiency*. However, (7) is almost true, for it was shown by Le Cam (1953) that for any sequence δ_n satisfying (8) the set S of points of superefficiency has Lebesgue measure zero. The following version of this result which we shall not prove, is due to Bahadur (1964). The assumptions are somewhat stronger but similar to those of Theorem 2.6.4.

Remark on notation. In the present chapter we shall use X_i, X, and x_i, x for real-valued random variables and the values they take on, and $\underline{X}$, $\underline{x}$ for the vectors $(X_1, \ldots, X_n)$, $(x_1, \ldots, x_n)$. This differs from the notation in earlier chapters, for example, from that in Section 2.6.

Theorem 1.1. *Let* $X_1, \ldots, X_n$ *be iid, each with density* $f(x, \theta)$ *with respect to a σ-finite measure* μ, *where* θ *is real-valued, and suppose the following regularity conditions hold.*

(i) *The parameter space* Ω *is an open interval (not necessarily finite).*

(ii) *The distributions* P_θ *of the* X_i *have common support, so that the set* $A = \{x: f(x, \theta) > 0\}$ *is independent of* θ.

(iii) *For every* $x \in A$, *the density* $f(x, \theta)$ *is twice differentiable with respect to* θ, *and the second derivative is continuous in* θ.

(iv) *The integral* $\int f(x, \theta)\, d\mu(x)$ *can be twice differentiated under the integral sign.*

(v) *The Fisher information* $I(\theta)$ *defined by* (2.6.10) *satisfies* $0 < I(\theta) < \infty$.

(vi) *For any given* $\theta_0 \in \Omega$ *there exists a positive number* c *and a function* $M(x)$ *(both of which may depend on* θ_0*) such that*

$$|\partial^2 \log f(x, \theta)/\partial\theta^2| \leqslant M(x)$$

$$\text{for all} \quad x \in A, \quad \theta_0 - c < \theta < \theta_0 + c$$

and

$$E_{\theta_0}[M(X)] < \infty.$$

Under these assumptions, if $\delta_n = \delta_n(X_1, \ldots, X_n)$ *is any estimator satisfying* (8), *then* $v(\theta)$ *satisfies* (7) *except on a set of Lebesgue measure zero.*

Note that by Lemma 2.6.1, condition (iv) ensures that for all $\theta \in \Omega$

(vii) $$E[\partial \log f(X, \theta)/\partial\theta] = 0$$

and

$$\text{(viii)}\ E\left[-\partial^2 \log f(X,\theta)/\partial\theta^2\right] = E\left[\partial \log f(X,\theta)/\partial\theta\right]^2 = I(\theta).$$

Condition (iv) can be replaced by conditions (vii) and (viii) in the statement of the theorem.

The example makes it clear that no regularity conditions on the densities $f(x,\theta)$ can prevent estimators from violating (7). This possibility can be avoided only by placing restrictions also on the sequence of estimators. In view of the information inequality (2.6.28), an obvious sufficient condition is (6) [with $g(\theta) = \theta$] together with

$$b_n'(\theta) \to 0 \tag{10}$$

where $b_n(\theta) = E_\theta(\delta_n) - \theta$ is the bias of δ_n.

If $I(\theta)$ is continuous, as will typically be the case, a more appealing assumption is perhaps that $v(\theta)$ also be continuous. Then (7) clearly cannot be violated at any point since otherwise it would be violated in an interval around this point in contradiction to Theorem 1.1. As an alternative, which under mild assumptions on f implies continuity of $v(\theta)$, Rao (1963) and Wolfowitz (1965) require the convergence in (2) to be uniform in θ. By working with coverage probabilities rather than asymptotic variance, the latter author also removes the unpleasant assumption that the limit distribution in (2) must be normal. An analogous result is proved by Pfanzagl (1970), who requires the estimators to be asymptotically median unbiased.

The search for restrictions on the sequence $\{\delta_n\}$, which would ensure (7) for all values of θ, is motivated in part by the hope of the existence, within the restricted class, of uniformly best estimators for which $v(\theta)$ attains the lower bound. It is further justified by the fact, brought out by Le Cam (1953), Huber (1966), and Hájek (1972), that violation of (7) at a point θ_0 entails certain unpleasant properties of the risk of the estimator in the neighborhood of θ_0.

This behavior can be illustrated on the Hodges example.

Example 1.1. *(Continued).* The normalized risk function

$$R_n(\theta) = nE(\delta_n - \theta)^2 \tag{11}$$

of the Hodges estimator δ_n can be written as

$$R_n(\theta) = 1 - (1-a^2)\int_{\underline{l}_n}^{\bar{l}_n} (x + \sqrt{n}\,\theta)^2 \phi(x)\,dx$$

$$+2\theta\sqrt{n}\,(1-a)\int_{\underline{l}_n}^{\bar{l}_n} (x + \sqrt{n}\,\theta)\phi(x)\,dx$$

where $\bar{l}_n = \sqrt[4]{n} - \sqrt{n}\,\theta$, and $\underline{l}_n = -\sqrt[4]{n} - \sqrt{n}\,\theta$. When the integrals are broken up into their three and two terms, respectively, and the relations

$$\Phi'(x) = \phi(x) \quad \text{and} \quad \phi'(x) = -x\phi(x)$$

are used, $R_n(\theta)$ reduces to

$$R_n(\theta) = \left[1 - (1 - a^2)\int_{\underline{l}_n}^{\bar{l}_n} x^2\phi(x)\,dx\right]$$

$$+ n\theta^2(1 - a)^2\left[\Phi(\bar{l}_n) - \Phi(\underline{l}_n)\right] + 2\sqrt{n}\,\theta a(1 - a)\left[\phi(\bar{l}_n) - \phi(\underline{l}_n)\right].$$

Consider now the sequence of parameter values $\theta_n = 1/\sqrt[4]{n}$, so that

$$\sqrt{n}\,\theta_n = n^{1/4}, \quad \underline{l}_n = -2n^{1/4}, \quad \bar{l}_n = 0.$$

Then

$$\sqrt{n}\,\theta_n\phi(\underline{l}_n) \to 0$$

so that the third term tends to infinity as $n \to \infty$. Since the second term is positive and the first term is bounded, it follows that

$$R_n(\theta_n) \to \infty \qquad \text{for} \quad \theta_n = 1/\sqrt[4]{n},$$

and hence a fortiori that

$$\sup_\theta R_n(\theta) \to \infty.$$

Let us now compare this result with the fact that (Problem 1.6) for any fixed θ

$$R_n(\theta) \to 1 \qquad \text{for} \quad \theta \neq 0, \qquad R_n(0) \to a^2.$$

(This shows that in the present case the limiting risk is equal to the asymptotic variance found in Example 1.1.) The functions $R_n(\theta)$ are continuous functions of θ with discontinuous limit function

$$L(\theta) = 1 \qquad \text{for} \quad \theta \neq 0, \qquad L(0) = a^2.$$

However, each of the functions with large values of n rises to a high above the limit value 1, at values of θ tending to the origin with n, and with the value of the peak tending to infinity. The improvement (over $\bar{X}$) from 1 to a^2 in the limiting risk at the origin and hence for large finite n also near the origin, therefore, leads to an enormous increase in risk at points slightly further away but nevertheless close to the origin. (In this connection, see Problem 1.8.)

2. EFFICIENT LIKELIHOOD ESTIMATION

Under smoothness assumptions similar to those of Theorem 1.1, we shall in the present section prove the existence of asymptotically efficient estimators and provide a method for determining such estimators which in many cases leads to an explicit solution.

We begin with the following assumptions:

(A0) The distributions P_θ of the observations are distinct (otherwise, θ cannot be estimated consistently; but see Redner (1981) for a different point of view).

(A1) The distributions P_θ have common support.

(A2) The observations are $\underline{X} = (X_1, \ldots, X_n)$, where the X_i are iid with probability density $f(x_i, \theta)$ with respect to μ.

(A3) The parameter space Ω contains an open interval ω of which the true parameter value θ_0 is an interior point.

Note: The true value of θ will be denoted by θ_0. The density $f(x_1, \theta) \cdots f(x_n, \theta)$, considered for fixed $\underline{x}$ as a function of θ, is called the *likelihood function*.

Theorem 2.1. *Under assumptions* $(A0)$–$(A2)$,

$$(1) \quad P_{\theta_0}\{f(X_1, \theta_0) \cdots f(X_n, \theta_0) > f(X_1, \theta) \cdots f(X_n, \theta)\} \to 1 \quad \text{as } n \to \infty$$

for any fixed $\theta \neq \theta_0$.

Proof. The inequality is equivalent to

$$\frac{1}{n}\Sigma \log[f(X_i, \theta)/f(X_i, \theta_0)] < 0.$$

By the law of large numbers, the left side tends in probability toward

$$E_{\theta_0}\log[f(X, \theta)/f(X, \theta_0)].$$

Since $-\log$ is strictly convex, Jensen's inequality shows that

$$E_{\theta_0}\log[f(X, \theta)/f(X, \theta_0)] < \log E_{\theta_0}[f(X, \theta)/f(X, \theta_0)] = 0,$$

and the results follows.

By (1), the density of $\underline{X}$ at the true θ_0 exceeds that at any other fixed θ with high probability when n is large. We do not know θ_0 but we can determine the value $\hat{\theta}$ of θ which maximizes the density of $\underline{X}$, that is, which maximizes the likelihood function at the observed $\underline{X}$. If this value exists and is unique, it is the *maximum likelihood estimator* (MLE) of θ.* The MLE of $g(\theta)$ is defined to be $g(\hat{\theta})$. If g is 1 : 1 and $\xi = g(\theta)$, this agrees with the definition of $\hat{\xi}$ as the value of ξ that maximizes the likelihood, and the definition is consistent also in the case that g is not 1 : 1. (In this connection, see Berk, 1967, and Zehna, 1966.)

Theorem 2.1 suggests that if the density of $\underline{X}$ varies smoothly with θ, the MLE of θ typically should be close to the true value of θ, and hence be a reasonable estimator.

Example 2.1. Let X have the binomial distribution $b(p, n)$. Then the MLE of p is obtained by maximizing $\binom{n}{x} p^x q^{n-x}$ and hence is $\hat{p} = X/n$ (Problem 2.1).

Example 2.2. If $X_1, \ldots, X_n$ are iid as $N(\xi, \sigma^2)$, it is convenient to obtain the MLE by maximizing the logarithm of the density, $-n \log \sigma - 1/2\sigma^2 \Sigma(x_i - \xi)^2 - c$. When (ξ, σ) are both unknown, the maximizing values are $\hat{\xi} = \bar{X}$, $\hat{\sigma}^2 = \Sigma(X_i - \bar{X})^2/n$ (Problem 2.3).

As a first question regarding the MLE for iid variables, let us ask whether it is consistent. We begin with the case in which Ω is finite, so that θ can take on only a finite number of values. In this case, a sequence δ_n is consistent if and only if

$$P_\theta(\delta_n = \theta) \to 1 \quad \text{for all} \quad \theta \in \Omega \tag{2}$$

(Problem 2.6).

Corollary 2.1. *Under assumptions (A0)–(A2) if Ω is finite, the MLE $\hat{\theta}_n$ exists, it is unique with probability tending to* 1, *and it is consistent.*

Proof. The result is an immediate consequence of Theorem 2.1 and the fact that if $P(A_{in}) \to 1$ for $i = 1, \ldots, k$, then also $P[A_{1n} \cap \cdots \cap A_{kn}] \to 1$ as $n \to \infty$.

The proof of Corollary 2.1 breaks down when Ω is not restricted to be finite. That the consistency conclusion itself can break down even if Ω is only countably infinite is shown by the following example due to Bahadur (1958) and Le Cam (1979b).

Example 2.3. Let h be a continuous function defined on (0, 1], which is strictly decreasing, with $h(x) \geqslant 1$ for all $0 < x \leqslant 1$ and satisfying

$$\int_0^1 h(x)\, dx = \infty \tag{3}$$

*For a more general definition, see Scholz (1980). A discussion of the MLE as a summarizer of the data rather than an estimator is given by Efron (1982).

Given a constant $0 < c < 1$, let a_k, $k = 0, 1, \ldots$ be a sequence of constants defined inductively as follows: $a_0 = 1$; given $a_0, \ldots, a_{k-1}$, the constant a_k is defined by

$$\int_{a_k}^{a_{k-1}} [h(x) - c]\, dx = 1 - c. \tag{4}$$

It is easy to see that there exists a unique value $0 < a_k < a_{k-1}$ satisfying (4) (Problem 2.7). Since the sequence $\{a_k\}$ is decreasing, it tends to a limit $a \geqslant 0$. If a were > 0, the left side of (4) would tend to zero which is impossible. Thus $a_k \to 0$ as $k \to \infty$.

Consider now the sequence of densities

$$f_k(x) = \begin{cases} c & \text{if } x \leqslant a_k \text{ or } x > a_{k-1} \\ h(x) & \text{if } a_k < x \leqslant a_{k-1} \end{cases},$$

and the problem of estimating the parameter k on the basis of independent observations $X_1, \ldots, X_n$ from f_k. We shall show that the MLE exists and that it tends to infinity in probability regardless of the true value k_0 of k and is therefore not consistent, provided $h(x) \to \infty$ sufficiently fast as $x \to 0$.

Let us denote the joint density of the X's by

$$p_k(\underline{x}) = f_k(x_1) \cdots f_k(x_n).$$

That the MLE exists follows from the fact that $p_k(\underline{x}) = c^n < 1$ for any value of k for which the interval $I_k = (a_k, a_{k-1}]$ contains none of the observations, so that the maximizing value of k must be one of the $\leqslant n$ values for which I_k contains at least one of the x's.

For $n = 1$, the MLE is the value of k for which $X_1 \in I_k$, and for $n = 2$ the MLE is the value of k for which $X_{(1)} \in I_k$. For $n = 3$, it may happen that one observation lies in I_k and two in I_l $(k < l)$, and whether the MLE is k or l then depends on whether $c \cdot h(x_{(1)})$ is greater than or less than $h(x_{(2)})h(x_{(3)})$.

We shall now prove that the MLE $\hat{K}_n$ (which is unique with probability 1) tends to infinity in probability, that is, that

$$P(\hat{K}_n > k) \to 1 \quad \text{for every } k, \tag{5}$$

provided h satisfies

$$h(x) \geqslant e^{1/x^2} \tag{6}$$

for all sufficiently small values of x.

To prove (5), we will show that for any fixed j

$$P\left[p_{K_n^*}(\underline{X}) > p_j(\underline{X})\right] \to 1 \quad \text{as } n \to \infty \tag{7}$$

where K_n^* is the value of k for which $X_{(1)} \in I_k$. Since $p_{\hat{K}_n}(\underline{X}) \geqslant p_{K_n^*}(\underline{X})$, it then follows that for any fixed k

$$P\left[\hat{K}_n > k\right] \geqslant P\left[p_{\hat{K}_n}(\underline{X}) > p_j(\underline{X}) \quad \text{for } j = 1, \ldots, k\right] \to 1.$$

To prove (7), consider

$$L_{jk} = \log\frac{f_k(x_1)\cdots f_k(x_n)}{f_j(x_1)\cdots f_j(x_n)} = \Sigma_i^{(1)}\log\frac{h(x_i)}{c} - \Sigma_i^{(2)}\log\frac{h(x_i)}{c}$$

where $\Sigma^{(1)}$ and $\Sigma^{(2)}$ extend over all i for which $x_i \in I_k$ and $x_i \in I_j$, respectively. Now $x_i \in I_j$ implies that $h(x_i) < h(a_j)$, so that

$$\Sigma^{(2)}\log[h(x_i)/c] < \nu_{jn}\log[h(a_j)/c]$$

where ν_{jn} is the number of x's in I_j. Similarly, for $k = K_n^*$,

$$\Sigma^{(1)}\log[h(x_i)/c] \geq \log[h(x_{(1)})/c]$$

since $\log[h(x)/c] \geq 0$ for all x. Thus,

$$\frac{1}{n}L_{j,K_n^*} \geq \frac{1}{n}\log\frac{h(x_{(1)})}{c} - \frac{1}{n}\nu_{jn}\log\frac{h(a_j)}{c}.$$

Since ν_{jn}/n tends in probability to $P(X_1 \in I_j) < 1$, it only remains to show that

$$\text{(8)} \qquad \frac{1}{n}\log h(X_{(1)}) \to \infty \qquad \text{in probability.}$$

Instead of $X_1, \ldots, X_n$ consider a sample $Y_1, \ldots, Y_n$ from the uniform distribution $U(0, 1/c)$. Then for any x, $P(Y_i > x) \geq P(X_i > x)$ and hence

$$P[h(Y_{(1)}) > x] \leq P[h(X_{(1)}) > x]$$

and it is therefore enough to prove that $(1/n)\log h(Y_{(1)}) \to \infty$ in probability. If h satisfies (6), $(1/n)\log h(Y_{(1)}) \geq 1/nY_{(1)}^2$, and the right side tends to infinity in probability since $nY_{(1)}$ tends to a limit distribution (Problem 5.2.16). This completes the proof.

For later reference, note that the proof has established not only (7) but the fact that for any fixed A (Problem 2.8)

$$\text{(9)} \qquad P[p_{K_n^*}(\underline{X}) > A^n p_j(\underline{X})] \to 1.$$

The example suggests (and this suggestion will be verified in the next section) that also for densities depending smoothly on a continuously varying parameter θ, the MLE need not be consistent. We shall now show, however, that a slightly weaker conclusion is possible under relatively mild conditions. Throughout the present section we shall assume θ to be real-valued. The case of several parameters will be taken up in Section 6.4.

The logarithm of the joint density

$$p(\underline{x}, \theta) = f(x_1, \theta) \cdots f(x_n, \theta) \tag{10}$$

when considered for fixed $\underline{x}$ as a function of θ,

$$L(\theta, \underline{x}) = \Sigma \log f(x_i, \theta) \tag{11}$$

is called the *log-likelihood*. We shall frequently use the shorthand notation $L(\theta)$ for $L(\theta, \underline{x})$ and $L'(\theta)$, $L''(\theta), \ldots$ for its derivatives with respect to θ.

Theorem 2.2. *Let* $X_1, \ldots, X_n$ *satisfy* (*A*0)–(*A*3) *and suppose that for almost all* x, $f(x, \theta)$ *is differentiable with respect to* θ *in* ω, *with derivative* $f'(x, \theta)$. *Then with probability tending to* 1 *as* $n \to \infty$, *the* likelihood equation

$$\frac{\partial}{\partial\theta}[f(x_1, \theta) \cdots f(x_n, \theta)] = 0 \tag{12}$$

or, equivalently, the equation

$$L'(\theta, \underline{x}) = \Sigma \frac{f'(x_i, \theta)}{f(x_i, \theta)} = 0 \tag{13}$$

has a root $\hat{\theta}_n = \hat{\theta}_n(x_1, \ldots, x_n)$ *such that* $\hat{\theta}_n(X_1, \ldots, X_n)$ *tends to the true value* θ_0 *in probability.*

Proof. Let a be small enough so that $(\theta_0 - a, \theta_0 + a) \subset \omega$, and let

$$S_n = \{\underline{x}: L(\theta_0, \underline{x}) > L(\theta_0 - a, \underline{x}) \text{ and } L(\theta_0, \underline{x}) > L(\theta_0 + a, \underline{x})\}. \tag{14}$$

By Theorem 2.1, $P_{\theta_0}(S_n) \to 1$. For any $\underline{x} \in S_n$ there thus exists a value $\theta_0 - a < \hat{\theta}_n < \theta_0 + a$ at which $L(\theta)$ has a local maximum, so that $L'(\hat{\theta}_n) = 0$. Hence for any $a > 0$ sufficiently small, there exists a sequence $\hat{\theta}_n = \hat{\theta}_n(a)$ of roots such that

$$P_{\theta_0}(|\hat{\theta}_n - \theta_0| < a) \to 1.$$

It remains to show that we can determine such a sequence, which does not depend on a.

Let θ_n^* be the root closest to θ_0, [This exists because the limit of a sequence of roots is again a root by the continuity of $L(\theta)$.] Then clearly $P_{\theta_0}(|\theta_n^* - \theta_0| < a) \to 1$ and this completes the proof.

In connection with this theorem, the following comments may be helpful.

1. The proof yields the additional fact that with probability tending to 1, the roots $\hat{\theta}_n(a)$ can be chosen to be local maxima and so, therefore, can the θ_n^* if we let θ_n^* be the closest root corresponding to a maximum.
2. On the other hand, the theorem does not establish the existence of a consistent estimator sequence since, with the true value θ_0 unknown, the data do not tell us which root to choose so as to obtain a consistent sequence. An exception, of course, is the case in which the root is unique.
3. It should also be emphasized that existence of a root $\hat{\theta}_n$ is not asserted for all $\underline{x}$ (or for a given n even for any $\underline{x}$). This does not affect consistency, which only requires $\hat{\theta}_n$ to be defined on a set S_n', the probability of which tends to 1 as $n \to \infty$.
4. Although the likelihood equation can have many roots, the consistent sequence of roots generated by Theorem 2.2 is essentially unique. For a more precise statement of this result which is due to Huzurbazar (1948), see Problem 2.27.

Corollary 2.2. *Under the assumptions of Theorem 2.2, if the likelihood equation has a unique root δ_n for each n and all $\underline{x}$, then $\{\delta_n\}$ is a consistent sequence of estimators of θ. If, in addition, the parameter space is an open interval $(\underline{\theta}, \bar{\theta})$ (not necessarily finite), then with probability tending to 1, δ_n maximizes the likelihood, that is, δ_n is the MLE, which is therefore consistent.*

Proof. The first statement is obvious. To prove the second, suppose the probability of δ_n being the MLE does not tend to 1. Then for sufficiently large values of n, the likelihood must tend to a supremum as θ tends toward $\underline{\theta}$ or $\bar{\theta}$ with positive probability. Now with probability tending to 1, δ_n is a local maximum of the likelihood, which must then also possess a local minimum. This contradicts the assumed uniqueness of the root.

The conclusion of Corollary 2.2 holds, of course, not only when the root of the likelihood equation is unique but also when the probability of multiple roots tends to zero as $n \to \infty$. On the other hand, even when the root is unique, the corollary says nothing about its properties for finite n.

Example 2.4. Let X take on the values 0, 1, 2 with probabilities $6\theta^2 - 4\theta + 1$, $\theta - 2\theta^2$, $3\theta - 4\theta^2$ $(0 < \theta < 1/2)$. Then the likelihood equation has a unique root for all x, which is a minimum for $x = 0$ and a maximum for $x = 1$ and 2 (Problem 2.10).

Theorem 2.2 establishes the existence of a consistent root of the likelihood equation. The next theorem asserts that any such sequence is asymptotically normal and efficient.

Theorem 2.3. *Suppose that* $X_1, \ldots, X_n$ *are iid and satisfy the assumptions of Theorem* 1.1, *with* (*iii*) *and* (*vi*) *replaced by the corresponding assumptions on the third* (*rather than the second*) *derivative, that is, by the existence of a third derivative satisfying*

$$\left|\frac{\partial^3}{\partial\theta^3}\log f(x, \theta)\right| \leqslant M(x) \tag{15}$$

$$\text{for all} \quad x \in A, \qquad \theta_0 - c < \theta < \theta_0 + c$$

with

$$E_{\theta_0}[M(x)] < \infty .$$

Then any consistent sequence $\hat{\theta}_n = \hat{\theta}_n(X_1, \ldots, X_n)$ *of roots of the likelihood equation satisfies*

$$\sqrt{n}\,(\hat{\theta}_n - \theta) \xrightarrow{\mathcal{L}} N\left(0, \frac{1}{I(\theta)}\right). \tag{17}$$

We shall call such a sequence $\hat{\theta}_n$ of roots an *efficient likelihood estimator* (ELE) of θ. It is typically (but need not be, see Example 3.1) provided by the MLE. More generally, any sequence $\hat{\theta}_n^*$ of estimators satisfying (17) will be said to be *asymptotically efficient*.

Proof. For any fixed x, expand $L'(\hat{\theta}_n)$ about θ_0

$$L'(\hat{\theta}_n) = L'(\theta_0) + (\hat{\theta}_n - \theta_0)L''(\theta_0) + \tfrac{1}{2}(\hat{\theta}_n - \theta_0)^2 L'''(\theta_n^*)$$

where θ_n^* lies between θ_0 and $\hat{\theta}_n$. By assumption, the left side is zero, so that

$$\sqrt{n}\,(\hat{\theta}_n - \theta_0) = \frac{(1/\sqrt{n})L'(\theta_0)}{-(1/n)L''(\theta_0) - (1/2n)(\hat{\theta}_n - \theta_0)L'''(\theta_n^*)}$$

where it should be remembered that $L(\theta)$, $L'(\theta)$, and so on are functions of

$\underline{X}$ as well as θ. We shall show that

(18) $$\frac{1}{\sqrt{n}} L'(\theta_0) \xrightarrow{\mathcal{L}} N[0, I(\theta_0)],$$

that

(19) $$-\frac{1}{n} L''(\theta_0) \xrightarrow{P} I(\theta_0)$$

and that

(20) $$\frac{1}{n} L'''(\theta_n^*) \quad \text{is bounded in probability.}$$

The desired result then follows from Theorem 5.1.4.

Of the above statements, (18) follows from the fact that

$$\frac{1}{\sqrt{n}} L'(\theta_0) = \sqrt{n} \frac{1}{n} \Sigma \left[\frac{f'(X_i, \theta_0)}{f(X_i, \theta_0)} - E_{\theta_0} \frac{f'(X_i, \theta_0)}{f(X_i, \theta_0)} \right]$$

since the expectation term is zero, and then from the CLT and the definition of $I(\theta)$.

Next,

$$-\frac{1}{n} L''(\theta_0) = \frac{1}{n} \Sigma \frac{f'^2(X_i, \theta_0) - f(X_i, \theta_0) f''(X_i, \theta_0)}{f^2(X_i, \theta_0)}.$$

By the law of large numbers, this tends in probability to

$$I(\theta_0) - E_{\theta_0} \frac{f''(X_i, \theta_0)}{f(X_i, \theta_0)} = I(\theta_0).$$

Finally,

$$\frac{1}{n} L'''(\theta) = \frac{1}{n} \Sigma \frac{\partial^3}{\partial \theta^3} \log f(X_i, \theta)$$

so that by (15)

$$\left| \frac{1}{n} L'''(\theta_n^*) \right| < \frac{1}{n} [M(X_1) + \cdots + M(X_n)]$$

with probability tending to 1. The right side tends in probability to $E_{\theta_0}[M(X)]$, and this completes the proof.

Although the conclusions of Theorem 2.3 are quite far-reaching, the proof is remarkably easy. The reason is that Theorem 2.2 already puts $\hat{\theta}_n$ into the neighborhood of the true value θ_0, so that an expansion about θ_0 essentially linearizes the problem and thereby prepares the way for application of the central limit theorem.

Corollary 2.3. *Under the assumptions of Theorem 2.3, if the likelihood equation has a unique root for all* n *and* $\underline{x}$, *and more generally if the probability of multiple roots tends to zero as* $n \to \infty$, *the MLE is asymptotically efficient.*

Before giving some applications of Theorem 2.3 and Corollary 2.3, let us briefly consider what is involved in checking its assumptions. The following are two conditions that may not be obvious:

1. Differentiability twice of $\int f(x, \theta)\, d\mu(x)$ with respect to θ by differentiating under the integral sign.
2. Condition (15), which states that the third derivative is uniformly bounded by an integrable function.

Conditions for 1. are given in books on calculus, although it is often easier simply to calculate the difference quotient and pass to the limit.

Condition 2 is usually easy to check after realizing that it is not necessary for (15) to hold for all θ, but that it is enough if there exist $\theta_1 < \theta_0 < \theta_2$ such that (15) holds for all $\theta_1 \leqq \theta \leqq \theta_2$.

Example 2.5. ***One-Parameter Exponential Family.*** Let $X_1, \ldots, X_n$ be iid according to a one-parameter exponential family with density

$$f(x_i, \eta) = e^{\eta T(x_i) - A(\eta)} \tag{21}$$

with respect to a σ-finite measure μ, and let the estimand be η. The likelihood equation is

$$\frac{1}{n}\Sigma T(x_i) = A'(\eta)$$

and hence by (1.4.11)

$$E_\eta\left[T(X_j)\right] = \frac{1}{n}\Sigma T(x_i). \tag{22}$$

The left side of (22) is a strictly increasing function of η since by (1.4.12)

$$\frac{d}{d\eta}\left[E_\eta T(X_j)\right] = \text{var}_\eta T(X_j) > 0.$$

It follows that equation (22) has at most one solution. The conditions of Theorem 2.3 are easily checked in the present case. In particular condition 1 follows from Theorem 1.4.1, and 2 from the fact that the third derivative of $\log f(x, \eta)$ is independent of x and a continuous function of η. With probability tending to 1, (22) therefore has a solution $\hat{\eta}$. This solution is unique, consistent, and asymptotically efficient, so that

$$\sqrt{n}(\hat{\eta} - \eta) \xrightarrow{\mathcal{L}} N(0, 1/\text{var}\, T) \tag{23}$$

where $T = T(X_i)$ and the asymptotic variance follows from (2.7.13).

Example 2.6. Truncated Normal. As an illustration of the preceding example, consider a sample of n observations from a normal distribution $N(\xi, 1)$, truncated at two fixed points $a < b$. The density of a single X is then

$$\frac{1}{\sqrt{2\pi}} \exp\left[-\frac{1}{2}(x - \xi)^2\right] / [\Phi(b - \xi) - \Phi(a - \xi)], \qquad a < x < b,$$

which satisfies (21) with $\eta = \xi, T(x) = x$. An ELE will therefore be the unique solution of $E_\xi(X) = \bar{x}$ if it exists. To see that this equation has a solution for any value $a < \bar{x} < b$, note that as $\xi \to -\infty$ or $+\infty$, X tends in probability to a or b, respectively (Problem 2.11). Since X is bounded, this implies that also $E_\xi(X)$ tends to a or b. Since $E_\xi(X)$ is continuous, existence of $\hat{\xi}$ follows.

As a second class of examples, consider some location problems. Since the MLE or ELE is equivariant (see Section 3.1), its risk cannot be smaller than that of the MRE estimator. One would thus typically expect the MRE (Pitman) estimator also to be asymptotically efficient. This was shown to be the case for a large class of loss functions by Stone (1974). Given a specific loss function L and finite n, the Pitman estimator δ_{Ln} has a risk advantage over the ELE $\hat{\xi}_n$. However, if one is vague about the loss function, the ELE may be viewed as an alternative of δ_{Ln}, with the two asymptotically equivalent in the sense that $\sqrt{n}(\hat{\xi}_n - \delta_{Ln}) \to 0$ in probability [Stone (1974)], but where for finite n and $L^* \neq L$, $\hat{\xi}_n$ may have smaller or larger risk than δ_{Ln}.

Example 2.7. Logistic. If $X_1, \ldots, X_n$ are iid according to the logistic density of Table 1.3.1 with $a = \theta, b = 1$, the likelihood equation after some simplification becomes (Problem 2.13)

$$\Sigma \frac{1}{1 + \exp(x_i - \theta)} = \frac{n}{2}. \tag{24}$$

The left side is an increasing function of θ which is zero at $-\infty$ and n at $+\infty$. (24) therefore has a unique root $\hat{\theta}$, which is the MLE since $L'(\theta)$ is > 0 or < 0 as θ is $< \hat{\theta}$ or $> \hat{\theta}$.

Consider now the general case of $X_1, \ldots, X_n$ iid each with density $f(x - \theta)$, where f is differentiable and $f(x) > 0$ for all x. The likelihood equation reduces to

$$\Sigma \frac{f'(x_i - \theta)}{f(x_i - \theta)} = 0. \tag{25}$$

If f is strictly strongly unimodal so that $f'(x)/f(x)$ is strictly decreasing, this equation has at most one root. On the other hand, the likelihood tends to zero as $\theta \to \pm\infty$ and therefore has a maximum $\hat{\theta}$ in the interior which must satisfy (25). When f is not strongly unimodal, (25) may have several roots. The determination of an ELE in this case will be considered in the next section.

Example 2.8. Double Exponential. For the double exponential density $DE(\theta, 1)$ given in Table 1.3.1, it is not true that for all (or almost all) x, $f(x - \theta)$ is differentiable with respect to θ, since for every x there exists a value ($\theta = x$) at which the derivative does not exist. Despite this failure, the MLE (which is the median of the X's) satisfies the conclusion of Theorem 2.3.

In the following example, $f(x, \theta)$ is neither an exponential, location, or scale family.

Example 2.9. Weibull. Let $X_1, \ldots, X_n$ be iid according to a *Weibull distribution* with density

$$f_\theta(x) = \theta x^{\theta - 1} \exp(-x^\theta), \qquad x > 0, \qquad \theta > 0. \tag{26}$$

The log likelihood of $x_1, \ldots, x_n$ is

$$n \log \theta + (\theta - 1)\Sigma \log x_i - \Sigma \exp(\theta \log x_i)$$

so that

$$L'(\theta) = \frac{n}{\theta} + \Sigma \log x_i - \Sigma \log x_i \exp(\theta \log x_i)$$

and

$$L''(\theta) = -\frac{n}{\theta^2} - \Sigma(\log x_i)^2 \exp(\theta \log x_i) < 0.$$

Thus, if the likelihood equation has a root, it is unique and equals the MLE. As $\theta \to 0$, $L'(\theta) \to \infty$ while for θ sufficiently large, $L'(\theta)$ is negative. This proves the existence of a root.

3. LIKELIHOOD ESTIMATION: MULTIPLE ROOTS

When the likelihood equation has multiple roots, the assumptions of Theorem 2.3 are no longer sufficient to guarantee consistency of the MLE, even when it exists for all n. This is shown by the following example due to Le Cam (1979b), which is obtained by embedding the sequence $\{f_k\}$ of Example 2.3 in a sufficiently smooth continuous-parameter family.

Example 3.1. For $k \leqslant \theta < k+1$, $k = 1, 2, \ldots$, let

$$f(x, \theta) = [1 - u(\theta - k)]f_k(x) + u(\theta - k)f_{k+1}(x), \tag{1}$$

with f_k defined as in Example 2.3 and u defined on $(-\infty, \infty)$ such that $u(x) = 0$ for $x \leqslant 0$, $u(x) = 1$ for $x \geqslant 1$, is strictly increasing on (0, 1) and infinitely differentiable on $(-\infty, \infty)$ (Problem 3.1). Let $X_1, \ldots, X_n$ be iid, each with density $f(x, \theta)$, and let $p(\underline{x}, \theta) = \Pi f(x_i, \theta)$.

Since for any given $\underline{x}$, the density $p(\underline{x}, \theta)$ is bounded and continuous in θ and is equal to c^n for all sufficiently large θ and greater than c^n for some θ, it takes on its maximum for some finite θ and the MLE $\hat{\theta}_n$ therefore exists.

To see that $\hat{\theta}_n \to \infty$ in probability, note that for $k \leqslant \theta < k+1$

$$p(\underline{x}, \theta) \leqslant \Pi \max[f_k(x_i), f_{k+1}(x_i)] = \bar{p}_k(\underline{x}). \tag{2}$$

If $\hat{K}_n$ and K_n^* are defined as in Example 2.3, the argument of that example shows that it is enough to prove that for any fixed j

$$P\left[p_{K_n^*}(\underline{X}) > \bar{p}_j(\underline{X})\right] \to 1 \quad \text{as} \quad n \to \infty, \tag{3}$$

where $p_k(\underline{x}) = p(\underline{x}, k)$. Now

$$\bar{L}_{jk} = \frac{p_k(\underline{x})}{\bar{p}_j(\underline{x})} = \Sigma^{(1)} \log \frac{h(x_i)}{c} - \Sigma^{(2)} \log \frac{h(x_i)}{c} - \Sigma^{(3)} \log \frac{h(x_i)}{c}$$

where $\Sigma^{(1)}$, $\Sigma^{(2)}$, and $\Sigma^{(3)}$ extend over all i for which $x_i \in I_k$, $x_i \in I_j$ and $x_i \in I_{j+1}$, respectively. The argument is now completed as before to show that $\hat{\theta}_n \to \infty$ in probability regardless of the true value of θ and is therefore not consistent.

The example is not yet completely satisfactory since $\partial f(x, \theta)/\partial \theta = 0$ and hence $I(\theta) = 0$ for $\theta = 1, 2, \ldots$. [The remaining conditions of Theorem 2.3 are easily checked (Problem 3.2).] To remove this difficulty, define

$$g(x, \theta) = \tfrac{1}{2}\left[f(x, \theta) + f(x, \theta + \alpha e^{-\theta^2})\right], \qquad \theta \geqslant 1 \tag{4}$$

for some fixed $\alpha < 1$.

If $X_1, \ldots, X_n$ are iid according to $g(x, \theta)$, we shall now show that the MLE $\hat{\theta}_n$ continues to tend to infinity for any fixed θ. We have, as before

$$P[\hat{\theta}_n > k] \geqslant P[\Pi g(x_i, K_n^*) > \Pi g(x_i, \theta) \quad \text{for all} \quad \theta \leqslant k]$$

$$\geqslant P\left\{\frac{1}{2^n}\Pi f(x_i, K_n^*) > \Pi\left(\frac{1}{2}\left[f(x_i, \theta) + f\left(x_i, \theta + \alpha e^{-\theta^2}\right)\right]\right)\right\}.$$

For $j \leqslant \theta < j+1$ it is seen from (1) that $[f(x_i, \theta) + f(x_i, \theta + \alpha e^{-\theta^2})]/2$ is a weighted average of $f_j(x_i), f_{j+1}(x_i)$ and possibly $f_{j+2}(x_i)$. By using $\bar{p}_j(\underline{x}) = \Pi \max[f_j(x_i), f_{j+1}(x_i), f_{j+2}(x_i)]$ in place of $p_j(\underline{x})$, the proof can now be completed as before. Since the densities $g(x_i, \theta)$ satisfy the conditions of Theorem 2.3 (Problem 3.3) these conditions therefore are not enough to ensure the consistency of the MLE. (For another example, see Ferguson, 1982.)

Even under the assumptions of Theorem 2.3 one is thus, in the case of multiple roots, still faced with the problem of identifying a consistent sequence of roots. Following are three possible approaches.

(i) In many cases, the maximum likelihood estimator is consistent. Conditions which ensure this were given by among others Wald (1949), Le Cam (1953, 1970), Kiefer and Wolfowitz (1956), Bahadur (1967), and Perlman (1972). A survey of the literature can be found in Perlman (1983). This material is technically difficult, and even when the conditions are satisfied, the determination of the MLE may present problems [see Barnett (1966)]. We shall therefore turn to somewhat simpler alternatives.

The following two methods require that some sequence of consistent (but not necessarily efficient) estimators be available. The existence of such a sequence is guaranteed under very weak assumptions by a theorem of Le Cam (1956); see also Kraft and Le Cam (1956). In any given situation, it is usually easy to construct a consistent sequence, as will be illustrated below and in the next section.

(ii) Suppose that δ_n is any consistent estimator of θ and that the assumptions of Theorem 2.3 hold. Then the root $\hat{\theta}_n$ of the likelihood equation closest to δ_n (which exists by the proof of Theorem 2.2) is also consistent, and hence is efficient by Theorem 2.3.

To see this, note that by Theorem 2.3 there exists a consistent sequence of roots, say $\hat{\theta}_n^*$. Since $\hat{\theta}_n^* - \delta_n \to 0$ in probability, so does $\hat{\theta}_n - \delta_n$.

The following approach, which does not require the determination of the closest root, and in which the estimators are no longer exact roots of the likelihood equation, is often more convenient.

(iii) The usual iterative methods for solving the likelihood equation

$$L'(\theta) = 0 \tag{5}$$

are based on replacing the left side by the linear terms of its Taylor expansion about an approximate solution $\tilde{\theta}$. If $\hat{\theta}$ denotes a root of (5), this leads to the approximation

$$0 = L'(\hat{\theta}) \doteq L'(\tilde{\theta}) + (\hat{\theta} - \tilde{\theta})L''(\tilde{\theta}), \tag{6}$$

and hence to

$$\hat{\theta} = \tilde{\theta} - \frac{L'(\tilde{\theta})}{L''(\tilde{\theta})}. \tag{7}$$

The procedure is then iterated by replacing $\tilde{\theta}$ by the value $\tilde{\tilde{\theta}}$ of the right side of (7), and so on. (For a discussion of the performance of this procedure, see for example Barnett, 1966.)

Here we are concerned only with the first step and with the performance of the one-step approximation (7) as an estimator of θ. The following result gives conditions on $\tilde{\theta}$ under which the resulting sequence of estimators is consistent, asymptotically normal, and efficient.

Theorem 3.1. *Suppose that the assumptions of Theorem* 2.3 *hold and that* $\tilde{\theta}_n$ *is not only a consistent but a* $\sqrt{n}$*-consistent* estimator of* θ, *that is, that* $\sqrt{n}(\tilde{\theta}_n - \theta)$ *is bounded in probability so that* $\tilde{\theta}_n$ *tends to* θ *at least at the rate of* $1/\sqrt{n}$. *Then the estimator sequence*

$$\delta_n = \tilde{\theta}_n - \frac{L'(\tilde{\theta}_n)}{L''(\tilde{\theta}_n)} \tag{8}$$

is asymptotically efficient, that is, it satisfies (2.17) *with* δ_n *in place of* $\hat{\theta}_n$.

Proof. As in the proof of Theorem 2.3, expand $L'(\tilde{\theta}_n)$ about θ_0 as

$$L'(\tilde{\theta}_n) = L'(\theta_0) + (\tilde{\theta}_n - \theta_0)L''(\theta_0) + \tfrac{1}{2}(\tilde{\theta}_n - \theta_0)^2 L'''(\theta_n^*)$$

where θ_n^* lies between θ_0 and $\tilde{\theta}_n$. Substituting this expression into (8) and simplifying, we find

$$\sqrt{n}(\delta_n - \theta_0) = \frac{(1/\sqrt{n})L'(\theta_0)}{-(1/n)L''(\tilde{\theta}_n)} + \sqrt{n}(\tilde{\theta}_n - \theta_0) \tag{9}$$

$$\times \left[1 - \frac{L''(\theta_0)}{L''(\tilde{\theta}_n)} - \frac{1}{2}(\tilde{\theta}_n - \theta_0)\frac{L'''(\theta_n^*)}{L''(\tilde{\theta}_n)}\right].$$

To prove the result, we shall use (2.18) and (2.19), which are consequences of the assumptions on $f(x, \theta)$, and (2.20) together with the fact that

*A general method for constructing $\sqrt{n}$-consistent estimators is given by Le Cam (1969, p. 103).

$\theta_n^* \to \theta_0$ in probability. We shall also need that

$$\frac{L''(\tilde{\theta}_n)/n}{L''(\theta_0)/n} \to 1 \quad \text{in probability.} \tag{10}$$

This follows from (2.19) and (2.20), and the expansion

$$\frac{1}{n}L''(\tilde{\theta}_n) = \frac{1}{n}L''(\theta_0) + \frac{1}{n}(\tilde{\theta}_n - \theta_0)L'''(\theta_n^{**})$$

where θ_n^{**} is between θ_0 and $\tilde{\theta}_n$.

Consider now the right side of (9). By (2.18), (2.19), and (10), the first term has the limit distribution asserted for the left side. It only remains to show that the expression in square brackets on the right side of (9) tends to zero in probability. This follows from (2.20) and the assumptions made about $\tilde{\theta}_n$.

Corollary 3.1. *Suppose that the assumptions of Theorem* 3.1 *hold and that* $I(\theta)$ *is a continuous function of* θ. *Then the estimator*

$$\delta_n' = \tilde{\theta}_n + \frac{L'(\tilde{\theta}_n)}{nI(\tilde{\theta}_n)} \tag{11}$$

is also asymptotically efficient.

Proof. By (10), condition (viii) of Theorem 1.1, and the law of large numbers, $-(1/n)L''(\tilde{\theta}_n) \to I(\theta_0)$ in probability. Since $I(\theta)$ is continuous, also $I(\tilde{\theta}_n) \to I(\theta_0)$ in probability, so that $-(1/n)L''(\tilde{\theta}_n)/I(\tilde{\theta}_n) \to 1$ in probability, and this completes the proof.

The estimators (8) and (11) are compared by Stuart (1958), who gives a heuristic argument why (11) might be expected to be closer to the ELE than (8) and provides a numerical example supporting this argument.

Example 3.2. Location Parameter. Consider the case of a symmetric location family in which the likelihood equation (2.25) has multiple roots. [For the Cauchy distribution, for example, it has been shown by Reeds (1980) that if (2.25) has $K + 1$ roots, then as $n \to \infty$, K tends in law to a Poisson distribution with expectation $1/\pi$. The Cauchy case has also been considered by Barnett (1966).] If $\text{var}(X) < \infty$, it follows from the CLT that the sample mean $\bar{X}_n$ is $\sqrt{n}$-consistent and that an asymptotically efficient estimator of θ is therefore provided by (8) or (11) with $\tilde{\theta}_n = \bar{X}$ if $f(x, \theta)$ satisfies the conditions of Theorem 2.3. For distributions such as the Cauchy for which $E(X^2) = \infty$, Theorem 5.3.2 shows that one can, instead, take for $\tilde{\theta}_n$ the sample median provided $f(0) > 0$; other L, M, or R-estimators provide still further possibilities.

Example 3.3. Grouped or Censored Observations. Suppose that $X_1, \ldots, X_n$ are iid according to a location family with cdf $F(x - \theta)$, with F known and with $0 < F(x) < 1$ for all x, but that it is only observed whether each X_i falls below a, between a and b, or above b where $a < b$ are two given constants. The n observations constitute n trinomial trials with probabilities $p_1 = p_1(\theta) = F(a - \theta)$, $p_2(\theta) = F(b - \theta) - F(a - \theta)$, $p_3(\theta) = 1 - F(b - \theta)$ for the three outcomes. If V denotes the number of observations less than a, then

$$\sqrt{n}\left[\frac{V}{n} - p_1\right] \xrightarrow{\mathcal{L}} N[0, p_1(1 - p_1)] \tag{12}$$

and, by Theorem 5.1.5,

$$\tilde{V}_n = a - F^{-1}\left(\frac{V}{n}\right) \tag{13}$$

is a $\sqrt{n}$-consistent estimator of θ. Since the estimator is not defined when $V = 0$ or $V = n$, some special definition has to be adopted in these cases whose probability however tends to zero as $n \to \infty$.

If the trinomial distribution for a single trial satisfies the assumptions of Theorem 2.3, as will be the case under mild assumptions on F, the estimator (8) is asymptotically efficient (but see the comment following Example 3.5). The approach applies, of course, equally to the case of more than three groups.

A very similar situation arises when the X's are *censored*, say at a fixed point a. For example, they might be lengths of life of light bulbs or patients, with observation discontinued at time a. The observations can then be represented as

$$Y_i = \begin{cases} X_i & \text{if } X_i < a \\ a & \text{if } X_i \geqslant a. \end{cases}$$

Here the value a of Y_i when $X_i \geqslant a$ has no significance; it simply indicates that the value of X_i is $\geqslant a$. The Y's are then iid with density

$$g(y, \theta) = \begin{cases} f(y - \theta) & \text{if } y < a \\ 1 - F(a - \theta) & \text{if } y = a \end{cases}$$

with respect to the measure μ which is Lebesgue measure on $(-\infty, a)$ and assigns measure 1 to the point $y = a$.

The estimator (13) continues to be $\sqrt{n}$-consistent in the present situation. An alternative starting point is, for example, the best linear combination of the ordered X's less than a (see, for example, Chan, 1967).

Estimation from grouped or censored observations arises not only for location families but also for any other family of distributions and has been treated in particular for exponential families (see, for example, Sundberg 1976). An algorithm for computing MLEs in such problems is discussed by Dempster, Laird, and Rubin (1977).

Example 3.4. Mixtures. Let $X_1, \ldots, X_n$ be a sample from a distribution $\theta G + (1 - \theta)H$, $0 < \theta < 1$, where G and H are two specified distributions with densities g

and h. The log likelihood of a single observation is a concave function of θ, and so therefore is the log likelihood of a sample (Problem 3.5). It follows that the likelihood equation has at most one solution. (The asymptotic performance of the estimator is studied by Hill, 1963.)

Even when the root is unique, Theorem 3.1 provides an alternative, which may be more convenient than the MLE. In the mixture problem, as in many other cases, a $\sqrt{n}$-consistent estimator can be obtained by the method of moments, which consists in equating the first k moments of X to the corresponding sample moments, say

$$E_\theta(X_i^r) = \frac{1}{n}\sum_{j=1}^{n} X_j^r, \qquad r = 1, \dots, k,$$

where k is the number of unknown parameters. [For further discussion, see, for example, Cramér (1946, Sect. 33.1) and Serfling (1980, Sect. 4.3.1),] In the present case, suppose that $E(X_i) = \xi$ or η when X is distributed as G or H where $\eta \neq \xi$ and G and H have finite variance. Since $k = 1$, the method of moments estimates θ as the solution of the equation

$$\xi\theta + \eta(1 - \theta) = \bar{X}_n$$

and hence by

$$\tilde{\theta}_n = \frac{\bar{X}_n - \eta}{\xi - \eta}.$$

[If $\eta = \xi$ but the second moments of X_i under H and G differ, one can instead equate $E(X_i^2)$ with $\Sigma X_j^2/n$ (Problem 3.6).] An asymptotically efficient estimator is then provided by (8).

In any application of Theorem 3.1, the question naturally arises which $\sqrt{n}$-consistent sequence $\{\tilde{\theta}_n\}$ to use as the starting point. It seems plausible that one would do best with a highly efficient starting sequence. More generally, this is a special case of the problem of choosing among the many alternative sequences of asymptotically efficient estimators. It had been pointed out earlier that an efficient estimator sequence can always be modified by terms of order $1/n$ without affecting the asymptotic efficiency. To distinguish among them requires taking into account the terms of the next order.

A number of authors (among them Rao, 1963; Pfanzagl, 1973; Ghosh and Subramanyam, 1974; Efron, 1975, 1978; and Akahira and Takeuchi, 1981), have investigated estimators that are "second-order efficient," that is, efficient and among efficient estimators have the greatest accuracy to terms of the next order, and in particular they have tried to determine to what extent the MLE is second-order efficient. Their findings, for sufficiently regular situations, can be summarized roughly as follows.

Typically, the MLE $\hat{\theta}_n$ has bias of order $1/n$, say

$$b_n(\theta) = \frac{B(\theta)}{n} + 0\left(\frac{1}{n^2}\right).$$

The estimator can be made approximately unbiased by subtracting from it an estimator of the bias based on the MLE. This leads to the bias-corrected estimator

$$\hat{\hat{\theta}}_n = \hat{\theta}_n - \frac{B(\hat{\theta}_n)}{n} \tag{14}$$

whose bias will be of order $1/n^2$. The basic result then states that among the estimators whose bias is of order $1/n^2$, the estimator (14) is second-order efficient.

That the MLE or bias-corrected MLE is not second-order efficient when compared with all other efficient estimators is illustrated by the following example in which, for the sake of simplicity, we shall consider expected squared error instead of asymptotic variance.

Example 3.5. Consider the estimation of σ^2 on the basis of a sample $X_1, \ldots, X_n$ from $N(0, \sigma^2)$. The MLE is then

$$\hat{\sigma}^2 = \frac{1}{n}\Sigma X_i^2,$$

which happens to be unbiased so that no correction is needed. Let us now consider the more general class of estimators

$$\delta_n = \left(\frac{1}{n} + \frac{a}{n^2}\right)\Sigma X_i^2. \tag{15}$$

It follows from formula (2.5.25) with $c = 1/n + a/n^2$ that

$$E(\delta_n - \sigma^2)^2 = \frac{2\sigma^4}{n} + \frac{(4a + a^2)\sigma^4}{n^2} + 0\left(\frac{1}{n^3}\right).$$

Thus the estimators δ_n are all asymptotically efficient in the sense that $nE(\delta_n - \theta)^2 \to 1/I(\theta)$ where $\theta = \sigma^2$. However, the MLE is not second-order efficient since the term of order $1/n^2$ is minimized not by $a = 0$ (MLE) but by $a = -2$, so that $(1/n - 2/n^2)\Sigma X_i^2$ has higher second-order efficiency than the MLE. (A measure of the loss due to the lower second-order efficiency is the asymptotic deficiency discussed in Section 5.2.) It follows from Theorem 5.2.4 that the limiting risk deficiency of the MLE ($a = 0$) relative to δ_n with $a = -2$ is 2 (Problem 3.7).

In fact, a uniformly best estimator (up to second-order terms) typically will not exist. The second-order situation is thus similar to that encountered in the exact (small-sample) theory. One can obtain uniform second-order optimality by imposing restrictions such as first-order unbiasedness, or must be content with weaker properties such as second-order admissibility or minimaxity. An admissibility result (somewhat similar to Theorem 4.3.3) is given by Ghosh and Sinha (1981); the minimax problem is treated by Levit (1980).

In the context of choosing a $\sqrt{n}$-consistent estimator $\tilde{\theta}_n$ for (8), it is of interest to note that in sufficiently regular situations good efficiency of $\tilde{\theta}_n$ is equivalent to high correlation with $\hat{\theta}_n$. This is made precise by the following result which is concerned only with first-order approximations.

Theorem 3.2. *Suppose $\hat{\theta}_n$ is an ELE estimator and $\tilde{\theta}_n$ a $\sqrt{n}$-consistent estimator, for which the joint distribution of*

$$T_n = \sqrt{n}\left(\hat{\theta}_n - \theta\right) \quad \text{and} \quad T'_n = \sqrt{n}\left(\tilde{\theta}_n - \theta\right)$$

tends to a bivariate limit distribution H *with zero means and covariance matrix* $\Sigma = \|\sigma_{ij}\|$. *Let* (T, T') *have distribution* H *and suppose that the means and covariance matrix of* (T_n, T'_n) *tend toward those of* (T, T') *as* $n \to \infty$. *Then the ARE of* T'_n *with respect to* T_n *is given by* $e = \rho^2$ *where* $\rho = \sigma_{12}/\sqrt{\sigma_{11}\sigma_{22}}$ *is the correlation coefficient of* (T, T').

Proof. Consider $\operatorname{var}[(1 - \alpha)T_n + \alpha T'_n]$ which tends to

$$(16) \quad \operatorname{var}[(1 - \alpha)T + \alpha T'] = (1 - \alpha)^2\sigma_{11} + 2\alpha(1 - \alpha)\sigma_{12} + \alpha^2\sigma_{22}.$$

This is non-negative for all values of α and takes on its minimum at $\alpha = 0$ since $\hat{\theta}_n$ is asymptotically efficient. Thus the coefficient of α in (16) is equal to zero and hence $\sigma_{11} = \sigma_{12}$ (Problem 3.8). Thus, $\rho = \sqrt{\sigma_{11}/\sigma_{22}} = \sqrt{e}$ as was to be proved.

4. THE MULTIPARAMETER CASE

In the preceding sections asymptotically efficient estimators were obtained when the distribution depends on a single parameter θ. When extending this theory to probability models involving several parameters $\theta_1, \ldots, \theta_s$, one may be interested either in the simultaneous estimation of these parameters (or certain functions of them) or with the estimation of one of the parameters at a time, the remaining parameters then playing the role of nuisance or incidental parameters. In the present section we shall primarily take the latter point of view.

Let $X_1, \ldots, X_n$ be iid with a distribution that depends on $\theta = (\theta_1, \ldots, \theta_s)$ and satisfies assumptions (A0)–(A3) of Section 6.2. For the time being we shall assume s to be fixed. Suppose we wish to estimate θ_j. Then it was seen in Section 2.7 that the variance of any unbiased estimator δ_n of θ_j, based on n observations, satisfies the inequality

$$\text{(1)} \qquad \operatorname{var}(\delta_n) \geqq [I(\theta)]_{jj}^{-1}/n$$

where the numerator on the right side is the jjth element of the inverse of the information matrix $I(\theta)$ with elements $I_{jk}(\theta)$, $j, k = 1, \ldots, s$, defined by

$$\text{(2)} \qquad I_{jk}(\theta) = \operatorname{cov}\left[\frac{\partial}{\partial\theta_j}\log f(X,\theta),\ \frac{\partial}{\partial\theta_k}\log f(X,\theta)\right].$$

It was further shown by Bahadur (1964) under conditions analogous to those of Theorem 1.1 that for any sequence of estimators δ_n of θ_j satisfying

$$\text{(3)} \qquad \sqrt{n}\,(\delta_n - \theta_j) \xrightarrow{\mathcal{L}} N[0, v(\theta)],$$

the asymptotic variance v satisfies

$$\text{(4)} \qquad v(\theta) \geqq [I(\theta)]_{jj}^{-1}$$

except on a set of values θ having measure zero.

We shall now show under assumptions generalizing those of Theorem 2.3 that with probability tending to 1 there exist solutions $\hat{\theta}_n = (\hat{\theta}_{1n}, \ldots, \hat{\theta}_{sn})$ of the likelihood equations

$$\text{(5)} \qquad \frac{\partial}{\partial\theta_j}[f(x_1,\theta)\cdots f(x_n,\theta)] = 0, \qquad j = 1,\ldots, s,$$

or equivalently

$$\text{(6)} \qquad \frac{\partial}{\partial\theta_j}[L(\theta)] = 0, \qquad j = 1,\ldots, s$$

such that $\hat{\theta}_{jn}$ is consistent for estimating θ_j and asymptotically efficient in the sense of satisfying (3) with

$$\text{(7)} \qquad v(\theta) = [I(\theta)]_{jj}^{-1}.$$

We state first some assumptions:

(A) There exists an open subset ω of Ω containing the true parameter point θ^0 such that for almost all x the density $f(x, \theta)$ admits all third derivatives $(\partial^3/\partial\theta_j\partial\theta_k\partial\theta_l)f(x, \theta)$ for all $\theta \in \omega$.

(B) the first and second logarithmic derivatives of f satisfy the equations

$$E_\theta\left[\frac{\partial}{\partial\theta_j}\log f(X, \theta)\right] = 0 \qquad \text{for } j = 1, \ldots, s, \tag{8}$$

and

$$I_{jk}(\theta) = E_\theta\left[\frac{\partial}{\partial\theta_j}\log f(X, \theta) \cdot \frac{\partial}{\partial\theta_k}\log f(X, \theta)\right] \tag{9}$$

$$= E_\theta\left[-\frac{\partial^2}{\partial\theta_j\partial\theta_k}\log f(X, \theta)\right].$$

Clearly (8) and (9) imply (2).

(C) Since the $s \times s$ matrix $I(\theta)$ is a covariance matrix, it is positive semidefinite. In generalization of condition (v) of Theorem 1.1 we shall assume that the $I_{jk}(\theta)$ are finite and that the matrix $I(\theta)$ is positive definite for all θ in ω, and hence that the s statistics

$$\frac{\partial}{\partial\theta_1}\log f(X, \theta), \ldots, \frac{\partial}{\partial\theta_s}\log f(X, \theta)$$

are affinely independent with probability 1.

(D) Finally, we shall suppose that there exist functions M_{jkl} such that

$$\left|\frac{\partial^3}{\partial\theta_j\partial\theta_k\partial\theta_l}\log f(x, \theta)\right| \leqq M_{jkl}(x) \qquad \text{for all} \quad \theta \in \omega$$

where

$$m_{jkl} = E_{\theta^0}\left[M_{jkl}(X)\right] < \infty \qquad \text{for all} \quad j, k, l.$$

Theorem 4.1. *Let* $X_1, \ldots, X_n$ *be iid each with a density* $f(x, \theta)$ *(with respect to* μ*) which satisfies (A0)–(A2) of Section 6.2 and assumptions*

(A)–(D) above. Then with probability tending to 1 *as* $n \to \infty$, *there exist solutions* $\hat{\theta}_n = \hat{\theta}_n(X_1, \ldots, X_n)$ *of the likelihood equations such that*

(i) $\hat{\theta}_{jn}$ *is consistent for estimating* θ_j,

(ii) $\sqrt{n}(\hat{\theta}_n - \theta)$ *is asymptotically normal with (vector) mean zero and covariance matrix* $[I(\theta)]^{-1}$ *and*

(iii) $\hat{\theta}_{jn}$ *is asymptotically efficient in the sense that*

$$\sqrt{n}\left(\hat{\theta}_{jn} - \theta_j\right) \xrightarrow{\mathcal{L}} N\{0, [I(\theta)]_{jj}^{-1}\}. \tag{10}$$

Proof. *(i) Existence and Consistency.** To prove the existence, with probability tending to 1, of a sequence of solutions of the likelihood equations which is consistent, we shall consider the behavior of the log likelihood $L(\theta)$ on the sphere Q_a with center at the true point θ^0 and radius a. We will show that for any sufficiently small a the probability tends to 1 that

$$L(\theta) < L(\theta^0)$$

at all points θ on the surface of Q_a, and hence that $L(\theta)$ has a local maximum in the interior of Q_a. Since at a local maximum the likelihood equations must be satisfied it will follow that for any $a > 0$, with probability tending to 1 as $n \to \infty$, the likelihood equations have a solution $\hat{\theta}_n(a)$ within Q_a and the proof can be completed as in the one-dimensional case.

To obtain the needed facts concerning the behavior of the likelihood on Q_a for small a, we expand the log likelihood about the true point θ^0 and divide by n to find

$$\begin{aligned} \frac{1}{n}L(\theta) - \frac{1}{n}L(\theta^0) &= \frac{1}{n}\Sigma A_j(x)(\theta_j - \theta_j^0) \\ &\quad + \frac{1}{2n}\Sigma\Sigma B_{jk}(x)(\theta_j - \theta_j^0)(\theta_k - \theta_k^0) \\ &\quad + \frac{1}{6n}\sum_j\sum_k\sum_l(\theta_j - \theta_j^0)(\theta_k - \theta_k^0)(\theta_l - \theta_l^0)\sum_{i=1}^n \gamma_{jkl}(x_i)M_{jkl}(x_i) \\ &= S_1 + S_2 + S_3 \end{aligned}$$

*Different proofs of existence and consistency under slightly different assumptions are given by Foutz (1977). Consistency proofs in more general settings were given by Wald (1949), Le Cam (1953), Bahadur (1967), Huber (1967), and Perlman (1972) among others. See also Pfanzagl (1969), Landers (1972), and Pfaff (1982).

where

$$A_j(x) = \frac{\partial}{\partial \theta_j} L(\theta)\bigg|_{\theta=\theta^0}$$

and

$$B_{jk}(x) = \frac{\partial^2}{\partial \theta_j \partial \theta_k} L(\theta)\bigg|_{\theta=\theta^0},$$

and where by assumption (D)

$$0 \leqslant |\gamma_{jkl}(x)| \leqslant 1.$$

To prove that the maximum of this difference for θ on Q_a is negative with probability tending to 1 if a is sufficiently small, we will show that with high probability the maximum of S_2 is negative while S_1 and S_3 are small compared to S_2. The basic tools for showing this are the facts that by (8), (9), and the law of large numbers

$$\text{(11)} \qquad \frac{1}{n} A_j(X) = \frac{1}{n} \frac{\partial}{\partial \theta_j} L(\theta)\bigg|_{\theta^0} \to 0 \qquad \text{in probability}$$

and

$$\text{(12)} \qquad \frac{1}{n} B_{jk}(X) = \frac{1}{n} \frac{\partial^2}{\partial \theta_j \partial \theta_k} L(\theta)\bigg|_{\theta^0} \to -I_{jk}(\theta^0) \qquad \text{in probability.}$$

Let us begin with S_1. On Q_a we have

$$|S_1| \leqslant \frac{1}{n} a \Sigma |A_j(X)|.$$

For any given a, it follows from (11) that $|A_j(X)|/n < a^2$ and hence that $|S_1| < sa^3$ with probability tending to 1. Next consider

$$\text{(13)} \qquad 2S_2 = \Sigma\Sigma\left[-I_{jk}(\theta^0)(\theta_j - \theta_j^0)(\theta_k - \theta_k^0)\right]$$

$$+ \Sigma\Sigma\left\{\frac{1}{n} B_{jk}(X) - \left[-I_{jk}(\theta^0)\right]\right\}(\theta_j - \theta_j^0)(\theta_k - \theta_k^0).$$

For the second term it follows from an argument analogous to that for S_1 that its absolute value is less than s^2a^3 with probability tending to 1. The first term is a negative (nonrandom) quadratic form in the variables $(\theta_j - \theta_j^0)$. By an orthogonal transformation this can be reduced to diagonal form $\Sigma\lambda_i\zeta_i^2$ with Q_a becoming $\Sigma\zeta_i^2 = a^2$. Suppose that the λ's that are negative are numbered so that $\lambda_s \leqslant \lambda_{s-1} \leqslant \cdots \leqslant \lambda_1 < 0$. Then $\Sigma\lambda_i\zeta_i^2 \leqslant \lambda_1\Sigma\zeta_i^2 = \lambda_1 a^2$. Combining the first and second terms, we see that there exist $c > 0$, $a_0 > 0$ such that for $a < a_0$

$$S_2 < -ca^2$$

with probability tending to 1.

Finally, with probability tending to 1,

$$\left|\frac{1}{n}\Sigma M_{jkl}(X_i)\right| < 2m_{jkl}$$

and hence $|S_3| < ba^3$ on Q_a where

$$b = \frac{s^3}{3}\Sigma\Sigma\Sigma m_{jkl}.$$

Combining the three inequalities, we see that

$$\max(S_1 + S_2 + S_3) < -ca^2 + (b + s)a^3 \tag{14}$$

which is less than zero if $a < c/(b + s)$, and this completes the proof of (i).

The proof of part (ii) of Theorem 4.1 is basically the same as that of Theorem 2.3. However, the single equation derived there from the expansion of $\hat{\theta}_n - \theta_0$ is now replaced by a system of s equations which must be solved for the differences $(\hat{\theta}_{jn} - \theta_j^0)$. This makes the details of the argument somewhat more cumbersome. In preparation, it will be convenient to consider quite generally a set of random linear equations in s unknowns

$$\sum_{k=1}^{s} A_{jkn}Y_{kn} = T_{jn} \qquad (j = 1, \ldots, s). \tag{15}$$

Lemma 4.1. *Let* $(T_{1n}, \ldots, T_{sn})$ *be a sequence of random vectors tending weakly to* $(T_1, \ldots, T_s)$ *and suppose that for each fixed* j *and* k, A_{jkn} *is a sequence of random variables tending in probability to constants* a_{jk} *for which the matrix* $A = \|a_{jk}\|$ *is nonsingular. Let* $B = \|b_{jk}\| = A^{-1}$. *Then if the* *distribution of* $(T_1, \ldots, T_s)$ *has a density with respect to Lebesgue measure over* E_s, *the solutions* $(Y_{1n}, \ldots, Y_{sn})$ *of* (15) *tend in probability to the solutions* $(Y_1, \ldots, Y_s)$ *of*

$$\sum_{k=1}^{s} a_{jk} Y_k = T_j \qquad (j = 1, \ldots, s) \tag{16}$$

given by

$$Y_j = \sum_{k=1}^{s} b_{jk} T_k. \tag{17}$$

Proof. With probability tending to 1, the matrices $\|A_{jkn}\|$ are nonsingular, and by Theorem 5.1.7 (Problem 4.1), the elements of the inverse of $\|A_{jkn}\|$ tend in probability to the elements of B. Therefore, by a slight extension of Theorem 5.1.4, the solutions of (15) have the same limit distribution as those of

$$Y_{jn} = \sum_{k=1}^{s} b_{jk} T_{kn}. \tag{18}$$

By applying Theorem 5.1.7 to the set S:

$$\Sigma b_{1k} T_k \leqslant y_1, \ldots, \Sigma b_{sk} T_k \leqslant y_s, \tag{19}$$

it is only necessary to show that the distribution of $(T_1, \ldots, T_s)$ assigns probability zero to the boundary of (19). Since this boundary is contained in the union of the hyperplanes $\Sigma b_{jk} T_k = y_j$, the result follows.

Proof of Part (ii) of Theorem 4.1. In generalization of the proof of Theorem 2.3, expand $\partial L(\theta)/\partial \theta_j = L_j'(\theta)$ about θ^0 to obtain

$$\begin{aligned} L_j'(\theta) = {} & L_j'(\theta^0) + \Sigma(\theta_k - \theta_k^0) L_{jk}''(\theta^0) \\ & + \tfrac{1}{2} \Sigma\Sigma(\theta_k - \theta_k^0)(\theta_l - \theta_l^0) L_{jkl}'''(\theta^*) \end{aligned} \tag{20}$$

where L_{jk}'' and L_{jkl}''' denote the indicated second and third derivatives of L and where θ^* is a point on the line segment connecting θ and θ^0. In this expansion, replace θ by a solution $\hat{\theta}_n$ of the likelihood equations, which by part (i) of the theorem can be assumed to exist with probability tending to 1 and to be consistent. The left side of (20) is then zero and the resulting equations can be written as

$$\text{(21)} \qquad \sqrt{n}\,\Sigma(\hat{\theta}_k - \theta_k^0)\left[\frac{1}{n}L''_{jk}(\theta^0) + \frac{1}{2n}\Sigma(\hat{\theta}_l - \theta_l^0)L'''_{jkl}(\theta^*)\right]$$

$$= -\frac{1}{\sqrt{n}}L'_j(\theta^0).$$

These have the form (15) with

$$\text{(22)} \qquad \begin{cases} Y_{kn} = \sqrt{n}\,(\hat{\theta}_k - \theta_k^0) \\ A_{jkn} = \frac{1}{n}L''_{jk}(\theta^0) + \frac{1}{2n}\Sigma(\hat{\theta}_l - \theta_l^0)L'''_{jkl}(\theta^*) \\ T_{jn} = -\frac{1}{\sqrt{n}}L'_j(\theta^0) = -\sqrt{n}\left[\frac{1}{n}\sum_{i=1}^{n}\frac{\partial}{\partial\theta_j}\log f(X_i, \theta)\right]_{\theta=\theta^0}. \end{cases}$$

Since $E_{\theta^0}[(\partial/\partial\theta_j)\log f(X_i, \theta)] = 0$, the multivariate central limit theorem (Theorem 5.1.8) shows that $(T_{1n}, \ldots, T_{sn})$ has a multivariate normal limit distribution with mean zero and covariance matrix $I(\theta^0)$.

On the other hand, it is easy to see—again in parallel to the proof of Theorem 2.3—that

$$\text{(23)} \qquad A_{jkn} \xrightarrow{P} a_{jk} = E\left[L''_{jk}(\theta^0)\right] = -I_{jk}(\theta^0).$$

The limit distribution of the Y's is therefore that of the solution $(Y_1, \ldots, Y_s)$ of the equations

$$\text{(24)} \qquad \sum_{k=1}^{s} I_{jk}(\theta^0)Y_k = T_j$$

where $T = (T_1, \ldots, T_s)$ is multivariate normal with mean zero and covariance matrix $I(\theta^0)$. It follows that the distribution of Y is that of

$$[I(\theta^0)]^{-1}T,$$

which is a multivariate distribution with zero mean and covariance matrix $[I(\theta^0)]^{-1}$.

This completes the proof of asymptotic normality and efficiency.

If the likelihood equations have a unique solution $\hat{\theta}_n$, then $\hat{\theta}_n$ is consistent, asymptotically normal, and efficient. It is, however, interesting to note that even if the parameter space is an open interval, it does not follow as in Corollary 2.2. that the MLE exists and hence is consistent (Problem 4.6). Sufficient conditions for existence and uniqueness are given in Mäkeläinen, Schmidt, and Styan (1981).

As in the one-parameter case, if the solution of the likelihood equations is not unique, Theorem 4.1 does not establish the existence of an efficient estimator of θ. However, the methods mentioned in Section 6.3 also work in the present case. In particular, if $\tilde{\theta}_n$ is a consistent sequence of estimators of θ, then the solutions $\hat{\theta}_n$ of the likelihood equations closest to $\tilde{\theta}_n$, for example, in the sense that $\Sigma(\hat{\theta}_{jn} - \tilde{\theta}_{jn})^2$ is smallest, is asymptotically efficient.

More convenient, typically, is the approach of Theorem 3.1, which we now generalize to the multiparameter case.

Theorem 4.2. *Suppose that the assumptions of Theorem* 4.1 *hold and that* $\tilde{\theta}_{jn}$ *is a* $\sqrt{n}$*-consistent estimator of* θ_j *for* $j = 1, \ldots, s$. *Let* $\{\delta_{kn}, k = 1, \ldots, s\}$ *be the solution of the equations*

$$\sum_{k=1}^{s} (\delta_{kn} - \tilde{\theta}_{kn}) L''_{jk}(\tilde{\theta}_n) = -L'_j(\tilde{\theta}_n). \tag{25}$$

Then $\delta_n = (\delta_{1n}, \ldots, \delta_{sn})$ *satisfies* (10) *with* δ_{jn} *in place of* $\hat{\theta}_{jn}$ *and thus is asymptotically efficient.*

Proof. The proof is a simple combination of the proofs of Theorem 3.1 and 4.1 and we shall only sketch it. Expanding the right side about θ^0 allows us to rewrite (25) as

$$\Sigma(\delta_{kn} - \tilde{\theta}_{kn}) L''_{jk}(\tilde{\theta}_n) = -L'_j(\theta^0) - \Sigma(\tilde{\theta}_{kn} - \theta_k^0) L''_{jk}(\theta^0) + R_n$$

where

$$R_n = -\tfrac{1}{2}\Sigma\Sigma(\tilde{\theta}_{kn} - \theta_k^0)(\tilde{\theta}_{ln} - \theta_l^0) L'''_{jkl}(\theta_n^*)$$

and hence as

$$\sqrt{n}\,\Sigma(\delta_{kn} - \theta_k^0)\frac{1}{n}L''_{jk}(\tilde{\theta}_n) = -\frac{1}{\sqrt{n}}L'_j(\theta^0) + \sqrt{n}\,\Sigma(\tilde{\theta}_{kn} - \theta_k^0)\left[\frac{1}{n}L''_{jk}(\tilde{\theta}_n) - \frac{1}{n}L''_{jk}(\theta^0)\right] + \frac{1}{\sqrt{n}}R_n. \tag{26}$$

This has the form (15), and it is easy to check (Problem 4.2) that the limits (in probability) of the A_{jkn} are the same a_{jk} as in (23) and that the second and third terms on the right side of (26) tend toward zero in probability. Thus the joint distribution of the right side is the same as that of the T_{jn} given by (22). It follows that the joint limit distribution of the $\sqrt{n}\,(\delta_{kn} - \theta_k^0)$ is the same as that of the $\sqrt{n}\,(\hat{\theta}_{kn} - \theta_k^0)$ in Theorem 4.1, and this completes the proof.

The following result generalizes Corollary 3.1 to the multiparameter case.

Corollary 4.1. *Suppose that the assumptions of Theorem* 4.2 *hold and that the elements* $I_{jk}(\theta)$ *of the information matrix of the* X*'s are continuous. Then the solutions* δ'_{kn} *of the equations*

$$n\Sigma(\delta'_{kn} - \tilde{\theta}_{kn})I_{jk}(\tilde{\theta}_n) = L'_j(\tilde{\theta}_n) \tag{27}$$

are asymptotically efficient.

The proof is left to Problem 4.5.

5. APPLICATIONS

Maximum likelihood (together with some of its variants) is the most widely used method of estimation, and a list of its applications would cover practically the whole field of statistics. [For a survey with a comprehensive set of references, see Norden (1972/73).] In this section we will discuss a few applications to illustrate some of the issues arising. The discussion, however, is not carried to the practical level, and in particular the problem of choosing among alternative asymptotically efficient methods is not addressed. Such a choice must be based not only on theoretical considerations but requires empirical evidence on the performance of the estimators at various sample sizes. For any specific example, the relevant literature should be consulted.

Example 5.1. Weibull Distribution. Let $X_1, \ldots, X_n$ be iid according to a two-parameter Weibull distribution, whose density it is convenient to write in a parametrization suggested by Cohen (1965) as

$$\frac{\gamma}{\beta} x^{\gamma-1} e^{-x^\gamma/\beta}, \qquad x > 0, \qquad \beta > 0, \qquad \gamma > 0, \tag{1}$$

where γ is a shape parameter and $\beta^{1/\gamma}$ a scale parameter. The likelihood equations, after some simplification, reduce to (Problem 5.1)

$$h(\gamma) = \frac{\Sigma x_i^\gamma \log x_i}{\Sigma x_i^\gamma} - \frac{1}{\gamma} = \frac{1}{n}\Sigma \log x_i \tag{2}$$

and

(3) $$\beta = \Sigma x_i^\gamma / n.$$

To show that (2) has at most one solution, note that $h'(\gamma)$ exceeds the derivative of the first term, which equals (Problem 5.2) $\Sigma a_i^2 p_i - (\Sigma a_i p_i)^2$ with

(4) $$a_i = \log x_i, \quad p_i = e^{\gamma a_i} / \sum_j e^{\gamma a_j}.$$

It follows that $h'(\gamma) > 0$ for all $\gamma > 0$. That (2) always has a solution follows from (Problem 5.3)

(5) $$-\infty = \lim_{\gamma \to 0} h(\gamma) < \frac{1}{n} \Sigma \log x_i < \log x_{(n)} = \lim_{\gamma \to \infty} h(\gamma).$$

This example, therefore, illustrates the simple situation in which the likelihood equations always have a unique solution.

Example 5.2. Location-Scale Families. Let $X_1, \ldots, X_n$ be iid, each with density $(1/a) f[(x - \xi)/a]$. The calculation of an ELE is easy when the likelihood equations have a unique root $(\hat{\xi}, \hat{a})$. It was shown by Barndorff–Nielsen and Blaesild (1980) and by Scholz (1981) that sufficient conditions for this to be the case is that $f(x)$ is positive, twice differentiable for all x, and strongly unimodal. Surprisingly, Copas (1975) showed that it is unique also when f is Cauchy, despite the fact that the Cauchy density is not strongly unimodal and that in this case the likelihood equation can have multiple roots when a is known.

In the presence of multiple roots, the simplest approach typically is that of Theorem 4.2. The $\sqrt{n}$-consistent estimators of ξ and a required by this theorem are easily obtained in the present case. As was pointed out in Example 3.2, the mean or median of the X's will usually have the desired property for ξ. (When f is asymmetric, this requires that ξ be specified to be some particular location measure such as the mean or median of the distribution of the X_i.) If $E(X_i^4) < \infty$, $\hat{a}_n = \sqrt{\Sigma (X_i - \bar{X})^2 / n}$ will be $\sqrt{n}$-consistent for a if the latter is taken to be the population standard deviation. If $E(X_i^4) = \infty$, one can instead, for example, take a suitable multiple of the interquartile range $X_{(k)} - X_{(j)}$, where $k = [3n/4]$ and $j = [n/4]$ (see, for example, Mosteller, 1946).

If f satisfies the assumptions of Theorem 4.1, then $[\sqrt{n}(\hat{\xi}_n - \xi), \sqrt{n}(\hat{a}_n - a)]$ have a joint bivariate normal distribution with zero means and covariance matrix $I^{-1}(a) = \|I_{ij}(a)\|^{-1}$, which is independent of ξ and where $I(a)$ is given by (2.7.25) and (2.7.26).

If the distribution of the X_i depends on $\theta = (\theta_1, \ldots, \theta_s)$, it is interesting to compare the estimation of θ_j when the other parameters are unknown with the situation in which they are known. The mathematical meaning of this distinction is that an estimator is permitted to depend on known parameters but not on unknown ones. Since the class of possible estimators is thus more restricted when the nuisance parameters are unknown, it follows from Theorems 2.3 and 4.1 that the asymptotic variance of an efficient estimator when some of the θ's are unknown can never fall below its value when they are known, so that

$$\frac{1}{I_{jj}(\theta)} \leqslant [I(\theta)]_{jj}^{-1}, \tag{6}$$

as was already shown in Section 2.7 as (2.7.30). There, it was also proved that equality holds in (6) whenever

$$\operatorname{cov}\left(\partial \log f(X,\theta)/\partial\theta_j, \partial \log f(X,\theta)/\partial\theta_k\right) = 0 \qquad \text{for all} \quad j \neq k, \tag{7}$$

and that this condition, which states that

$$I(\theta) \qquad \text{is diagonal,} \tag{8}$$

is also necessary for equality. For the location-scale families of Example 5.2, it follows from (2.7.26) that $I_{12} = 0$ whenever f is symmetric about zero but not necessarily otherwise. For symmetric f there is therefore no loss of asymptotic efficiency in estimating ξ or a when the other parameter is unknown.

Suppose the efficient estimator of θ_j depends on the remaining parameters and yet θ_j can be estimated without loss of efficiency when these parameters are unknown. The situation can then be viewed as a rather trivial example of the idea of adaptive estimation of which more typical examples were mentioned in Sections 5.4 and 5.6. It was viewed this way by Stein (1956) in a paper that initiated the development of adaptive estimation.

As another illustration of efficient likelihood estimation, consider a multiparameter exponential family. Here UMVU estimators often are satisfactory solutions of the estimation problem. However, the estimand may not be U-estimable and then another approach is needed. In some cases, even when a UMVU estimator exists, the MLE has the advantage of not taking on values outside the range of the estimand.

Example 5.3. Multiparameter Exponential Families. Let $\underline{X} = (X_1, \ldots, X_n)$ be distributed according to an s-parameter exponential family with density (1.4.2) with respect to a σ-finite measure μ, where $\underline{x}$ takes the place of x and where it is assumed that $T_1(\underline{X}), \ldots, T_s(\underline{X})$ are affinely independent with probability 1. Using the fact that

$$\frac{\partial}{\partial\eta_j}[L(\eta)] = -\frac{\partial}{\partial\eta_j}[A(\eta)] + T_j(\underline{x}) \tag{9}$$

and other properties of the densities (1.4.2), one sees that the conditions of Theorem 4.1 are satisfied when the X's are iid. By (1.4.11), the likelihood equations for the η's reduce to

$$T_j(\underline{x}) = E_\eta\left[T_j(\underline{X})\right]. \tag{10}$$

If these equations have a solution, it is unique (and is the MLE) since $L(\eta)$ is a strictly concave function of η. This follows from Theorem 1.6.6 and the fact that by (1.4.12)

$$-\frac{\partial^2}{\partial\eta_j\partial\eta_k}[L(\eta)] = \frac{\partial^2}{\partial\eta_j\partial\eta_k}[A(\eta)] = \text{cov}\left[T_j(\underline{X}), T_k(\underline{X})\right] \tag{11}$$

so that by assumption the matrix with entries (11) is positive definite.

Sufficient conditions for the existence of a solution of the likelihood equations are given by Crain (1976) and Barndorff-Nielsen (1978, Sect. 9.3, 9.4), where they are shown to be satisfied for the two-parameter gamma family of Table 1.4.1.

An alternative method for obtaining asymptotically efficient estimators for the parameters of an exponential family is based on the mean-value parametrization (2.7.22). Slightly changing the formulation of the model, consider a sample $(X_1, \ldots, X_n)$ of size n from the family (1.4.2), let $\bar{T}_j = [T_j(X_1) + \cdots + T_j(X_n)]/n$ and $\theta_j = E(T_j)$. By the CLT, the joint distribution of the $\sqrt{n}(\bar{T}_j - \theta_j)$ is multivariate normal with zero means and covariance matrix $\sigma_{ij} = \text{cov}[T_i(X), T_j(X)]$. This proves the $\bar{T}_j$ to be asymptotically efficient estimators by (2.7.23).

For further discussion of maximum likelihood estimation in exponential families, see Berk (1972), Sundberg (1974), Barndorff-Nielsen (1978), and Johansen (1979). In the next two examples, we shall consider in somewhat more detail the most important case of Example 5.3, the multivariate and, in particular, the bivariate normal distribution.

Example 5.4. Multivariate Normal Distribution. Let $(X_{1\nu}, \ldots, X_{p\nu})$, $\nu = 1, \ldots, n$ be a sample from a nonsingular normal distribution with means $E(X_{i\nu}) = \xi_i$ and covariances $\text{cov}(X_{i\nu}, X_{j\nu}) = \sigma_{ij}$. By (1.3.15) the density of the X's is given by

$$|\Xi|^{n/2}(2\pi)^{-pn/2}\exp\left(-\tfrac{1}{2}\Sigma\Sigma\eta_{jk}S_{jk}\right) \tag{12}$$

where

$$S_{jk} = \Sigma(X_{j\nu} - \xi_j)(X_{k\nu} - \xi_k), \qquad j, k = 1, \ldots, p, \tag{13}$$

and where $\Xi = \|\eta_{jk}\|$ is the inverse of the covariance matrix $\|\sigma_{jk}\|$.

Consider first the case in which the ξ's are known. Then (12) is an exponential family with $T_{jk} = -\frac{1}{2}S_{jk}$. If the matrix $\|\sigma_{jk}\|$ is nonsingular, the T_{jk} are affinely independent with probability 1, so that the result of the preceding example applies. Since $E(S_{jk}) = n\sigma_{jk}$, the likelihood equations (10) reduce to $n\sigma_{jk} = S_{jk}$ and thus have the solutions

$$\hat{\sigma}_{jk} = \frac{1}{n}S_{jk}. \tag{14}$$

The sample moments and correlations are therefore ELEs of the population variances, covariances, and correlation coefficients. Also, the (jk)th element of $\|\hat{\sigma}_{jk}\|^{-1}$ is an asymptotically efficient estimator of η_{jk}. In addition to being the MLE, $\hat{\sigma}_{jk}$ is the UMVU estimator of σ_{jk} (Example 2.2.4).

If the ξ's are unknown,

$$\hat{\xi}_j = \frac{1}{n}\Sigma X_{j\nu} = X_{j\cdot}$$

and $\hat{\sigma}_{jk}$, given by (14) but with S_{jk} now defined as

$$S_{jk} = \Sigma(X_{j\nu} - X_{j\cdot})(X_{k\nu} - X_{k\cdot}), \tag{15}$$

continue to be ELEs for ξ_j and σ_{jk} (Problem 5.5).

If ξ is known, the asymptotic distribution of S_{jk} given by (13) is immediate from the central limit theorem since S_{jk} is the sum of n iid variables with expectation

$$E(X_{j\nu} - \xi_j)(X_{k\nu} - \xi_k) = \sigma_{jk}$$

and variance

$$E\left[(X_{j\nu} - \xi_j)^2(X_{k\nu} - \xi_k)^2\right] - \sigma_{jk}^2.$$

If $j \neq k$, it follows from Problem 1.4.19 that

$$E\left[(X_{j\nu} - \xi_j)^2(X_{k\nu} - \xi_k)^2\right] = \sigma_{jj}\sigma_{kk} + 2\sigma_{jk}^2$$

so that

$$\text{var}\left[(X_{j\nu} - \xi_j)(X_{k\nu} - \xi_k)\right] = \sigma_{jj}\sigma_{kk} + \sigma_{jk}^2$$

and

$$\sqrt{n}\left(\frac{S_{jk}}{n} - \sigma_{jk}\right) \xrightarrow{\mathcal{L}} N\left(0, \sigma_{jj}\sigma_{kk} + \sigma_{jk}^2\right). \tag{16}$$

If ξ is unknown, the S_{jk} given by (15) are independent of the $X_{i\cdot}$ and the asymptotic distribution of (15) is the same as that of (13) (Problem 5.6).

Example 5.5. Bivariate Normal Distribution. In the preceding example it was seen that knowing the means does not affect the efficiency with which the covariances can be estimated. Let us now restrict attention to the covariances and, for the sake of simplicity, suppose that $p = 2$. With an obvious change of notation, let (X_i, Y_i), $i = 1, \ldots, n$ be iid, each with density (1.3.16). Since the asymptotic distribution of $\hat{\sigma}$, $\hat{\tau}$, and $\hat{\rho}$ are not affected by whether or not ξ and η are known, let us assume $\xi = \eta = 0$. For the information matrix $I(\theta)$ [where $\theta = (\sigma^2, \tau^2, \rho)$], we find

[Problem 5.7(i)]

$$(17)\qquad (1-\rho^2)I(\theta) = \begin{bmatrix} \dfrac{2-\rho^2}{4\sigma^4} & \dfrac{-\rho^2}{4\sigma^2\tau^2} & \dfrac{-\rho}{2\sigma^2} \\ \dfrac{-\rho^2}{4\sigma^2\tau^2} & \dfrac{2-\rho^2}{4\tau^4} & \dfrac{-\rho}{2\tau^2} \\ \dfrac{-\rho}{2\sigma^2} & \dfrac{-\rho}{2\tau^2} & \dfrac{1+\rho^2}{1-\rho^2} \end{bmatrix}.$$

Inversion of this matrix gives the covariance matrix of the $\sqrt{n}(\hat{\theta}_j - \theta_j)$ as [Problem 5.7(ii)]

$$(18)\qquad \begin{bmatrix} 2\sigma^4 & 2\rho^2\sigma^2\tau^2 & \rho(1-\rho^2)\sigma^2 \\ 2\rho^2\sigma^2\tau^2 & 2\tau^4 & \rho(1-\rho^2)\tau^2 \\ \rho(1-\rho^2)\sigma^2 & \rho(1-\rho^2)\tau^2 & (1-\rho^2)^2 \end{bmatrix}.$$

Thus, we find that

$$\sqrt{n}(\hat{\sigma}^2 - \sigma^2) \xrightarrow{\mathcal{L}} N(0, 2\sigma^4)$$

$$\sqrt{n}(\hat{\tau}^2 - \tau^2) \xrightarrow{\mathcal{L}} N(0, 2\tau^4)$$

$$\sqrt{n}(\hat{\rho} - \rho) \xrightarrow{\mathcal{L}} N\left[0, (1-\rho^2)^2\right]$$

On the other hand, if σ and τ are known to be equal to 1, the MLE $\hat{\hat{\rho}}$ of ρ satisfies (Problem 5.8)

$$(19)\qquad \sqrt{n}(\hat{\hat{\rho}} - \rho) \xrightarrow{\mathcal{L}} N\left(0, \frac{(1-\rho^2)^2}{1+\rho^2}\right)$$

whereas if ρ and τ are known, the MLE $\hat{\hat{\sigma}}$ of σ satisfies

$$(20)\qquad \sqrt{n}(\hat{\hat{\sigma}}^2 - \sigma^2) \xrightarrow{\mathcal{L}} N\left(0, \frac{4\sigma^4(1-\rho^2)}{2-\rho^2}\right).$$

The efficiency of $\hat{\rho}$ to $\hat{\hat{\rho}}$ is thus

$$(21)\qquad e_{\hat{\rho}, \hat{\hat{\rho}}} = \frac{1}{1+\rho^2}.$$

This is 1 when $\rho = 0$ but can be close to $1/2$ when $|\rho|$ is close to 1. Similarly,

$$e_{\hat{\sigma}^2, \hat{\hat{\sigma}}^2} = \frac{2(1-\rho)^2}{2-\rho^2}. \tag{22}$$

This efficiency is again 1 when $\rho = 0$ but tends to zero as $|\rho| \to 1$. This last result, which at first may seem surprising, actually is easy to explain. If ρ were equal to 1, and $\tau = 1$ say, we would have $X_i = \sigma Y_i$. Since both X_i and Y_i are observed, we could then determine σ without error from a single observation.

The next example deals with an important class of problems in which the MLE does not exist, although the conditions of Theorem 4.1 are satisfied.

Example 5.6. Normal Mixtures. Let $X_1, \dots, X_n$ be iid each with probability p as $N(\xi, \sigma^2)$ and probability $q = 1 - p$ as $N(\eta, \tau^2)$. (The Tukey models are examples of such distributions with $\eta = \xi$.) The joint density of the X's is then given by

$$\prod_{i=1}^{n} \left\{ \frac{p}{\sqrt{2\pi}\,\sigma} \exp\left[-\frac{1}{2\sigma^2}(x_i - \xi)^2 \right] + \frac{q}{\sqrt{2\pi}\,\tau} \exp\left[\frac{1}{2\tau^2}(x_i - \eta)^2 \right] \right\}. \tag{23}$$

This is a sum of non-negative terms of which one, for example, is proportional to

$$\frac{1}{\sigma\tau^{n-1}} \exp\left[-\frac{1}{2\sigma^2}(x_1 - \xi)^2 - \frac{1}{2\tau^2}\sum_{i=2}^{n}(x_i - \eta)^2 \right].$$

When $\xi = x_1$ and $\sigma \to 0$ this term tends to infinity for any fixed values of η, τ, and $x_2, \dots, x_n$. The likelihood is therefore unbounded and the MLE does not exist. (The corresponding result holds for any other mixture with density $\Pi\langle(p/\sigma)f[(x_i - \xi)/\sigma] + (q/\tau)f[(x_i - \eta)/\tau]\rangle$ when $f(0) \neq 0$.)

On the other hand, the conditions of Theorem 4.1 are satisfied (Problem 5.10) so that efficient solutions of the likelihood equations exist and asymptotically efficient estimators can be obtained through Theorem 4.2. One approach to the determination of the required $\sqrt{n}$-consistent estimators is the method of moments. In the present case this means equating the first five moments of the X's with the corresponding sample moments and then solving for the five parameters. For the normal mixture problem, these estimators were proposed in their own right by K. Pearson (1894). For a more recent discussion and possible simplifications see Cohen (1967).

A study of the improvement of an asymptotically efficient estimator over that obtained by the method of moments has been carried out for the case that it is known that $\tau = \sigma$ (Tan and Chang, 1972). If $\Delta = (\eta - \xi)/\sigma$, the AREs for the estimation of all four parameters depend only on Δ and p. As an example, consider the estimation of p. Here the ARE is < 0.01 if $\Delta < 1/2$ and $p < 0.2$; it is < 0.1 if $\Delta < 1/2$ and $0.2 < p < 0.4$, and is > 0.9 if $\Delta > 0.5$. [For an alternative starting point for the application of Theorem 4.2, see Quandt and Ramsey (1978), particularly the discussion by N. Kiefer.]

Example 5.7. Multinomial Experiments. Let $(X_0, X_1, \dots, X_s)$ have the multinomial distribution (1.4.4). In the full-rank exponential representation (3.5.12), the statistics T_j can be taken to be the X_j. Using the mean-value parametrization, the likelihood equations (10) reduce to $np_j = X_j$ so that the MLE of p_j is $\hat{p}_j = X_j/n$ ($j = 1,\dots, s$). If X_j is 0 or n, the likelihood equations have no solution in the parameter space $0 < p_j < 1$, $\Sigma_{j=1}^{s} p_j < 1$. However, the probability of any X_j taking on either of these values tends to zero as $n \to \infty$. That the MLEs $\hat{p}_j$ are asymptotically efficient is seen by introducing the indicator variables $X_{j\nu}$, $\nu = 1,\dots, n$ which are 1 when the νth trial results in outcome j and are 0 otherwise. Then the vectors $(X_{0\nu},\dots, X_{s\nu})$ are iid and $T_j = X_{j1} + \cdots + X_{jn}$, so that asymptotic efficiency follows from Example 5.3.

As discussed in Section 3.5, in applications of the multinomial distribution to contingency tables, the p's are usually subject to additional restrictions. Theorem 4.1 typically continues to apply although the computation of the estimators may be less obvious. This class of problems is treated comprehensively in Bishop, Fienberg, and Holland (1975) and Haberman (1974). Empty cells often present special problems.

6. EXTENSIONS

The discussion of efficient likelihood estimation so far has been restricted to the iid case. In the present section we mention briefly extensions to some more general situations, which permit results analogous to those of Sections 6.2–6.4. Treatments not requiring the stringent (but frequently applicable) assumptions of Theorems 2.3 and 4.1 have been developed by Le Cam (1953, 1969, 1970) and others. For further work in this direction see Pfanzagl (1970), Weiss and Wolfowitz (1974) and Blyth (1982), and Ibragimov and Has'minskii (1981).

The theory easily extends to the case of two or more samples. Suppose that the variables $X_{\alpha 1},\dots, X_{\alpha n_\alpha}$ in the αth sample are iid according to the distribution with density $f_{\alpha,\theta}$ ($\alpha = 1,\dots, r$) and that the r samples are independent. In applications, it will typically turn out that the vector parameter $\theta = (\theta_1,\dots, \theta_s)$ has some components occurring in more than one of the r distributions while others may be specific to just one distribution. However, for the present discussion we shall permit each of the distributions to depend on all of the parameters.

The limit situation we shall consider supposes that each of the sample sizes n_α tends to infinity, all at the same rate, but that r remains fixed. Consider, therefore, sequences of sample sizes n_{α_k} ($k = 1,\dots, \infty$) with total sample size $N_k = \Sigma_{\alpha=1}^{r} n_{\alpha_k}$ such that

$$n_{\alpha_k}/N_k \to \lambda_\alpha \quad \text{as} \quad k \to \infty \tag{1}$$

where $\Sigma\lambda_\alpha = 1$, and the λ_α are > 0.

Theorem 6.1. *Suppose the assumptions of Theorem* 4.1 *hold for each of the densities* $f_{\alpha,\theta}$. *Let* $I^{(\alpha)}(\theta)$ *denote the information matrix corresponding to* $f_{\alpha,\theta}$ *and let*

$$I(\theta) = \Sigma\lambda_\alpha I^{(\alpha)}(\theta). \tag{2}$$

The log likelihood $L(\theta)$ *is given by*

$$L(\theta) = \sum_{\alpha=1}^{r} \sum_{j=1}^{n_\alpha} \log f_{\alpha,\theta}(x_{\alpha j})$$

and the likelihood equations by

$$\frac{\partial}{\partial\theta_j} L(\theta) = 0 \qquad (j = 1,\ldots, s). \tag{3}$$

With these identifications, the conclusions of Theorem 4.1 *remain valid.*

The proof is an easy extension of that of Theorem 4.1 since $L(\theta)$, and therefore each term of its Taylor expansion, is a sum of r independent terms of the kind considered in the proof of Theorem 4.1 (Problem 6.1). (For further discussion of this situation, see Bradley and Gart, 1962.)

That asymptotic efficiency continues to have the meaning it had in Theorems 2.3 and 4.1 follows from the fact that Theorem 1.1 and its extension to the multiparameter case also extends to the present situation (see Bahadur, 1964, Section 4).

Corollary 6.1. *Under the assumptions of Theorem* 6.1 *suppose that for each* α *all off-diagonal elements in the* j*th row and* j*th column of* $I^{(\alpha)}(\theta)$ *are zero. Then the asymptotic variance of* $\hat{\theta}_j$ *is the same when the remaining* θ*'s are unknown as when they are known.*

Proof. If the property in question holds for each $I^{(\alpha)}(\theta)$, it also holds for $I(\theta)$ and the result thus follows from Problem 5.4.

The following four examples illustrate some applications of Theorem 6.1.

Example 6.1. Estimation of a Common Mean. Let $X_1,\ldots, X_m$ and $Y_1,\ldots, Y_n$ be independently distributed according to $N(\xi, \sigma^2)$ and $N(\xi, \tau^2)$, respectively, with ξ, σ, and τ unknown. The problem of estimating ξ was considered briefly in Example 2.2.3 where it was found that a UMVU estimator for ξ does not exist. On the other hand, the problem of asymptotically efficient estimation of ξ is easy in the present case. Since the MLEs of the mean and variance of a single normal distribution are asymptotically independent, Corollary 6.1 applies and shows that ξ can be estimated with the efficiency that is attainable when σ and τ are known. Now in that case, the MLE—which is also UMVU— is

$$\frac{(m/\sigma^2)\bar{X} + (n/\tau^2)\bar{Y}}{m/\sigma^2 + n/\tau^2}.$$

It follows from Theorem 5.1.4 that the asymptotic distribution of $\hat{\xi}$ is not changed when σ^2 and τ^2 are replaced by

$$\hat{\sigma}^2 = \frac{1}{m-1}\Sigma(X_i - \bar{X})^2 \quad \text{and} \quad \hat{\tau}^2 = \frac{1}{n-1}\Sigma(Y_j - \bar{Y})^2$$

and the resulting estimator $\hat{\xi}$ is therefore asymptotically efficient.

Example 6.2. Random Effects: Balanced One-way Layout. Consider the estimation of the variance components σ_A^2 and σ^2 in model (3.5.1). In the canonical form (3.5.2), we are dealing with independent normal variables Z_{11}, Z_{i1} $(i = 2, \ldots, s)$, Z_{ij} $(i = 1, \ldots, s; j = 2, \ldots, n)$. We shall here restrict attention to the second and third group, as suggested by Thompson (1962), and are then dealing with samples of sizes $s - 1$ and $(n-1)s$ from $N(0, \tau^2)$ and $N(0, \sigma^2)$ where $\tau^2 = \sigma^2 + n\sigma_A^2$. The assumptions of Theorem 6.1 are satisfied with $r = 2$, $\theta = (\sigma^2, \tau^2)$ and the parameter space $\Omega = \{(\sigma, \tau)\colon 0 < \sigma^2 < \tau^2\}$. For fixed n, the sample sizes $n_1 = s - 1$, $n_2 = s(n-1)$ tend to infinity as $s \to \infty$, with $\lambda_1 = 1/n$, $\lambda_2 = (n-1)/n$.

The joint density of the second and third group of Z's constitutes a two-parameter exponential family; the log likelihood is given by

$$(4) \qquad -L(\theta) = n_2 \log \sigma + n_1 \log \tau + \frac{S^2}{2\sigma^2} + \frac{S_A^2}{2\tau^2} + c,$$

where $S^2 = \Sigma_{i=1}^s \Sigma_{j=2}^n Z_{ij}^2$ and $S_A^2 = \Sigma_{i=2}^s Z_{i1}^2$. By Example 5.3, the likelihood equations have at most one solution. Solving the equations yields

$$(5) \qquad \hat{\sigma}^2 = S^2/n_2, \qquad \hat{\tau}^2 = S_A^2/n_1,$$

and these are the desired (unique, ML) solutions, provided they are in Ω, that is, they satisfy

$$(6) \qquad \hat{\sigma}^2 < \hat{\tau}^2.$$

It follows from Theorem 4.1 that the probability of (6) tends to 1 as $s \to \infty$ for any $\theta \in \Omega$; this can also be seen directly from the fact that $\hat{\sigma}^2$ and $\hat{\tau}^2$ tend to σ^2 and τ^2 in probability.

What can be said when (6) is violated? The likelihood equations then have no root in Ω and an MLE does not exist (the likelihood attains its maximum at the boundary point $\hat{\sigma}^2 = \hat{\tau}^2 = (S_A^2 + S^2)/(n_1 + n_2)$ which is not in Ω). However, none of this matters from the present point of view since the asymptotic theory has nothing to say about a set of values whose probability tends to zero. [For small-sample computations of the mean square errors of a number of estimators of σ^2 and σ_A^2, see Klotz, Milton, and Zacks (1969) and Portnoy (1971). Note also the earlier discussion of this problem in Sections 3.5 and 4.1.]

The joint asymptotic distribution of $\hat{\sigma}^2$ and $\hat{\tau}^2$ can be obtained from Theorem 4.1 or directly from the definition of S^2 and S_A^2 and the CLT, and a linear transformation of the limit distribution then gives the joint asymptotic distribution of $\hat{\sigma}^2$ and $\hat{\sigma}_A^2$ (Problem 6.2).

Example 6.3. Random Effects: Balanced Two-Way Layout. A new issue arises as we go from the one-way to the two-way layout with the model given by (3.5.4). After elimination of Z_{111} (in the notation of Example 3.5.2), the data in canonical form consists of four samples Z_{i11} ($i = 2, \ldots, I$), Z_{1j1} ($j = 2, \ldots, J$), Z_{ij1} ($i = 2, \ldots, I$; $j = 2, \ldots, J$), and Z_{ijk} ($i = 1, \ldots, I$; $j = 1, \ldots, J$; $k = 2, \ldots, n$), and the parameter is $\theta = (\sigma, \tau_A, \tau_B, \tau_C)$ where

$$(7) \quad \tau_C^2 = \sigma^2 + n\sigma_C^2, \qquad \tau_B^2 = nI\sigma_B^2 + n\sigma_C^2 + \sigma^2, \qquad \tau_A^2 = nJ\sigma_A^2 + n\sigma_C^2 + \sigma^2$$

so that $\Omega = \{\theta\colon \sigma^2 < \tau_C^2 < \tau_A^2, \tau_B^2\}$. The joint density of these variables constitutes a four-parameter exponential family. The likelihood equations thus again have at most one root, and this is given by

$$\hat{\sigma}^2 = S^2/(n-1)IJ, \qquad \hat{\tau}_C^2 = S_C^2/(I-1)(J-1),$$

$$\hat{\tau}_B^2 = S_B^2/(J-1), \qquad \hat{\tau}_A^2 = S_A^2/(I-1)$$

when $\hat{\sigma}^2 < \hat{\tau}_C^2 < \hat{\tau}_A^2, \hat{\tau}_B^2$. No root exists when these inequalities fail.

In this case, asymptotic theory requires that both I and J tend to infinity, and assumption (1) of Theorem 6.1 then does not hold. Asymptotic efficiency of the MLEs follows, however, from Theorem 4.1 since each of the samples depends only on one of the parameters $\sigma^2, \tau_C^2, \tau_A^2, \tau_B^2$. The apparent linkage of these parameters through the inequalities $\sigma^2 < \tau_C^2 < \tau_A^2, \tau_B^2$ is immaterial. The true point $\theta^0 = (\sigma^0, \tau_C^0, \tau_A^0, \tau_B^0)$ is assumed to satisfy these restrictions, and each parameter can then independently vary about the true value, which is all that is needed for Theorem 4.1. It therefore follows as in the preceding example that the MLEs are asymptotically efficient, and that $\sqrt{(n-1)IJ}\,(\hat{\sigma}^2 - \sigma^2)$, etc. have the limit distributions given by Theorem 4.1 or are directly obtainable from the definitions of these estimators.

A general large-sample treatment both of components of variance and the more general case of mixed models, without assuming the models to be balanced, is given by Miller (1977).

Example 6.4. Independent Binomial Experiments. As in Section 3.5, let X_i ($i = 1, \ldots, s$) be independently distributed according to the binomial distributions $b(p_i, n_i)$, with the p_i being functions of a smaller number of parameters. If the n_i tend to infinity at the same rate, the situation is of the type considered in Theorem 6.1, which will in typical cases ensure the existence of an efficient solution of the likelihood equations with probability tending to 1.

As an illustration, suppose, as in (3.5.22), that the p's are given in terms of the logistic distribution, and more specifically that

$$(8) \qquad q_i = \frac{1}{1 + e^{-(\alpha + \beta t_i)}}$$

where the t's are known numbers and α and β the parameters to be estimated. The likelihood equations

$$(9) \qquad \Sigma n_i p_i = \Sigma x_i, \qquad \Sigma n_i t_i p_i = \Sigma t_i x_i$$

have at most one solution (Problem 6.3) which will exist with probability tending to 1 (but may not exist for some particular finite values) and which can be obtained by standard iterative methods.

That the likelihood equations have at most one solution is true not only for the model (8) but more generally when

$$q_i = F(\Sigma \beta_j t_j) \tag{10}$$

where the t's are known, the β's are being estimated, and F is a known distribution function with $\log F(x)$ and $\log[1 - F(x)]$ strictly concave. [See Haberman, 1974, Chapter 8; and Problem 6.7.] For further discussion of this and more general logistic regression models see Pregibon (1981).

For the multinomial problem mentioned in the preceding section and those of Example 6.4, alternative methods have been developed, which are asymptotically equivalent to the ELEs and hence also asymptotically efficient. These methods are based on minimizing χ^2 or some other functions measuring the distance of the vector of probabilities from that of the observed frequencies. [See, for example, Neyman (1949), Taylor (1953), Le Cam (1956), Wijsman (1959), Berkson (1980), Amemiya (1980), and Ghosh and Sinha (1981) for entries to the literature on choosing between these different estimators.]

The situation of Theorem 6.1 shares with that of Theorem 2.3 the crucial property that the total amount of information $T(\theta)$ asymptotically becomes arbitrarily large. In the general case of independent but not identically distributed variables, this need no longer be the case.

Example 6.5. Let X_i $(i = 1, \dots, n)$ be independent Poisson variables with $E(X_i) = \gamma_i \lambda$ where the γ's are known numbers. Consider two cases.

(i) $\Sigma_{i=1}^{\infty} \gamma_i < \infty$. The amount of information X_i contains about λ is γ_i/λ by (2.6.11) and Table 2.6.1, and the total amount $T_n(\lambda)$ that $(X_1, \dots, X_n)$ contains about λ is therefore

$$T_n(\lambda) = \frac{1}{\lambda} \sum_{i=1}^{n} \gamma_i. \tag{11}$$

It is intuitively plausible that in these circumstances λ cannot be estimated consistently because only the early observations provide an appreciable amount of information. To prove this formally, note that $Y_n = \Sigma_{i=1}^{n} X_i$ is a sufficient statistic for λ on the basis of $(X_1, \dots, X_n)$ and that Y_n has a Poisson distribution with mean $\lambda \Sigma_{i=1}^{n} \gamma_i$. Thus all the Y's are less informative than a random variable Y with distribution $P(\lambda \Sigma_{i=1}^{\infty} \gamma_i)$ in the sense that the distribution of any estimator based on Y_n can be duplicated by one based on Y (Problem 6.9). Since λ cannot be estimated exactly on the basis of Y, the result follows.

(ii) $\Sigma_{i=1}^{\infty}\gamma_i = \infty$. Here the MLE $\delta_n = \Sigma_{i=1}^n X_i / \Sigma_{i=1}^n \gamma_i$ is consistent and asymptotically normal (Problem 6.10) with

$$\sqrt{\sum_{i=1}^{n} \gamma_i}\,(\delta_n - \lambda) \xrightarrow{\mathcal{L}} N(0, \lambda). \tag{12}$$

Thus δ_n is approximately distributed as $N[\lambda, 1/T_n(\lambda)]$ and an extension of Theorem 1.1 to the present case (see Bahadur, 1964) permits the conclusion that δ_n is asymptotically efficient.

Note: The norming constant required for asymptotic normality must be proportional to $\sqrt{\Sigma_{i=1}^n \gamma_i}$. Depending on the nature of the γ's this can be any function of n tending to infinity rather than the customary $\sqrt{n}$. In general, it is the total amount of information rather than the sample size which governs the asymptotic distribution of an asymptotically efficient estimator. In the iid case, $T_n(\theta) = nI(\theta)$ so that $\sqrt{T_n(\theta)}$ is proportional to $\sqrt{n}$.

A general treatment of the case of independent random variables with densities $f_j(x_j, \theta)$, $\theta = (\theta_1, \ldots, \theta_r)$ along the lines of Theorems 2.3 and 4.1 has been given by Bradley and Gart (1962) and Hoadley (1971) (see also Nordberg, 1980). The proof (for $r = 1$) is based on generalizations of (2.18)–(2.20) (see Problem 6.20) and hence depends on a suitable law of large numbers and central limit theorem for sums of independent nonidentical random variables. In the multiparameter case, of course, it may happen that some of the parameters can be consistently estimated and others not.

The theory for iid variables summarized by Theorems 1.1, 2.3, and 4.1 can be generalized not only to the case of independent nonidentical variables, but also to dependent variables whose joint distribution depends on a fixed number of parameters $\theta = (\theta_1, \ldots, \theta_r)$. The log likelihood $L(\theta)$ is now the sum of the logarithms of the conditional densities $f_j(x_j, \theta | x_1, \ldots, x_{j-1})$ and the total amount of information $T_n(\theta)$ is the sum of the expected conditional amounts of information $I_j(\theta)$ in X_j, given $X_1, \ldots, X_{j-1}$:

$$I_j(\theta) = E\left\{ E\left[\frac{\partial}{\partial\theta} \log f_j(X_j, \theta | X_1, \ldots, X_{j-1})\right]^2 \right\} = E\left[\frac{\partial}{\partial\theta} \log f_j(X_j, \theta)\right]^2.$$

Under regularity conditions on the f_j's analogous to those of Theorems 2.3 and 4.1 together with additional conditions to ensure that the total amount of information tends to infinity as $n \to \infty$ and that the appropriate CLT for dependent variables is applicable, it can be shown that with probability tending to 1 there exists a root $\hat{\theta}_n$ of the likelihood equations such that $\sqrt{T_n(\theta)}\,(\hat{\theta}_n - \theta) \xrightarrow{\mathcal{L}} N(0, 1)$. This program has been carried out in a series of papers by Bar-Shalom (1971), Bhat (1974), and Crowder (1976).* [The

*A review of the literature of maximum likelihood estimation in both discrete and continuous parameter stochastic processes can be found in Basawa and Prakasa Rao (1980).

required extension of Theorem 1.1 can be obtained from Bahadur (1964); see also Kabaila (1980).] The following illustrates the theory with a simple classic example involving only one parameter.

Example 6.6. Normal Autoregressive Markov Series. Let

$$X_j = \beta X_{j-1} + U_j, \qquad j = 2, \ldots, n \tag{13}$$

where the U_j are iid as $N(0, 1)$, where β is an unknown parameter satisfying $|\beta| < 1$,* and where X_1 is $N(0, \sigma^2)$. The X's all have marginal normal distributions with mean zero. The variance of X_j satisfies

$$\text{var}(X_j) = \beta^2 \text{var}(X_{j-1}) + 1 \tag{14}$$

and hence $\text{var}(X_j) = \sigma^2$ for all j provided

$$\sigma^2 = 1/(1 - \beta^2). \tag{15}$$

This is the stationary case in which $(X_{j_1}, \ldots, X_{j_k})$ has the same distribution as $(X_{j_1+r}, \ldots, X_{j_k+r})$ for all $r = 1, 2, \ldots$ (Problem 6.14).

The amount of information that each X_j $(j > 1)$ contains about β is (Problem 6.28) $I_j(\theta) = 1/(1 - \beta^2)$ so that $T_n(\beta) \sim n/(1 - \beta^2)$. The general theory therefore suggests the existence of a root $\hat{\beta}_n$ of the likelihood equation such that

$$\sqrt{n}\,(\hat{\beta}_n - \beta) \xrightarrow{\mathcal{L}} N(0, 1 - \beta^2). \tag{16}$$

That (16) does hold, can also be checked directly (see, for example, Anderson, 1959).

The conclusions of this section up to this point can be summarized by saying that the asymptotic theory developed for the iid case in Sections 6.1–6.5 continues to hold—under appropriate safeguards—even if the iid assumption is dropped, provided the number of parameters is fixed and the total amount of information goes to infinity.

We shall now briefly consider two generalizations of the earlier situation to which this conclusion does not apply. The first concerns the case in which the number of parameters tends to infinity with the total sample size.

In Theorem 6.1 the number r of samples was considered fixed while the sample sizes n_α were assumed to tend to infinity. Such a model is appropriate when one is dealing with a small number of moderately large samples. A quite different asymptotic situation arises in the reverse case of a large number (considered as tending to infinity) of finite samples. Here an important distinction arises between *structural parameters* such as ξ in Example 6.1, which are common to all the samples and which are the parameters of interest, and *incidental parameters* such as σ^2 and τ^2 in

*For a discussion without this restriction, see Anderson (1959) and Heyde and Feigin (1975).

Example 6.1, which occur in only one of the samples. That Theorem 4.1 does not extend to this case is illustrated by the following two examples.

Example 6.7. Estimation of a Common Variance. Let $X_{\alpha j}$ $(j = 1, \ldots, r)$ be independently distributed according to $N(\theta_\alpha, \sigma^2)$, $\alpha = 1, \ldots, n$. The MLEs are

$$\hat{\theta}_\alpha = X_{\alpha\cdot}, \qquad \hat{\sigma}^2 = \frac{1}{rn}\Sigma\Sigma(X_{\alpha j} - X_{\alpha\cdot})^2. \tag{17}$$

Furthermore, these are the unique solutions of the likelihood equations.

However, in the present case, the MLE of σ^2 is not even consistent. To see this, note that the statistics

$$S_\alpha^2 = \Sigma(X_{\alpha j} - X_{\alpha\cdot})^2$$

are identically independently distributed with expectation

$$E(S_\alpha^2) = (r-1)\sigma^2$$

so that $\Sigma S_\alpha^2/n \to (r-1)\sigma^2$ and hence

$$\hat{\sigma}^2 \to \frac{r-1}{r}\sigma^2 \quad \text{in probability.} \tag{18}$$

[A consistent and efficient estimator sequence of σ^2 is available in the present case, namely

$$\hat{\hat{\sigma}}^2 = \frac{1}{(r-1)n}\Sigma S_\alpha^2.\Big]$$

The study of this class of problems (including Example 6.7) was initiated by Neyman and Scott (1948), who also considered a number of other examples including one in which an MLE is consistent but not efficient.

A reformulation of the problem of structural parameters was proposed by Kiefer and Wolfowitz (1956), who considered the case in which the incidental parameters are themselves random variables, identically independently distributed according to some distribution, but of course unobservable. This will often bring the situation into the area of applicability of Theorem 4.1 or 6.1.

Example 6.8. Regression with Both Variables Subject to Error. Let X_i, Y_i $(i = 1, \ldots, n)$ be independent normal with means $E(X_i) = \xi_i$, $E(Y_i) = \eta_i$ and variances σ^2, τ^2, where $\eta_i = \alpha + \beta\xi_i$. There is thus a linear relationship between ξ and η, both of which are observed with independent, normally distributed errors. We are interested in estimating β and, for the sake of simplicity, shall take α as known to be zero. Then $\theta = (\beta, \sigma^2, \tau^2, \xi_1, \ldots, \xi_n)$, with the first three parameters being structural and the ξ's incidental. The likelihood is proportional to

$$(19) \qquad \frac{1}{\sigma^n \tau^n} \exp\left[-\frac{1}{2\sigma^2}\Sigma(x_i - \xi_i)^2 - \frac{1}{2\tau^2}\Sigma(y_i - \beta\xi_i)^2\right].$$

The likelihood equations have two roots, given by (Problem 6.18)

$$(20) \quad \hat{\beta} = \pm\sqrt{\frac{\Sigma y_i^2}{\Sigma x_i^2}}, \qquad 2n\hat{\sigma}^2 = \Sigma x_j^2 - \frac{1}{\hat{\beta}}\Sigma x_j y_j, \qquad 2n\hat{\tau}^2 = \Sigma y_j^2 - \hat{\beta}\Sigma x_j y_j,$$

$$2\hat{\xi}_i = x_i + \frac{1}{\hat{\beta}} y_i, \qquad i = 1, \dots, n$$

and the likelihood is larger at the root for which $\hat{\beta}\Sigma x_i y_i > 0$. If Theorem 4.1 applies, one of these roots must be consistent and hence tend to β in probability. Since $S_X^2 = \Sigma X_j^2$ and $S_Y^2 = \Sigma Y_j^2$ are independently distributed according to noncentral χ^2-distributions with noncentrality parameters $\lambda_n^2 = \Sigma_{j=1}^n \xi_j^2$ and $\beta^2\lambda_n^2$, their limit behavior depends on that of λ_n. (Note, incidentally, that for $\lambda_n = 0$, the parameter β becomes unidentifiable.) Suppose that $\lambda_n^2/n \to \lambda^2 > 0$. The distribution of S_X^2 and S_Y^2 is unchanged if we replace each ξ_i^2 by λ_n^2/n, and by the law of large numbers $\Sigma X_j^2/n$ therefore has the same limit as

$$E(X_1^2) = \sigma^2 + \frac{1}{n}\lambda_n^2 \to \sigma^2 + \lambda^2.$$

Similarly, $\Sigma Y_j^2/n$ tends in probability to $\tau^2 + \beta^2\lambda^2$ and hence $\hat{\beta}_n^2 \xrightarrow{P} (\tau^2 + \beta^2\lambda^2)/(\sigma^2 + \lambda^2)$. Thus neither of the roots is consistent. [It was pointed out by Solari (1969) that the likelihood in this problem is unbounded so that an MLE does not exist (Problem 6.19).]

If in (19) it is assumed that $\tau = \sigma$, it is easily seen that the MLE of β is consistent (Problem 6.16). For a discussion of this problem and some of its generalizations, see Anderson (1976) and Anderson and Sawa (1982). Another modification leading to a consistent MLE is suggested by Copas (1972).

Instead of (19) it is sometimes assumed that the ξ's are themselves iid according to a normal distribution $N(\mu, \gamma^2)$. The pairs (X_i, Y_i) then constitute a sample from a bivariate normal distribution, and asymptotically efficient estimators of the parameters μ, γ, β, σ, and τ can be obtained from the MLEs of Example 5.4. An analogous treatment is possible for Example 6.7.

Kiefer and Wolfowitz (1956) have considered not only this problem and that of Example 6.7 but a large class of problems of this type by postulating that the ξ's are iid according to a distribution G, but treating G as unknown, subject only to some rather general regularity assumptions. Alternative approaches to the estimation of structural parameters in the presence of a large number of incidental parameters are discussed by

Kalbfleisch and Sprott (1970) and Andersen (1970). A detailed discussion of Example 6.8 and its extension to more general regression models can be found in Kendall and Stuart (1979, Chap. 29), and of Example 6.7 in Jewell and Raab (1981).

The results of this chapter have all been derived in the so-called *regular case*, that is, when the densities satisfy regularity assumptions such as those of Theorems 1.1, 2.3, and 4.1. Of particular importance for the validity of the conclusions is that the support of the distributions P_θ does not vary with θ. Varying support brings with it information that often makes it possible to estimate some of the parameters with greater accuracy than that attainable in the regular case.

Example 6.9. Uniform. Let $X_1, \ldots, X_n$ be iid as $U(0, \theta)$. Then the MLE of θ is $\hat{\theta}_n = X_{(n)}$, and satisfies (Problem 5.2.16)

$$n(\theta - \hat{\theta}_n) \xrightarrow{\mathcal{L}} E(0, \theta). \tag{21}$$

Since $\hat{\theta}_n$ always underestimates θ and has a bias of order $1/n$, the order of the error $\hat{\theta}_n - \theta$, consider as an alternative the UMVU estimator $\delta_n = [(n+1)/n]X_{(n)}$, which satisfies

$$n(\theta - \delta_n) \xrightarrow{\mathcal{L}} E(-\theta, \theta). \tag{22}$$

The two asymptotic distributions have the same variance, but the first has expectation θ while the second has expectation zero and is thus much better centered.

The improvement of δ_n over $\hat{\theta}_n$ is perhaps seen more clearly by considering expected squared error. We have (Problem 5.2.17)

$$E\left[n(\hat{\theta}_n - \theta)\right]^2 \to 2\theta^2, \qquad E\left[n(\delta_n - \theta)\right]^2 \to \theta^2. \tag{23}$$

Thus the risk efficiency of $\hat{\theta}_n$ with respect to δ_n is $1/2$.

The example illustrates two ways in which such situations differ from the regular iid cases. First, the appropriate normalizing factor is n rather than $\sqrt{n}$ reflecting the fact that the error of the MLE is of order $1/n$ instead of $1/\sqrt{n}$. Second, the MLE need no longer be asymptotically optimal even when it is consistent. Typically, it can be improved by a bias correction.

Example 6.10. Exponential. Let $X_1, \ldots, X_n$ be iid according to the exponential distribution $E(\xi, b)$. Then the MLEs of ξ and b are

$$\hat{\xi} = X_{(1)} \quad \text{and} \quad \hat{b} = \frac{1}{n}\Sigma\left[X_i - X_{(1)}\right]. \tag{24}$$

It follows from Problem 1.5.18 that $n[X_{(1)} - \xi]/b$ is exactly (and hence asymptotically) distributed as $E(0, 1)$. As was the case for $\hat{\theta}$ in the preceding example, $\hat{\xi}$ is therefore asymptotically biased. More satisfactory is the UMVU estimator δ_n given by (2.2.23), which is obtained from $\hat{\xi}$ by subtracting an estimator of the bias (Problem 6.21).

It was further seen in Problem 1.5.18 that $2n\hat{b}/b$ is distributed as χ^2_{2n-2}. Since $(\chi^2_n - n)/\sqrt{2n} \to N(0, 1)$ in law, it is seen that $\sqrt{n}(\hat{b} - b) \to N(0, b^2)$. We shall now show that $\hat{b}$ is asymptotically efficient. For this purpose consider the case that ξ is known. The resulting one-parameter family of the X's is an exponential family and the MLE $\hat{\hat{b}}$ of b is asymptotically efficient and satisfies $\sqrt{n}(\hat{\hat{b}} - b) \to N(0, b^2)$ (Problem 6.22). Since $\hat{b}$ and $\hat{\hat{b}}$ have the same asymptotic distribution, $\hat{b}$ is a fortiori also asymptotically efficient, as was to be proved.

Example 6.11. Pareto. Let $X_1, \ldots, X_n$ be iid according to the Pareto distribution $P(a, c)$ with density

$$f(x) = ac^a/x^{a+1}, \qquad 0 < c < x, \qquad 0 < a. \tag{25}$$

The distribution is widely used, for example, in economics [see Johnson and Kotz (1970, Chap. 19)] and is closely connected with the exponential distribution of the preceding example through the fact that if X has density (25), then $Y = \log X$ has the exponential distribution $E(\xi, b)$ with (Problem 1.4.18)

$$\xi = \log c, \qquad b = 1/a. \tag{26}$$

From this fact it is seen that the MLEs of a and c are

$$\hat{a} = \frac{n}{\Sigma \log(X_i/X_{(1)})} \quad \text{and} \quad \hat{c} = X_{(1)} \tag{27}$$

and that these estimators are independently distributed, $\hat{c}$ as $P(na, c)$ and $2na/\hat{a}$ as χ^2_{2n-2} (Problem 6.23).

Since $\hat{b}$ is asymptotically efficient in the exponential case, the same is true of $1/\hat{b}$ and hence of $\hat{a}$. On the other hand, $n(X_{(1)} - c)$ has the limit distribution $E(0, c/a)$ and hence is biased. As was the case with the MLE of ξ in Example 6.10, an improvement over the MLE $\hat{c}$ of c is obtained by removing its bias and replacing $\hat{c}$ by the UMVU estimator

$$X_{(1)}\left[1 - \frac{1}{(n-1)\hat{a}}\right]. \tag{28}$$

For the details of these calculations, see Problems 6.23–6.25.

Example 6.12. Lognormal. As a last situation with variable support, consider a sample $X_1, \ldots, X_n$ from a three-parameter lognormal distribution, defined by the requirement that $Z_i = \log(X_i - \xi)$ are iid $N(\gamma, \sigma^2)$ so that

$$f(x; \xi, \gamma, \sigma^2) = \frac{1}{(x - \xi)\sqrt{2\pi}\,\sigma} \exp\left\{-\frac{1}{2\sigma^2}[\log(x - \xi) - \gamma]^2\right\} \tag{29}$$

when $x > \xi$, and $f = 0$ otherwise. When ξ is known, the problem reduces to that of estimating the mean γ and variance σ^2 from the normal sample $Z_1, \dots, Z_n$. However, when ξ is unknown, the support varies with ξ. Although in this case the density (29) tends to zero very smoothly at ξ (Problem 6.27), the theory of Section 6.4 is not applicable, and the problem requires a more powerful approach such as that of Le Cam (1969).

Note: For a discussion of the literature on this problem, see, for example, Johnson and Kotz (1970, Chap. 14).

The difficulty can be circumvented by a device used in other contexts by Kempthorne (1966), Lambert (1970), and Copas (1972), and suggested for the present problem by Giesbrecht and Kempthorne (1976). These authors argue that observations are never recorded exactly but only to the nearest unit of measurement. This formulation leads to a multinomial model of the kind considered for one parameter in Example 3.3, and Theorem 4.1 is directly applicable.

The corresponding problem for the three-parameter Weibull distribution is reviewed in Scholz (1983). For further discussion of such irregular cases, see for example Polfeldt (1970) and Woodrofe (1972).

7. ASYMPTOTIC EFFICIENCY OF BAYES ESTIMATORS

Bayes estimators were defined and some of their properties illustrated in Section 4.1. For the case of n binomial trials with a beta prior and squared error loss, it was found in Example 5.2.2 that the Bayes estimator of the success probability p, suitably normalized, has a normal limit distribution which is independent of the parameters of the prior distribution. The limit distribution is, in fact, the same as that of the normalized success frequency X/n, so that the Bayes estimators are asymptotically efficient. An exactly analogous result holds in Example 4.1.3 for the Bayes estimator of a normal mean when the prior distribution is normal. This raises the question whether the same limit distribution also obtains when the conjugate priors in these two examples are replaced by more general prior distributions, and whether the phenomenon persists in more general situations. The principal result of the present section (Theorem 7.2) shows that under suitable conditions the distribution of Bayes estimators based on n iid random variables tends to become independent of the prior distribution as $n \to \infty$ and that the Bayes estimators are asymptotically efficient.

Recent versions of such a theorem were given by Bickel and Yahav (1969) and by Ibragimov and Has'minskii (1972, 1981). The present proof, which combines elements from these papers, is due to Bickel. We begin by stating some assumptions.

Let $X_1, \dots, X_n$ be iid with density $f(x_i, \theta)$ (with respect to μ), where θ is real-valued and the parameter space Ω is an open interval. The true value of θ will be denoted by θ_0.

(B1) The function $L(\theta)$ satisfies the assumptions of Theorem 1.1.

To motivate the next assumption, note that under the assumptions of Theorem 1.1, if $\tilde{\theta} = \tilde{\theta}_n$ is any sequence for which $\tilde{\theta} \xrightarrow{P} \theta$, then

$$L(\theta) = L(\theta_0) + (\theta - \theta_0)L'(\theta_0) - \tfrac{1}{2}(\theta - \theta_0)^2[nI(\theta_0) + R_n(\theta)] \tag{1}$$

where

$$\frac{1}{n}R_n(\theta) \xrightarrow{P} 0 \quad \text{as} \quad n \to \infty \tag{2}$$

(Problem 7.2). We require here the following stronger assumption.

(B2) Given any $\varepsilon > 0$ there exists $\delta > 0$ such that in the expansion (1) the probability of the event

$$\sup\left\{\left|\frac{1}{n}R_n(\theta)\right| : |\theta - \theta_0| \leqslant \delta\right\} \geqslant \varepsilon \tag{3}$$

tends to zero as $n \to \infty$.

In the present case it is not enough to impose conditions on $L(\theta)$ in the neighborhood of θ_0 as is typically the case in asymptotic results. Since the Bayes estimators involve integration over the whole range of θ values, it is also necessary to control the behavior of $L(\theta)$ at a distance from θ_0.

(B3) For any $\delta > 0$ there exists $\varepsilon > 0$ such that the probability of the event

$$\sup\left\{\frac{1}{n}[L(\theta) - L(\theta_0)] : |\theta - \theta_0| \geqslant \delta\right\} \leqslant -\varepsilon \tag{4}$$

tends to 1 as $n \to \infty$.

(B4) The prior density π of θ is continuous and positive for all $\theta \in \Omega$.

(B5) The expectation of θ under π exists, that is,

$$\int |\theta| \pi(\theta)\, d\theta < \infty. \tag{5}$$

To establish the asymptotic efficiency of Bayes estimators under these assumptions, we shall first prove that for large values of n the posterior distribution of θ given the X's is approximately normal with

$$\text{mean} = \theta_0 + \frac{1}{nI(\theta_0)}L'(\theta_0) \quad \text{and} \quad \text{variance} = 1/nI(\theta_0). \tag{6}$$

Theorem 7.1. *If* $\pi^*(t|\underline{x})$ *is the posterior density of* $\sqrt{n}(\theta - T_n)$ *where*

$$T_n = \theta_0 + \frac{1}{nI(\theta_0)} L'(\theta_0), \tag{7}$$

(i) *then if* $(B1)$–$(B4)$ *hold*,

$$\int \left| \pi^*(t|\underline{x}) - \sqrt{I(\theta_0)}\, \phi\left[t\sqrt{I(\theta_0)}\right] \right| dt \xrightarrow{P} 0. \tag{8}$$

(ii) *If, in addition,* $(B5)$ *holds, then*

$$\int (1 + |t|) \left| \pi^*(t|\underline{x}) - \sqrt{I(\theta_0)}\, \phi\left[t\sqrt{I(\theta_0)}\right] \right| dt \xrightarrow{P} 0. \tag{9}$$

Proof. (i) By the definition of T_n,

$$\pi^*(t|\underline{x}) = \frac{\pi\left(T_n + \frac{t}{\sqrt{n}}\right) \exp\left[L\left(T_n + \frac{t}{\sqrt{n}}\right)\right]}{\int \pi\left(T_n + \frac{u}{\sqrt{n}}\right) \exp\left[L\left(T_n + \frac{u}{\sqrt{n}}\right)\right] du} \tag{10}$$

$$= e^{\omega(t)} \pi\left(T_n + \frac{t}{\sqrt{n}}\right) \Big/ C_n$$

where

$$\omega(t) = L\left(T_n + \frac{t}{\sqrt{n}}\right) - L(\theta_0) - \frac{1}{2nI(\theta_0)} [L'(\theta_0)]^2 \tag{11}$$

and

$$C_n = \int e^{\omega(u)} \pi\left(T_n + \frac{u}{\sqrt{n}}\right) du. \tag{12}$$

We shall prove at the end of the section that

$$J_1 = \int \left| e^{\omega(t)} \pi\left(T_n + \frac{t}{\sqrt{n}}\right) - e^{-t^2 I(\theta_0)/2} \pi(\theta_0) \right| dt \xrightarrow{P} 0 \tag{13}$$

so that

$$C_n \xrightarrow{P} \int e^{-t^2 I(\theta_0)/2} \pi(\theta_0)\, dt = \pi(\theta_0) \sqrt{2\pi / I(\theta_0)}\,. \tag{14}$$

The left side of (8) is equal to J/C_n where

$$(15) \qquad J = \int \left| e^{\omega(t)}\pi\left(T_n + \frac{t}{\sqrt{n}}\right) - C_n\sqrt{I(\theta_0)}\,\phi\left[t\sqrt{I(\theta_0)}\right] \right| dt$$

and by (14) it is enough to show that $J \xrightarrow{P} 0$.

Now $J \leqslant J_1 + J_2$ where J_1 is given by (13) and

$$J_2 = \int \left| C_n\sqrt{I(\theta_0)}\,\phi\left[t\sqrt{I(\theta_0)}\right] - \exp\left[-\frac{t^2}{2}I(\theta_0)\right]\pi(\theta_0) \right| dt$$

$$= \left| \frac{C_n\sqrt{I(\theta_0)}}{\sqrt{2\pi}} - \pi(\theta_0) \right| \cdot \int \exp\left[-\frac{t^2}{2}I(\theta_0)\right] dt.$$

By (13) and (14), J_1 and J_2 tend to zero in probability, and this completes the proof.

(ii) The left side of (9) is equal to

$$\frac{1}{C_n}J' \leqslant \frac{1}{C_n}(J_1' + J_2')$$

where J', J_1', J_2' are obtained from J, J_1, J_2 by inserting the factor $(1 + |t|)$ under the integral signs. It is therefore enough to prove that J_1' and J_2' both tend to zero in probability. The proof for J_2' is the same as that for J_2; the proof for J_1' will be given at the end of the section together with that for J_1.

On the basis of Theorem 7.1, we are now able to prove the principal result of this section.

Theorem 7.2. *If* $(B1)$–$(B5)$ *hold, and if* $\tilde{\theta}_n$ *is the Bayes estimator when the prior density is* π *and the loss is squared error, then*

$$(16) \qquad \sqrt{n}(\tilde{\theta}_n - \theta_0) \xrightarrow{\mathcal{L}} N[0, 1/I(\theta_0)],$$

so that $\tilde{\theta}_n$ *is consistent* and asymptotically efficient.*

Proof. We have

$$\sqrt{n}(\tilde{\theta}_n - \theta_0) = \sqrt{n}(\tilde{\theta}_n - T_n) + \sqrt{n}(T_n - \theta_0).$$

*A general relationship between the consistency of MLEs and Bayes estimators is discussed by Strasser (1981).

By the CLT, the second term has the limit distribution $N[0, 1/I(\theta_0)]$, so that it only remains to show that

$$\sqrt{n}\left(\tilde{\theta}_n - T_n\right) \xrightarrow{P} 0. \tag{17}$$

Now

$$\tilde{\theta}_n = \int \theta \pi(\theta | \underline{x})\, d\theta = \int \left(\frac{t}{\sqrt{n}} + T_n \right) \pi^*(t | \underline{x})\, dt$$

and hence

$$\sqrt{n}\left(\tilde{\theta}_n - T_n\right) = \int t \pi^*(t | \underline{x})\, dt.$$

Thus

$$\sqrt{n}\,|\tilde{\theta}_n - T_n| = \left| \int t \pi^*(t|\underline{x})\, dt - \int t \sqrt{I(\theta_0)}\, \phi\left[t\sqrt{I(\theta_0)} \right] dt \right|$$

$$\leqslant \int |t| \cdot \left| \pi^*(t|\underline{x}) - \sqrt{I(\theta_0)}\, \phi\left[t\sqrt{I(\theta_0)} \right] \right| dt,$$

which tends to zero in probability by Theorem 7.1(ii).

Before discussing the implications of Theorem 7.2, we shall show that assumptions (B1)–(B5) are satisfied in exponential families.

Example 7.1. Exponential Families. Let

$$f(x_i, \theta) = e^{\theta T(x_i) - A(\theta)}$$

so that

$$A(\theta) = \log \int e^{\theta T(x)}\, d\mu(x).$$

Recall from Section 1.4 that A is differentiable to all orders and that

$$A'(\theta) = E_\theta[T(X)];$$

$$A''(\theta) = \text{var}_\theta[T(X)] = I(\theta).$$

Suppose $I(\theta) > 0$. Then,

$$L(\theta) - L(\theta_0) = (\theta - \theta_0)\Sigma T(X_i) - n[A(\theta) - A(\theta_0)]$$
$$= (\theta - \theta_0)\Sigma[T(X_i) - A'(\theta_0)]$$
$$-n\{[A(\theta) - A(\theta_0)] - [(\theta - \theta_0)A'(\theta_0)]\}.$$

The first term is equal to $(\theta - \theta_0)L'(\theta_0)$. Apply Taylor's theorem to $A(\theta)$ to find

$$A(\theta) = A(\theta_0) + (\theta - \theta_0)A'(\theta_0) + \tfrac{1}{2}(\theta - \theta_0)^2 A''(\theta^*),$$

so that the second term is equal to $(-n/2)(\theta - \theta_0)^2 A''(\theta^*)$. Hence

$$L(\theta) - L(\theta_0) = (\theta - \theta_0)L'(\theta_0) - \frac{n}{2}(\theta - \theta_0)^2 A''(\theta^*).$$

To prove (B2) we must show that

$$A''(\theta^*) = I(\theta_0) + \frac{1}{n}R_n(\theta)$$

where

$$R_n(\theta) = n[A''(\theta^*) - I(\theta_0)]$$

satisfies (3); that is, we must show that given ε there exists δ such that the probability of

$$\sup\{|A''(\theta^*) - I(\theta_0)| : |\theta - \theta_0| \leqslant \delta\} \geqslant \varepsilon$$

tends to zero. This follows from the facts that $I(\theta) = A''(\theta)$ is continuous and that $\theta^* \to \theta_0$ as $\theta \to \theta_0$.

To see that (B3) holds, write

$$\frac{1}{n}[L(\theta) - L(\theta_0)] = (\theta - \theta_0)\left\{\frac{1}{n}\Sigma[T(X_i) - A'(\theta_0)] - \left[\frac{A(\theta) - A(\theta_0)}{\theta - \theta_0} - A'(\theta_0)\right]\right\}$$

and suppose without loss of generality that $\theta > \theta_0$.

Since $A''(\theta) > 0$, so that $A(\theta)$ is strictly convex, it is seen that $\theta > \theta_0$ implies $[A(\theta) - A(\theta_0)]/(\theta - \theta_0) > A'(\theta_0)$. On the other hand, $\Sigma[T(X_i) - A'(\theta_0)]/n \xrightarrow{P} 0$

and hence with probability tending to 1, the factor of $(\theta - \theta_0)$ is negative. It follows that

$$\sup\left\{\frac{1}{n}[L(\theta) - L(\theta_0)]: \theta - \theta_0 \geqslant \delta\right\}$$

$$\leqslant \delta\left\{\frac{\Sigma[T(X_i) - A'(\theta_0)]}{n} - \inf\left[\frac{A(\theta) - A(\theta_0)}{\theta - \theta_0} - A'(\theta_0): \theta - \theta_0 \geqslant \delta\right]\right\}.$$

and hence that (B3) is satisfied.

Theorems 7.1 and 7.2 were stated under the assumption that π is the density of a proper distribution, so that its integral is equal to 1. There is a trivial but useful extension to the case in which $\int \pi(\theta)\, d\theta = \infty$ but where there exists n_0 so that the posterior density

$$\tilde{\pi}(\theta | x_1, \ldots, x_{n_0}) = \frac{\prod_{i=1}^{n_0} f(x_i, \theta)\pi(\theta)}{\int \prod_{i=1}^{n_0} f(x_i, \theta)\pi(\theta)\, d\theta}$$

of θ given $x_1, \ldots, x_{n_0}$ is, with probability 1, a proper density satisfying assumptions (B4) and (B5). The posterior density of θ given $X_1, \ldots, X_n$ $(n > n_0)$ when θ has prior density π is then the same as the posterior density of θ given $X_{n_0+1}, \ldots, X_n$ when θ has prior density $\tilde{\pi}$, and the result now follows.

Example 7.2. ***Location Families.*** The Pitman estimator derived in Theorem 3.1.5 is the Bayes estimator corresponding to the improper prior density $\pi(\theta) \equiv 1$. If $X_1, \ldots, X_n$ are iid with density $f(x_i - \theta)$ satisfying (B1)–(B3) the posterior density after one observation $X_1 = x_1$ is $f(x_1 - \theta)$ and hence a proper density satisfying assumption (B5) provided $E_\theta|X_1| < \infty$ (Problem 7.3). Under these assumptions, the Pitman estimator is therefore asymptotically efficient.* An analogous result holds in the scale case (Problem 7.4).

Theorem 7.1 can be generalized further. Rather than requiring the posterior density $\tilde{\pi}$ to be proper with finite expectation after a fixed number n_0 of observations, it is enough to assume that it satisfies these conditions for all $n \geqslant n_0$ when $(X_1, \ldots, X_{n_0}) \in S_{n_0}$ where $P(S_{n_0}) \to 1$ as $n_0 \to \infty$ (Problem 7.6).

Example 7.3. Let X_i be independent taking on the values 1 and 0 with probability p and $q = 1 - p$, respectively, and let $\pi(p) = 1/pq$. Then the posterior

*For a more general treatment of this result, see Stone (1974).

distribution of p will be proper (and will then automatically have finite expectation) as soon as $0 < \Sigma X_i < n$, but not before. Since for any $0 < p < 1$ the probability of this event tends to 1 as $n \to \infty$, the asymptotic efficiency of the Bayes estimator follows.

Theorem 7.2 provides additional support for the suggestion, made in Section 4.1, that Bayes estimation constitutes a useful method for generating estimators. However, the theorem is unfortunately of no help in choosing among different Bayes estimators, since all prior distributions satisfying assumptions (B4) and (B5) lead to the same asymptotic behavior. In fact, if $\tilde{\theta}_n$ and $\tilde{\theta}'_n$ are Bayes estimators corresponding to two different prior distributions Λ and Λ' satisfying (B4) and (B5), (17) implies the even stronger statement

$$\sqrt{n}\left(\tilde{\theta}'_n - \tilde{\theta}_n\right) \xrightarrow{P} 0. \tag{18}$$

Nevertheless, the interpretation of θ as a random variable with density $\pi(\theta)$ leads to some suggestions concerning the choice of π. Theorem 7.1 showed that the posterior distribution of θ, given the observations, eventually becomes a normal distribution which is concentrated near the true θ_0 and which is independent of π. It is intuitively plausible that a close approximation to the asymptotic result will tend to be achieved more quickly (i.e., for smaller n) if π assigns a relatively high probability to the neighborhood of θ_0 than if this probability is very small. A minimax approach thus leads to the suggestion of a uniform assignment of prior density. It is clear what this means for a location parameter but not in general since the parametrization is arbitrary and reparametrization destroys uniformity. In addition, it seems plausible that account should also be taken of the relative informativeness of the observations corresponding to different parameter values.

As discussed in Section 4.1, proposals for prior distributions satisfying such criteria have been made (from a somewhat different point of view) by Jeffreys and others. For details, further suggestions, and references, see Box and Tiao (1973) and Jaynes (1979).

When the likelihood equation has a unique root $\hat{\theta}_n$ (which with probability tending to 1 is then the MLE), this estimator has a great practical advantage over the Bayes estimators which share its asymptotic properties. It provides a unique estimating procedure, applicable to a large class of problems, which is supported (partly because of its intuitive plausibility and partly for historical reasons) by a substantial proportion of the statistical profession. This advantage is less clear in the case of multiple roots where asymptotically efficient likelihood estimators such as the one-step estimator

(3.11) depend on a somewhat arbitrary initial estimator and need no longer agree with the MLE even for large n. A disadvantage of Bayes estimators in the multiparameter case is that they require the computationally inconvenient evaluation of multiple integrals.

To resolve the difficulty raised by the profusion of asymptotically efficient estimators, it seems natural to carry the analysis one step further and to take into account terms (for example, in an asymptotic expansion of the distribution of the estimator) of order $1/n$. Investigations along these lines have been undertaken by Rao (1961), Ghosh and Subramanyam (1974), Efron (1975), Pfanzagl and Wefelmeyer (1978/79), and others (see also Section 6.3). They are complicated by the fact that to this order the estimators tend to be biased and their efficiencies can be improved by removing these biases. For an interesting discussion of these issues, see Berkson (1980). The subject still requires further study.

We conclude this section by proving that the quantities J_1 [defined by (13)] and J_1' tend to zero in probability. For this purpose it is useful to obtain the following alternative expression for $\omega(t)$.

Lemma 7.1. *The quantity* $\omega(t)$, *defined by* (11), *is equal to*

$$\omega(t) = -I(\theta_0)\frac{t^2}{2} - \frac{1}{2n}R_n\left(T_n + \frac{t}{\sqrt{n}}\right)\left[t + \frac{1}{I(\theta_0)\sqrt{n}}L'(\theta_0)\right]^2 \tag{19}$$

where R_n *is the function defined in* (1) (*Problem* 7.8).

Proof for J_1. To prove that the integral (13) tends to zero in probability, divide the range of integration into the three parts: (i) $|t| \leq M$, (ii) $|t| \geq \delta\sqrt{n}$, and (iii) $M < |t| < \delta\sqrt{n}$, and show that the integral over each of the three tends to zero in probability.

(i) $|t| \leq M$. To prove this result, we shall show that for every $0 < M < \infty$,

$$\sup\left|e^{\omega(t)}\pi\left(T_n + \frac{t}{\sqrt{n}}\right) - e^{-I(\theta_0)t^2/2}\pi(\theta_0)\right| \xrightarrow{P} 0 \tag{20}$$

where here and throughout the proof of (i) the sup is taken over $|t| \leq M$. The result will follow from (20) since the range of integration is bounded. Substituting the expression (19) for $\omega(t)$, (20) is seen to follow from the following two facts (Problem 7.9):

$$\sup\left\{\left|\frac{1}{n}R_n\left(T_n + \frac{t}{\sqrt{n}}\right)\right|\left[t + \frac{1}{I(\theta_0)\sqrt{n}}L'(\theta_0)\right]^2\right\} \xrightarrow{P} 0 \tag{21}$$

and

$$(22)\qquad \sup\left|\pi\left(T_n + \frac{t}{\sqrt{n}}\right) - \pi(\theta_0)\right| \xrightarrow{P} 0.$$

The second of these is obvious from the continuity of π and the fact that (Problem 7.10)

$$(23)\qquad T_n \xrightarrow{P} \theta_0.$$

To prove (21), it is enough to show that

$$(24)\qquad \sup\left|\frac{1}{n}R_n\left(T_n + \frac{t}{\sqrt{n}}\right)\right| \xrightarrow{P} 0$$

and

$$(25)\qquad \frac{1}{I(\theta_0)}\frac{1}{\sqrt{n}}L'(\theta_0) \qquad \text{is bounded in probability.}$$

Of these, (25) is clear from (B1) and the central limit theorem. To see (24), note that $|t| \leqslant M$ implies

$$T_n - \frac{M}{\sqrt{n}} \leqslant T_n + \frac{t}{\sqrt{n}} \leqslant T_n + \frac{M}{\sqrt{n}}$$

and hence by (23) that for any $\delta > 0$ the probability of

$$\theta_0 - \delta \leqslant T_n + \frac{t}{\sqrt{n}} \leqslant \theta_0 + \delta$$

will be arbitrarily close to 1 for sufficiently large n. The result now follows from (B2).

(ii) $M \leqslant |t| \leqslant \delta\sqrt{n}$. For this part it is enough to prove that for $|t| \leqslant \delta\sqrt{n}$ the integrand of J_1 is bounded by an integrable function with probability $\geqslant 1 - \varepsilon$. Then the integral can be made arbitrarily small by choosing a sufficiently large M. Since the second term of the integrand of (13) is integrable, it is enough to show that such an integrable bound exists for the first term. More precisely, we shall show that given $\varepsilon > 0$ there exists $\delta > 0$, $C < \infty$ such that for sufficiently large n,

$$(26)\qquad P\left[e^{\omega(t)}\pi\left(T_n + \frac{t}{\sqrt{n}}\right) \leqslant Ce^{-t^2 I(\theta_0)/4} \qquad \text{for all} \quad |t| \leqslant \delta\sqrt{n}\right] \geqslant 1 - \varepsilon.$$

The factor $\pi(T_n + t/\sqrt{n})$ causes no difficulty by (23) and the continuity of π, so that it remains to establish such a bound for

$$\text{(27)} \quad \exp\omega(t) \leqslant \exp\left\{-\frac{t^2}{2}I(\theta_0) + \frac{1}{n}\left|R_n\left(T_n + \frac{t}{\sqrt{n}}\right)\right|\left[t^2 + \frac{(L'(\theta_0))^2}{nI^2(\theta_0)}\right]\right\}.$$

For this purpose, note that

$$|t| \leqslant \delta'\sqrt{n} \qquad \text{implies} \qquad T_n - \delta' \leqslant T_n + \frac{t}{\sqrt{n}} \leqslant T_n + \delta'$$

and hence by (23) that with probability arbitrarily close to 1, for n sufficiently large

$$|t| \leqslant \delta'\sqrt{n} \qquad \text{implies} \qquad \left|T_n + \frac{t}{\sqrt{n}} - \theta_0\right| \leqslant 2\delta'.$$

By (B2) there exists δ' such that the latter inequality implies

$$P\left\{\sup_{|t| \leqslant \delta'\sqrt{n}}\left|\frac{1}{n}R_n\left(T_n + \frac{t}{\sqrt{n}}\right)\right| \leqslant \frac{1}{4}I(\theta_0)\right\} \geqslant 1 - \varepsilon.$$

Combining this fact with (25), we see that the right side of (27) is $\leqslant C'e^{-t^2 I(\theta_0)/4}$ for all t satisfying (ii), with probability arbitrarily close to 1, and this establishes (26).

(iii) $|t| \geqslant \delta\sqrt{n}$. As in (ii), the second term in the integrand of (13) can be neglected, and it is enough to show that for all δ

$$\text{(28)} \quad \int_{|t| \geqslant \delta\sqrt{n}} \exp[\omega(t)]\pi\left(T_n + \frac{t}{\sqrt{n}}\right) dt$$
$$= \sqrt{n}\int_{|\theta - T_n| \geqslant \delta} \pi(\theta)\exp\left\{L(\theta) - L(\theta_0) - \frac{1}{2nI(\theta_0)}[L'(\theta_0)]^2\right\} d\theta \xrightarrow{P} 0.$$

From (23) and (B3) it is seen that given δ there exists ε such that

$$\sup_{|\theta - T_n| \geqslant \delta}[e^{L(\theta) - L(\theta_0)}] \leqslant e^{-n\varepsilon}$$

with probability tending to 1. By (25), the right side of (28) is therefore bounded above by

$$C\sqrt{n}\, e^{-n\varepsilon} \int \pi(\theta)\, d\theta = C\sqrt{n}\, e^{-n\varepsilon} \tag{29}$$

with probability tending to 1, and this completes the proof of (iii).

To prove (13), let us now combine (i)–(iii). Given $\varepsilon > 0$, $\delta > 0$, choose M so large that

$$\int_M^\infty \left[C \exp\left[-\frac{t^2}{4} I(\theta_0) \right] + \exp\left[-\frac{t^2}{2} I(\theta_0) \right] \pi(\theta_0) \right] dt \leqslant \varepsilon/3, \tag{30}$$

and hence that for sufficiently large n, the integral (13) over (ii) is $\leqslant \varepsilon/3$ with probability $\geqslant 1 - \varepsilon$. Next choose n so large that the integrals (13) over (i) and over (iii) are also $\leqslant \varepsilon/3$ with probability $\geqslant 1 - \varepsilon$. Then $P[J_1 \leqslant \varepsilon] \geqslant 1 - 3\varepsilon$, and this completes the proof of (13).

The proof for J_1' requires only trivial changes. In part (i), the factor $[1 + |t|]$ is bounded, so that the proof continues to apply. In part (ii), multiplication of the integrand of (30) by $[1 + |t|]$ does not affect its integrability, and the proof goes through as before. Finally, in part (iii), the integral in (29) must be replaced by $Cne^{-n\varepsilon}\int |\theta| \pi(\theta)\, d\theta$, which is finite by (B5).

8. LOCAL ASYMPTOTIC OPTIMALITY

In the preceding sections, it was seen that under suitable regularity conditions there exist estimator sequences $\hat{\theta}_n$ of a real-valued parameter θ which are asymptotically efficient in the sense that

$$\sqrt{n}\,(\hat{\theta}_n - \theta) \xrightarrow{\mathcal{L}} N[0, 1/I(\theta)]. \tag{1}$$

Unfortunately, the optimality claim based on (1) is marred by the fact that the competition has been restricted to estimators that are asymptotically normal, by the phenomenon of superefficiency, and by the possibility that an estimator that behaves well at every point θ may misbehave quite badly in the neighborhood of some values of θ.

The last of these difficulties suggests a change of point of view which, as it turns out, avoids all three of the problems mentioned, and which leads to a different type of optimality result, developed by Le Cam and Hájek. [For an exposition of this work and some history, see Hájek (1972).] Instead of considering the performance of an estimator *at* the true θ_0, we shall study it in the *neighborhood* of a given fixed point. It is convenient not to let the latter be the true value but any given θ_0 and then to consider the performance of the estimator in a $1/\sqrt{n}$-neighborhood of θ_0, that is, when the true parameter point θ_n is equal to

$$\theta_n = \theta_0 + \frac{\eta}{\sqrt{n}}. \tag{2}$$

We will therefore define *local asymptotic* (loc. as.) properties such as admissibility or minimaxity of an estimator sequence *at a point* θ_0. These properties may then hold for some values of θ_0 and not for others. Eventually, a global point of view can be restored by considering the local properties at θ_0, for varying θ_0.

For a sequence $\{\theta_n\}$ of values satisfying (2), consider the normalized loss of an estimator sequence $\delta = \{\delta_n\}$,

$$n(\delta_n - \theta_n)^2 = n\left(\delta_n - \theta_0 - \frac{\eta}{\sqrt{n}}\right)^2.$$

The variable η plays the role of a *local parameter* (at θ_0) and replaces the global parameter θ.

We should like to evaluate the performance of δ in terms of the limit distribution of $\sqrt{n}(\delta_n - \theta_n)$. If the latter exists, it was shown in Lemma 5.1.2 that its asymptotic expected squared error is equal to the limit of the truncated squared error risk

$$R_{n,A}(\eta, \delta) = E_{\theta_n}\left\{\min\left[n\left(\delta_n - \theta_0 - \frac{\eta}{\sqrt{n}}\right)^2, A^2\right]\right\} \tag{3}$$

as we let first n and then A tend to infinity. One difficulty remains. We are not assuming the existence of a limit distribution and hence not the existence of $\lim R_{n,A}(\eta, \delta)$ as $n \to \infty$. Since we shall be concerned with a minimax criterion, we shall consider the worst possible limiting risk and define the *loc. as. risk function* of δ at θ_0 as

$$R(\eta, \delta) = R(\eta, \{\delta_n\}) = \lim_{A\to\infty} \overline{\lim_{n\to\infty}} R_{n,A}(\eta, \delta). \tag{4}$$

This definition is sufficiently general to cover any sequence of estimators (but see Example 8.4 below). However, for specific estimators, the inner limit in (4) will typically exist so that $\overline{\lim}$ can be replaced by lim, and the limit as $A \to \infty$ will coincide with the expected untruncated squared error, in which case (4) reduces to the usual squared error risk

$$\lim_{n\to\infty} E_{\theta_0+\eta/\sqrt{n}} n\left[\delta_n - \theta_0 - \frac{\eta}{\sqrt{n}}\right]^2.$$

A sequence $\delta = \{\delta_n\}$ is said to be *loc. as. minimax* at θ_0 if it minimizes $\sup_\eta R(\eta, \delta)$. It is loc. as. inadmissible if there exists a sequence $\delta' = \{\delta_n'\}$ such that

$$(5) \qquad R(\eta, \delta') \leqslant R(\eta, \delta) \qquad \text{for all} \quad \eta$$

with strict inequality for at least some η, and *loc. as. admissible* if no such δ' exists. If $R(\eta, \delta)$ is constant (i.e., independent of η) and loc. as. admissible, it is also loc. as. minimax.

Example 8.1. Let X have the binomial distribution $b(p, n)$.

(i) Consider the sequence of estimators $\delta_n = c_n X/n$ with $c_n = 1 - c/n$. To obtain the loc. as. risk function of δ at p_0, put $p_n = p_0 + \eta/\sqrt{n}$ and calculate

$$(6) \qquad nE(\delta_n - p_n)^2 = c_n^2 p_n q_n + (1 - c_n)^2 n p_n^2 \to p_0 q_0.$$

This result is not affected by truncating the loss at A^2 before passing to the limit in n, and then letting $A \to \infty$ (Problem 8.1). Thus (6) shows that in this case the loc. as. risk $R(\eta, \delta) = p_0 q_0$ at any given p_0 is constant, but that it varies with p_0.

(ii) Consider next the minimax estimator $\delta_n' = (X + \frac{1}{2}\sqrt{n})/(n + \sqrt{n})$. A corresponding calculation gives $R(\eta, \delta') = p_0 q_0 + (p_0 - \frac{1}{2})^2$, and comparison with $R(\eta, \delta) = p_0 q_0$ shows that δ', though unique minimax and admissible for every n, is loc. as. neither minimax nor admissible except at the point $p_0 = 1/2$.

Example 8.2. Let $\sqrt{n}(\delta_n - \theta) \xrightarrow{\mathcal{L}} N(0, \tau^2)$. Then it follows from Lemma 5.1.2 that $R(0, \delta) = \tau^2$ and that for $\eta = 0$, the $\overline{\lim}$ in the definition of (4) can be replaced by lim.

Example 8.3. If δ is the sequence of estimators δ_n defined in Example 1.1, then [Problem 8.3(i)] at $\theta = 0$,

$$R(\eta, \delta) = a^2 + \eta^2(1 - a)^2,$$

so that the loc. as. risk is unbounded. Since the loc. as. risk of $\bar{X}$ is $\equiv 1$ for all θ [Problem 8.3(ii)], it follows that δ is not loc. as. minimax at $\theta = 0$.

Example 8.4. Let $X_1, \ldots, X_n$ be iid according to $U(0, \theta)$. Since good estimators of θ converge at rate $1/n$ rather than $1/\sqrt{n}$, we shall in this example define the loc. as. risk of δ by (4) with

$$(7) \qquad R_{n,A}(\eta, \delta) = E_{\theta_n}\left\{\min\left[n^2\left(\delta_n - \theta_0 - \frac{\eta}{n}\right]^2, A^2\right)\right\}$$

where $\theta_n = \theta_0 + \eta/n$. The MLE of θ is $X_{(n)}$ and one finds its loc. as. risk at θ_0 to be constant $= 2\theta_0^2$, while the loc. as. risk of the UMVU estimator $[(n+1/n)]X_{(n)}$ at θ_0 is equal to θ_0^2 for all η (Problem 8.4). In the present case, the MLE is thus neither loc. as. admissible nor minimax.

In the following it will be convenient to say that two sequences $\{Y_n\}$ and $\{Z_n\}$ of random variables are asymptotically equivalent, to be denoted by $Y_n \sim Z_n$ if $Z_n - Y_n \to 0$ in probability, that is, if $Z_n = Y_n + R_n$ where $R_n \to 0$ in probability. We shall be concerned with local asymptotic properties of the ELE $\hat{\theta}_n$ under the assumptions of Theorem 2.3. A basic property of these estimators is that under θ_0

$$L\left(\theta_0 + \frac{\eta}{\sqrt{n}}\right) - L(\theta_0) \sim I\eta\sqrt{n}\,(\hat{\theta}_n - \theta_0) - \tfrac{1}{2}I\eta^2 \tag{8}$$

where $I = I(\theta_0)$. This follows from the fact that

$$L\left(\theta_0 + \frac{\eta}{\sqrt{n}}\right) - L(\theta_0) = \frac{\eta}{\sqrt{n}}L'(\theta_0) + \frac{\eta^2}{2n}L''(\theta_0) + \frac{\eta^3}{6n^{3/2}}L'''(\theta_n^*)$$

and from (2.19), (2.20), and the equation preceding (2.18).

The following proof of loc. as. admissibility of a sequence of estimators $\hat{\theta}_n$ satisfying (1) and (8) under the assumptions of Theorem 2.3 is due to C. Stone.

In preparation, we require the following definition and lemma. A sequence $\{P_n\}$ of distributions on E_s is *tight* if given any $\varepsilon > 0$ there exists a bounded rectangle R such that $P_n(R) > 1 - \varepsilon$ for all n. The condition essentially states that no probability mass escapes to infinity as $n \to \infty$.

Lemma 8.1. *The sequence of distributions of* $\underline{Y}_n = (Y_{1n}, \ldots, Y_{sn})$ *is tight provided each of the sequences* $\{Y_{in}\}$, $i = 1, \ldots, s$ *is tight.*

(ii) If $\{Y_n\}$ *is a sequence of random variables for which*

$$\lim_{A\to\infty}\ \overline{\lim_{n\to\infty}}\ E(Y_{nA}^2) < \infty,$$

where Y_{nA}^2 *is* Y_n^2 *truncated at* A^2, *then the sequence* $\{Y_n\}$ *is tight.*

Proof. (i) This follows from the fact that

$$P\{|Y_{in}| \geqslant A_i \text{ for at least one } i\} \leqslant \Sigma P(|Y_{in}| \geqslant A_i).$$

(ii) For all A and n,

$$P(|Y_n| \geqslant A) \leqslant \frac{1}{A^2} E(Y_{nA}^2),$$

and the result follows by letting $A \to \infty$.

A basic consequence of tightness is the following property, which is proved, for example, in Billingsley (1979, Th. 29.3).

Theorem 8.1. *If $\{P_n\}$ is a tight sequence of probability distributions, there exists a subsequence $\{P_{n_j}\}$ and a probability distribution P such that $P_{n_j} \to P$ as $j \to \infty$.*

Consider now the estimator $\hat{\theta}_n$ satisfying (1) and (8) and a competing estimator $\hat{\theta}_n'$, let

$$T_n = \sqrt{n}\left(\hat{\theta}_n - \theta_0\right), \qquad T_n' = \sqrt{n}\left(\hat{\theta}_n' - \theta_0\right) \tag{9}$$

and $\delta = \{\hat{\theta}_n\}$, $\delta' = \{\hat{\theta}_n'\}$. We shall show in Theorem 8.2 that the loc. as. risk of $\hat{\theta}_n$ at θ_0 is constant, $= 1/I$ where $I = I(\theta_0)$, and that any sequence $\{\hat{\theta}_n'\}$ satisfying $R(\eta, \delta') \leqslant 1/I$ for all η must have $R(\eta, \delta') \equiv 1/I$, so that $\hat{\theta}_n$ is loc. as. admissible and minimax.

We shall begin with three lemmas concerning the limiting behavior of the distribution of (T_n, T_n').

Lemma 8.2. *Let $\hat{\theta}_n$ satisfy (1) and (8), and let $\hat{\theta}_n'$ be a competing sequence of estimators. If the joint distribution of (T_n, T_n') tends to a limit distribution H_0 under θ_0, then the joint distribution of (T_n, T_n') under $\theta_n = \theta_0 + \eta/\sqrt{n}$ tends to the limit distribution H_η given by*

$$\frac{dH_\eta(t, t')}{dH_0(t, t')} = \exp\left(I\eta t - \frac{1}{2} I\eta^2\right) = g(t, \eta). \tag{10}$$

Proof. Let (T, T') be two random variables with joint distribution H_0. Then since $g(t, \eta)$ is continuous in t, it follows from Theorem 5.1.6 that

$$(T_n, T_n', g(T_n, \eta))_{\theta_0} \xrightarrow{\mathcal{L}} (T, T', g(T, \eta))_{H_0}.$$

Therefore by (8),

$$\left(T_n, T_n', \exp\left[L\left(\theta_0 + \frac{\eta}{\sqrt{n}}\right) - L(\theta_0)\right]\right)_{\theta_0} \xrightarrow{\mathcal{L}} (T, T', g(T, \eta))_{H_0}.$$

If ψ is any bounded continuous function of two arguments, we have

$$(11) \qquad E_{\theta_n}\psi(T_n, T_n') = E_{\theta_0}\left[\psi(T_n, T_n')e^{L(\theta_0+\eta/\sqrt{n})-L(\theta_0)}\right]$$
$$\to E_{H_0}\left[\psi(T, T')g(T, \eta)\right].$$

This last step does not follow from Theorem 5.1.6 since the exponential factor is not bounded. It follows, however, from the standard theorem about uniform integrability. To see that the integrand in the middle term is uniformly integrable, note that since ψ is bounded, it is enough to show that $X_n = \exp[L(\theta_0 + \eta/\sqrt{n}) - L(\theta_0)]$ is uniformly integrable. This follows from the facts that $X_n \xrightarrow{\mathcal{L}} g(T, \eta)$, $E_{\theta_0}(X_n) = E_{H_0}[g(T, \eta)]$, and $X_n, g(T, \eta) \geqslant 0$ by Theorem 5.4 of Billingsley (1968).

From (11) it follows by Theorem 5.1.7 that

$$(T_n, T_n')_{\theta_n} \xrightarrow{\mathcal{L}} H_\eta(t, t')$$

where H_η is given by (10).

Lemma 8.3. *If $\hat{\theta}_n$ satisfies* (1) *and* (8), *then its loc. as. risk is constant,* $= 1/I$.

Proof. Apply Lemma 8.2 with $\hat{\theta}_n' = \hat{\theta}_n$ so that $T_n' = T_n$. Then by (1), $(T_n, T_n)_{\theta_0} \xrightarrow{\mathcal{L}} (T, T)$, $(T_n, T_n)_{\theta_n} \xrightarrow{\mathcal{L}} (T, T)_{H_\eta}$. Thus, T_n under θ_n tends in law to a distribution with density

$$\frac{I}{\sqrt{2\pi}}\exp\left(-\frac{1}{2}It^2 + I\eta t - \frac{1}{2}I\eta^2\right)$$

which is the density of $N(\eta, 1/I)$. The distribution of $\sqrt{n}(\hat{\theta}_n - \theta_0 - \eta/\sqrt{n})$ therefore tends to $N(0, 1/I)$ and it follows from Lemma 5.1.2 that the loc. as. risk of $\hat{\theta}_n$ is constant $= 1/I$ and that for it $\overline{\lim}$ can be replaced in (4) by lim.

To utilize Lemma 8.2, we require convergence of (T_n, T_n') under θ_0.

Lemma 8.4. *Let $\hat{\theta}_n$ satisfy* (1) *and* (8), *and let $\hat{\theta}_n'$ be a competing sequence of estimators which satisfies* $R(\eta, \delta') \leqslant 1/I$ *for all* η.

(i) *Then* (T_n, T_n') *tends to a limit distribution* H_0 *under* θ_0 *and hence to* H_η *under* θ_n.

(ii) *If* (T, T') *are distributed according to the family* $\{H_\eta, -\infty < \eta < \infty\}$, *then* T *is sufficient for* η *on the basis of* T *and* T'.

(iii) $R(\eta, \delta') \equiv 1/I$.

Proof. Since $R(0, \delta) = 1/I$ and $R(0, \delta') \leqslant 1/I$, it follows from Lemma 8.1 that the sequence $\{(T_n, T_n')\}$ is tight under θ_0 and hence that there exists a subsequence $\{(T_{n_j}, T_{n_j}')\}$ tending to a limit distribution H_0 under θ_0. By Lemma 8.2, (T_{n_j}, T_{n_j}') then converges under θ_n toward a limit distribution H_η satisfying (10).

If (T, T') is distributed according to H_η, it follows from (10) and Theorem 1.5.2 that T is sufficient for η on the basis of (T, T').

Consider, now, the problem of estimating η on the basis of (T, T'). This is essentially the situation of Example 4.3.2 with T, distributed as $N(\eta, 1/I)$, playing the role of the sufficient statistic $\bar{X}$. In both cases, the full data set (T, T') can be generated by a normal random variable with unknown mean and known variance, together with variables that have a known distribution. The result of Example 4.3.2 is thus applicable and shows that T is unique minimax, and hence that $T' = T$ with probability 1. Therefore $(T_{n_j}, T_{n_j}')_{\theta_0} \xrightarrow{\mathfrak{L}} (T, T')_{H_0}$. Since this is true for any convergent subsequence, it follows that $(T_n, T_n')_{\theta_0} \xrightarrow{\mathfrak{L}} (T, T)_{H_0}$, and this completes the proof by Lemma 8.2.

Theorem 8.2. *If* $\hat{\theta}_n$ *satisfies* (1) *and* (8), *it is loc. as. admissible and minimax.*

Proof. If $\hat{\theta}_n$ is not loc. as. admissible, there exists a sequence $\hat{\theta}_n'$ with $R(\eta, \delta') \leqslant 1/I$ for all η and $< 1/I$ for some η. This contradicts Lemma 8.4(iii).

In the finite sample normal situation to which the local asymptotic problem was reduced in the proof of Lemma 8.4, $\bar{X}$ is not only minimax but is the unique estimator with this property.

Theorem 8.3. *Under the assumptions of Theorem* 8.2, *if* $\hat{\theta}_n'$ *is any estimator satisfying* $R(\eta, \delta') \leqslant 1/I$ *for all* η, *then* $\hat{\theta}_n' \sim \hat{\theta}_n$.

Proof. It follows from the proof of Lemma 8.4 that $(T_n, T_n')_{\theta_0} \to (T, T)_{H_0}$. For any $\varepsilon > 0$

$$P_{\theta_0}[|T_n' - T_n| \geqslant \varepsilon] \to P_{\theta_0}(|T' - T| \geqslant \varepsilon) = 0$$

by Theorem 5.1.7, and this completes the proof.

Local asymptotic risk functions generalize with only the obvious changes to the estimation of vector-valued parameters and the local asymptotic minimax property of Theorem 8.2 extends to that case. However, this is not true for the corresponding admissibility property (and hence for Theorem 8.3) when the dimensionality of the parameter exceeds 2, in view of the inadmissibility results of Sections 4.5 and 4.6.

9. PROBLEMS

Section 1

1.1 Let $X_1, \dots, X_n$ be iid as $N(\theta, 1)$. Consider the two estimators

$$T_n = \begin{cases} \bar{X}_n & \text{if } S_n \leqslant a_n \\ n & \text{if } S_n > a_n \end{cases}$$

where $S_n = \Sigma(X_i - \bar{X})^2$, $P(S_n > a_n) = 1/n$, and

$$T_n' = \frac{X_1 + \cdots + X_{k_n}}{k_n}$$

with k_n the largest integer $\leqq \sqrt{n}$.

(i) Show that the asymptotic efficiency of T_n' relative to T_n is zero.
(ii) Show that for any *fixed* $\varepsilon > 0$

$$P[|T_n - \theta| > \varepsilon] = \frac{1}{n} + o\left(\frac{1}{n}\right) \tag{1}$$

but

$$P[|T_n' - \theta| > \varepsilon] = o\left(\frac{1}{n}\right). \tag{2}$$

(iii) For large values of n, what can you say about the two probabilities (1) and (2) when ε is replaced by $a/\sqrt{n}$? (Basu, 1956)

1.2 Verify the asymptotic distribution claimed for δ_n in Example 1.1.

1.3 Let δ_n be any estimator satisfying (1.2) with $g(\theta) = \theta$. Construct a sequence δ_n' such that $\sqrt{n}(\delta_n' - \theta) \xrightarrow{\mathcal{L}} N[0, w^2(\theta)]$ with $w(\theta) = v(\theta)$ for $\theta \neq \theta_0$ and $w(\theta_0) = 0$.

1.4 In the preceding problem, construct δ_n' such that $w(\theta) = v(\theta)$ for all $\theta \neq \theta_0, \theta_1$ and $< v(\theta)$ for $\theta = \theta_0, \theta_1$.

1.5 Construct a sequence $\{\delta_n\}$ satisfying (1.2) but for which the bias $b_n(\theta)$ does not tend to zero.

1.6 In Example 1.1 with $R_n(\theta)$ given by (1.11), show that $R_n(\theta) \to 1$ for $\theta \neq 0$, and that $R_n(0) \to a^2$.

1.7 Let $b_n(\theta) = E_\theta(\delta_n) - \theta$ be the bias of the estimator δ_n of Example 1.1. Show

(i) that

$$b_n(\theta) = \frac{-(1-a)}{\sqrt{n}} \int_{-\sqrt[4]{n}}^{\sqrt[4]{n}} x\phi(x - \sqrt{n}\theta)\, dx;$$

(ii) that

$$b_n'(\theta) \to 0 \qquad \text{for any} \quad \theta \neq 0, \qquad b_n'(0) \to -(1-a).$$

(iii) Use (ii) to explain how the Hodges estimator δ_n can violate (1.7) without violating the information inequality.

1.8 In Example 1.1 (continued), show that if $\theta_n = c/\sqrt{n}$, then $R_n(\theta_n) \to a^2 + c^2(1-a)^2$.

Section 2

2.1 Let X have the binomial distribution $b(p, n)$, $0 \leqslant p \leqslant 1$. Determine the MLE of p

(i) by the usual calculus method for determining the maximum of a function;

(ii) by showing that $p^x q^{n-x} \leqslant (x/n)^x[(n-x)/n]^{n-x}$.

[*Hint*: (ii) Apply the fact that the geometric mean is equal to or less than the arithmetic mean, to n numbers of which x are equal to np/x and $n-x$ equal to $nq/(n-x)$.]

2.2 In the preceding problem show that the MLE does not exist when p is restricted to $0 < p < 1$ and when $x = 0$ or $= n$.

2.3 Let $X_1, \ldots, X_n$ be iid according to $N(\xi, \sigma^2)$. Determine the MLE

(i) of ξ when σ is known;

(ii) of σ when ξ is known;

(iii) of (ξ, σ) when both are unknown.

2.4 Suppose $X_1, \ldots, X_n$ are iid as $N(\xi, 1)$ with $\xi > 0$. Show that the MLE is $\bar{X}$ when $\bar{X} > 0$ and does not exist when $\bar{X} \leqslant 0$.

2.5 Let X take on the values 0 and 1 with probabilities p and q, respectively. When it is known that $\frac{1}{3} \leqslant p \leqslant \frac{2}{3}$,

(i) find the MLE;

(ii) show that the expected squared error of the MLE is uniformly larger than that of $\delta(x) \equiv \frac{1}{2}$.

2.6 When Ω is finite, show that the MLE is consistent if and only if it satisfies (2.2).

2.7 Prove the existence of unique $0 < a_k < a_{k-1}$, $k = 1, 2, \ldots$, satisfying (2.4).

2.8 Prove (2.9).

2.9 In Example 2.3 with $0 < c < 1/2$ determine a consistent estimator of k.

[*Hint*: (i) The smallest value K of j for which I_j contains at least as many of the X's as any other I is consistent.
(ii) The value of j for which I_j contains the median of the X's is consistent since the median of f_k is in I_k.]

2.10 Verify the nature of the roots in Example 2.4.

2.11 Let X be distributed as $N(\theta, 1)$. Show that conditionally given $a < X < b$, the variable X tends in probability to b as $\theta \to \infty$.

2.12 Consider a sample $X_1, \dots, X_n$ from a Poisson distribution conditioned to be positive, so that

$$P(X_i = x) = \theta^x e^{-\theta}/x!(1 - e^{-\theta}) \qquad \text{for} \quad x = 1, 2, \dots .$$

Show that the likelihood equation has a unique root for all values of x.

2.13 Verify equation (2.24).

2.14 Let X have the negative binomial distribution (2.3.3). Find an ELE of p.

2.15 (i) Let $X_1, \dots, X_n$ be positive random variables (or symmetrically distributed about zero) with joint density $a^n \Pi f(ax_i)$, $a > 0$. The likelihood equation has a unique root if $xf'(x)/f(x)$ is strictly decreasing for $x > 0$.

(ii) If $X_1, \dots, X_n$ are iid with density $f(x_i - \theta)$ where f is unimodal and if the likelihood equation has a unique root, then the likelihood equation also has a unique root when the density of each X_i is $af[a(x_i - \theta)]$, with a known.

2.16 If $X_1, \dots, X_n$ are iid with density $f(x_i - \theta)$ or $af(ax_i)$ and f is the logistic density $L(0, 1)$, the likelihood equation has unique solutions $\hat{\theta}$ and $\hat{a}$ both in the location and the scale case. Determine the limit distribution of $\sqrt{n}(\hat{\theta} - \theta)$ and $\sqrt{n}(\hat{a} - a)$.

2.17 In Problem 2.15(i), with f the Cauchy density $C(0, 1)$, the likelihood equation has a unique root $\hat{a}$ and $\sqrt{n}(\hat{a} - a) \xrightarrow{\mathcal{L}} N(0, 2a^2)$.

2.18 If $X_1, \dots, X_n$ are iid as $C(\theta, 1)$, then for any fixed n there is positive probability

(i) that the likelihood equation has $2n - 1$ roots;

(ii) that the likelihood equation has a unique root.

[*Hint*: (i) If the x's are sufficiently widely separated, the value of $L'(\theta)$ in the neighborhood of x_i is dominated by the term $(x_i - \theta)/[1 + (x_i - \theta)^2]$. As θ passes through x_i, this term changes signs so that the log likelihood has a local maximum near x_i. (ii) Let the x's be close together.]

2.19 Let (X_i, Y_i), $i = 1, \dots, n$, be iid according to a bivariate normal distribution with $E(X_i) = E(Y_i) = 0$, $\text{var}(X_i) = \text{var}(Y_i) = 1$, and unknown correlation coefficient ρ. The likelihood equation is a cubic for which the probability of a unique root tends to 1 as $n \to \infty$.

[*Hint*: For a cubic equation $ax^3 + 3bx^2 + 3cx + d = 0$ let $G = a^2d - 3abc + 2b^3$ and $H = ac - b^2$. Then the condition for a unique real root is $G^2 + 4H^3 > 0$.]

2.20 In the preceding problem, if $\hat{\rho}_n$ is a consistent solution of the likelihood equation, then

$$\sqrt{n}\,(\hat{\rho}_n - \rho) \xrightarrow{\mathcal{L}} N\left[0, (1-\rho^2)^2/(1+\rho^2)\right].$$

[*Hint*: To evaluate $I(\rho)$, note that

$$E(X_iY_i) = \rho, \quad E\left(X_i^2Y_i^2\right) = 1 + 2\rho^2, \quad E\left(X_iY_i^2\right) = 0, E\left(X_iY_i^3\right) = 3\rho.]$$

2.21 If $X_1, \ldots, X_n$ are iid according to the gamma distribution $\Gamma(\theta, 1)$, the likelihood equation has a unique root.

[*Hint*: Use Example 2.5. Alternatively, write down the likelihood and use the fact that $\Gamma'(\theta)/\Gamma(\theta)$ is an increasing function of θ. (See, for example, Artin, "The Gamma Function" (pp. 16–17), Holt, Rinehart and Winston, 1964.]

2.22 Under the assumptions of Theorem 1.1 show that

$$\left[L\left(\theta_0 + \frac{1}{\sqrt{n}}\right) - L(\theta_0) + \tfrac{1}{2}I(\theta_0)\right]\Big/\sqrt{I(\theta_0)} \tag{3}$$

tends in law to $N(0, 1)$.

2.23 Let $X_1, \ldots, X_n$ be iid according to $N(\theta, a\theta^2)$, $\theta > 0$, where a is a known positive constant.

(i) Find an explicit expression for an ELE.

(ii) Determine whether there exists an MRE estimator under a suitable group of transformations.

2.24 Check that the assumptions of Theorem 2.3 are satisfied in the following examples: (i) 2.4, (ii) 2.7, and (iii) 2.9.

2.25 In Example 2.5, show directly that $(1/n)\Sigma T(X_i)$ is an asymptotically efficient estimator of $\theta = E_\eta[T(X)]$ by considering its limit distribution.

2.26 Let $X_1, \ldots, X_n$ be iid according to $\theta g(x) + (1-\theta)h(x)$, where (g, h) is a pair of specified probability densities with respect to μ, and where $0 < \theta < 1$.

(i) Give one example of (g, h) for which the assumptions of Theorem 2.3 are satisfied and one for which they are not.

(ii) Discuss the existence and nature of the roots of the likelihood equation for $n = 1, 2, 3$.

2.27 Under the assumptions of Theorem 2.2, suppose that $\hat{\theta}_{1n}$ and $\hat{\theta}_{2n}$ are two consistent sequences of roots of the likelihood equation. Prove that $P_{\theta_0}(\hat{\theta}_{1n} = \hat{\theta}_{2n}) \to 1$ as $n \to \infty$.

[*Hint*: (i) Let $S_n = \{\underline{x} = (x_1, \ldots, x_n)$ such that $\hat{\theta}_{1n}(\underline{x}) \neq \hat{\theta}_{2n}(\underline{x})\}$. For all $\underline{x} \in S_n$ there exists θ_n^* between $\hat{\theta}_{1n}$ and $\hat{\theta}_{2n}$ such that $L''(\theta_n^*) = 0$. For all $\underline{x} \notin S_n$, let θ_n^* be the common value of $\hat{\theta}_{1n}$ and $\hat{\theta}_{2n}$. Then θ_n^* is a consistent sequence of roots of the likelihood equation.

(ii) $(1/n)L''(\theta_n^*) - (1/n)L''(\theta_0) \to 0$ in probability and therefore $(1/n)L''(\theta_n^*) \to -I(\theta_0)$ in probability.

(iii) Let $0 < \varepsilon < I(\theta_0)$ and let

$$S_n' = \left\{\underline{x}: \frac{1}{n}L''(\theta_n^*) < -I(\theta_0) + \varepsilon\right\}.$$

Then $P_{\theta_0}(S_n') \to 1$. On the other hand, $L''(\theta_n^*) = 0$ on S_n so that S_n is contained in the complement of S_n'. (Huzurbazar, 1948).]

Section 3

3.1 Let

$$u(t) = \begin{cases} c\int_0^t e^{-1/x(1-x)}\,dx & \text{for} \quad 0 < t < 1, \\ 0 & \text{for} \quad t \leqslant 0 \\ 1 & \text{for} \quad t \geqslant 1. \end{cases}$$

Show that for a suitable c, the function u is continuous and infinitely differentiable for $-\infty < t < \infty$.

3.2 Show that the density (3.1) with $\Omega = (0, \infty)$ satisfies all conditions of Theorem 2.3 with the exception of (v) of Theorem 1.1.

3.3 Show that the density (3.4) with $\Omega = (0, \infty)$ satisfies all conditions of Theorem 2.3.

3.4 In Example 3.2, evaluate the estimators (3.8) and (3.11) for the Cauchy case, using for $\tilde{\theta}_n$ the sample median.

3.5 In Example 3.4 show that $L(\theta)$ is concave.

3.6 In Example 3.4, if $\eta = \xi$, show how to obtain a $\sqrt{n}$-consistent estimator by equating sample and population second moments.

3.7 In Example 3.5 show that the LRD of the MLE with respect to the estimator (3.15) with $a = -2$ is 2.

3.8 In Theorem 3.2, show that $\sigma_{11} = \sigma_{12}$.

Section 4

4.1 (i) If a vector $\underline{Y}_n$ in E_s converges in probability to a constant vector $\underline{a}$, and if h is a continuous function defined over E_s, then $h(\underline{Y}_n) \to h(\underline{a})$ in probability.

(ii) Use (i) to show that the elements of $\|A_{jkn}\|^{-1}$ tend in probability to the elements of B as claimed in the proof of Lemma 4.1.

[*Hint*: (i) Apply Theorem 5.1.7 and Problem 5.1.13.]

4.2 (i) Show that (4.26) with the remainder term neglected has the same form as (4.15) and identify the A_{jkn}.

(ii) Show that the resulting a_{jk} of Lemma 4.1 are the same as those of (4.23).

(iii) Show that the remainder term in (4.26) can be neglected in the proof of Theorem 4.2.

4.3 Let $X_1, \dots, X_n$ be iid according to $N(\xi, \sigma^2)$.

(i) Show that the likelihood equations have a unique root.

(ii) Show directly (i.e., without recourse to Theorem 4.1) that the MLEs $\hat{\xi}$ and $\hat{\sigma}$ are asymptotically efficient.

4.4 Let $(X_0, \dots, X_s)$ have the multinomial distribution $M(p_0, \dots, p_s; n)$.

(i) Show that the likelihood equations have a unique root.

(ii) Show directly that the MLEs $\hat{p}_i$ are asymptotically efficient.

4.5 Prove Corollary 4.1.

4.6 Show that there exists a function f of two variables for which the equations $\partial f(x, y)/\partial x = 0$, $\partial f(x, y)/\partial y = 0$ have a unique solution, and this solution is a local but not a global maximum of f.

Section 5

5.1 In Example 5.1, show that the likelihood equations are given by (5.2) and (5.3).

5.2 In Example 5.1, verify equation (5.4).

5.3 Verify (5.5).

5.4 If $\theta = (\theta_1, \dots, \theta_r, \theta_{r+1}, \dots, \theta_s)$ and if

$$\operatorname{cov}\left[\frac{\partial}{\partial \theta_i} L(\theta), \frac{\partial}{\partial \theta_j} L(\theta)\right] = 0 \qquad \text{for any} \quad i \leqslant r < j,$$

then the asymptotic distribution of $(\hat{\theta}_1, \dots, \hat{\theta}_r)$ under the assumptions of Theorem 4.1 is unaffected by whether or not $\theta_{r+1}, \dots, \theta_s$ are known.

5.5 In Example 5.4 verify the MLEs $\hat{\xi}_i$ and $\hat{\sigma}_{jk}$ when the ξ's are unknown.

5.6 In Example 5.4 show that the S_{jk} given by (5.15) are independent of $(X_{1\cdot}, \dots, X_{p\cdot})$ and have the same joint distribution as the statistics (5.13) with n replaced by $n - 1$.

[*Hint*: Subject each of the p vectors $(X_{i1}, \dots, X_{in})$ to the same orthogonal transformation, where the first row of the orthogonal matrix is $(1/\sqrt{n}, \dots, 1/\sqrt{n})$.]

5.7 Verify the matrices (i) (5.17), and (ii) (5.18).

5.8 Verify (5.19) and (5.20).

5.9 In Example 5.5, if it is known that $\xi = \eta = 0$ and $\sigma = \tau = 1$, then $\delta = \Sigma X_i Y_i / n$ is a consistent estimator of $\rho = \text{cov}(X_i, Y_i)$. Show that $\sqrt{n}(\delta - \rho) \xrightarrow{\mathcal{L}} N(0, 1 + \rho^2)$ and hence that δ is less efficient than the MLE of ρ.

5.10 In Example 5.6 show that the conditions of Theorem 4.1 are satisfied.

Section 6

6.1 Prove Theorem 6.1.

6.2 In Example 6.2, determine the joint asymptotic distribution of (i) $(\hat{\sigma}^2, \hat{\tau}^2)$ and (ii) $(\hat{\sigma}^2, \hat{\sigma}_A^2)$.

6.3 Show that the likelihood equations (6.9) have at most one solution.

6.4 Consider samples $(X_1, Y_1), \ldots, (X_m, Y_m)$ and $(X_1', Y_1'), \ldots, (X_n', Y_n')$ from two bivariate normal distributions with means zero and variance–covariances $(\sigma^2, \tau^2, \rho\sigma\tau)$ and $(\sigma'^2, \tau'^2, \rho'\sigma'\tau')$, respectively. Use Theorem 6.1 and Example 5.5 to find the limit distribution

(i) of $\hat{\sigma}^2$ and $\hat{\tau}^2$ when it is known that $\rho' = \rho$;
(ii) of $\hat{\rho}$ when it is known that $\sigma' = \sigma$, $\tau' = \tau$.

6.5 In the preceding problem, find the efficiency gain (if any)

(i) in part (i) resulting from the knowledge that $\rho' = \rho$;
(ii) in part (ii) resulting from the knowledge that $\sigma' = \sigma$, $\tau' = \tau$.

6.6 In Example 6.1, determine the MLE of ξ.

[*Hint*: Begin by finding the MLE of σ and τ given ξ.]

6.7 In Example 6.4 suppose that $q_i = F(\alpha + \beta t_i)$ and that both $\log F(x)$ and $\log[1 - F(x)]$ are strictly concave. Then the likelihood equations have at most one solution.

6.8 (i) If the cdf F is symmetric and if $\log F(x)$ is strictly concave, so is $\log[1 - F(x)]$.
(ii) Show that $\log F(x)$ is strictly concave when F is strongly unimodal but not when F is Cauchy.

6.9 In Example 6.5 show that Y_n is less informative than Y.

[*Hint*: Let Z_n be distributed as $P(\lambda\Sigma_{i=n+1}^{\infty}\gamma_i)$ independently of Y_n. Then $Y_n + Z_n$ is a sufficient statistic for λ on the basis of (Y_n, Z_n) and $Y_n + Z_n$ has the same distribution as Y.]

6.10 Show that the estimator δ_n of Example 6.5 satisfies (6.12).

6.11 Find suitable normalizing constants for δ_n of Example 6.5 when (i) $\gamma_i = i$, (ii) $\gamma_i = i^2$, and (iii) $\gamma_i = \dfrac{1}{i}$.

6.12 Let X_i $(i = 1, \ldots, n)$ be independent normal with variance 1 and mean βt_i (with t_i known). Discuss the estimation of β along the lines of Example 6.5.

6.13 Generalize the preceding problem to the situation in which (i) $E(X_i) = \alpha + \beta t_i$, $\text{var}(X_i) = 1$; (ii) $E(X_i) = \alpha + \beta t_i$, $\text{var}(X_i) = \sigma^2$ where α, β, and σ^2 are unknown parameters to be estimated.

6.14 Prove that the sequence $X_1, X_2, \ldots$ of Example 6.6 is stationary provided it satisfies (6.15).

6.15 (i) In Example 6.6, show that the likelihood equation has a unique solution, that it is the MLE, and that it has the same asymptotic distribution as $\delta_n' = \Sigma_{i=1}^n X_i X_{i+1} / \Sigma_{i=1}^n X_i^2$.

(ii) Show directly that δ_n' is a consistent estimator of β.

6.16 When $\tau = \sigma$ in (6.19), show that the MLE exists and is consistent.

6.17 Suppose that in (6.19) the ξ's are themselves random variables, which are iid as $N(\mu, \gamma^2)$.

(i) Show that the joint density of the (X_i, Y_i) is that of a sample from a bivariate normal distribution, and identify the parameters of that distribution.

(ii) In the model of part (i) find asymptotically efficient estimators of the parameters μ, γ, β, σ, and τ.

6.18 Verify the roots (6.20).

6.19 Show that the likelihood (6.19) is unbounded.

6.20 Let X_j $(j = 1, \ldots, n)$ be independently distributed with densities $f_j(x_j, \theta)$ (θ real-valued), let $I_j(\theta)$ be the information X_j contains about θ, and let $T_n(\theta) = \Sigma_{j=1}^n I_j(\theta)$ be the total information about θ in the sample. Suppose that $\hat{\theta}_n$ is a consistent root of the likelihood equation $L'(\theta) = 0$ and that, in generalization of (2.18)–(2.20),

$$\frac{1}{\sqrt{T_n(\theta_0)}} L'(\theta_0) \xrightarrow{\mathcal{L}} N(0, 1),$$

and

$$-\frac{L''(\theta_0)}{T_n(\theta_0)} \xrightarrow{P} 1 \quad \text{and} \quad \frac{L'''(\theta_n^*)}{T_n(\theta_0)} \text{ is bounded in probability.}$$

Then

$$\sqrt{T_n(\theta_0)}\,(\hat{\theta}_n - \theta_0) \xrightarrow{\mathcal{L}} N(0, 1).$$

6.21 In Example 6.10 compare

(i) the asymptotic distributions of $\hat{\xi}$ and δ_n;

(ii) the normalized expected squared error of $\hat{\xi}$ and δ_n.

6.22 In Example 6.10 show that

(i) $\sqrt{n}(\hat{\hat{b}} - b) \xrightarrow{\mathcal{L}} N(0, b^2)$;

(ii) $\sqrt{n}(\hat{b} - b) \xrightarrow{\mathcal{L}} N(0, b^2)$.

6.23 In Example 6.11 show that

(i) $\hat{c}$ and $\hat{a}$ are independent and have the stated distributions;
(ii) $X_{(1)}$ and $\Sigma \log[X_i/X_{(1)}]$ are complete sufficient statistics on the basis of a sample from (6.25).

6.24 In Example 6.11 determine the UMVU estimators of a and c, and the asymptotic distributions of these estimators.

6.25 In the preceding problem compare

(i) the asymptotic distribution of the MLE and the UMVU estimator of c;
(ii) the normalized expected squared error of these two estimators.

6.26 Let $X_1, \ldots, X_n$ be iid according to the three-parameter lognormal distribution (6.29). Show that

(i) $p^*(\underline{x}, \xi) = \sup_{\gamma, \sigma^2} p(\underline{x}; \xi, \gamma, \sigma^2) = c/[\hat{\sigma}(\xi)]^n \Pi[1/(x_i - \xi)]$ where

$$p(\underline{x}; \xi, \gamma, \sigma^2) = \prod_{i=1}^{n} f(x_i; \xi, \gamma, \sigma^2),$$

$$\hat{\sigma}^2(\xi) = \frac{1}{n}\Sigma[\log(x_i - \xi) - \hat{\gamma}(\xi)]^2 \quad \text{and} \quad \hat{\gamma}(\xi) = \frac{1}{n}\Sigma \log(x_i - \xi).$$

(ii) $p^*(\underline{x}, \xi) \to \infty$ as $\xi \to x_{(1)}$.

[*Hint*: (ii) For ξ sufficiently near $x_{(1)}$,

$$\hat{\sigma}^2(\xi) \leqslant \frac{1}{n}\Sigma[\log(x_i - \xi)]^2 \leqslant [\log(x_{(1)} - \xi)]^2$$

and hence

$$p^*(\underline{x}, \xi) \geqslant |\log(x_{(1)} - \xi)|^{-n} \Pi(x_{(i)} - \xi)^{-1}.$$

The right side tends to infinity as $\xi \to x_{(1)}$. (Hill, 1963).]

6.27 The derivatives of all orders of the density (6.29) tend to zero as $x \to \xi$.

6.28 In Example 6.6,

(i) show that for $j > 1$ the expected value of the conditional information (given X_{j-1}) that X_j contains about β is $1/(1 - \beta^2)$;
(ii) determine the information X_1 contains about β.

Section 7

7.1 Determine the limit distribution of the Bayes estimator corresponding to squared error loss, and verify that it is asymptotically efficient, in each of the following cases:

(i) The observations Y_i have the gamma distribution $\Gamma(\gamma, 1/\tau)$, the estimand is $1/\tau$, and τ has the conjugate prior density $\Gamma(g, \alpha)$.

(ii) The observations and prior are as in Problem 4.1.9 and the estimand is λ.

(iii) The observations Y_i have the negative binomial distribution (2.3.3), p has the prior density $B(a, b)$, and the estimand is $(a) p, (b) 1/p$.

7.2 The assumptions of Theorem 1.1 imply (7.1) and (7.2).

7.3 In Example 7.2, the posterior density of θ after one observation is $f(x_1 - \theta)$; it is a proper density, and it satisfies (B5) provided $E_\theta|X_1| < \infty$.

7.4 Let $X_1, \ldots, X_n$ be independent, positive variables, each with density $(1/\tau)f(x_i/\tau)$, and let τ have the improper density $\pi(\tau) = 1/\tau$ ($\tau > 0$). The posterior density after one observation is a proper density, and it satisfies (B5) provided $E_\tau(1/X_1) < \infty$.

7.5 Give an example in which the posterior density is proper (with probability one) after two observations but not after one.

[*Hint*: In the preceding example let $\pi(\tau) = 1/\tau^2$.]

7.6 Prove the result stated preceding Example 7.3.

7.7 Let $X_1, \ldots, X_n$ be iid as $N(\theta, 1)$ and consider the improper density $\pi(\theta) = e^{\theta^4}$. Then the posterior prior will be improper for all n.

7.8 Prove Lemma 7.1.

7.9 (i) If $\sup|Y_n(t)| \xrightarrow{P} 0$ and $\sup|X_n(t) - c| \xrightarrow{P} 0$ as $n \to \infty$, then $\sup|X_n(t) - ce^{Y_n(t)}| \xrightarrow{P} 0$, where the sup is taken over a common set $t \in T$.

(ii) Use (i) to show that (7.21) and (7.22) imply (7.20).

7.10 Show that (B1) implies (i) (7.23) and (ii) (7.25).

Section 8

8.1 (i) Verify (8.6).

(ii) Show that (8.6) is not affected by truncating the loss at A^2 before passing to the limit in n, and then letting $A \to \infty$.

8.2 In Example 8.1 verify $R(\eta, \delta')$.

8.3 In Example 8.3,

(i) verify $R(\eta, \delta)$ at $\theta = 0$;

(ii) show that the loc. as. risk of $\bar{X}$ is $\equiv 1$ for all θ.

[*Hint*: (i) Problem 1.8.]

8.4 In Example 8.4, show that the loc. as. risk at θ_0 (i) of $X_{(n)}$ is $2\theta_0^2$ and (ii) of $(n + 1)X_{(n)}/n$ is θ_0^2.

8.5 Let $X_1, \ldots, X_n$ be iid as $N(0, \sigma^2)$. Determine the loc. as. risk of $\Sigma X_i^2/(n + a)$ for any given constant a.

8.6 Let $X_1, \ldots, X_n$ be iid according to (1.4.1) with $s = 1$. Evaluate the loc. as. risk of $\Sigma T_1(X_i)/n$ as an estimator of $E[T_1(X_i)]$, and show that it is constant for all values of η_1.

10. REFERENCES

The origins of the concept of maximum likelihood go back to the work of Lambert, Daniel Bernoulli, and Lagrange in the second half of the 18th century, and of Gauss and Laplace at the beginning of the 19th. [For details and references, see Edwards (1974).] The modern history begins with Edgeworth (1908–09) and Fisher (1922, 1925), whose contributions are discussed by Savage (1976) and Pratt (1976).

Fisher's work was followed by a euphoric belief in the universal consistency and asymptotic efficiency of maximum likelihood estimators, at least in the iid case. The true situation was sorted out only gradually. Landmarks are Wald (1949), who provided fairly general conditions for consistency, Cramér (1946), who defined the "regular" case in which the likelihood equation has a consistent asymptotically efficient root, the counterexamples of Bahadur (1958) and Hodges (Le Cam, 1953), and Le Cam's resulting theorem on superefficiency (1953).

Convergence (under suitable restrictions and appropriately normalized) of the posterior distribution of a real-valued parameter with a prior distribution to its normal limit was first discovered by Laplace (1820) and later reobtained by Bernstein (1917) and von Mises (1931). More general versions of this result are given in Le Cam (1958). The asymptotic efficiency of Bayes solutions was established by Le Cam (1958), Bickel and Yahav (1969), and Ibragimov and Has'minskii (1972). [See also Ibragimov and Has'minskii (1981).]

Local asymptotic minimaxity as a formulation of asymptotic optimality that avoids restrictions on the competing estimators and the possibility of superefficiency was first discussed by Chernoff (1956), who credits the idea to C. Stein and H. Rubin. The theory was developed further by Hájek (1972) and Le Cam (1979a).

Aickin, M.
(1979). "Existence of MLES for discrete linear exponential models." *Ann. Inst. Statist. Math.* **31**, 103–113.

Aitchison, J. and Silvey, S. D.
(1958). "Maximum-likelihood estimation of parameters subject to restraints." *Ann. Math. Statist.* **29**, 813–828.

Akahira, M. and Takeuchi, K.
(1981). *Asymptotic Efficiency of Statistical Estimators*. Springer, New York.

Amemiya, T.
(1980). "The n^{-2}-order mean squared errors of the maximum likelihood and the minimum logit chi-square estimator." *Ann. Statist.* **8**, 488–505.

Andersen, E. B.
(1970). "Asymptotic properties of conditional maximum likelihood estimators," *J. Roy. Statist. Soc.* (*B*) **32**, 283–301.

Anderson, T. W.
(1958). *Introduction to Multivariate Statistical Analysis*. Wiley, New York.

Anderson, T. W.
(1959). "On asymptotic distributions of estimates of parameters of stochastic difference equations." *Ann. Math. Statist.* **30**, 676–687.

Anderson, T. W.
(1976). "Estimation of linear functional relationships: Approximate distributions and connections with simultaneous equations in economics." *J. Roy. Statist. Soc.* (*B*) **38**, 1–19.

Anderson, T. W. and Sawa, T.
(1982). "Exact and approximate distributions of the AARHUS maximum likelihood estimator of a slope coefficient." *J. Roy. Statist. Soc.* (*B*) **44**, 52–62.

Bahadur, R. R.
(1958). "Examples of inconsistency of maximum likelihood estimates." *Sankhyā* **20**, 207–210.

Bahadur, R. R.
(1964). "On Fisher's bound for asymptotic variances." *Ann. Math. Statist.* **35**, 1545–1552.

Bahadur, R. R.
(1967). "Rates of convergence of estimates and test statistics." *Ann. Math. Statist.* **38**, 303–324.

Barndorff–Nielsen, O.
(1978). *Information and Exponential Families in Statistical Theory*. Wiley, New York.

Barndorff–Nielsen, O. and Blaesild, P.
(1980). "Global maxima, and likelihood in linear models." Research Rep. 57, Department of Theoretical Statistics, University of Aarhus.

Barnett, V. D.
(1966). "Evaluation of the maximum likelihood estimator where the likelihood equation has multiple roots." *Biometrika* **53**, 151–166.

Bar-Shalom, Y.
(1971). "Asymptotic properties of maximum likelihood estimates." *J. Roy. Statist. Soc.* (*B*) **33**, 72–77.

Basawa, I. V. and Prakasa Rao, B. L. S.
(1980). *Statistical Inference in Stochastic Processes*. Academic Press, London.

Basu, D.
(1956). "The concept of asymptotic efficiency." *Sankhyā* **17**, 193–196.

Berk, R. H.
(1967). "Review of Zehna (1966)." *Math. Rev.* **33**, No. 1922.

Berk, R. H.
(1972). "Consistency and asymptotic normality of MLE's for exponential models." *Ann. Math. Statist.* **43**, 193–204.

Berkson, J.
(1980). "Minimum chi-square, not maximum likelihood!" *Ann. Statist.* **81**, 457–469.

Bernstein, S.
(1917). *Theory of Probability*. (Russian).

Bhat, B. R.
(1974). "On the method of maximum likelihood for dependent observations," *J. Roy. Statist. Soc.* (*B*) **36**, 48–53.

Bickel, P. J. and Yahav, J. A.
(1969). "Some contributions to the asymptotic theory of Bayes solutions." *Z. Wahrsch. verw. Geb.* **11**, 257–276.

Billingsley, P.
(1968). Convergence of probability measures. Wiley, New York.

Billingsley, P.
(1979). *Probability and Measure*. Wiley, New York.

Bishop, Y. M., Fienberg, S. E., and Holland, P. W.
(1975). *Discrete Multivariate Analysis* M.I.T. Press, Cambridge.

Blyth, C. R.
(1982). "Maximum probability estimation in small samples." In: A Festschrift for Erich L. Lehmann. (Bickel, Doksum, Hodges, Eds). Wadsworth, Belmont.

Box, G. E. and Tiao, G. C.
(1973). *Bayesian Inference in Statistical Analysis*. Addison-Wesley, Reading, Mass.

Bradley, R. A. and Gart, J. J.
(1962). "The asymptotic properties of ML estimators when sampling for associated populations." *Biometrika* **49**, 205–214.

Chan, L. K.
(1967). "Remark on the linearized maximum likelihood estimate." *Ann. Math. Statist.* **38**, 1876–1881.

Chernoff, H.
(1956). "Large-sample theory: Parametric case." *Ann. Math. Statist.* **27**, 1–22.

Cohen, A. C.
(1965). "Maximum likelihood estimation in the Weibull distribution based on complete and on censored samples." *Technometrics* **7**, 579–588.

Cohen, A. C.
(1967). "Estimation in mixtures of two normal distributions." *Technometrics* **9**, 15–28.

Cohen, A. and Sackrowitz, H. B.
(1974). "On estimating the common mean of two normal distributions." *Ann. Statist.* **2**, 1274–1282.

Copas, J. B.
(1972). "The likelihood surface in the linear functional relationship problem." *J. Roy. Statist. Soc.* B, **34**, 274–278.

Copas, J. B.
(1975). "On the unimodality of the likelihood for the Cauchy distribution." *Biometrika* **62**, 701–704.

Cox, D. R. and Hinkley, D. V.
(1974). *Theoretical Statistics*. Chapman and Hall, London.

Cox, N. R.
(1976). "A note on the determination of the nature of turning points of likelihoods." *Biometrika* **63**, 199–201.

Crain, B. R.
(1976). "Exponential models, maximum likelihood estimation, and the Haar condition." *J. Amer. Statist. Assoc.* **71**, 737–740.

Cramér, H.
(1946). *Mathematical Methods of Statistics*. Princeton University Press, Princeton, N.J.

Crowder, M. J.
(1976). "Maximum likelihood estimation for dependent observations." *J. Roy. Statist. Soc. (B)* **38**, 45–53.

Dempster, A. P., Laird, N. M., and Rubin, D. B.
(1977). "Maximum likelihood from incomplete data via the EM algorithm." *J. Roy Statist. Soc. (B)* **39**, 1–22.

Edgeworth, F. Y.
(1908–09). "On the probable errors of frequency constants." *J. Roy. Statist. Soc.* **71**, 381–397, 499–512; *J. Roy. Statist. Soc.* **72**, 81–90.

Edwards, A. W. F.
(1974). "The history of likelihood," *Internat. Statist. Rev.* **42**, 4–15.

Efron, B.
(1975). "Defining the curvature of a statistical problem (with applications to second order efficiency)." *Ann. Statist.* **3**, 1189–1242.

Efron, B.
(1978). "The geometry of exponential families." *Ann Statist.* **6**, 362–376.

Efron, B.
(1982). "Maximum likelihood and decision theory" *Ann. Statist.* **10**, 340–356.

Ferguson, T. S.
(1982). "An inconsistent maximum likelihood estimate." *J. Amer. Statist. Assoc.* **77**, 831–834.

Fisher, R. A.
(1922). "On the mathematical foundations of theoretical statistics." *Philos. Trans. Roy. Soc. London, Ser. A* **222**, 309–368.

Fisher, R. A.
(1925). "Theory of statistical estimation." *Proc. Camb. Phil. Soc.* **22**, 700–725.

Foutz, R. V.
(1977). "On the unique consistent solution to the likelihood equations." *J. Amer. Statist. Assoc.* **72**, 147–148.

Ghosh, J. K. and Sinha, B. K.
(1981). "A necessary and sufficient condition for second order admissibility with applications to Berkson's bioassay problem." *Ann. Statist.* **9**, 1334–1338.

Ghosh, J. K., Sinha, B. K., and Wieand, H. S.
(1980). "Second order efficiency of the MLE with respect to any bounded bowl shaped loss function." *Ann. Statist.* **8**, 506–521.

Ghosh, J. K. and Subramanyam, K.
(1974). "Second order efficiency of maximum likelihood estimators." *Sankhyā (A)* **36**, 325–358.

Giesbrecht, F. and Kempthorne, O.
(1976). "Maximum likelihood estimation in the three-parameter lognormal distribution." *J. Roy. Statist. Soc. (B)* **38**, 257–264.

Haas, G., Bain, L., and Antle, C.
(1970). "Inferences for the Cauchy distribution based on maximum likelihood estimators." *Biometrika* **57**, 403–408.

Haberman, S. J.
(1974). *The Analysis of Frequency Data*. University of Chicago Press, Chicago.

Hájek, J.
(1972). "Local asymptotic minimax and admissibility in estimation." *Proc. Sixth Berkeley Symp. on Math. Statist. Probab.* **1**, 175–194.

Heyde, C. C. and Feigin, P. D.
(1975). "On efficiency and exponential families in stochastic process estimation." In *Statistical Distributions in Scientific Work*, vol. 1, Patil, Kotz, and Ord, Eds. D. Reidel, Dordrecht, Holland, pp. 227–240.

Hill, B. M.
(1963). "The three-parameter lognormal distribution and Bayesian analysis of a point-source epidemic." *J. Amer. Statist. Assoc.* **58**, 72–84.

Hoadley, B.
(1971). "Asymptotic properties of maximum likelihood estimators for the independent not identically distributed case." *Ann. Math. Statist.* **42**, 1977–1991.

Huber, P. J.
(1966). "Strict efficiency excludes superefficiency" (Abstract). *Ann. Math. Statist.* **37**, 1425.

Huber, P. J.
(1967). "The behaviour of maximum likelihood estimates under non-standard conditions." *Proc. Fifth. Berkeley Symp. on Math. Statist. Probab.* **1**, 221–233.

Huzurbazar, V. S.
(1948). "The likelihood equation, consistency and the maxima of the likelihood function." *Ann. Eugenics* **14**, 185–200.

Ibragimov, I. A. and Has'minskii, R. Z.
(1972). "Asymptotic behavior of statistical estimators. II. Limit theorems for the a posteriori density and Bayes' estimators." Theory of Probability and Its Applications. **18**, 76–91.

Ibragimov, I. A. and Has'minskii, R. Z.
(1981). *Statistical Estimation: Asymptotic Theory*. Springer, New York.

Jaynes, E. T.
(1979). "Where do we stand on maximum entropy?" In *The Maximum Entropy Formalism*. Levine and Tribus, Eds. Cambridge, pp. 15–118. M.I.T. Press.

Jewell, N. P. and Raab, G. M.
(1981). "Difficulties in obtaining consistent estimators of variance parameters." *Biometrika* **68**, 221–226.

Johansen, S.
(1979). "Introduction to the theory of regular exponential families." *Institute of Mathematical Statistics Lecture Notes*, Vol. 3. University of Copenhagen.

Johnson, N. L. and Kotz, S.
(1970). *Continuous Univariate Distributions*. 2 Vols. Houghton Mifflin, Boston.

Kabaila, P. V.
(1980). "On the asymptotic variance of estimators based on time series." To be published.

Kalbfleisch, J. D. and Sprott, D. A.
(1970). "Application of likelihood methods to models involving large numbers of parameters." *J. Roy. Statist. Soc.* (*B*) **32**, 175–208.

Kempthorne, O.
(1966). "Some aspects of experimental inference." *J. Amer. Statist. Assoc.* **61**, 11–34.

Kendall, M. G. and Stuart, A.
(1979). *The Advanced Theory of Statistics.* Vol. 2, 4th ed. MacMillan, New York.

Kiefer, J. and Wolfowitz, J.
(1956). "Consistency of the maximum likelihood estimator in the presence of infinitely many incidental parameters." *Ann. Math. Statist.* **27**, 887–906.

Klotz, J. H., Milton, R. C., and Zacks, S.
(1969). "Mean square efficiency of estimators of variance components." *J. Amer. Statist. Assoc.* **64**, 1383–1402.

Kraft, C. and LeCam, L.
(1956). "A remark on the roots of the likelihood equation." *Ann. Math. Statist.* **27**, 1174–1177.

Lambert, J. A.
(1970). "Estimation of parameters in the four-parameter lognormal." *Austr. J. Statist.* **12**, 33–44.

Landers, D.
(1972). "Existence and consistency of modified minimum contrast estimates." *Ann. Math. Statist.* **43**, 74–83.

Laplace, P. S.
(1820). *Theorie Analytique des Probabilités.* 3rd ed. Paris.

Le Cam, L.
(1953). On some asymptotic properties of maximum likelihood estimates and related Bayes' estimates." *Univ. of Calif. Publ. in Statist.* **1**, 277–330.

Le Cam, L.
(1956). "On the asymptotic theory of estimation and testing hypotheses." *Proc. Third Berkeley Symp. on Math. Statist. Probab.* **1**, 129–156.

Le Cam, L.
(1958). "Les propriétés asymptotiques des solutions de Bayes." *Publ. Inst. Statist. l'Univ. de Paris* **VII**, Fasc. 3–4, 17–35.

Le Cam, L.
(1969). *Théorie Asymptotique de la Décision Statistique.* Les Presses de l'Université de Montréal.

Le Cam, L.
(1970). "On the assumptions used to prove asymptotic normality of maximum likelihood estimates." *Ann. Math. Statist.* **41**, 802–828.

Le Cam, L.
(1979a). "On a theorem of J. Hajek." In *Contributions to Statistics.* J. Hajek Memorial Volume. Jureckova, Ed. Academia, Prague.

Le Cam, L.
(1979b). *Maximum Likelihood: An Introduction.* Lecture Notes in Statistics No. 18. University of Maryland, College Park, Md.

Levit, B. Ya.
(1980). "On asymptotic minimax estimators of the second order." *Theor. Probabl. Appl.* **25**, 552–568.

Mäkeläinen, T., Schmidt, K., and Styan, G.
(1981). "On the existence and uniqueness of the maximum likelihood estimate of a vector-valued parameter in fixed-size samples." *Ann. Statist.* **9**, 758–767.

Miller, J. J.
(1977). "Asymptotic properties of maximum likelihood estimates in the mixed model of the analysis of variance." *Ann. Statist.* **5**, 746–762.

Mosteller, F.
(1946). "On some useful inefficient statistics." *Ann. Math. Statist.* **17**, 377–408.

Neyman, J.
(1949). "Contributions to the theory of the χ^2 test." *Proc. Berk. Symp. Math. Statist. Probab.* 239–273.

Neyman, J. and Scott, E. L.
(1948). "Consistent estimates based on partially consistent observations." *Econometrica* **16**, 1–32.

Nordberg, L.
(1980). "Asymptotic normality of maximum likelihood estimators based on independent, unequally distributed observations in exponential family models." *Scand. J. Statist.* **7**, 27–32.

Norden, R. H.
(1972–73). "A survey of maximum likelihood estimation." *Internat. Statist. Rev.* **40**, 329–354; **41**, 39–58.

Pearson, K.
(1894). "Contributions to the mathematical theory of evolution." *Phil. Trans. Royal Soc., Ser. A.* **185**, 71–110.

Perlman, M.
(1972). "On the strong consistency of approximate maximum likelihood estimators." *Proc. Sixth Berkeley Symp. Math. Statist. Probab.* **1**, 263–281.

Perlman, M.
(1983). "The limiting behavior of multiple roots of the likelihood equation."

Pfaff, Th.
(1982). "Quick consistency of quasi maximum likelihood estimators." *Ann. Statist.* **10**, 990–1005.

Pfanzagl, J.
(1969). "On the measurability and consistency of minimum constrast estimators." Metrika **14**, 249–272.

Pfanzagl, J.
(1970). "On the asymptotic efficiency of median unbiased estimates." *Ann. Math. Statist.* **41**, 1500–1509.

Pfanzagl, J.
(1973). "Asymptotic expansions related to minimum contrast estimators." *Ann. Statist.* **1**, 993–1026.

Pfanzagl, J. and Wefelmeyer, W.
(1978/79). "A third order optimum property of the maximum likelihood estimator." *J. Multiv. Anal.* **8**, 1–29; **9**, 179–182.

Polfeldt, T.
(1970). "Asymptotic results in non-regular estimation." Skand. Akt. Tidskr. Suppl. 1–2.

Portnoy, S.
(1971). "Formal Bayes estimation with application to a random effects model." *Ann. Math. Statist.* **42**, 1379–1402.

Pratt, J. W.
(1976). "F. Y. Edgeworth and R. A. Fisher on the efficiency of maximum likelihood estimation." *Ann. Statist.* **4**, 501–514.

Pregibon, D.
(1981). "Logistic regression diagnostics." *Ann. Statist.* **9**, 705–724.

Quandt, R. E. and Ramsey, J. B.
(1978). "Estimating mixtures of normal distributions and switching regressions." *J. Amer. Statist. Assoc.* **73**, 730–752.

Rao, C. R.
(1961). "Asymptotic efficiency and limiting information." *Proc. Fourth Berkeley Symp. Math. Statist. Prob.* **1**, 531–546.

Rao, C. R.
(1962). "Apparent anomalies and irregularities in maximum likelihood estimation." *Sankhyā* **24**, 73–102.

Rao, C. R.
(1963). "Criteria of estimation in large samples." *Sankhyā* **25**, 189–206.

Redner, R.
(1981). "Note on the consistency of the maximum likelihood estimate for nonidentifiable distributions." *Ann. Statist.* **9**, 225–228.

Reeds, J.
(1980). "The asymptotic distribution of the number of roots of the Cauchy likelihood equation." Unpublished.

Richards, F. S. G.
(1967). "On finding local maxima of functions of a real variable." *Biometrika* **54**, 310–311.

Savage, L. J.
(1976). "On rereading R. A. Fisher." *Ann. Statist.* **4**, 441–500.

Scholz, F. W.
(1980). "Towards a unified definition of maximum likelihood." *Canad. J. Statist.* **8**, 193–203.

Scholz, F. W.
(1981). "On the uniqueness of roots of the likelihood equation." Technical Report No. 14. Department of Statistics, University of Washington, Seattle.

Scholz, F. W.
(1983). "Maximum likelihood estimation." In: *Encyclopedia of Statistical Sciences*, vol. 5. Wiley, New York.

Serfling, R. J.
(1980). *Approximation Theorems of Mathematical Statistics.* Wiley, New York.

Solari, M. E.
(1969). "The 'maximum likelihood solution' of the problem of estimating a linear functional relationship." *J. Roy. Statist. Soc. Ser. B.* **31**, 372–375.

Stein, C.
(1956). "Efficient nonparametric testing and estimation." *Proc. Third Berkeley Symp. Math. Statist. Probab.* **1**, 187–195.

Stone, C. J.
(1974). "Asymptotic properties of estimators of a location parameter." *Ann. Statist.* **2**, 1127–1137.

Strasser, H.
(1981). "Consistency of maximum likelihood and Bayes estimates." *Ann. Statist.* **9**, 1107–1113.

Stuart, A.
(1958). "Note 129: Iterative solutions of likelihood equations." *Biometrics* **14**, 128–130.

Sundberg, R.
(1974). "Maximum likelihood theory for incomplete data from an exponential family." *Scand. J. Statist.* **2**, 49–58.

Sundberg, R.
(1976). "An iterative method for solution of the likelihood equations for incomplete data from exponential families." *Comm. Statist.* (*B*) **5**, 55–64.

Tan, W. Y. and Chang, W. C.
(1972). "Comparisons of method of moments and method of maximum likelihood in estimating parameters of a mixture of two normal densities." *J. Amer. Statist. Assoc.* **67**, 702–708.

Tarone, R. E. and Gruenhage, G.
(1975). "A note on the uniqueness of roots of the likelihood equations for vector-valued parameters." *J. Amer. Statist. Assoc.* **70**, 903–904.

Taylor, W. F.
(1953). "Distance functions and regular best asymptotically normal estimates." *Ann. Math. Statist.* **24**, 85–92.

Thompson, W. A., Jr.
(1962). "The problem of negative estimates of variance components." *Ann. Math. Statist.* **33**, 273–289.

von Mises, R.
(1931). *Wahrscheinlichkeitsrechnung.* Springer, Berlin.

Wald, A.
(1949). "Note on the consistency of the maximum likelihood estimate." *Ann. Math. Statist.* **20**, 595–601.

Weiss, L. and Wolfowitz, J.
(1966). "Generalized maximum likelihood estimators." *Theory Prob. Applic.* **11**, 58–81.

Weiss, L. and Wolfowitz, J.
(1974). *Maximum Probability Estimators and Related Topics.* Springer, Berlin.

Wijsman, R.
(1959). "On the theory of BAN estimates." *Ann. Math. Statist.* **30**, 185–191, 1268–1270.

Wolfowitz, J.
(1965). "Asymptotic efficiency of the maximum likelihood estimator." *Theory Prob. Applications* **10**, 247–260.

Woodrofe, M.
(1972). "Maximum likelihood estimation of a translation parameter of a truncated distribution." *Ann. Math. Statist.* **43**, 113–122.

Zehna, P. W.
(1966). "Invariance of maximum likelihood estimation." *Ann. Math. Statist.* **37**, 744.

Author Index

Subject Index

Testing Statistical Hypotheses
Second Edition

Contents